Sherpa Owners Workshop Manual

R G O Hawes

Models covered
All Sherpa models with in-line petrol engines
185, 200, 215, 230, 240, 250, 255, 280, 285, 310 and 350 models
1622 cc, 1700 cc, 1798 cc and 1994 cc

Does not cover V8 or Diesel-engine vehicles, or bodies and conversions produced by specialist manufacturers

ISBN 1 85010 297 X

Printed in England *(463-6R3)*

ABCDE
FGHIJ
KLMNO
P
2

THE BOOK

Haynes Publishing Group
Sparkford Nr Yeovil
Somerset BA22 7JJ England

Haynes Publications, Inc
861 Lawrence Drive
Newbury Park
California 91320 USA

British Library Cataloguing in Publication Data

Hawes, R.G.O.
Freight Rover/BL Sherpa (petrol) owners workshop manual. – (Owners Workshop Manuals).
1. British Leyland Sherpa Truck
I. Title II. Series
629.28'73 TL 230.5.B/
ISBN 1-85010-297-X

Acknowledgements

Thanks are due to BL Cars Ltd for the supply of technical information used in compilation of the original manual, and to Freight Rover Ltd, who supplied the details required for compiling the Supplement. Certain illustrations are the copyright of BL Cars Ltd or Freight Rover Ltd, and are used with their permission. The Champion Sparking Plug Company Ltd supplied the illustrations showing the spark plug conditions. Thanks are also due to Sykes-Pickavant, who supplied some of the workshop tools, and also to the staff at Sparkford who assisted in the production of this manual.

About this manual

Its aim

The aim of this manual is to help you get the best value from your vehicle. It can do so in several ways. It can help you decide what work must be done (even should you choose to get it done by a garage), provide information on routine maintenance and servicing, and give a logical course of action and diagnosis when random faults occur. However, it is hoped that you will make use of the manual by tackling the work yourself. On simpler jobs it may be even quicker than booking the vehicle into a garage, and having to go there twice, to leave and collect it. Perhaps most important, a lot of money can be saved by avoiding the costs the garage must charge to cover its labour and overheads.

The manual has drawings and descriptions to show the function of the various components so that their layout can be understood. Then the tasks are described and photographed in a step-by-step sequence so that even a novice can do the work.

Its arrangement

The manual is divided into thirteen Chapters, each covering a logical sub-division of the vehicle. The Chapters are each divided into consecutively numbered Sections and the Sections into paragraphs (or sub-sections), with decimal numbers following on from the Section they are in, eg 5.1, 5.2, 5.3 etc.

It is freely illustrated, especially in those parts where there is a detailed sequence of operations to be carried out. There are two forms of ilustration; figures and photographs. The figures are numbered in sequence with decimal numbers, according to their position in the Chapter: eg, Fig. 6.4 is the 4th drawing/illustration in Chapter 6. Photographs are numbered (either individually or in related groups) the same as the Section of sub-section of the text where the operation they show is described.

There is an alphabetical index at the back of the manual as well as a contents list at the front.

References to the 'left' or 'right' of the vehicle are in the sense of a person in a seat facing towards the front of the vehicle.

Unless otherwise stated, nuts and bolts are removed by turning anti-clockwise, and tightened by turning clockwise.

Vehicle manufacturers continually make changes to specifications and recommendations, and these, when notified, are incorporated into our manuals at the earliest opportunity.

Whilst every care is taken to ensure that the information in this manual is correct no liability can be accepted by the authors or publishers for loss, damage or injury caused by any errors in, or omissions from, the information given.

Introduction to the Sherpa

In the Sherpa BL have produced a rugged, light commercial vehicle which has secured a wide reputation for versatility and reliability. With a volume capacity of 190 cu ft usable loadspace the vehicle is produced in three nominal payload versions of 13/14 cwt, 18/19 cwt and 22/23 cwt. There are five body options, namely, van, pick-up, minibus, crewbus and chassis-cab. In addition there is a wide range of optional extras including Basic or De-luxe equipment standards, automatic transmission and overdrive. The wide range of possible variations is perhaps exemplified by the twelve different combinations of sliding and hinged doors that are possible.

Originally produced with the well-tried and tested 1622cc or 1798cc B-series overhead valve petrol engines, the latest Sherpa now has the new O-series 1700cc overhead camshaft engine. (Diesel-engined versions of the Sherpa have always been available, but they are not covered in this manual.)

The model designations – 185, 215, 240, 250 (ohv) and 200, 230, 250, 255 (ohc) – are derived from the gross vehicle weight (GVW), 1.85, 2.15, 2.40 and 2.50 tons etc.

Modifications to the range since 1980 include fitment of larger engine (1994 cc), front disc brakes and a 5-speed gearbox. Gross vehicle weight now extends to 3.50 tonnes nominal in the 350 model. Details of these and other changes will be found in Chapter 13.

References to 'BL' or 'BL dealer' made throughout the manual should now read 'Freight Rover' or 'Freight Rover dealer'.

Contents

Sherpa 250 De-luxe van with the 1700 cc (17V) overhead camshaft engine

Sherpa 250 De-luxe minibus (12-seater) with the 1700 cc (17V) overhead camshaft engine

The Sherpa minibus used as the project vehicle for this manual

Buying spare parts and vehicle identification numbers

For modifications, and information applicable to later models, see Supplement at end of manual

Buying spare parts

Spare parts are available from many sources, for example: BL dealers, other garages and accessory shops, and motor factors. Our advice regarding spare part sources is as follows:

Officially appointed BL garages – This is the best source of parts which are peculiar to your vehicle and are otherwise not generally available (eg complete cylinder heads, internal gearbox components, badges, interior trim etc). It is also the only place at which you should have repairs carried out if your vehicle is still under warranty – non-BL components may invalidate the warranty. To be sure of obtaining the correct parts it will always be necessary to give the storeman your vehicle's identification number, and if possible, to take the old part along for positive identification. It obviously makes good sense to go straight to the specialists on your vehicle for this type of part for they are best equipped to supply you.

Other garages and accessory shops – These are often very good places to buy materials and components needed for the maintenance of your vehicle (eg spark plugs, bulbs, fan belts, oils and greases, filler paste etc). They also sell general accessories, usually have convenient opening hours, charge reasonable prices and can often be found not far from home.

Motor factors – Good factors will stock all the more important components which wear out relatively quickly (eg clutch components, pistons, valves, exhaust systems, brake cylinders/pipes/hoses/seals/shoes and pads etc.) Motor factors will often provide new or reconditioned components on a part exchange basis – this can save a considerable amount of money.

Vehicle identification numbers

Although many individual parts, and in some cases sub-assemblies, fit a number of different models it is dangerous to assume that just because they look the same, they are the same. Differences are not always easy to detect except by serial numbers. Make sure therefore, that the appropriate identity number for the model or sub-assembly is known and quoted when a spare part is ordered.

The engine number, commission number and vehicle number are stamped on plates secured to the panel in front of the door pillar on the driver's side. The engine number is also stamped on a plate secured to the right-hand side of the cylinder block.

The body number is stamped on a plate secured to the left-hand wheel arch platform.

The manual gearbox number is stamped on the under-side of the gearbox casing.

The automatic transmission number is stamped on a plate fixed to the left-hand side of the transmission.

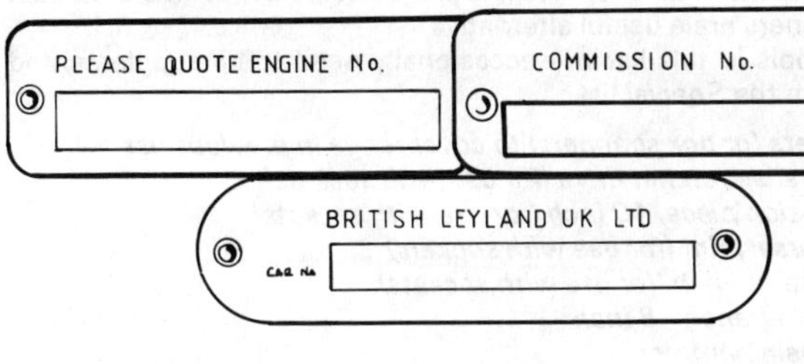

The engine number, commission number, and vehicle number are stamped on plates in front of the door pillar

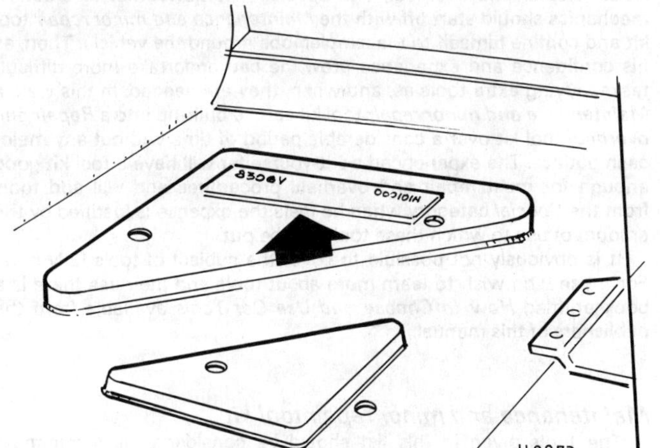

The body number is stamped on a plate on the left-hand wheel arch platform

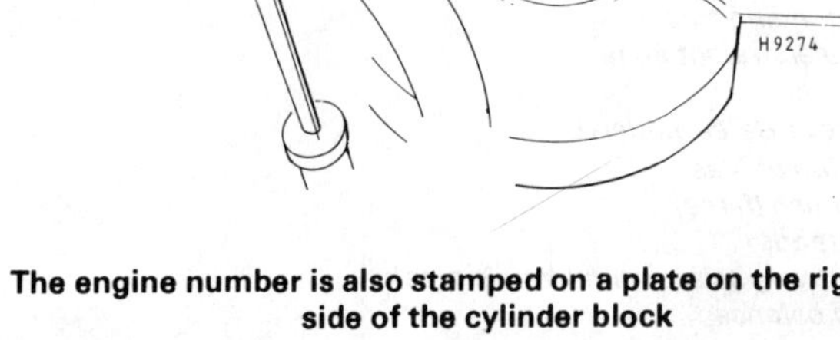

The engine number is also stamped on a plate on the right-hand side of the cylinder block

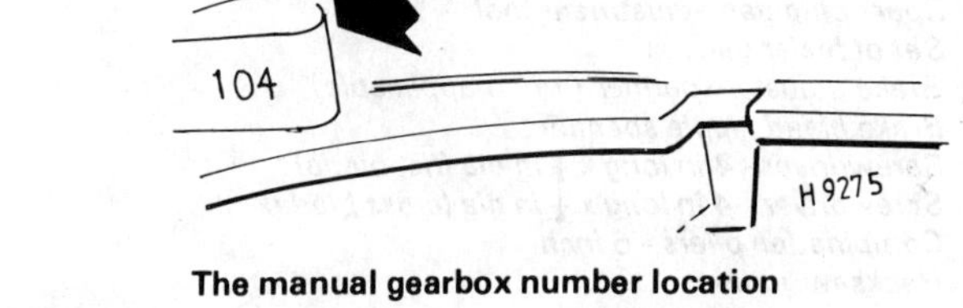

The manual gearbox number location

Tools and working facilities

Introduction

A selection of good tools is a fundamental requirement for anyone contemplating the maintenance and repair of a motor vehicle. For the owner who does not possess any, their purchase will prove a considerable expense, offsetting some of the savings made by doing-it-yourself. However, provided that the tools purchased meet the relevant national safety standards and are of good quality, they will last for many years and prove an extremely worthwhile investment.

To help the average owner to decide which tools are needed to carry out the various tasks detailed in this manual, we have compiled three lists of tools under the following headings: *Maintenance and minor repair, Repair and overhaul,* and *Special.* The newcomer to practical mechanics should start off with the *Maintenance and minor repair* tool kit and confine himself to the simpler jobs around the vehicle. Then, as his confidence and experience grow, he can undertake more difficult tasks, buying extra tools as, and when, they are needed. In this way, a *Maintenance and minor repair* tool kit can be built-up into a *Repair and overhaul* tool kit over a considerable period of time without any major cash outlays. The experienced do-it-yourselfer will have a tool kit good enough for most repair and overhaul procedures and will add tools from the *Special* category when he feels the expense is justified by the amount of use to which these tools will be put.

It is obviously not possible to cover the subject of tools fully here. For those who wish to learn more about tools and their use there is a book entitled *How to Choose and Use Car Tools* available from the publishers of this manual.

Maintenance and minor repair tool kit

The tools given in this list should be considered as a minimum requirement if routine maintenance, servicing and minor repair operations are to be undertaken. We recommend the purchase of combination spanners (ring one end, open-ended the other); although more expensive than open-ended ones, they do give the advantages of both types of spanner.

Nuts and bolts used on the Sherpa range are a mixture of metric and Imperial (UNF) sizes. **Make sure** that threads are compatible before tightening fasteners, or stripped threads may result.

Combination spanners - $\frac{7}{16}$, $\frac{1}{2}$, $\frac{9}{16}$, $\frac{5}{8}$, $\frac{3}{4}$, $\frac{13}{16}$, and $\frac{15}{16}$ in AF, and 8, 10, 12, 13, 14, 17 22, 24 and 27mm
Adjustable spanner - 9 inch
Engine sump/gearbox/rear axle drain plug key (where applicable)
Spark plug spanner (with rubber insert)
Spark plug gap adjustment tool
Set of feeler gauges
Brake adjuster spanner (when applicable)
Brake bleed nipple spanner
Screwdriver - 4 in long x $\frac{1}{4}$ in dia (flat blade)
Screwdriver - 4 in long x $\frac{1}{4}$ in dia (cross blade)
Combination pliers - 6 inch
Hacksaw, junior
Tyre pump
Tyre pressure gauge
Grease gun
Oil can
Fine emery cloth (1 sheet)
Wire brush (small)
Funnel (medium size)

Repair and overhaul tool kit

These tools are virtually essential for anyone undertaking any major repairs to a motor vehicle, and are additional to those given in the *Maintenance and minor repair* list. Included in this list is a comprehensive set of sockets. Although these are expensive they will be found invaluable as they are so versatile - particularly if various drives are included in the set. We recommend the $\frac{1}{2}$ in square-drive type, as this can be used with most proprietary torque wrenches. If you cannot afford a socket set, even bought piecemeal, then inexpensive tubular box spanners are a useful alternative.

The tools in this list will occasionally need to be supplemented by tools from the *Special* list.

Sockets (or box spanners) to cover range in previous list
Reversible ratchet drive (for use with sockets)
Extension piece, 10 inch (for use with sockets)
Universal joint (for use with sockets)
Torque wrench (for use with sockets)
'Mole' wrench - 8 inch
Ball pein hammer
Soft-faced hammer, plastic or rubber
Screwdriver - 6 in long x $\frac{5}{16}$ in dia (flat blade)
Screwdriver - 2 in long x $\frac{5}{16}$ in square (flat blade)
Screwdriver - $1\frac{1}{2}$ in long x $\frac{1}{4}$ in dia (cross blade)
Screwdriver - 3 in long x $\frac{1}{8}$ in dia (electricians)
Pliers - electricians side cutters
Pliers - needle nosed
Pliers - circlip (internal and external)
Cold chisel - $\frac{1}{2}$ inch
Scriber (this can be made by grinding the end of a broken hacksaw blade)
Scraper (this can be made by flattening and sharpening one end of a piece of copper pipe)
Centre punch
Pin punch
Hacksaw
Valve grinding tool
Steel rule/straight edge
Allen keys
'Torx' keys (later models)
Selection of files
Wire brush (large)
Axle-stands
Jack (strong scissor or hydraulic type)
Spring balance

Special tools

The tools in this list are those which are not used regularly, are expensive to buy, or which need to be used in accordance with their manufacturers' instructions. Unless relatively difficult mechanical jobs are undertaken frequently, it will not be economic to buy many of these tools. Where this is the case, you could consider clubbing together with friends (or a motorists' club) to make a joint purchase, or borrowing the tools against a deposit from a local garage or tool hire specialist.

The following list contains only those tools and instruments freely available to the public, and not those special tools produced by the vehicle manufacturer specifically for its dealer network. You will find occasional references to these manufacturers' special tools in the text of this manual. Generally, an alternative method of doing the job without the vehicle manufacturer's special tool is given. However, sometimes, there is no alternative to using them. Where this is the case and the relevant tool cannot be bought or borrowed you will have to entrust the work to a franchised garage.

Valve spring compressor
Piston ring compressor
Balljoint separator
Universal hub/bearing puller
Impact screwdriver
Micrometer and/or vernier gauge
Dial gauge
Stroboscopic timing light
Dwell angle meter/tachometer
Universal electrical multi-meter
Cylinder compression gauge
Lifting tackle
Trolley jack
Light with extension lead

Buying tools

For practically all tools, a tool factor is the best source since he will have a very comprehensive range compared with the average garage or accessory shop. Having said that, accessory shops often offer excellent quality tools at discount prices, so it pays to shop around.

There are plenty of good tools around at reasonable prices, but always aim to purchase items which meet the relevant national safety standards. If in doubt, ask the proprietor or manager of the shop for advice before making a purchase.

Care and maintenance of tools

Having purchased a reasonable tool kit, it is necessary to keep the tools in a clean serviceable condition. After use, always wipe off any dirt, grease and metal particles using a clean, dry cloth, before putting the tools away. Never leave them lying around after they have been used. A simple tool rack on the garage or workshop wall, for items such as screwdrivers and pliers is a good idea. Store all normal spanners and sockets in a metal box. Any measuring instruments, gauges, meters, etc, must be carefully stored where they cannot be damaged or become rusty.

Take a little care when tools are used. Hammer heads inevitably become marked and screwdrivers lose the keen edge on their blades from time to time. A little timely attention with emery cloth or a file will soon restore items like this to a good serviceable finish.

Working facilities

Not to be forgotten when discussing tools, is the workshop itself. If anything more than routine maintenance is to be carried out, some form of suitable working area becomes essential.

It is appreciated that many an owner mechanic is forced by circumstances to remove an engine or similar item, without the benefit of a garage or workshop. Having done this, any repairs should always be done under the cover of a roof.

Wherever possible, any dismantling should be done on a clean flat workbench or table at a suitable working height.

Any workbench needs a vice: one with a jaw opening of 4 in (100 mm) is suitable for most jobs. As mentioned previously, some clean dry storage space is also required for tools, as well as the lubricants, cleaning fluids, touch-up paints and so on which become necessary.

Another item which may be required, and which has a much more general usage, is an electric drill with a chuck capacity of at least 5/16 in (8 mm). This, together with a good range of twist drills, is virtually essential for fitting accessories such as wing mirrors and reversing lights.

Last, but not least, always keep a supply of old newspapers and clean, lint-free rags available, and try to keep any working area as clean as possible.

Spanner jaw gap comparison table

Jaw gap (in)	Spanner size
0·250	1/4 in AF
0·276	7 mm
0·313	5/16 in AF
0·315	8 mm
0·344	11/32 in AF; 1/8 in Whitworth
0·354	9 mm
0·375	3/8 in AF
0·394	10 mm
0·433	11 mm
0·438	7/16 in AF
0·445	3/16 in Whitworth; 1/4 in BSF
0·472	12 mm
0·500	1/2 in AF
0·512	13 mm
0·525	1/4 in Whitworth; 5/16 in BSF
0·551	14 mm
0·563	9/16 in AF
0·591	15 mm
0·600	5/16 in Whitworth; 3/8 in BSF
0·625	5/8 in AF
0·630	16 mm
0·669	17 mm
0·686	11/16 in AF
0·709	18 mm
0·710	3/8 in Whitworth; 7/16 in BSF
0·748	19 mm
0·750	3/4 in AF
0·813	13/16 in AF
0·820	7/16 in Whitworth; 1/2 in BSF
0·866	22 mm
0.875	7/8 in AF
0·920	1/2 in Whitworth; 9/16 in BSF
0·938	15/16 in AF
0·945	24 mm
1·000	1 in AF
1·010	9/16 in Whitworth; 5/8 in BSF
1·024	26 mm
1·063	1 1/16 in AF; 27 mm
1·100	5/8 in Whitworth; 11/16 in BSF
1·125	1 1/8 in AF
1·181	30 mm
1·200	11/16 in Whitworth; 3/4 in BSF
1·250	1 1/4 in AF
1·260	32 mm
1·300	3/4 in Whitworth; 7/8 in BSF
1·313	1 5/16 in AF
1·390	13/16 in Whitworth; 15/16 in BSF
1·417	36 mm
1·438	1 7/16 in AF
1·480	7/8 in Whitworth; 1 in BSF
1·500	1 1/2 in AF
1·575	40 mm; 15/16 in Whitworth
1·614	41 mm
1·625	1 5/8 in AF
1·670	1 in Whitworth; 1 1/8 in BSF
1·688	1 11/16 in AF
1·811	46 mm
1·813	1 13/16 in AF
1·860	1 1/8 in Whitworth; 1 1/4 in BSF
1·875	1 7/8 in AF
1·969	50 mm
2·000	2 in AF
2·050	1 1/4 in Whitworth; 1 3/8 in BSF
2·165	55 mm
2·362	60 mm

Jacking and towing

For modifications, and information applicable to later models, see Supplement at end of manual

Jacking points

To change a wheel in an emergency use the jack supplied with the vehicle. Before jacking-up, apply the handbrake and chock the wheel diagonally opposite the one to be removed. Never work under a vehicle that is supported solely on a jack. Safety supports such as axle stands must be used to avoid the possibility of personal injury or damage.

Front wheel: When changing a wheel with a flat tyre, position the jack beneath the front spring with the lifting platform of the jack under the spring in front of and as close as possible to the axle beam. At all other times the jack lifting platform should be positioned under the axle beam in line with the spring.

Rear wheel: When changing a wheel with a flat tyre, position the jack beneath the rear axle with the lifting platform as near to the U-bolts as possible. At all other times the jack lifting platform should be positioned under the axle tubes between the U-bolts.

Towing (automatic transmission)

When a vehicle with automatic transmission is on tow for recovery the following precautions must be observed.

Do not tow the vehicle with the rear wheels on the ground unless an extra 2 pints (1.2 litres) of fluid are added to the transmission.

Do not permit the vehicle to be towed faster than 30 mph (50 km/h) or for a distance greater than 40 miles (65 km).

If the above conditions cannot be met, or if excessive or unusual transmission noise is evident, disconnect the propeller shaft or tow with the rear wheels off the ground.

When being towed put the gear selector in the N position and the ignition/master switch to 1.

If the vehicle is to be transported on a trailer the handbrake must be applied and P must be selected, except for rail freight when N must be selected.

When making arrangements for recovery make sure that all concerned are aware of these precautions.

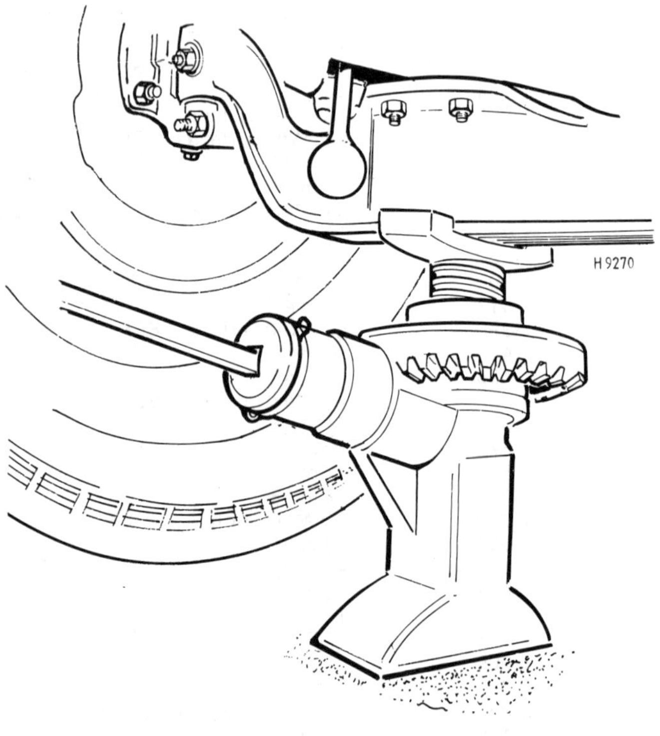

Jacking-up a front wheel

Jacking-up a rear wheel

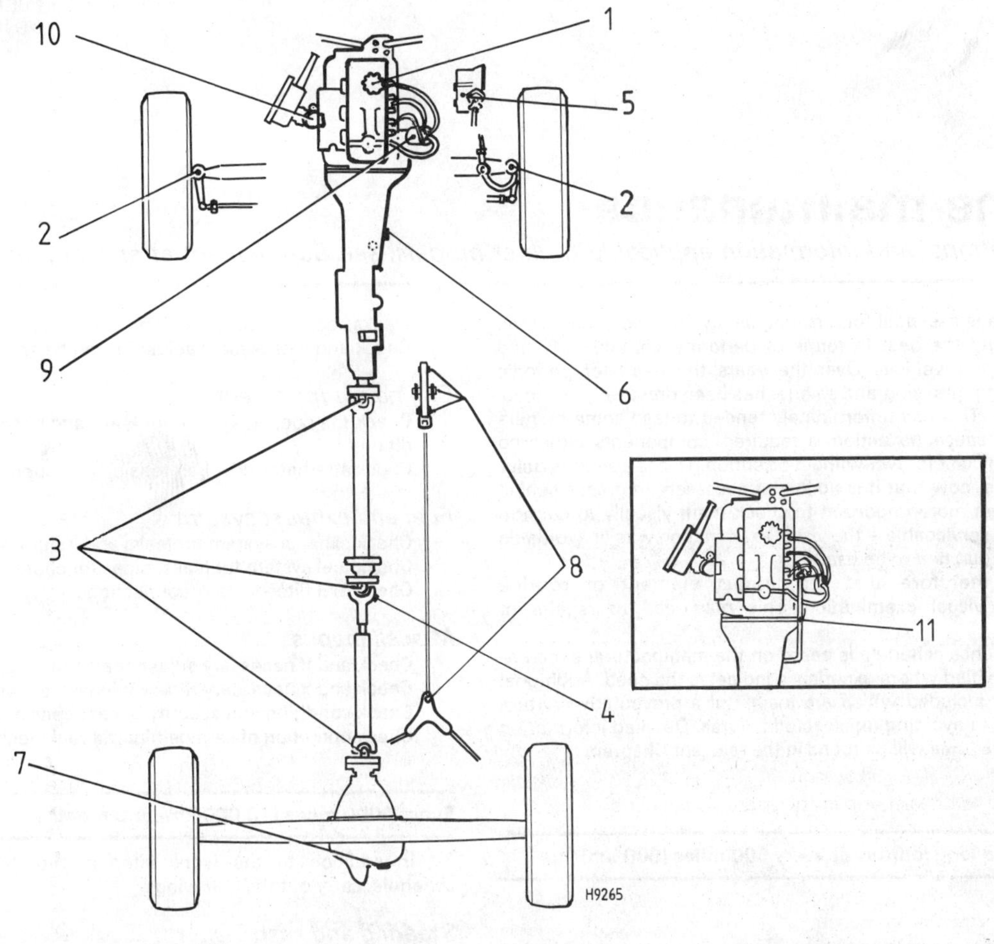

Recommended lubricants and fluids

Component or system	Lubricant type or specification
1 Engine	Multigrade engine oil, viscosity range 10W/40 to 15W/50, to API SE or SF
2 Swivel pins	Castrol PH grease or equivalent
3 Propeller shaft universal joints (not later models)	Multi-purpose lithium based grease
4 Propeller shaft splined joint (not later models)	Multi-purpose lithium based grease
5 Steering box	SAE EP90 hypoid gear oil
6 Manual gearbox:	
4-speed (with or without overdrive*)	SAE EP80 or EP90 hypoid gear oil
5-speed	Automatic transmission fluid type G
7 Rear axle	SAE EP90 hypoid gear oil
8 Handbrake linkage	Engine oil
9 Distributor (contact breaker type)	Engine oil (automatic advance mechanism) and multi-purpose grease (contact breaker cam and pivot post)
10 Carburettor dashpot	Engine oil
11 Automatic transmission	Automatic transmission fluid type G

**Do not use any anti-friction oil additives if overdrive is fitted*

Note: *Lubrication requirement may vary from territory to territory and with different operating conditions. Refer to the operator's handbook supplied with the vehicle or consult your local BL dealer.*

Routine maintenance

For modifications, and information applicable to later models, see Supplement at end of manual

Maintenance is essential for ensuring safety, and desirable for the purpose of getting the best in terms of performance, economy and reliability from your vehicle. Over the years the need for periodic lubrication – oiling, greasing and so on – has been drastically reduced, if not eliminated. This has unfortunately tended to lead some owners to think that, because no action is required, components either no longer exist or will last forever without attention. This is a serious delusion. For example, now that it is no longer necessary to grease sealed bearings it is even more important to check them visually to confirm that they remain serviceable – the visual examination was unavoidable when the grease gun had to be used!

It follows therefore that the largest element of routine maintenance is visual examination. This may lead to repairs or renewals.

The maintenance schedule is based on the manufacturer's recommendations, amplified where experience indicates the need. Additional tasks have been included which are mainly of a preventative nature and which assist in avoiding unexpected failures. Detailed information on servicing procedures will be found in the relevant Chapters.

Weekly, before a long journey or every 500 miles (800 km)

Steering

Check tyre pressures, including spare (when cold) and adjust if necessary (photo)
Examine tyres for tread depth and damage (photo)
Check tightness of wheel securing nuts
Check steering for smooth and accurate operation

Brakes

Check reservoir fluid level (photo). If this is low, top-up and check for fluid leakage
Check for satisfactory brake operation

Engine compartment

Check engine oil level and top-up if required (photos)
Check battery electrolyte level and top-up if required (photo)
Check expansion tank coolant level and top-up if required (photo)
Check washer reservoir lever and top-up if required (photo)

Lights, wipers, horn, instruments

Check operation of all lights
Check operation of windscreen wipers and washer
Check that the horn functions satisfactorily
Check that all instruments and warning indicators operate

Every 3000 miles (5000 km) or six weeks

In addition to the work listed in the Weekly schedule, carry out the following operations:

Steering and suspension

Lubricate the steering swivel pins (photos)
Check shock absorbers for fluid leaks
Check steering box for oil leaks
Check condition and security of steering joints

Brakes

Check visually hydraulic hoses, pipes and unions for chafing, cracks, leakage and corrosion
Check and if necessary adjust footbrake and handbrake

Engine compartment

Check the cooling system for leaks and hoses for security and condition
Check alternator drivebelt tension and condition (photo)

Fuel and exhaust system

Check exhaust system for leaks and mountings for security
Check fuel system for leaks, pipes for chafing and corrosion
Check fuel filter-to-tank connections

Miscellaneous

Check and if necessary adjust headlamp alignment
Check and if necessary renew windscreen wiper blades
Check condition and security of seat belts and seats
Check operation of seat belt inertia reel mechanism

Every 6000 miles (10 000 km) or 3 months

In addition to the work listed in the 3000 mile (5000 km) schedule, carry out the following:

Steering and suspension

Check steering box oil level and top-up if necessary
Have the front wheel alignment checked and adjusted if necessary by your BL dealer
Have the wheels balanced if necessary to eliminate vibration

Transmission

Lubricate the propeller shaft universal joints and splined joint
Check rear axle oil level and top-up if necessary (photo)
Manual gearbox: check oil level and top-up if necessary (photo)
Automatic transmission: check fluid level as described in Chapter 6 and top-up if necessary. Check correct operation of parking pawl. Lubricate the exposed selector linkage

Brakes

Lubricate the handbrake linkage

Engine

Drain engine oil when hot, renew filter and refill (photos)

Ignition system

Clean the spark plugs and adjust if necessary
Clean and adjust contact breaker points, or renew if necessary
Lubricate distributor cam and automatic advance mechanism
Check ignition timing and adjust if necessary (see Chapter 4)

Fuel and exhaust system

Top-up carburettor dashpot (photo)
Empty dust trap and clean air filter element (hot climate models)
Check carburettor idle settings and adjust if necessary (see Chapter 3)

Miscellaneous

Lubricate accelerator cable, linkage and pedal fulcrum
Clean the battery connections and smear with petroleum jelly
Lubricate all locks (except steering lock) and hinges
Lubricate base of windscreen wiper spindles with glycerine
Wash and wax polish bodywork, including chromium trim, forcing

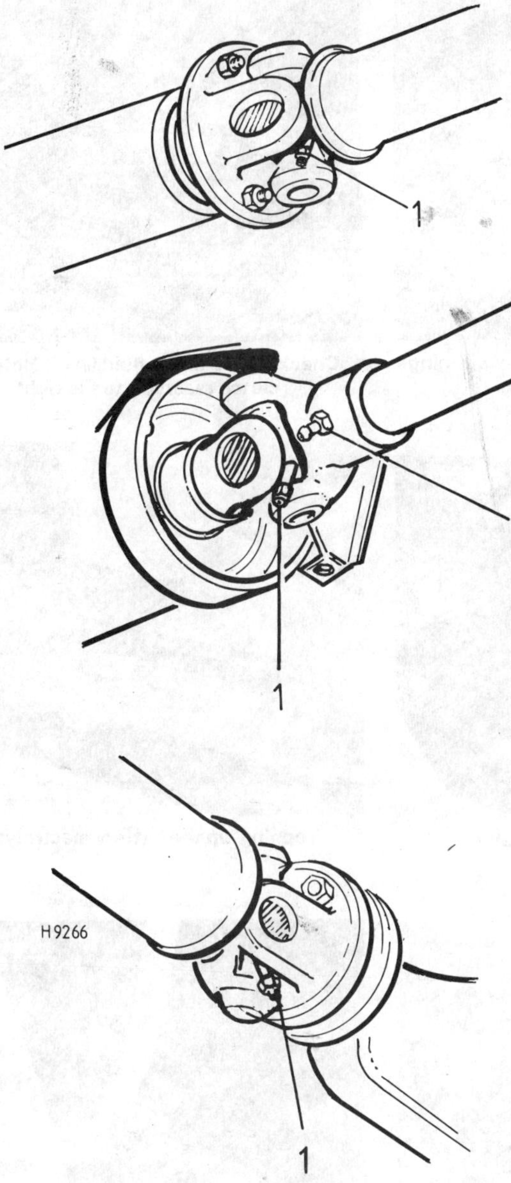

The propeller shaft grease nipples

1 Universal joint supply
2 Sliding joint supply

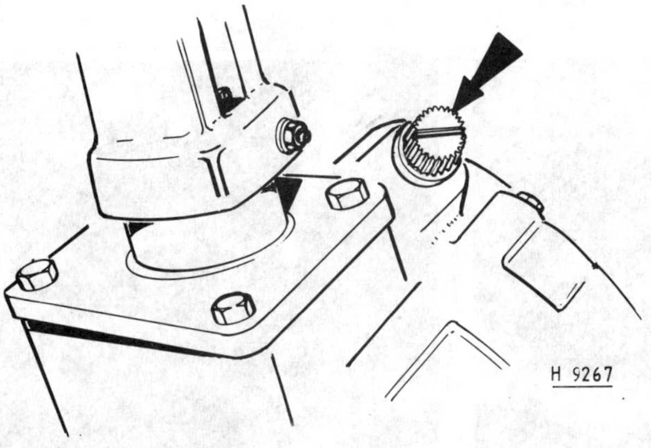

The steering box level/filler plug (arrowed)

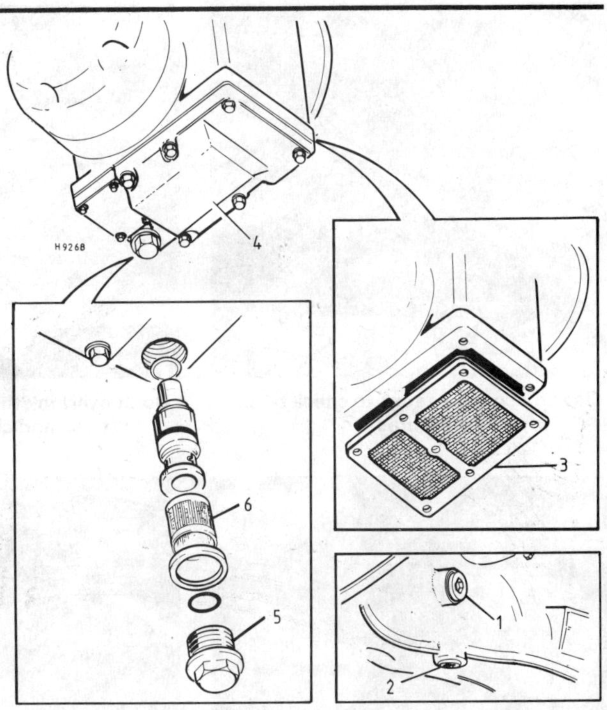

Servicing the overdrive gearbox

1 Level/filler plug
2 Drain plug
3 Filter
4 Sump cover
5 Relief valve filter plug
6 Relief valve filter

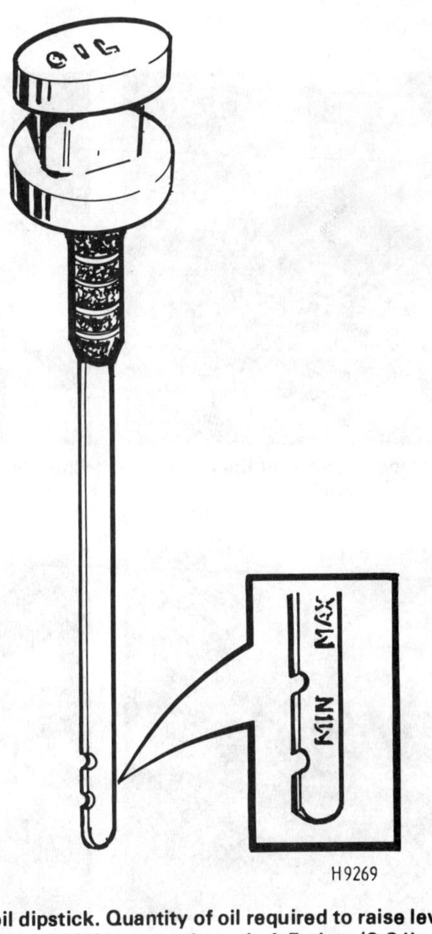

Engine oil dipstick. Quantity of oil required to raise level from MIN to MAX is approximately 1.5 pints (0.9 litre)

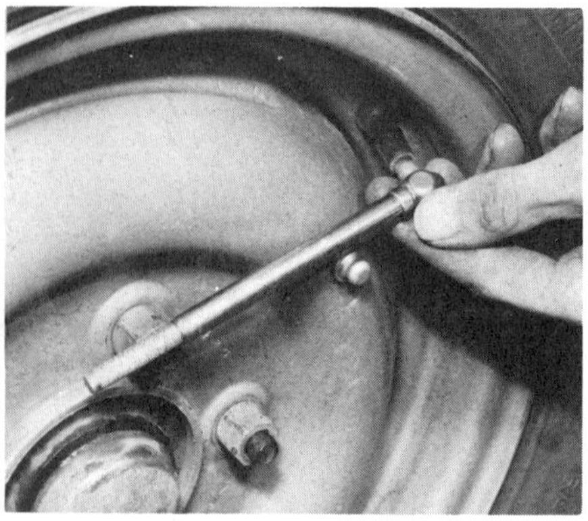

Use an accurate gauge to check tyre pressures

Don't overlook the spare when examining tyres and checking pressures

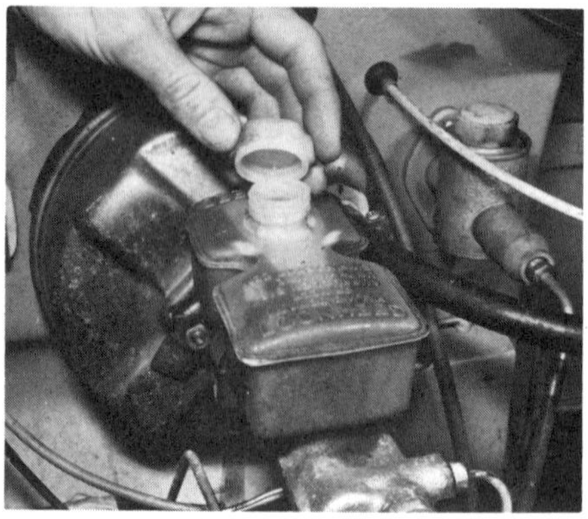

Checking the brake fluid level. Note the clutch reservoir to the right

Checking the engine oil level...

... and topping-up (17V engine)

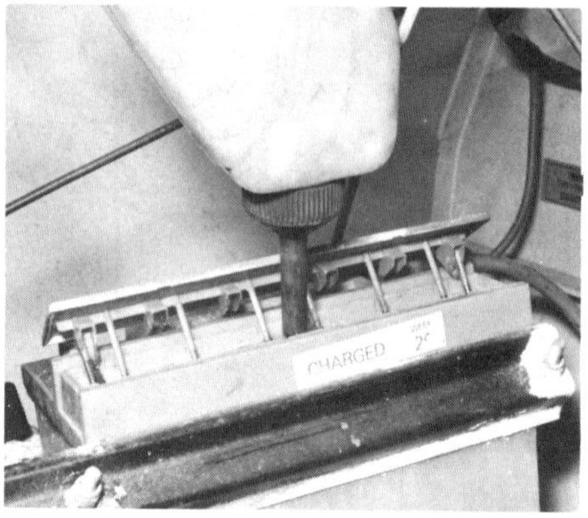

Topping-up the battery electrolyte

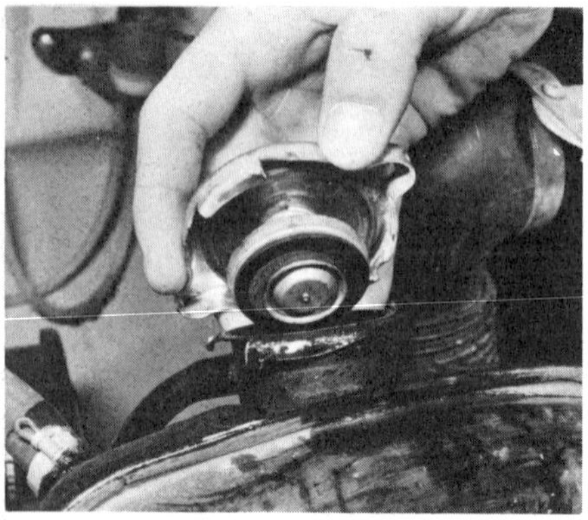

Checking the coolant level at the expansion tank

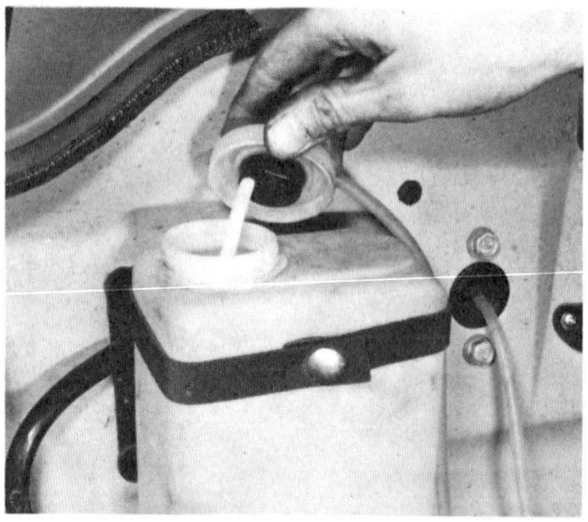

Checking the windscreen washer fluid level

A swivel pin upper greasing nipple and ...

... a lower one

Checking the alternator drivebelt tension

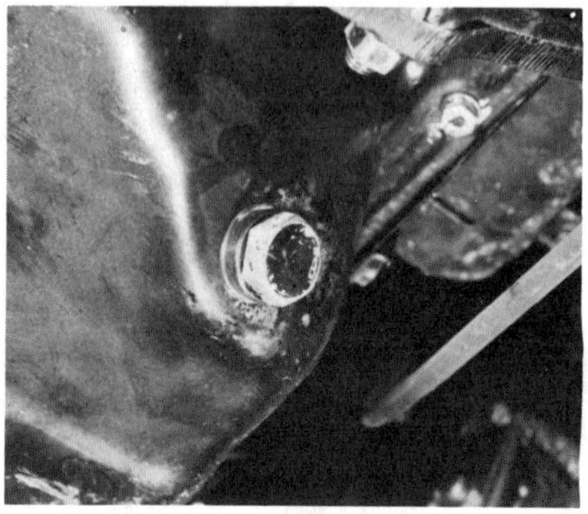

The engine oil sump drain plug

The engine oil filter on 16V and 18V engine and ...

... the engine oil filter on 17V engines (the radiator has been removed)

Removing the carburettor dashpot oil cap

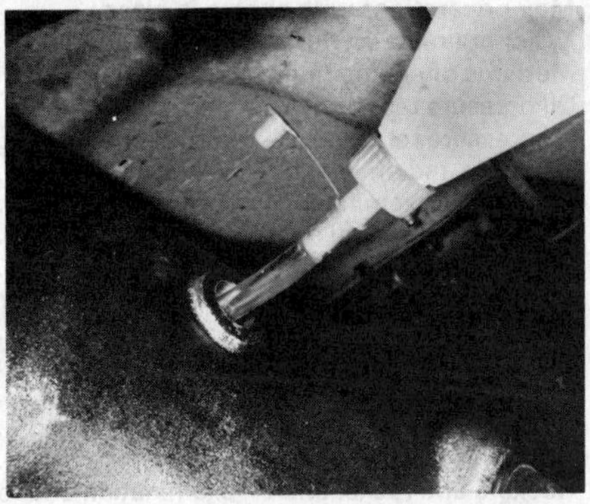

Topping-up the gearbox oil

The rear axle level/filler plug

wax into joints in bodywork to prevent rust. Clean out interior of vehicle

Every 12 000 miles (20 000 km) or 6 months

In addition to the procedures scheduled for the 6000 mile (10 000 km) servicing, carry out the following:

Steering and suspension

Check steering box and adjust if necessary, see Chapter 11
Adjust front wheel bearing endfloat, see Chapter 11

Brakes

Inspect brake linings for wear and brake drums for damage and distortion, see Chapter 9

Engine

Renew the engine oil filler cap
Check valve clearances and adjust if necessary, see Chapter 1

Ignition system

Renew spark plugs, setting correct gap before fitting

Fuel and exhaust system

Renew air cleaner element
Check air temperature control valve (17V engines), see Chapter 3

Every 24 000 miles (40 000 km) or 12 months

In addition to the procedures scheduled for the 12 000 mile (20 000 km) servicing, carry out the following:

Transmission

Drain and renew rear axle oil
Drain and renew gearbox oil. Overdrive models : clean filters and magnetic plugs before refilling, see Chapter 6

Engine

Check condition of timing belt (17V engines) and retension or renew, see Chapter 1

Every 36 000 miles (60 000 km) or 18 months

In addition to the procedures scheduled for the 12 000 mile (20 000 km) servicing, carry out the following:

Brakes

Renew the servo air filter element, see Chapter 9

Fuel and exhaust system

Renew the air cleaner element (hot climate models)

Every 48 000 miles (80 000 km) or 2 years

In addition to the procedures scheduled for the 24 000 mile (40 000 km) servicing, carry out the following:

Brakes

Renew all flexible hydraulic hoses, rubber seals and hydraulic fluid, see Chapter 9

Engine

Renew the timing belt (17V engines), see Chapter 1

Chapter 1 Part A
1622 cc (16V) and 1798 cc (18V) engines

Contents

Specifications

1622 cc (16V) engine

General

Type	Four-cylinder, in-line, ohv
Bore	3.0 in (76.2 mm)
Stroke	3.5 in (88.9 mm)
Capacity	99.55 cu in (1622 cc)
Firing order	1–3–4–2
Compression ratio	7.5:1

Crankshaft

Main journal diameter	2.0005 to 2.0010 in (50.80 to 50.82 mm)
Minimum regrind diameter	1.9605 to 1.9610 in (49.78 to 49.82 mm)
Crankpin journal diameter	1.8759 to 1.8764 in (47.64 to 47.66 mm)
Minimum regrind diameter	1.8359 to 1.8364 in (46.632 to 46.645 mm)
Endfloat	0.002 to 0.003 in (0.051 to 0.076 mm)
Oversize thrust washers	+ 0.003 in (0.076 mm)

Main and big-end bearings

Running clearance	0.001 to 0.0027 in (0.03 to 0.07 mm)
Undersizes available	- 0.010, 0.020, 0.030 and 0.040 in (0.25, 0.51, 0.76 and 1.02 mm)

Gudgeon pins

Type	Clamped
Fit in piston	0.00035 in (0.008 mm) clearance to 0.00005 in (0.0012 mm) interference

Pistons

Type	Solid skirt
Clearance in bore:	
Bottom of skirt	0.0015 to 0.0021 in (0.04 to 0.05 mm)
Top of skirt	0.0028 to 0.0037 in (0.07 to 0.09 mm)
Oversizes available	+ 0.020 and 0.040 in (0.51 and 1.02 mm)

Piston rings

Compression:	
Type – top	Chrome-faced
Type – second and third	Tapered, cast iron alloy
Fitted gap	0.012 to 0.017 in (0.30 to 0.43 mm)
Ring to groove clearance:	
Top	0.002 to 0.0035 in (0.05 to 0.09 mm)
Second and third	0.0015 to 0.0035 in (0.04 to 0.09 mm)
Oil control:	
Type	Slotted scraper
Fitted gap	0.012 to 0.017 in (0.30 to 0.43 mm)

Camshaft

Journal diameters:	
Front	1.78875 to 1.78725 in (45.43 to 45.45 mm)
Centre	1.72875 to 1.72925 in (43.91 to 43.92 mm)
Rear	1.62275 to 1.62325 in (41.22 to 41.23 mm)
Bearing liner inside diameter (reamed after fitting):	
Front	1.79025 to 1.79075 in (44.47 to 44.48 mm)
Centre	1.73025 to 1.73075 in (43.95 to 43.96 mm)
Rear	1.62425 to 1.62475 in (41.26 to 41.27 mm)
Running clearance	0.001 to 0.002 in (0.02 to 0.05 mm)
Endfloat	0.003 to 0.007 in (0.08 to 0.18 mm)

Valves

Valve to rocker clearance (cold)	0.015 in (0.38 mm)
Seat angle (inlet and exhaust)	45°
Head diameter:	
Inlet	1.500 to 1.505 in (38.10 to 38.23 mm)
Exhaust	1.281 to 1.286 in (32.54 to 32.66 mm)
Stem diameter:	
Inlet	0.3422 to 0.3427 in (8.69 to 8.70 mm)
Exhaust	0.34175 to 0.34225 in (8.68 to 8.69 mm)
Stem to guide clearance:	
Inlet	0.0014 to 0.0026 in (0.035 to 0.066 mm)
Exhaust	0.0018 to 0.0030 in (0.047 to 0.077 mm)

Valve guides

Fitted height above spring seat (inlet and exhaust)	0.625 in (15.86 mm)

Valve springs

Free length	2.031 to 2.051 in (51.59 to 52.09 mm)
Load at top of lift	131 lbf (59.4 kgf)

Valve timing

Inlet valve:	
Opens	TDC
Closes	50° ABDC
Exhaust valve:	
Opens	35° BBDC
Closes	15° ATDC

Lubrication system

System type	Wet sump, pressure fed
Oil pressure:	
Running	50 lbf/in^2 (3.52 kgf/cm^2)
Idling	15 lbf/in^2 (1.05 kgf/cm^2)
Oil pump	Rotor type
Oil filter	Full flow, disposable cartridge
Relief valve operating pressure	50 lbf/in^2 (3.52 kgf/cm^2)
Relief valve spring free length	2.86 in (72.62 mm)
Sump capacity	7 pints (3.98 litres)

1798 cc (18V) engine

Specifications are as listed above for the 16V engine, with the following exceptions

General

Bore	3.16 in (80.26 mm)
Capacity	109.7 cu in (1798 cc)
Compression ratio:	
Low compression	6.9:1
High compression	9.0:1

Crankshaft

Main journal diameter	2.1262 to 2.1270 in (54.00 to 54.02 mm)
Minimum regrind diameter	2.0862 to 2.0870 in (52.99 to 53.01 mm)

Gudgeon pin

Type	Press-fit in connecting rod
Fit in piston	Hand push-fit at 60°F (16°C)
Fit in connecting rod	0.001 to 0.0017 in (0.02 to 0.04 mm) interference

Piston clearances

Up to engine no 18V/764E/L2294:	
Bottom of skirt	0.0006 to 0.0016 in (0.015 to 0.040 mm)
Top of skirt	0.0021 to 0.0037 in (0.053 to 0.094 mm)
From engine no 18V/764E/L2295:	
Bottom of skirt	0.0008 to 0.0018 in (0.020 to 0.046 mm)
Top of skirt	0.0012 to 0.0028 in (0.030 to 0.071 mm)

Piston rings

Compression:	
Type-top	Chrome-faced
Type-second	Tapered or stepped, cast iron alloy
Fitted gap (except stepped ring)	0.012 to 0.022 in (0.304 to 0.559 mm)
Fitted gap (stepped ring)	0.012 to 0.017 in (0.304 to 0.432 mm)
Ring to groove clearance	0.0015 to 0.0035 in (0.028 to 0.088 mm)
Oil control:	
Type	Twin chrome-faced rails with expander
Fitted gap	0.015 to 0.045 in (0.38 to 1.14 mm)

Valves

Valve to rocker clearance (cold)	0.013 in (0.33 mm)
Seat angle (inlet and exhaust)	45½°
Head diameter:	
Inlet	1.562 to 1.567 in (39.67 to 39.80 mm)
Exhaust	1.343 to 1.348 in (34.11 to 34.23 mm)
Stem diameter:	
Inlet	0.3429 to 0.3434 in (8.70 to 8.72 mm)
Exhaust	0.3423 to 0.3428 in (8.69 to 8.70 mm)
Stem to guide clearance:	
Inlet	0.0007 to 0.0019 in (0.02 to 0.05 mm)
Exhaust	0.0013 to 0.0025 in (0.03 to 0.06 mm)

Valve guides

Fitted height above spring seat:	
Inlet	0.75 in (19.05 mm)
Exhaust	0.625 in (15.37 mm)

Valve springs

Free length	1.92 to 1.94 in (48.77 to 49.27 mm)
Load at top of lift	142 lbf (64.4 kgf)

Lubrication system

Oil pressure (running)	70 lbf/in^2 (4.92 kgf/cm^2)
Relief valve operating pressure	60 lbf/in^2 (4.21 kgf/cm^2)
Relief valve spring free length	3.0 in (76.0 mm)
Sump capacity	7.65 pints (4.34 litres)

Torque wrench settings

	lbf ft	Nm
1622 cc engines		
Cylinder head nuts	50	68
Rocker bracket nuts	25	34
Main bearing setscrews	70	95
Big-end bolts	35	47
Gudgeon pin clamp bolts	25	34
Spark plugs	18	24
Flywheel nuts	35	47

	lbf ft	Nm
Distributor clamp bolt:		
Fixed nut type	4	6
Fixed bolt type	2.5	3
Timing cover set screws:		
$\frac{1}{4}$ in	6	8
$\frac{5}{16}$ in	14	19
Engine rear plate setscrews:		
$\frac{5}{16}$ in	20	27
$\frac{3}{8}$ in	30	41
Water pump bolts	17	23
Water outlet elbow	8	11
Manifold nuts	16	22
Crankshaft pulley nut	75	101
Sump setscrews	6	8
Camshaft nut	65	88
Oil release valve domed nut	43	58
Water pump pullet set screws	18	24
Thermal transmitter	16	22
1798 cc engine		
Main bearing nuts/setscrews	70	95
Flywheel setscrews	40	54
Big-end bolts	35	47
Big-end nuts (12-sided) – Oil thread	33 ± 2	45 ± 3
Cylinder head nuts	50	68
Crankshaft rear oil seal retainer bolts	20	27
Rocker bracket nuts	25	34
Oil pump to crankcase	14	19
Sump securing bolts	6	8
Cylinder side cover screws	4	5
Timing cover:		
$\frac{1}{4}$ in screws	6	8
$\frac{5}{16}$ in screws	14	19
Crankshaft pulley nut	70	95
Camshaft nut	65	88
Front plate:		
$\frac{5}{16}$ in setscrews	20	27
Rear plate:		
$\frac{5}{16}$ in screws	20	27
$\frac{3}{8}$ in screws	30	41
Water pump to crankcase	17	23
Pulley screws	18	24
Water outlet elbow nuts	8	11
Rocker cover nuts	4	5
Manifold nuts	15	20
Oil filter centre-bolt	15	20
Clutch to flywheel	28	37
Automatic gearbox torque converter to driveplate bolts	28	37
Distributor clamp nut (bolt trapped)	2.5	3
Spark plugs	18	24
Oil pipe banjo	37 max	50 max
Oil release valve – dome nut	40	54

1 General description

1 The 1622 cc (16V) and 1798 cc (18V) petrol engines installed in the Sherpa range are almost identical in external appearance but differ considerably internally. Both are conventional four cylinder, overhead valve type engines fitted with a single SU carburettor. The 16V engine has three main bearings on the crankshaft whereas the 18V engine has five main bearings. The 18V e gine is also available in either high compression or low compression versions.

2 In these engines, two valves per cylinder are mounted vertically in the cast iron cylinder head and run in pressed-in valve guides. They are operated by rocker arms and pushrods from the camshaft which is located at the base of the cylinder bores in the left hand side of the engine.

3 The cylinder head has all five inlet and exhaust ports on the left hand side. Cylinders 1 and 2 have a siamesed inlet port as have cylinders 3 and 4. Cylinders 1 and 4 have individual exhaust ports and cylinders 2 and 3 share a siamesed exhaust port.

4 The cylinder block and upper half of the crankcase are cast together and a pressed steel oil sump is bolted to the underside. Attached to the rear of the engine backplate is the clutch bellhousing and gearbox.

5 The dished crown pistons are made from anodised aluminium and they have a solid skirt. Two or three compression and one oil control ring are fitted to each piston depending on type of pistons fitted. The gudgeon pin is clamped in the 16V connecting rod but is a press fit in the 18V.

6 At the front of the engine is a chain driving the camshaft via the camshaft and crankshaft sprockets. The chain is tensioned by a spring loaded slipper type tensioner which automatically adjusts for chain stretch. The camshaft is supported by three steel backed white metal bearings. If these are renewed it is necessary to ream the bearings in position.

7 The overhead valves are operated by means of rocker arms mounted on the rocker shaft running along the top of the cylinder head. The rocker arms are activated by pushrods and tappets which in turn rise and fall in accordance with the lobes on the camshaft. The valves are held closed by single springs.

8 The static and dynamically balanced forged steel crankshaft is supported by three (16V engines) or five (18V engines) renewable main bearings. Crankshaft endfloat is controlled by four semi-circular thrust washers, two of which are located on either side of the centre bearing.

9 The centrifugal water pump and radiator cooling fan are driven, together with the alternator, from the crankshaft pulley wheel by a rub-

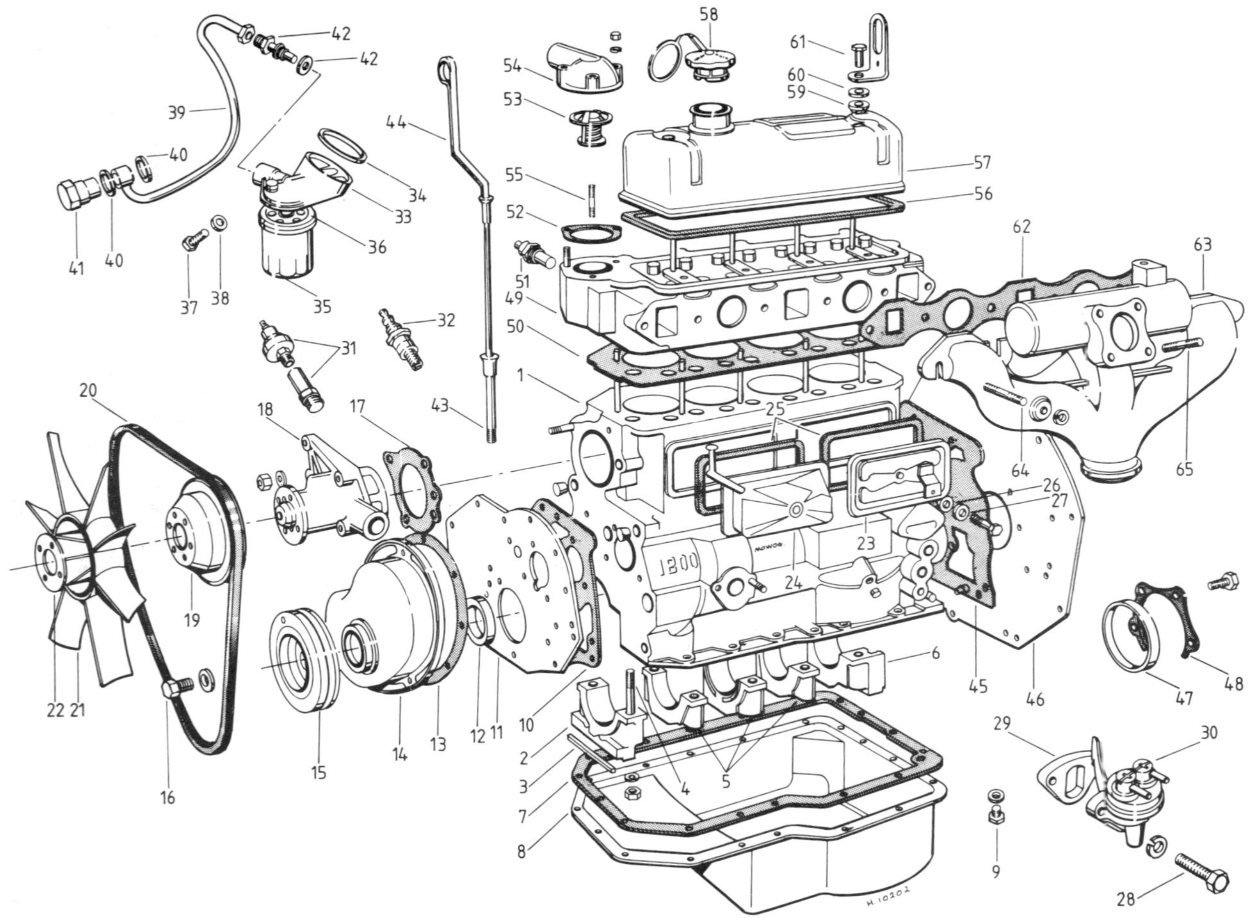

Fig. 1.1 18V engine external components (Sec 1)

1 Cylinder block
2 Front main bearing cap
3 Cap joint
4 Fixing stud
5 Main bearing caps
6 Rear main bearing cap
7 Sump gasket
8 Sump
9 Sump drain plug
10 Front mounting plate gasket
11 Front mounting plate
12 Oil seal
13 Timing cover gasket
14 Timing cover
15 Crankshaft pulley
16 Pulley bolt
17 Water pump gasket
18 Water pump
19 Pulley
20 Fan belt
21 Fan
22 Lockwasher
23 Cylinder side cover
24 Oil separator
25 Rear cover gaskets
26 Rubber seal
27 Cup washer
28 Petrol pump fixing bolt
29 Distance piece
30 Petrol pump
31 Oil pressure switch
32 Sparking plug
33 Oil filter head
34 Oil filter head seal
35 Oil filter cartridge
36 Oil filter cartridge seal
37 Oil filter head bolt
38 Bolt sealing washer
39 Lubrication pipe
40 Sealing washers
41 Banjo bolt
42 Adaptor and seal
43 Dipstick guide tube
44 Dipstick
45 Gasket
46 Engine backplate
47 Oil seal
48 Oil seal retainer
49 Cylinder head
50 Cylinder head gasket
51 Thermal transmitter
52 Thermostat housing gasket
53 Thermostat
54 Water outlet
55 Fixing stud
56 Rocker cover gasket
57 Rocker cover
58 Oil filler cap
59 Rubber seal
60 Cup washer
61 Rocker cover fixing
62 Manifold gasket
63 Manifold
64 Manifold fixing stud
65 Carburettor fixing stud

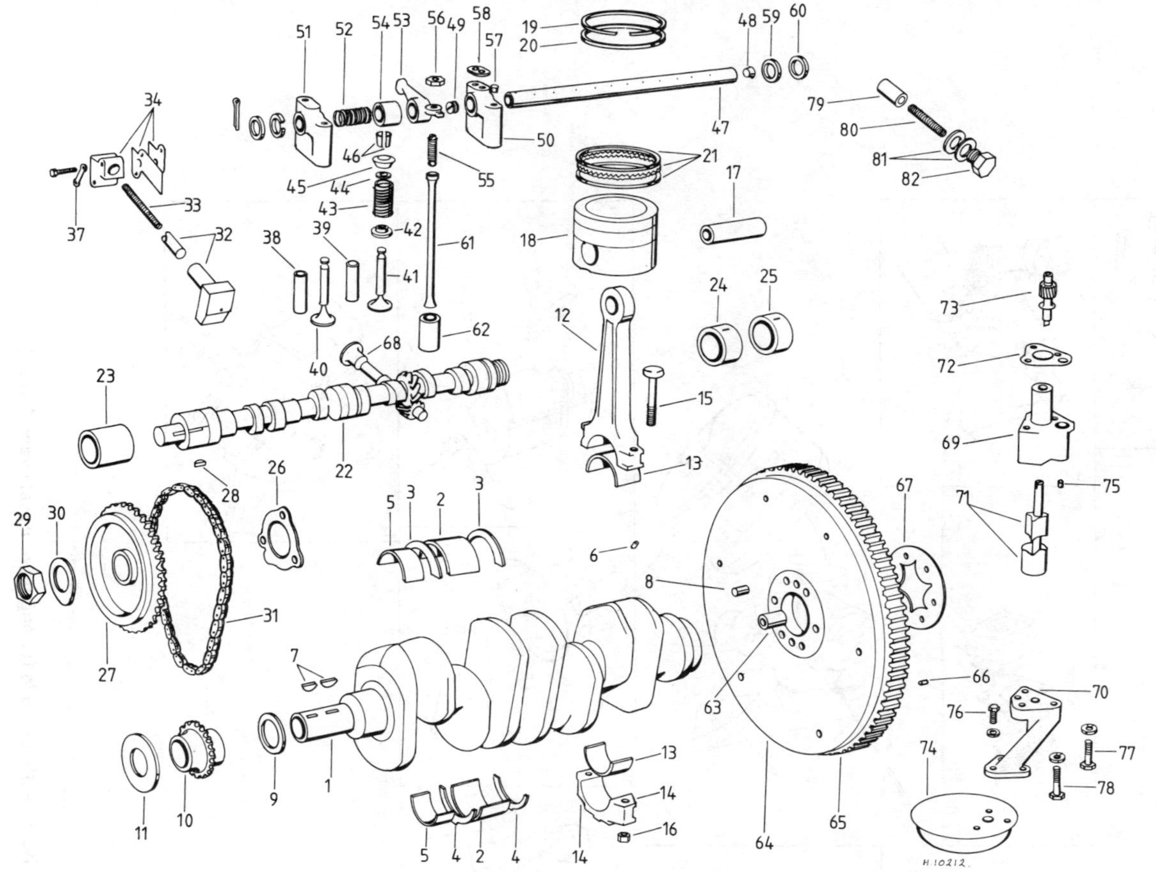

Fig. 1.2 18V engine internal components (Sec 1)

1 Crankshaft
2 Main bearing
3 Thrust washer – upper
4 Thrust washer – lower
5 Main bearing
6 Oil restrictor
7 Key for gear
8 Crankshaft dowel
9 Gear packing washer
10 Timing gear
11 Oil thrower
12 Connecting rod
13 Big end bearing
14 Big end bearing cap
15 Big end bearing cap bolt
16 Big end nuts
17 Gudgeon pin
18 Piston
19 Compression ring – top
20 Compression ring – 2nd
21 Scraper ring
22 Camshaft
23 Camshaft bearing – front
24 Camshaft bearing – centre
25 Camshaft bearing – rear
26 Camshaft locking plate
27 Timing gear
28 Timing gear key
29 Camshaft nut
30 Lock washer
31 Timing chain
32 Slipper head & cylinder
33 Spring
34 Tensioner body, plate & gasket
37 Lock washer
38 Exhaust valve guide
39 Inlet valve guide
40 Exhaust valve
41 Inlet valve
42 Valve spring collar
43 Valve spring
44 Packing ring
45 Spring cup
46 Valve cotter
47 Valve rocker shaft
48 Plain plug
49 Screwed plug
50 Bracket with tapped hole
51 Plain bracket
52 Spring
53 Valve rocker
54 Bush
55 Adjusting screw
56 Adjusting screw locknut
57 Locking screw
58 Locking plate
59 Spring washer
60 Washer for rocker shaft
61 Pushrod
62 Tappet
63 Crankshaft spigot bush
64 Flywheel
65 Starter ring
66 Dowel for clutch
67 Lock washer
68 Distributor drive shaft
69 Oil pump body
70 Oil pump cover
71 Oil pump rotors
72 Oil pump gasket
73 Oil pump drive spindle
74 Oil pump strainer
75 Pump cover to body dowel
76 Oil pump cover to strainer screw
77 Oil pump cover screw – short
78 Oil pump cover screw – long
79 Oil pressure relief valve
80 Spring for relief valve
81 Oil pressure relief valve washer
82 Oil pressure cap nut

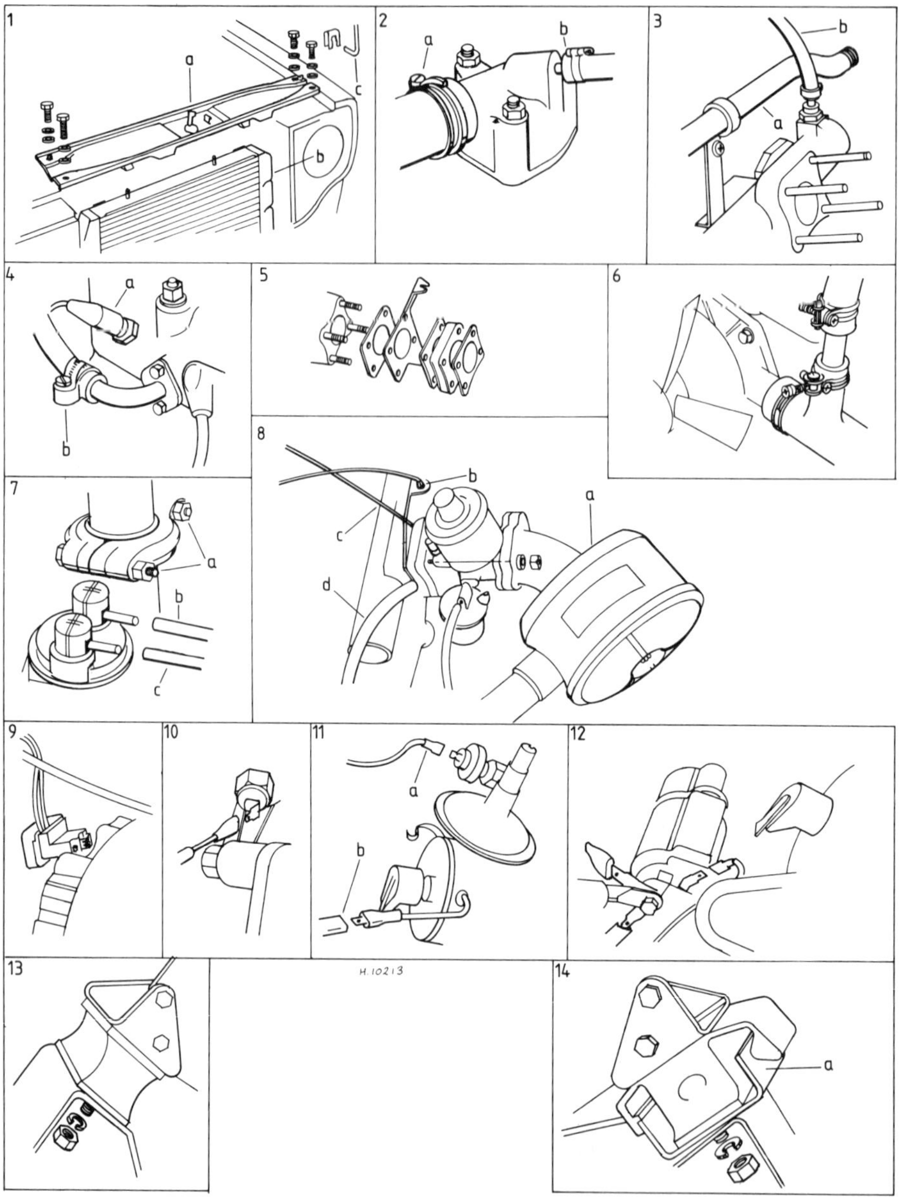

Fig. 1.3 Disconnection details for engine removal (Secs 5 and 6)

1 Removing the radiator
a Bonnet lock platform
b Radiator
c Bonnet stay and clip
2 Coolant hose disconnection
a Radiator top hose
b Air bleed hose (later models)
3 Heater (a) and brake servo (b) hoses
4 Spark plug leads (a), and heater hose (b) at elbow on rear of cylinder head
5 Carburettor fittings
6 Radiator bottom hose (later model shown)
7 Exhaust clamp (a), fuel-pump outlet (b) and inlet (c)
8 Air cleaner (a), throttle cable (b), vacuum pipe (c) and breather hose (d)
9 Alternator
10 Thermal transmitter (coolant temperature)
11 Oil pressure switch (a) and distributor lead (b)
12 Starter solenoid
13 Right-hand mounting
14 Left hand mounting and deflection stop bracket (a)

ber/fabric 'fan' belt. The distributor is mounted towards the rear of the right hand side of the cylinder block and advances and retards the ignition timing by mechanical and vacuum means. The distributor is driven at half crankshaft speed by a short shaft and skew gear from a skew gear on the camshaft. The oil pump is mounted inside the crankcase and driven from the camshaft by a short drive spindle.

10 Attached to the rear of the crankshaft by six bolts and two dowels is the flywheel to which is bolted the diaphragm spring clutch. Mounted on the circumference of the flywheel is the starter ring gear into which the starter motor drive engages when starting the engine.

11 To avoid repetition this part of Chapter 1 is mainly concerned with the 18V engine but where significant differences on the 16V engine arise these are specifically mentioned.

2 Major operations with engine in place

The following major operations can be carried out to the engine with it in place in the vehicle:

(a) Removal and refitting of the cylinder head assembly
(b) Removal and refitting of the sump
(c) Removal and refitting of the big-end bearings
(d) Removal and refitting of the pistons and connecting rods
(e) Removal and refitting of the timing chain and gears
(f) Removal and refitting of the camshaft
(g) Removal and refitting of the oil pump

3 Major operations with engine removed

The following major operations must be carried out with the engine out of the vehicle and on a bench or the floor:

(a) Removal and refitting of the main bearings
(b) Removal and refitting of the crankshaft
(c) Removal and refitting of the flywheel

4 Methods of engine removal

1 The engine can be removed either attached to the gearbox or on its own. Both methods are described.

2 It is easier if a hydraulic type trolley jack is used in conjunction with two pairs of axle stands so that the vehicle can be raised sufficiently to allow easy access underneath it. Overhead lifting tackle will be necessary in both cases.

3 Because of the weight consideration and the very steep angle to which the engine must be tilted, the do-it-yourself mechanic without the use of a pit or ramp should remove the gearbox first. A third method can be used and this is to detach the engine and gearbox from vehicle mountings and lower the unit onto the floor. The front of the vehicle can then be lifted up and the power unit drawn forwards. Using this method (possibly removing the cylinder head as well) can eliminate the need for an overhead hoist and may be of use in a confined space.

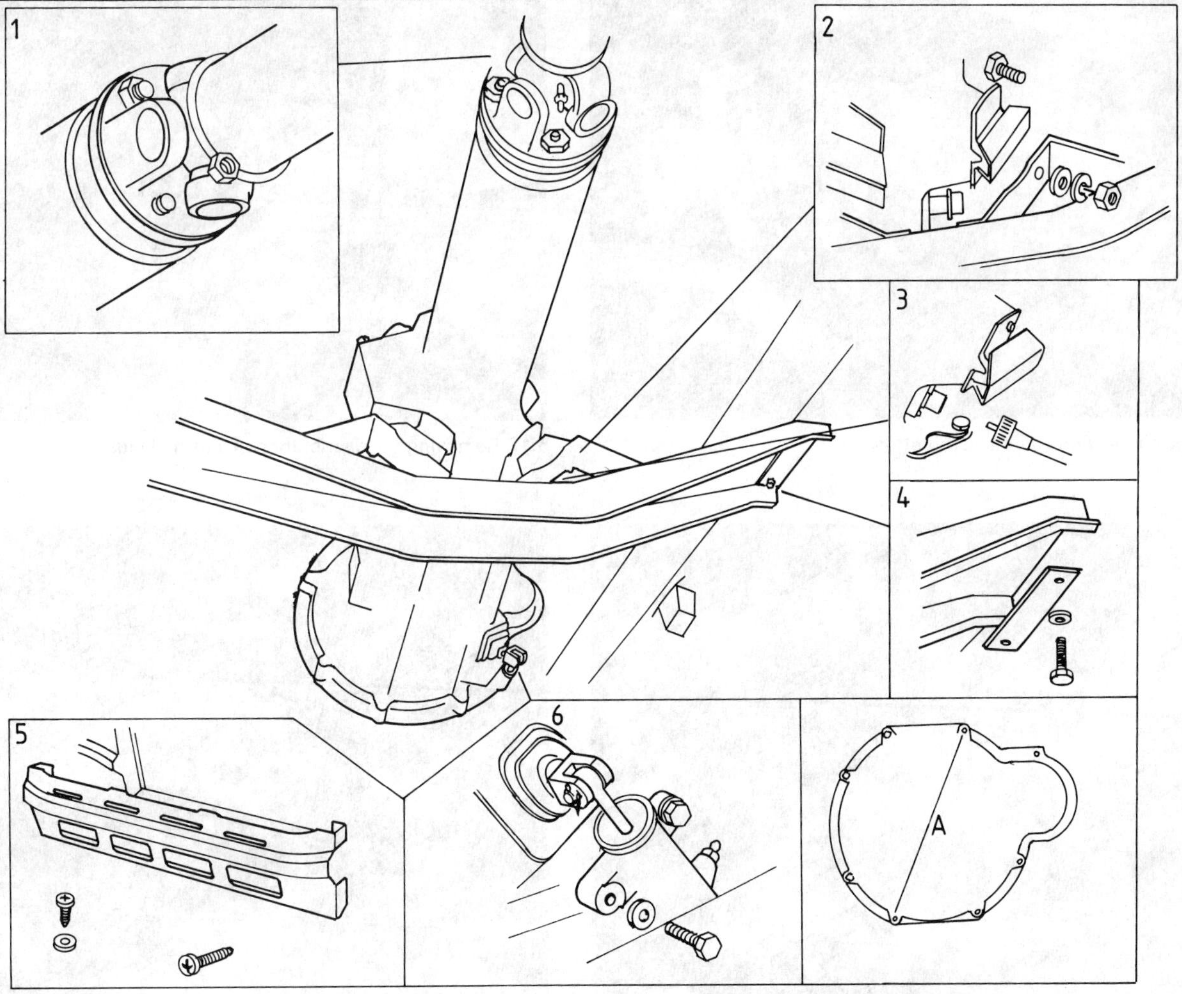

Fig. 1.4 Disconnection details for removing manual gearbox with engine (Sec 5)

1 *Gearbox/propeller shaft drive flanges; alignment marks (a)*
2 *Rear mounting*
3 *Speedometer cable*
4 *Rear crossmember reinforcing plates*
5 *Lower surround grille panel*
6 *Clutch slave cylinder*
A *First bolts to be fitted on reassembly*

5.2 Cylinder block drain plug

5.3 Battery earth terminal detachment

5.5 Pipe connections on the carburettor

5.16 Removing the distributor cap and HT leads

5.17A Detaching the cable terminal from the oil pressure switch

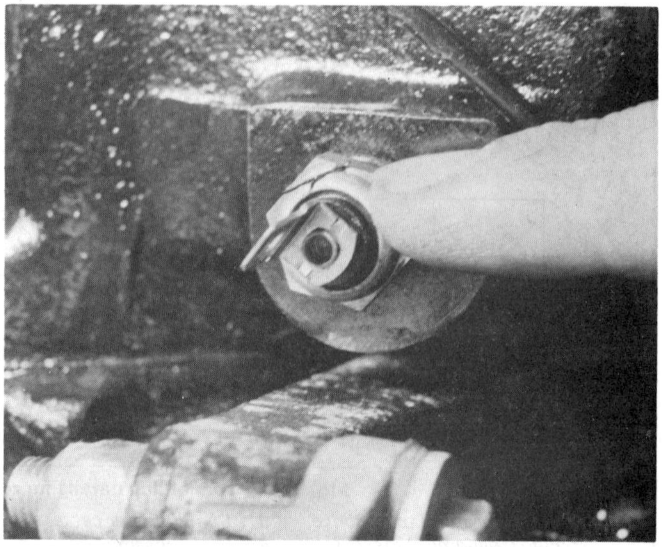

5.17B The thermal transmitter located in the side of the cylinder head

Note: *Vehicles fitted with automatic transmission which require engine and transmission removal should preferably have the transmission removed first as described in Chapter 6, Section 16, followed by the engine. This is because of the size and weight of the transmission.*

5 Engine – removal with gearbox

1 With the help of an assistant take the weight of the bonnet and undo and remove the four nuts and spring and plain washers securing the bonnet to the two hinges. Carefully lift the bonnet, releasing the prop, and carry it over the front of the vehicle. Store it in a safe place where it will not get damaged.
2 Refer to Chapter 2 and drain the cooling system (photo).
3 Disconnect the negative and then the positive battery cables and tuck them out of the way (photo).
4 Refer to Chapter 2 again and remove the radiator.
5 Carefully pull off the ignition vacuum pipe and the breather pipe from the carburettor (photo).
6 Loosen or cut the clip that secures the fuel tank feed pipe to the fuel pump. Pull off the hose and plug the bore with a pencil stub or similar peg. Disconnect the outlet pipe from the pump.
7 Disconnect the throttle outer cable and the throttle return spring from the throttle bracket.
8 On 18V engines with an automatic gearbox, disconnect the downshift cable from the carburettor linkage and throttle bracket.
9 Slacken the clips and pull off the brake servo pipe from the manifold.
10 Unclip the choke cable from the throttle cable.
11 Remove the four nuts and washers securing the carburettor to the manifold and remove the carburettor and the air cleaner as an assembly. Remove the carburettor insulator block, the throttle control bracket and the joint washer(s) from the manifold, noting the order in which they were assembled.
12 *Early models:* Disconnect the heater hose from the engine pipe at the rear of the manifold. *Later models:* Disconnect the heater hose from the heater and support bracket.
13 *Later models:* Disconnect the radiator bottom hose from the water pump and move the heater hose and bottom hose assembly out of the way to one side.
14 *Later models:* Disconnect the bleed hose from the water outlet elbows.
15 Undo the exhaust pipe clamp nuts, remove the clamp and the exhaust pipe from the manifold.
16 Mark the spark plug cables to ensure correct refitting and disconnect them from the plugs. Release the HT cable from the ignition coil, spring back the distributor clips and remove the cap and leads as an assembly (photo).
17 Disconnect the LT wiring from the distributor and the wiring from the oil pressure switch and the coolant temperature transmitter (photos). Make a note of the connections on the alternator and the starter motor and disconnect the wiring.
18 Disconnect the heater hose from the elbow at the rear of the cylinder head.
19 Apply the handbrake, jack-up the front of the vehicle and support it on stands.
20 Remove the nut and bolt which secure the starter motor earth strap and also, on automatic transmission models, the selector illumination lamp earth wire.
21 Place a rope sling or chain around the engine and support its weight with an overhead hoist. Make certain that there is no chance of this slipping as the unit has to be lifted out at a steep angle. To help the lift, try to arrange the lifting point to be just in front of the line of the No 3 cylinder stud (RH side) for manual cylinder models, or just behind the line for automatic transmission models.
22 Remove the seven screws and washers securing the lower surround front grille panel and remove the panel.
23 *Manual gearbox models:* Refer to Chapter 6 and remove the gear change lever from the gearbox.
24 Remove the propeller shaft and gearbox drive flanges for correct alignment when reassembling. Remove the four nuts and bolts from the flanges and tie the propeller shaft up to one side.
25 Undo the finger nut or remove the clamp and disconnect the speedometer cable from the gearbox (photo). Tuck it to one side.
26 *Manual gearbox models:* Remove the two bolts and washers securing the clutch slave cylinder to the gearbox and withdraw the cylinder from the pushrod. Move the cylinder clear of the gearbox assembly and tie it up out of the way. There is no need to disconnect the hydraulic hose.
27 *Manual gearbox models:* Remove the two bolts and nuts securing each rear mounting to the rear crossmember.
28 *Manual gearbox models:* Support the gearbox on a trolley jack and remove the two bolts and reinforcing plate securing each end of the rear crossmember to the body. Remove the crossmember.
29 *Automatic transmission models:* Disconnect the selector rod from both levers and remove the rod.
30 *Automatic transmission models:* Disconnect the oil cooler pipes from the transmission. Plug the pipes and unions to stop dirt getting in and to avoid loss of fluid.
31 *Automatic transmission models:* Disconnect the inhibitor switch leads at the harness multi-connector.
32 *Automatic transmission models:* Slacklen the central nut securing the rear mounting to the crossmember and remove the two nuts and bolts securing the rear mounting to the transmission.
33 Remove the two nuts which secure each engine mounting to the front crossmember brackets.
34 Lift the engine assembly a small amount to free the front mountings and remove the deflection stop bracket from the left-hand mounting.
35 Check all round the engine and gearbox to make certain that nothing has been left connected and then carefully raise, pull and tilt the assembly up and forward out of the vehicle. Keep checking that nothing gets caught up until the unit is well out and then park it where it can be cleaned before anything else is done to it.
36 Refitting an engine and gearbox assembly is the reverse of the removal procedure. Fit the gearbox rear mounting bolts and nuts before locating the engine front mountings. Make sure that the engine and gearbox are both filled with the correct quantity and grades of oil or fluid after fitting.

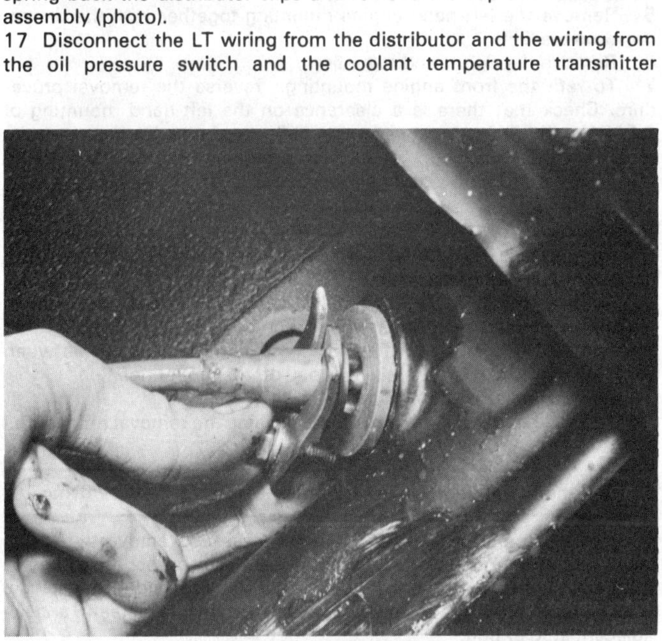
5.25 Removing the speedometer cable, clamp type retainer

6 Engine – removal without gearbox

1 Where it is not intended to work on the gearbox the engine can be removed on its own and this is somewhat easier than removing it with the gearbox.
2 Follow the procedure listed in the previous Section until the starter motor earth strap is removed (paragraph 20).
3 *Automatic transmission models:* Remove the cover from the mounting plate and then remove the four bolts retaining the torque converter to the driveplate. The crankshaft will have to be turned to bring the bolts into view.
4 Remove the two nuts and washers securing each engine mounting to the front crossmember brackets.
5 Place a rope sling or chain around the engine and support its weight with an overhead hoist. Make certain that there is no chance of

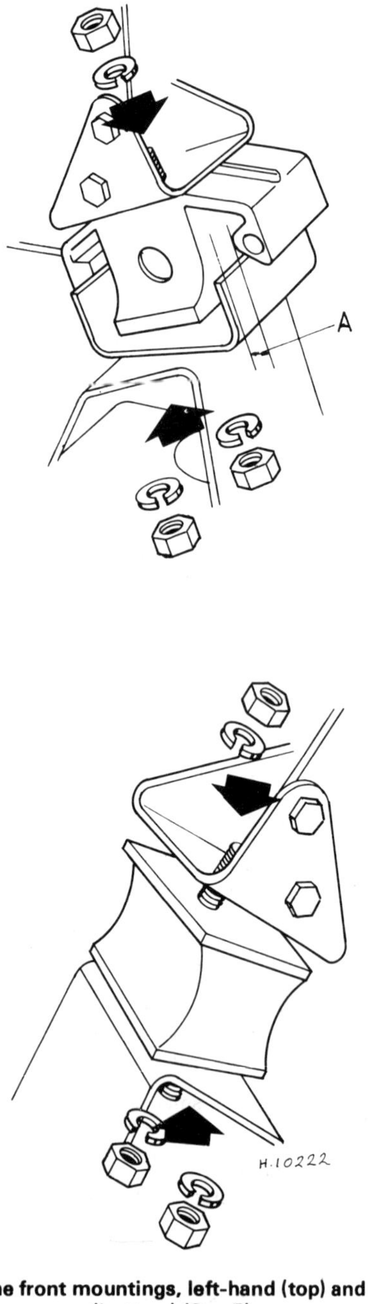

Fig. 1.5 Engine front mountings, left-hand (top) and right-hand (bottom) (Sec 7)
A minimum static clearance 0.312 in (7.92 mm)

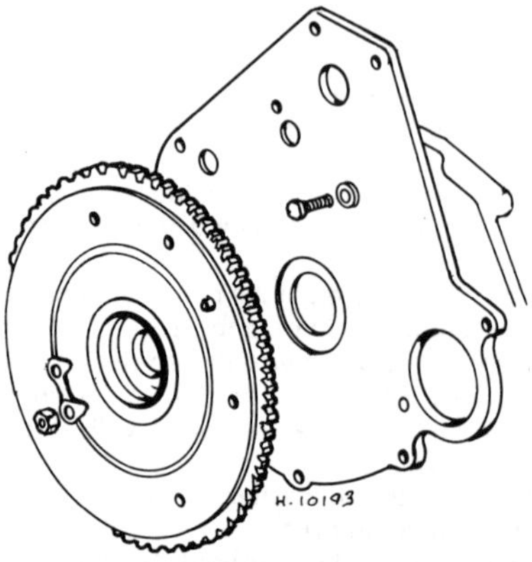

Fig. 1.6 Gearbox adaptor plate and flywheel (16V engine) (Sec 8)

this slipping as the engine has to be tilted during removal. To help the lift try to arrange the lifting point to be in line with No 3 cylinder head stud (RH side) for manual gearbox models or just in front of the line for automatic gearbox models.

6 Remove the two top bolts retaining the gearbox to the mounting plate.

7 Lift the engine a small amount to free its front mountings and remove the deflection stop bracket from the left-hand mounting.

8 Support the gearbox with a jack and a piece of wood between the two. Remove the remaining bolts securing the gearbox to the mounting plate.

9 *Manual gearbox models:* Check that all disconnections have been made and then carefully move the engine forward clear of the gearbox input shaft. Slowly lift the engine, turning it to the left to clear the bulkhead, and remove the engine from the vehicle.

10 *Automatic transmission models:* Check that all disconnections have been made. Carefully move the engine forward to clear the torque converter spigot and slowly remove the engine from the vehicle.

11 Refitting an engine is the reverse of the removal procedure, but the following additional points require attention:

- *(a) Manual gearbox: Lubricate the bore of the crankshaft spigot bush with high melting point grease before fitting. If the clutch has been dismantled it must be centralised (Chapter 5)*
- *(b) Manual gearbox: When fitting the gearbox to mounting plate bolts fit the two bolts marked A in Fig. 1.4 first*
- *(c) Automatic transmission: Tighten the torque converter to driveplate bolts to the specified torque*

12 Finally, make sure that the engine is filled with the correct quantity and grade of oil after refitting.

7 Engine mountings – removal and refitting

1 Abnormal vibration could be due to loose or decayed engine mountings. First check the mounting nuts for tightness. If the mounting rubbers are split, misshapen or otherwise damaged they can be changed without removing the engine. Proceed as follows.

Front mountings

2 Support the engine with a jack and a piece of wood, to spread the load, between the jack and the engine.

3 Remove the nut and spring washer which secure the engine mountings to the bracket attached to the cylinder block.

4 Remove the two nuts and spring washers which secure the engine mounting to the mounting bracket on the front crossmember.

5 Remove the left-hand engine mounting together with the deflection stop bracket.

6 Remove the right-hand mounting.

7 To refit the front engine mountings, reverse the removal procedure. Check that there is a clearance on the left-hand mounting of 0.312 in (7.92 mm) between the deflection stop bracket and the mounting shoe (Fig. 1.5). If this clearance cannot be maintained with the engine static, a new mounting must be fitted.

Rear mountings

8 Remove the two bolts and spring washers which secure each mounting to the gearbox casing.

9 Remove the two nuts and bolts with plain and spring washers which secure each mounting to the rear crossmember.

10 Support the gearbox with a jack, with a piece of wood between the two, and raise it to just take the load off the mountings.

11 Remove the rear mountings.

12 To refit the rear engine mountings, reverse the removal procedure.

8 Flywheel and gearbox adaptor plate – removal and refitting

1 If they are still assembled to the engine, first remove the clutch (see Chapter 5) or the gearbox (see Chapter 6) as appropriate.

2 Lock the flywheel to prevent it turning. This can be done by using a suitable bolt or piece of metal between the flywheel teeth and an adjacent stud or bolt.

3 Bend back the tabs on the lockwasher(s) and undo and remove the bolts (18V) or nuts (16V) securing the flywheel to the crankshaft.

4 If it is intended to refit the same flywheel, make ajdacent marks on the crankshaft and the flywheel to ensure correct reassembly.
5 Withdraw the flywheel.
6 Remove the screws securing the gearbox adaptor plate to the cylinder block. Make a record of the different sizes of screw and washers so that they can be refitted in their correct positions. Lift away the backplate from the block together with its gasket – two in the case of the 18V engine.
7 Refitting the gearbox adaptor plate and flywheel is the reverse of the removal sequence, but the following points should be noted:

(a) Use new gaskets on reassembly
(b) Use new lockwashers for the flywheel securing nuts or bolts (three lockwashers for 16V engines, one for 18V engines)

16V engines

(c) Tighten the plate retaining screws to their specified torques
(d) Tighten the flywheel securing nuts to the specified torque

18V engines

(e) Smear the drive flange of the crankshaft with clean SAE 90 EP oil
(f) Consider the value of fitting a new oil seal for the crankshaft rear end. At this stage of dismantling it is false economy not to. Prise the old seal out of the plate. Lubricate the new seal with SAE 90 EP oil and carefully tap it into the plate. Take great care not to damage it, make sure that the seal lips face into the crankshaft and that the back of the seal is flush with the adaptor plate
(g) When fitting the adaptor plate, carefully manoeuvre the seal into position as it passes over the end of the crankshaft
(h) Tighten the plate bolts to the specified torque. Note that the two larger bolts locate at the top of the plate in the non-countersunk holes
(j) Fit a new retainer for the rear oil seal and tighten the seal retainer securing bolts to the specified torque
(k) Tighten the flywheel securing bolts to the specified torque

9 Dismantling the engine – general

1 It is best to mount the engine on a dismantling stand, but if one is not available, stand the engine on a strong bench, to be at a comfortable working height. It can be dismantled on the floor but it is not easy.
2 During the dismantling process greatest care should be taken to keep the exposed parts free from dirt. As an aid to achieving this, thoroughly clean down the outside of the engine, removing all traces of oil and congealed dirt.
3 Use paraffin or water soluble solvent. The latter compound will make the job much easier for, after the solvent has been applied and allowed to stand for a time, a vigorous jet of water will wash off the solvent with all the grease and dirt. If the dirt is thick and deeply embedded, work the solvent into it with a wire brush.
4 Finally wipe down the exterior of the engine with a rag and only then, when it is quite clean, should the dismantling process begin. As the engine is stripped, clean each part in a bath of paraffin.

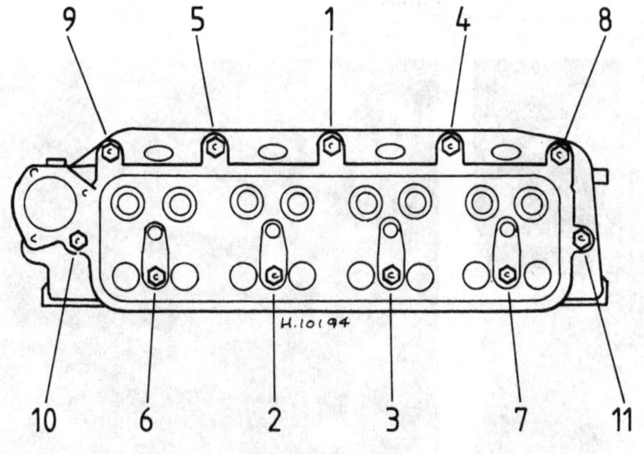

Fig. 1.7 Cylinder head nut slackening and tightening sequence (Secs 11, 12 and 59)

5 Never immerse parts with oilways (for example the crankshaft) in paraffin but, to clean, wipe down carefully with a petrol dampened cloth. Oilways can be cleaned out with wire. If an air line is available, all parts can be blown dry and the oilways blown through as an added precaution.
6 Re-use of old engine gaskets is false economy and will lead to oil and water leaks, if nothing worse. Always use new gaskets throughout.
7 Do not throw the old gasket away, for it sometimes happens that an immediate replacement cannot be found and the old gasket is then very useful as a template. Hang up the old gaskets as they are removed.
8 To strip the engine it is best to work from the top down. The underside of the crankcase when supported on wood blocks acts as a firm base. When the stage is reached where the crankshaft and connecting rods have to be removed, the engine can be turned on its side and all other work carried out with it in this position.
9 Whenever possible, refit nuts, bolts and washers finger tight from wherever they were removed. This helps avoid loss and muddle later. If they cannot be refitted lay them out in such a fashion that it is clear from whence they came.

10 Engine ancillary components – removal

Before basic engine dismantling begins it is necessary to strip it of ancillary components as follows:

(a) Alternator
(b) Distributor
(c) Thermostat
(d) Oil filter cartridge
(e) Inlet and exhaust manifold

It is possible to strip all these items with the engine in the vehicle if it is merely the individual items that require attention. Presuming the engine to be out of the vehicle and on the bench and that the item mentioned is still on the engine, follow the procedure described below.
1 Slacken off the alternator retaining bolts and nuts and remove the unit together with its adjustment link.
2 To remove the distributor first disconnect the vacuum advance/retard pipe from the side of the distributor. Undo and remove the two screws with spring and plain washers that secure the distributor clamp flange to the cylinder block. Lift away the distributor and clamp flange.
3 Remove the thermostat cover by undoing and removing the three nuts and spring washers which hold it in position. Lift away the cover and gasket, followed by the thermostat itself.
4 Remove the oil filter cartridge by simply unscrewing it from the oil filter head on the side of the cylinder block.
5 Undo and remove the six nuts and washers that secure the inlet and exhaust manifold assembly to the side of the cylinder head. If the carburettor is still mounted on the inlet manifold release it from the petrol feed pipe from the pump. Lift off the manifold assembly and recover the gasket.
6 Remove the mechanical fuel pump by unscrewing the two retaining nuts and spring washers which hold it to the block. Release it from the petrol feed pipe to the carburettor float chamber and lift away the fuel pump.
7 The engine is now stripped of ancillary components and is ready for major dismantling to begin.

11 Cylinder head removal – engine in vehicle

1 Drain the cooling system as described in Chapter 2.
2 For safety reasons disconnect the negative and then the positive battery terminals and tuck the leads to the rear of the battery.
3 Pull off the feed pipe from the carburettor float chamber union and plug the end of the pipe with a pencil (photo).
4 Slacken the hose clip and remove the breather hose from the carburettor body (photo).
5 Pull the distributor vacuum advance/retard hose union from the carburettor body (photo).
6 Slacken the accelerator cable to linkage securing nut.
7 Press in the two accelerator cable retainer ears on the underside of the support bracket and pull the cable through the bracket.

11.3 Detaching the fuel feed from the carburettor

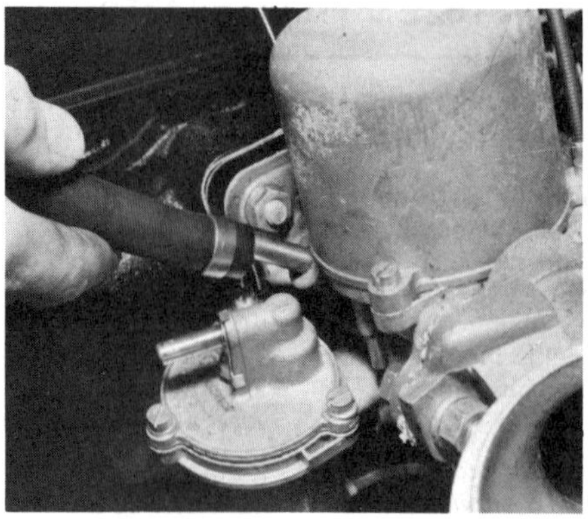

11.4 Detaching the breather hose from the carburettor

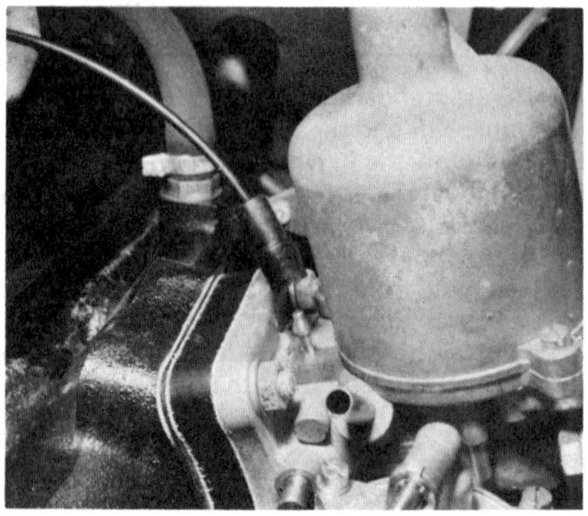

11.5 Detaching the vacuum hose from the carburettor

11.9 Detaching the radiator top hose from the thermostat housing

11.10 Disconnecting the thermal transmitter cable

11.11 Removing the HT leads from the spark plugs

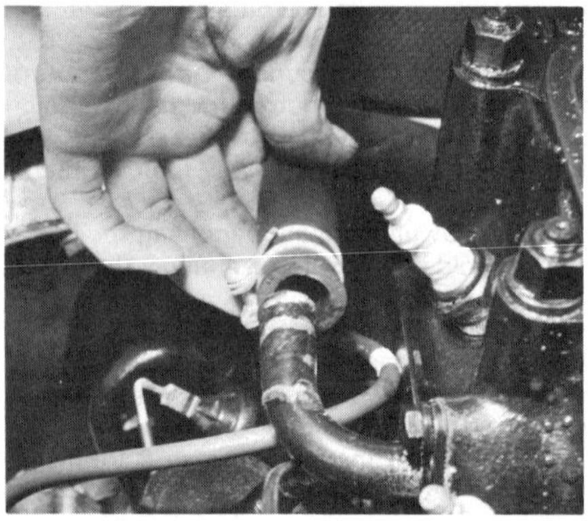

11.12 Removing the heater hose from the cylinder head

11.13 Removing the vacuum hose from the manifold

11.14 The rocker cover securing bolts being removed

11.15 Lifting off the rocker cover

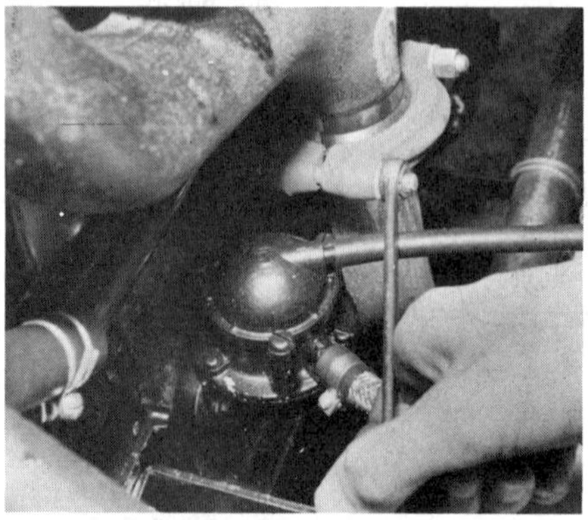

11.16 Undoing the exhaust manifold to pipe clamp

11.17 Lifting off the thermostat cover followed by ...

11.18 ... the thermostat

11.19 Undoing a rocker shaft pedestal securing nut

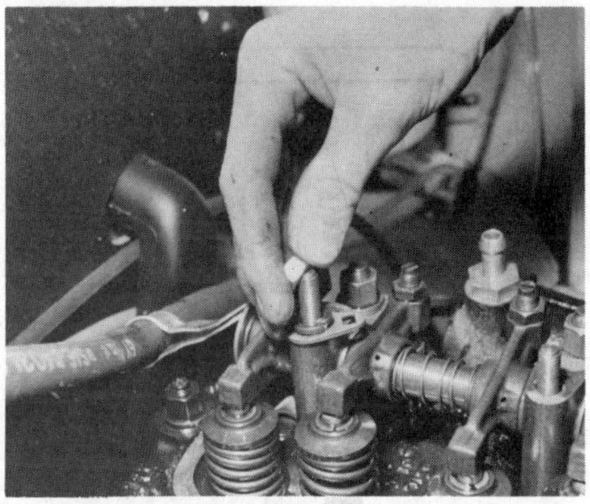
11.20 Taking off the lockwasher from the rocker pedestal

11.21 Lifting the rocker shaft assembly from the cylinder head

11.22 Taking out the pushrods

11.23 Special shims located under the rocker shaft pedestals

11.24 Lifting off the cylinder head

11.25 Removing the cylinder head gasket

8 Slacken the nut securing the choke control cable to the choke lever and pull the complete cable through its location on the side of the carburettor.

9 Slacken the top hose to thermostat housing securing clip and pull off the hose (photo).

10 Detach the Lucar terminal from the thermal transmitter located beneath the thermostat housing (photo).

11 Mark the spark plug HT leads to ensure correct refitting and detach the leads from the spark plugs (photo).

12 Slacken the clip securing the heater hose to the angled pipe on the left hand side of the cylinder head. Pull the hose from the pipe (photo).

13 Slacken the clip securing the vacuum servo unit hose to the union on the rear branch of the inlet manifold. Pull off the hose and tuck back on the bulkhead (photo).

14 Undo and remove the two rocker cover securing bolts together with the spacer, plain washer and seal (photo).

15 Lift away the rocker cover and its gasket (photo).

16 Undo and remove the two nuts, plain washers and bolts that secure the clamp around the exhaust downpipe to manifold joint (photo). Lift away the clamp halves.

17 Undo and remove the three nuts and spring washers securing the thermostat cover to the top of the cylinder head (photo). Lift the cover from the three studs and recover the gasket.

18 The thermostat may now be lifted out from its location in the cylinder head (photo).

19 Slacken the eleven cylinder head nuts, half a turn at a time, in the sequence shown in Fig. 1.7. Remove the nuts and washers. Progressively slacken the remaining four nuts which secure the rocker shaft pedestals; remove the nuts and spring washers (photo).

20 Recover the locking washer from the rear rocker shaft pedestal (photo).

21 Lift the rocker shaft assembly from the top of the cylinder head (photo).

22 Remove the pushrods, keeping them in the relative order in which they were removed. The easiest way to do this is to push them through a sheet of thin card in the correct sequence (photo).

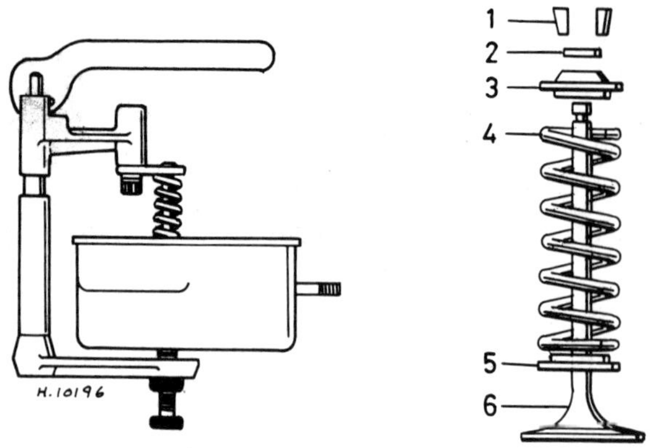

Fig. 1.8 Valve spring removal (Sec 13)

1	*Collets*	*3*	*Top cup*	*5*	*Lower cup*
2	*Oil seal*	*4*	*Spring*	*6*	*Valve*

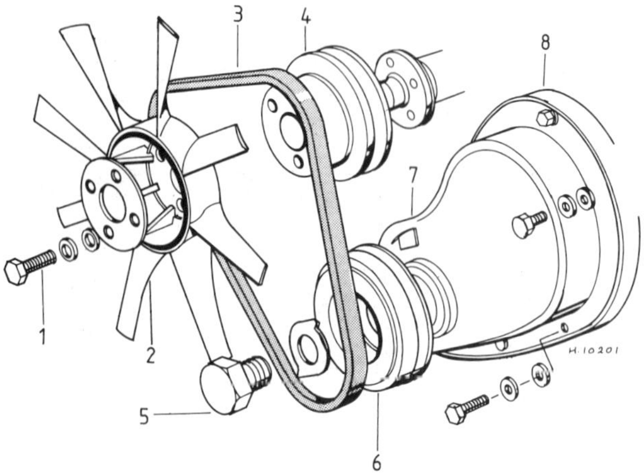

Fig. 1.10 Timing cover removal (Sec 16)

1	*Fan and pulley securing bolt, spring and plain washer*	*5*	*Crankshaft pulley securing bolt and lock washer*
2	*Fan*	*6*	*Crankshaft pulley*
3	*Fan belt*	*7*	*Timing cover*
4	*Fan pulley*	*8*	*Gasket*

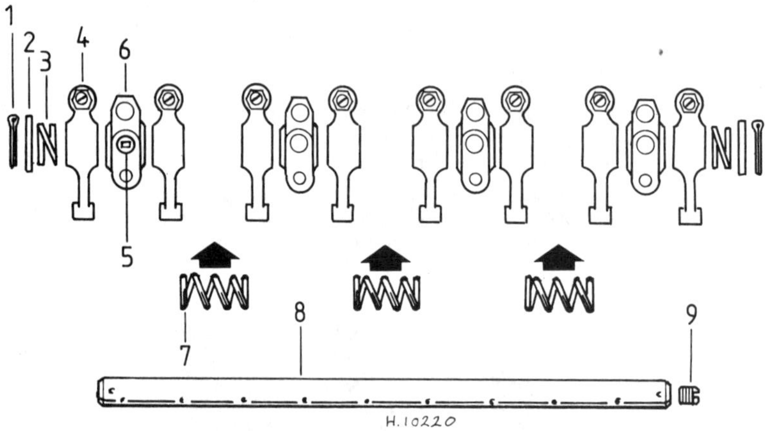

Fig. 1.9 Valve rocker assembly components (Sec 15)

1 Split-pin
2 Plain washer
3 Spring washer
4 Rocker arm
5 Locking screw
6 Pedestal
7 Spring
8 Rocker shaft
9 End plug

23 Look for any shims located on the rocker shaft retaining studs and carefully lift them from the studs (photo).

24 The cylinder head can now be removed by lifting upwards. If the head is jammed, try to rock it to break the seal. Under no circumstances try to prise it apart from the block with a screwdriver or cold chisel as damage may be done to the faces of the head and block. If the head will not free readily, turn the engine over using the starter motor as the compression in the cylinders will often break the cylinder head joint. If this fails to work, strike the head sharply with a plastic or wooden hammer, or with a metal hammer onto a piece of wood on the side of the head. Under no circumstances hit the head directly with a metal hammer as this may cause the iron casting to fracture. Several sharp taps with a hammer, at the same time pulling upwards, should free the head. Lift the head off and place to one side (photo).

25 Recover the old cylinder head gasket (photo).

12 Cylinder head removal – engine on bench

The sequence for removal of the cylinder head with the engine on the bench is basically identical to the latter operations for removal with the engine in the car. Refer to Section 11 and follow the instructions given in paragraphs 14, 15 and 17 to 25.

13 Valves – removal

1 The valves are easily removed from the cylinder head by the following method. Compress each spring in turn with a universal valve spring compressor until the two halves of the collets can be removed. Release the compressor and lift away the spring top cup, the spring, oil seal, lower cup and the valve (Fig. 1.8).

2 If, when the valve spring compressor is screwed down, the valve spring top cup refuses to free and expose the split collet, do not continue to screw down on the compressor as there is a likelihood of damaging it.

3 Gently tap the top of the tool directly over the cup with a light hammer. This should free the cup. To avoid the compressor jumping off the valve retaining cup when it is tapped, hold the compressor firmly in position with one hand.

4 It is essential that the valves are kept in their correct sequence unless they are so badly worn that they are to be renewed. If they are going to be re-used place them in a sheet of card having eight holes numbered 1 to 8 corresponding to the relative positions the valves were in when fitted. Also keep the valve springs, cups etc in this same correct order.

14 Valve guides – removal

Valve guide removal is a simple task but it is not recommended that you should do this because their refitting is too difficult to do accurately. It is far better to leave their removal and insertion to a BL garage (see also Section 41).

15 Rocker assembly – dismantling

1 To dismantle the rocker assembly, release the rocker shaft locating screw from the rear pedestal, remove the split-pin, flat and spring washers from each end of the shaft and slide from the shaft the

pedestals, rocker arms and rocker spacer springs (Fig. 1.9).
2 From the end of the shaft undo the plug which gives access to the inside of the rocker which can now be cleaned of sludge etc. Ensure the rocker arm lubricating holes are clear.

16 Timing cover, tensioner, gears and chain – removal

The timing cover, gears and chain can be removed with the engine in the vehicle provided that the radiator is removed first. The procedure for removing the timing cover, tensioner, gears and chain is otherwise the same irrespective of whether the engine is in the vehicle or on the bench.
1 Bend back the locking tab of the crankshaft pulley locking washer under the crankshaft pulley retaining bolt.
2 Using a large socket undo and remove the bolt and lock washer.
3 Placing two large screwdrivers or tyre levers behind the crankshaft pulley wheel at 180° to each other, carefully lever off the pulley. It is preferable to use a proper extractor if this is available but the large screwdrivers or tyre levers are quite suitable providing care is taken not to damage the pulley flange.
4 Remove the Woodruff key from the crankshaft nose with a pair of pliers and note how the groove in the pulley is designed to fit over it. Place the Woodruff key in a glass jar for it is a very small part and can easily become lost.
5 Unscrew the bolts holding the timing cover to the block. NOTE that three different sizes of bolt are used, and that each bolt makes use of a large flat washer as well as a spring washer.
6 Pull off the timing cover and its gasket.
7 With the timing cover off, take off the oil thrower noting which way round it is fitted.
8 Unlock the two screws retaining the tensioner by bending back the lock tabs and remove the screws. Note that the slipper head is under spring tension and components could be lost unless care is taken when removing the tensioner assembly.
9 Allow the spring loading against the slipper head to relax and withdraw the slipper head, the spring and the inner cylinder from the tensioner.
10 Bend back the locking tab on the washer under the camshaft retaining nut and unscrew the nut. Lift away the nut and lockwasher. Ease each timing gear forwards a little at a time by levering behind each gearwheel in turn with two large screwdrivers or tyre levers at 180° to each other. If the gearwheels are locked solid then it will be necessary to use a proper gearwheel and pulley extractor, and if one is available this should be used anyway in preference to levers. With both

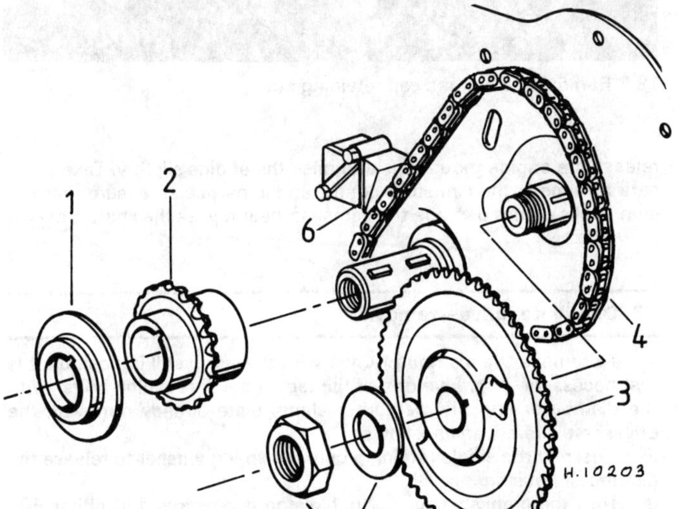

Fig. 1.11 Tensioner, timing gears and chain (Sec 16)

1 *Oil thrower*
2 *Crankshaft gearwheel*
3 *Camshaft gearwheel*
4 *Timing chain*
5 *Camshaft gearwheel retaining nut and washer*
6 *Tensioner*

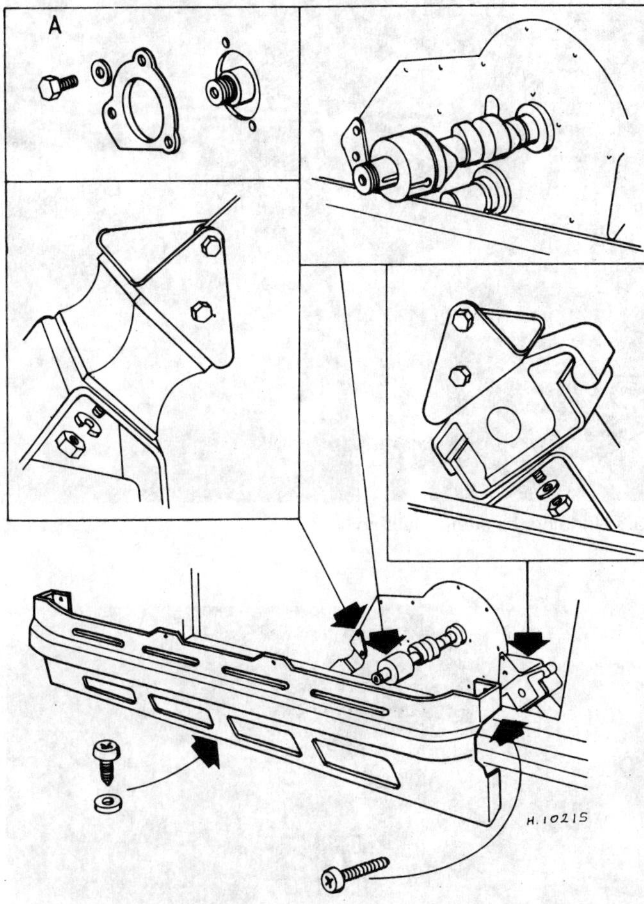

Fig. 1.12 Camshaft removal with engine in vehicle (Sec 17)

A Camshaft locating plate

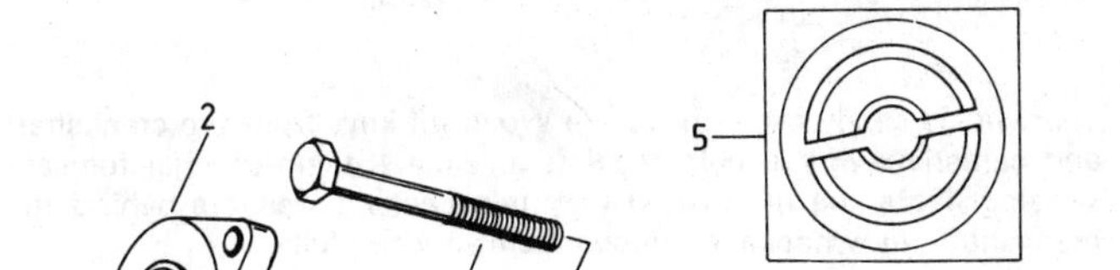

Fig. 1.13 Distributor drive (Sec 18)

1 *Distributor clamp securing bolt*
2 *Distributor housing*
3 *Screw – distributor housing to cylinder block*
4 *Distributor drive shaft*
5 *Inset: Correct position of drive shaft slot*

19.3 Undoing the sump securing bolts

19.5 Removing the oil pump strainer

19.6 Removing the oil pump from its mounting studs

19.7 Removing a big-end cap retaining nut

gearwheels safely off, remove the Woodruff keys from the crankshaft and camshaft with a pair of pliers and place them in a jar for safe keeping. Note the number of very thin packing washers behind the crankshaft gearwheel and remove them very carefully.

17 Camshaft – removal

1 The camshaft can be removed with the engine on the bench or in the vehicle. If the camshaft is to be removed with the engine in the vehicle, partly drain the cooling system and remove the hose from the water pump and heater pipe. The timing cover, gears and chain must be removed as described in Section 16. The distributor driveshaft must be removed as described in Section 18, and the fuel pump must also be removed. The front lower surround grille panel will have to be removed to allow the camshaft to be removed forwards through the front of the bodywork. With the drivegear out of the way proceed as follows.
2 Undo and remove the three bolts and spring washers which hold the camshaft locating plate to the block. The bolts are normally covered by the camshaft gearwheel.
3 Remove the plate. Recover the tappets if they are still in place and keep in the correct order as fitted in the engine.
4 The camshaft can now be withdrawn. It will be found necessary to release the engine mountings and raise the engine slightly. Take great care to remove the camshaft gently, and in particular, ensure that the cam peaks do not damage the camshaft bearings as the shaft is drawn forwards.

18 Distributor drive – removal

1 To remove the distributor drive with the sump still in position it is first necessary to remove one of the tappet cover retaining bolts. With the distributor and the distributor clamp plate already removed, the driveshaft is removed as follows.
2 Unscrew the single retaining bolt and spring washer to release the distributor housing.
3 Turn the crankshaft until No 1 piston is positioned at either 90° before or after the TDC position so that the pistons are halfway up the bores.
4 With the distributor housing removed, if the sump is still in position, screw into the end of the distributor drive shaft a 5/16 UNF bolt. A tappet cover bolt is ideal for this purpose. The drive shaft can then be lifted out, the shaft being turned slightly in the process to free the shaft skew gear from the camshaft skew gear.
5 If the sump has already been removed then it is a simple matter to push the driveshaft out from inside the crankcase.

Are your plugs trying to tell you something?

Normal.
Grey-brown deposits, lightly coated core nose. Plugs ideally suited to engine, and engine in good condition.

Heavy Deposits.
A build up of crusty deposits, light-grey sandy colour in appearance.
Fault: Often caused by worn valve guides, excessive use of upper cylinder lubricant, or idling for long periods.

Lead Glazing.
Plug insulator firing tip appears yellow or green/yellow and shiny in appearance.
Fault: Often caused by incorrect carburation, excessive idling followed by sharp acceleration. Also check ignition timing.

Carbon fouling.
Dry, black, sooty deposits.
Fault: over-rich fuel mixture.
Check: carburettor mixture settings, float level, choke operation, air filter.

Oil fouling.
Wet, oily deposits. Fault: worn bores/piston rings or valve guides; sometimes occurs (temporarily) during running-in period.

Overheating.
Electrodes have glazed appearance, core nose very white – few deposits. Fault: plug overheating. Check: plug value, ignition timing, fuel octane rating (too low) and fuel mixture (too weak).

Electrode damage.
Electrodes burned away; core nose has burned, glazed appearance. Fault: pre-ignition. Check: for correct heat range and as for 'overheating'.

Split core nose.
(May appear initially as a crack). Fault: detonation or wrong gap-setting technique. Check: ignition timing, cooling system, fuel mixture (too weak).

WHY DOUBLE COPPER IS BETTER FOR YOUR ENGINE.

Champion Double Copper plugs are the first in the world to have copper core in both centre and earth electrode: This innovative design means that they run cooler by up to 100°C – giving greater efficiency and longer life. These double copper cores transfer heat away from the tip of the plug faster and more efficiently. Therefore, Double Copper runs at cooler temperatures than conventional plugs giving improved acceleration response and high speed performance with no fear of pre-ignition.

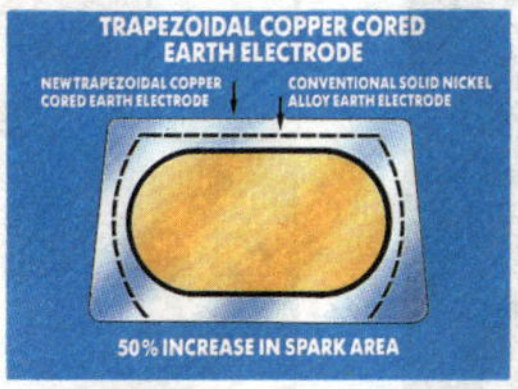

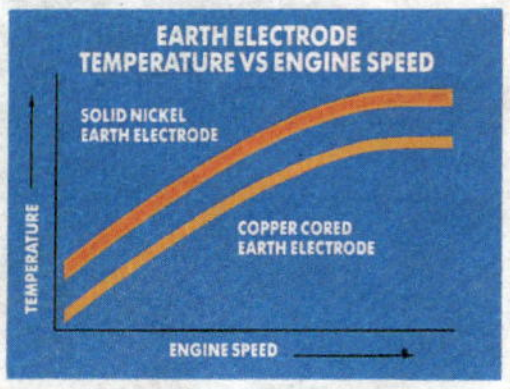

Champion Double Copper plugs also feature a unique trapezoidal earth electrode giving a 50% increase in spark area. This, together with the double copper cores, offers greatly reduced electrode wear, so the spark stays stronger for longer.

- **FASTER COLD STARTING**
- **FOR UNLEADED OR LEADED FUEL**
- **ELECTRODES UP TO 100°C COOLER**
- **BETTER ACCELERATION RESPONSE**
- **LOWER EMISSIONS**
- **50% BIGGER SPARK AREA**
- **THE LONGER LIFE PLUG**

Plug Tips/Hot and Cold.
Spark plugs must operate within well-defined temperature limits to avoid cold fouling at one extreme and overheating at the other.
Champion and the car manufacturers work out the best plugs for an engine to give optimum performance under all conditions, from freezing cold starts to sustained high speed motorway cruising.
Plugs are often referred to as hot or cold. With Champion, the higher the number on its body, the hotter the plug, and the lower the number the cooler the plug.

Plug Cleaning
Modern plug design and materials mean that Champion no longer recommends periodic plug cleaning. Certainly don't clean your plugs with a wire brush as this can cause metal conductive paths across the nose of the insulator so impairing its performance and resulting in loss of acceleration and reduced m.p.g.
However, if plugs are removed, always carefully clean the area where the plug seats in the cylinder head as grit and dirt can sometimes cause gas leakage.
Also wipe any traces of oil or grease from plug leads as this may lead to arcing.

CHAMPION
DOUBLE COPPER

This photographic sequence shows the steps taken to repair the dent and paintwork damage shown above. In general, the procedure for repairing a hole will be similar; where there are substantial differences, the procedure is clearly described and shown in a separate photograph.

First remove any trim around the dent, then hammer out the dent where access is possible. This will minimise filling. Here, after the large dent has been hammered out, the damaged area is being made slightly concave.

Next, remove all paint from the damaged area by rubbing with coarse abrasive paper or using a power drill fitted with a wire brush or abrasive pad. 'Feather' the edge of the boundary with good paintwork using a finer grade of abrasive paper.

Where there are holes or other damage, the sheet metal should be cut away before proceeding further. The damaged area and any signs of rust should be treated with Turtle Wax Hi-Tech Rust Eater, which will also inhibit further rust formation.

For a large dent or hole mix Holts Body Plus Resin and Hardener according to the manufacturer's instructions and apply around the edge of the repair. Press Glass Fibre Matting over the repair area and leave for 20-30 minutes to harden. Then ...

... brush more Holts Body Plus Resin and Hardener onto the matting and leave to harden. Repeat the sequence with two or three layers of matting, checking that the final layer is lower than the surrounding area. Apply Holts Body Plus Filler Paste as shown in Step 5B.

For a medium dent, mix Holts Body Plus Filler Paste and Hardener according to the manufacturer's instructions and apply it with a flexible applicator. Apply thin layers of filler at 20-minute intervals, until the filler surface is slightly proud of the surrounding bodywork.

For small dents and scratches use Holts No Mix Filler Paste straight from the tube. Apply it according to the instructions in thin layers, using the spatula provided. It will harden in minutes if applied outdoors and may then be used as its own knifing putty.

Use a plane or file for initial shaping. Then, using progressively finer grades of wet-and-dry paper, wrapped round a sanding block, and copious amounts of clean water, rub down the filler until glass smooth. 'Feather' the edges of adjoining paintwork.

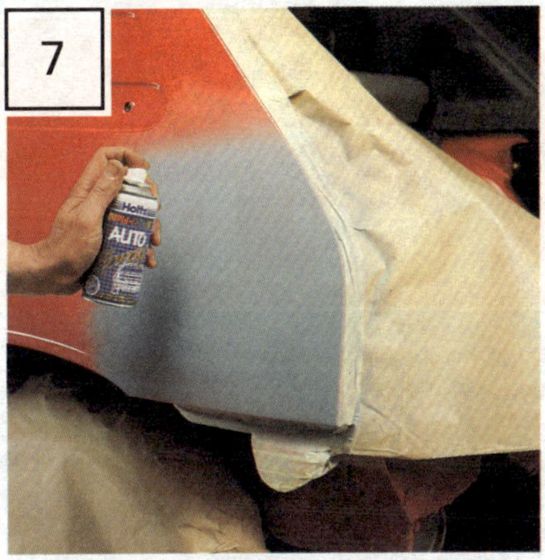

7

Protect adjoining areas before spraying the whole repair area and at least one inch of the surrounding sound paintwork with Holts Dupli-Color primer.

8

Fill any imperfections in the filler surface with a small amount of Holts Body Plus Knifing Putty. Using plenty of clean water, rub down the surface with a fine grade wet-and-dry paper – 400 grade is recommended – until it is really smooth.

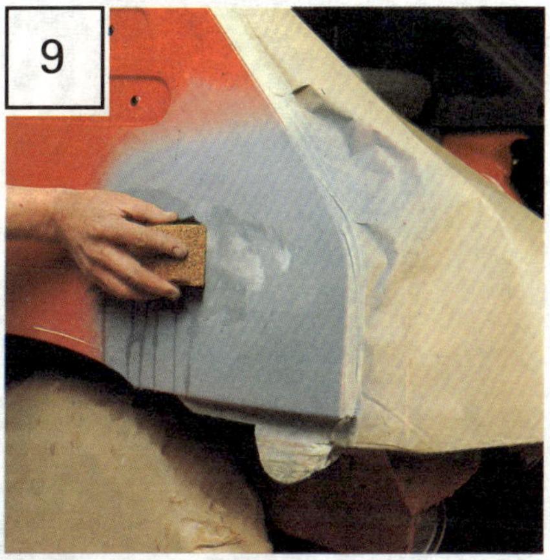

9

Carefully fill any remaining imperfections with knifing putty before applying the last coat of primer. Then rub down the surface with Holts Body Plus Rubbing Compound to ensure a really smooth surface.

10

Protect surrounding areas from overspray before applying the topcoat in several thin layers. Agitate Holts Dupli-Color aerosol thoroughly. Start at the repair centre, spraying outwards with a side-to-side motion.

10A

If the exact colour is not available off the shelf, local Holts Professional Spraymatch Centres will custom fill an aerosol to match perfectly.

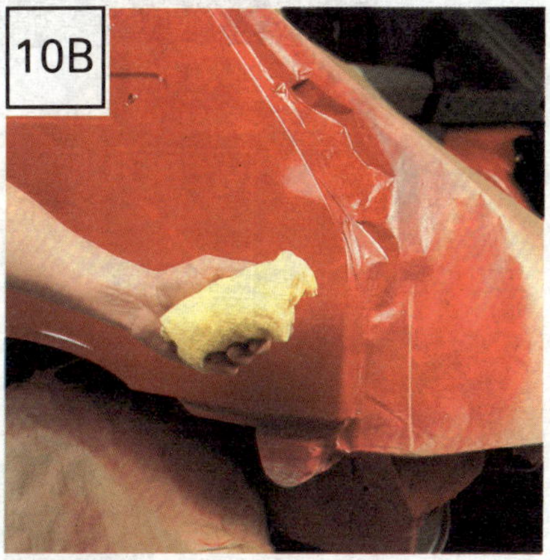

10B

To identify whether a lacquer finish is required, rub a painted unrepaired part of the body with wax and a clean cloth.

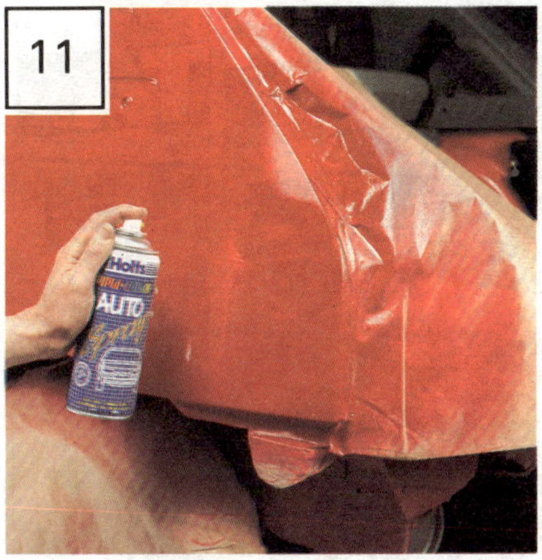

11

If *no* traces of paint appear on the cloth, spray Holts Dupli-Color clear lacquer over the repaired area to achieve the correct gloss level.

12

13

The paint will take about two weeks to harden fully. After this time it can be 'cut' with a mild cutting compound such as Turtle Wax Minute Cut prior to polishing with a final coating of Turtle Wax Extra.

14

When carrying out bodywork repairs, remember that the quality of the finished job is proportional to the time and effort expended.

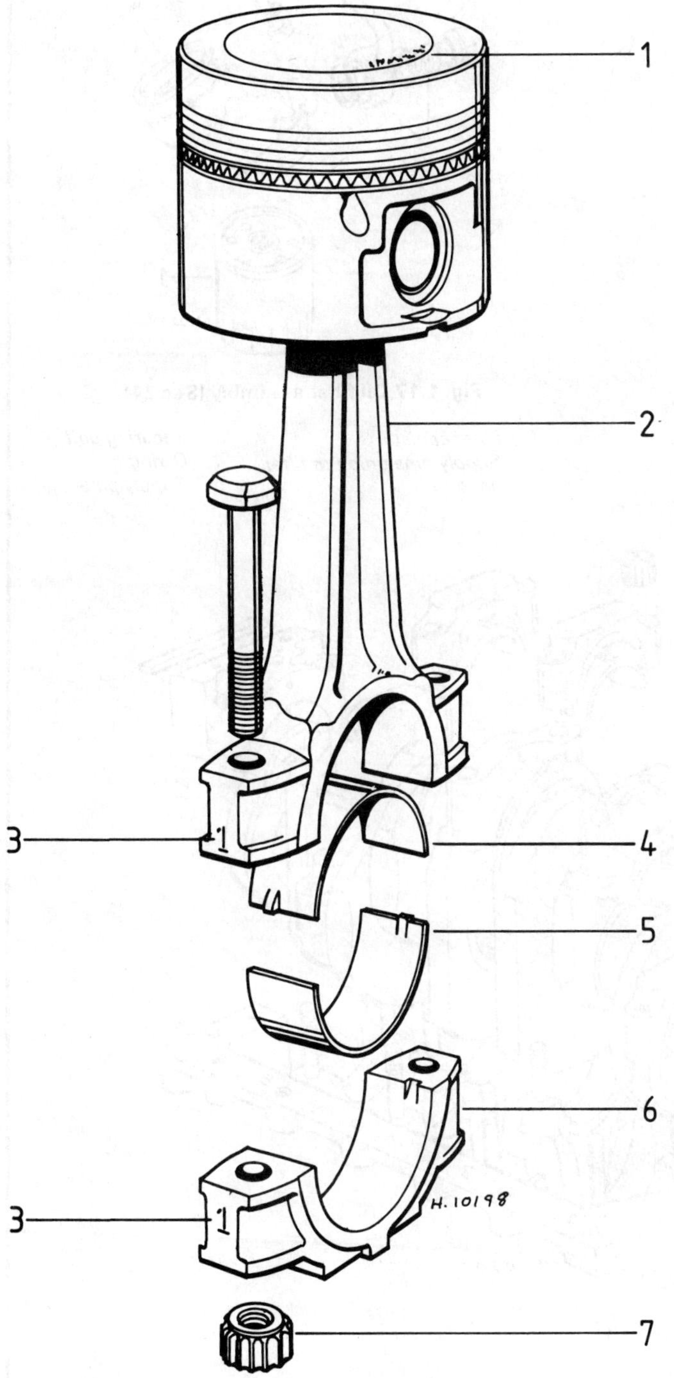

Fig. 1.14 Piston and connecting rod assembly (18V engine) (Sec 19)

1 *Piston – note identification marks*
2 *Connecting rod*
3 *Identification mark*
4 *Upper shell bearing half*
5 *Lower shell bearing half*
6 *End cap*
7 *Multi-sided nut*

19 Sump, pistons, connecting rods and big end bearings – removal

1 The sump, pistons and connecting rods can be removed with the engine still in the vehicle or with the engine on the bench. Proceed with the appropriate methods in either case for removing the cylinder head. The pistons and connecting rods are drawn up out of the top of the cylinder bores.
2 Undo and remove the two bolts, spring and plain washers securing the sump connecting plate to the gearbox bellhousing if the engine is still in the vehicle.
3 Undo and remove the bolts and spring washers holding the sump in position (photo). Lift away the sump and gasket.
4 To gain access to the bearings it is now necessary to remove the oil pump and strainer.
5 Undo and remove the three bolts and spring washers that secure the strainer to the oil pump cover. Lift away the strainer (photo).
6 Undo and remove the three nuts and spring washers that secure the oil pump body to the studs. Carefully withdraw the oil pump (photo).
7 Undo and remove the big-end cap retaining nuts using a socket and remove the big-end caps one at a time, taking care to keep them in the right order and the correct way round (photo). Ensure that the shell bearings are also kept with their correct connecting rods and caps unless they are to be renewed.
8 Normally the numbers 1 to 4 are stamped on adjacent sides of the big-end caps and connecting rods, indicating which cap fits on which rod and which way round the cap fits. If no numbers or lines can be found then scratch mating marks across the joint from the rod to the cap with a sharp screwdriver: one line for connecting rod No 1, two for connecting rod No 2 and so on. This will ensure there is no confusion later as it is most important that the caps go back in the position on the connecting rods from which they were removed.
9 If the big-end caps are difficult to remove they may be gently tapped with a soft hammer.
10 To remove the shell bearings press the bearing opposite the groove in both connecting rod and the connecting rod cap, and the bearing shell will slide out easily.
11 Withdraw the pistons and connecting rods upwards and ensure they are kept in the correct order, for refitting in the same bore. Refit the connecting rod caps and bearings to the rods if the bearings do not require renewal to minimise the risk of getting the caps and rods muddled.

20 Gudgeon pin – removal

16V engines

1 The gudgeon pin is clamped to the connecting rod and to remove the pin it is first necessary to remove the clamping bolt and spring washer. With the bolt removed press out the gudgeon pin to free the piston from the connecting rod. Take note which way round the piston fits on the connecting rod and which way the pin fits. Keep all mating components together for correct reassembly.

18V engines

2 These engines have a pressed-in gudgeon pin which requires a special BL tool (No 18G 1150 with adaptor 18G 1150D) to remove and refit the pin. This tool is shown in Fig. 1.15 and must be used in the following manner.
3 Securely hold the hexagonal body in a firm vice and screw back the large nut until it is flush with the end of the main centre screw. Well lubricate the screw and large nut as they have to withstand high loading. Now push the centre screw in until the nut just touches the thrust race.
4 Fit the adaptor number 18G 1150D onto the main centre screw with the piston ring cutaway positioned uppermost. Then slide the parallel sleeve with the groove end first onto the centre screw.
5 Fit the piston with the FRONT or ▲ mark on towards the adaptor on the centre screw. This is important because the gudgeon pin is spot faced on only one side of the piston (forward) and irreparable damage will result in fitting the wrong way round. Next fit the remover/replacer bush on the centre screw with the flange end away from the gudgeon pin.
6 Screw the stop nut onto the main centre screw and adjust it until

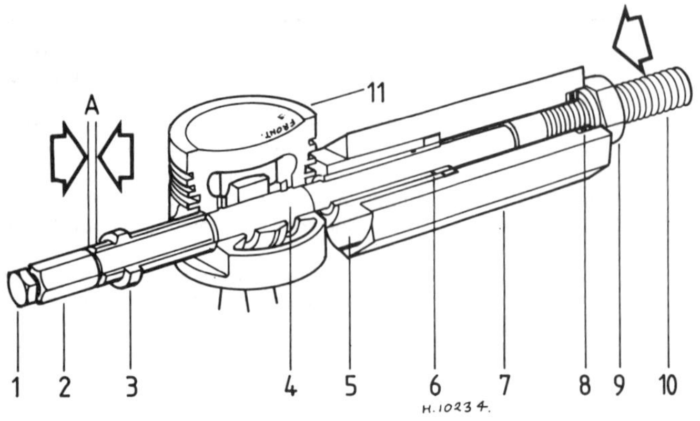

Fig. 1.15 Gudgeon pin removal (18V engines) using BL tool 18G 1150 and adaptor 18G 1150D (Sec 20)

1 *Lock screw*
2 *Stop nut*
3 *Flange away from gudgeon pin remover/replacer bush*
4 *Gudgeon pin*
5 *Piston support adaptor*
6 *Groove in sleeve away from gudgeon pin*
7 *Service tool body*
8 *Thrust race*
9 *Large nut*
10 *Centre screw*
11 *Piston*
A *See text*

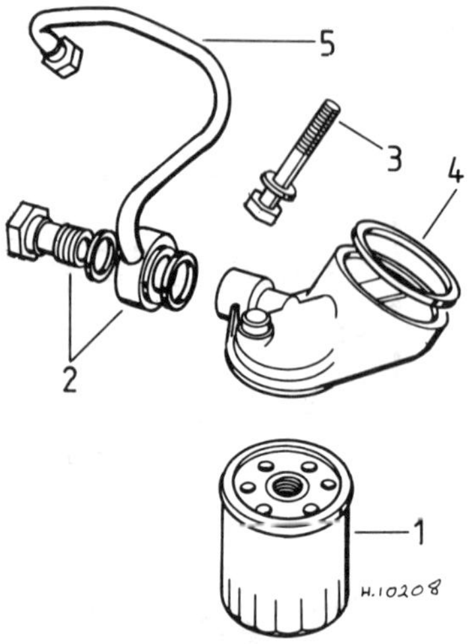

Fig. 1.17 Oil filter assembly (Sec 24)

1 *Filter canister*
2 *Supply pipe union to filter head*
3 *Securing bolt*
4 *O-ring*
5 *Supply pipe*

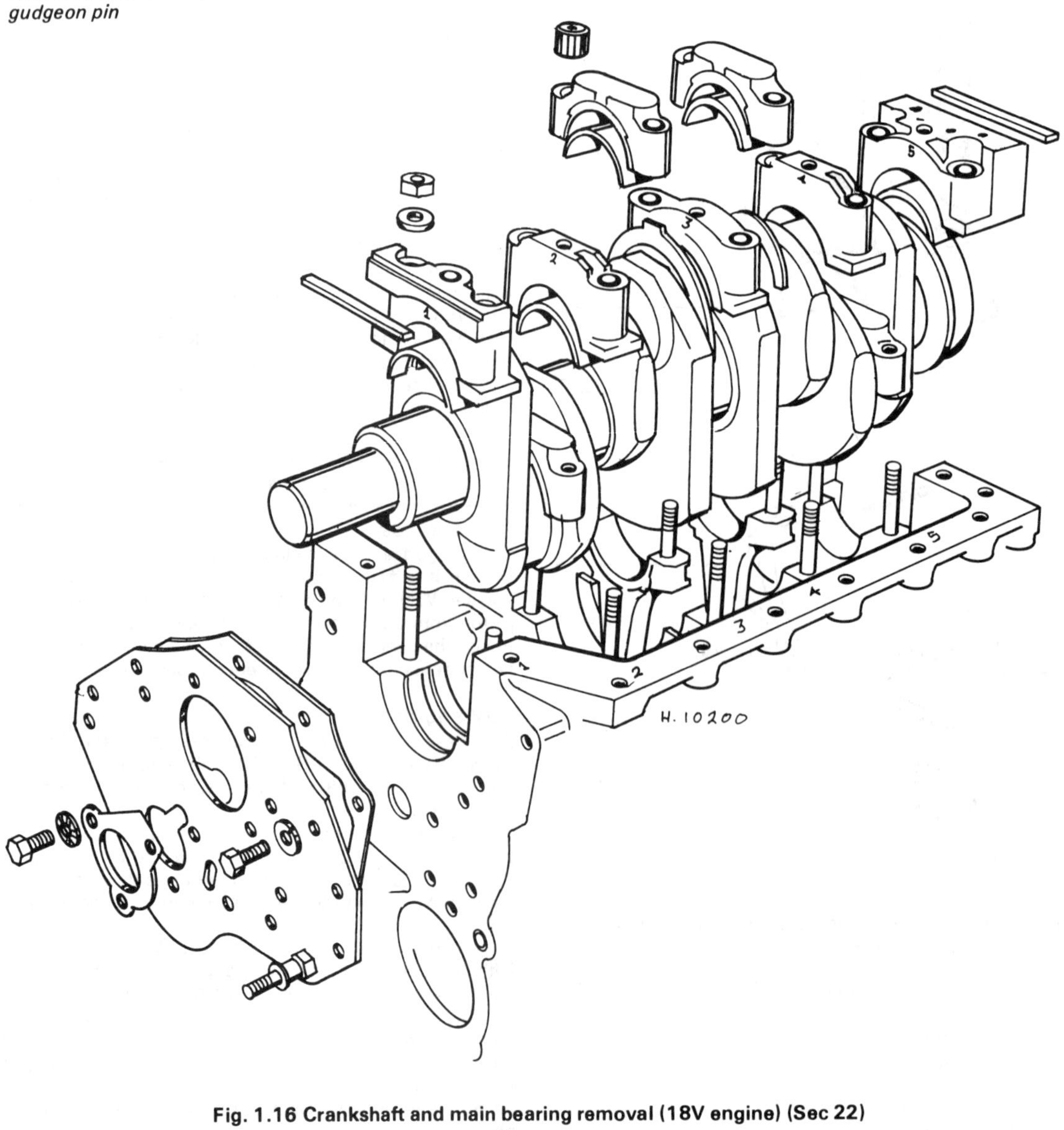

Fig. 1.16 Crankshaft and main bearing removal (18V engine) (Sec 22)

approximately 0.040 inch (1.0 mm) end play (A in Fig. 1.15) exists, and lock the stop nut securely with the lock screw. Now check that the remover/replacer and parallel sleeve are positioned correctly in the bore on both sides of the piston. Also check that the curved face of the adaptor is clean and slide the piston onto the tool, so it fits into the curved face of the adaptor with the piston rings over the cutaway.

7 Screw the large nut up to the thrust race and holding the lock screw turn the large nut with a ring spanner or long socket until the gudgeon pin is withdrawn from the piston.

21 Piston rings – removal

1 To remove the piston rings, slide them carefully over the top of the piston, taking care not to scratch the aluminium alloy of the piston. Never slide them off the bottom of the piston skirt. It is very easy to break piston rings if they are pulled off roughly so this operation should be done with extreme caution. It is helpful to use an old 0.020 inch feeler gauge to facilitate their removal.

2 Lift one end of the piston ring to be removed out of its groove and insert the end of the feeler gauge under it.

3 Turn the feeler gauge slowly round the piston and as the ring comes out of its groove it rests on the land above. It can then be eased off the piston with the feeler gauge stopping it from slipping into any empty grooves if it is any but the top piston ring that is being removed.

22 Crankshaft and main bearings – removal

1 Drain the engine oil, remove the timing gears and remove the sump, oil pump, big-end bearings, flywheel and engine backplate as already described. Removal can only be attempted with the engine on the bench. (The following description concerns the 18V engine but it can be used for the 16V engine allowing for the fact that there are five main bearings on the 18V and three on the 16V.)

2 Undo and remove the ten nuts securing the main bearing caps to the cylinder block. Then undo and remove the two bolts and spring washers which hold the front main bearing cap against the engine front plate. It will be beneficial if the front plate is completely removed, so remove the remaining securing bolts and spring washers and lift away together with the paper gasket.

3 Make sure that the main bearing caps are numbered 1 to 5 on the front faces and also have the word FRONT towards the front. If they are not so marked, identify them with punch-marks or by scribing lines so that they may be refitted correctly.

4 Remove the main bearing caps and the bottom half of each bearing shell, taking care to keep the bearing shells in the right caps.

5 When removing the centre bearing cap, note the bottom semi-circular halves of the thrust washers, one half lying on either side of the main bearing. Keep them with the centre bearing along the correct side.

6 Remove the crankshaft by lifting it away from the crankcase.

7 Remove the upper halves of the main bearing shells and thrust washers and keep them with their bearing caps.

23 Lubrication system – description

1 A force feed system of lubrication is fitted with oil circulated around the engine from the sump below the cylinder block. The level of engine oil in the sump is indicated by the dipstick which is fitted on the right-hand side of the engine. The optimum level is indicated by the maximum mark. The level of oil in the sump, ideally, should be neither above or below this line. Oil is replenished via the filler cap towards the front of the rocker cover.

2 The eccentric rotor type oil pump is bolted within the left-hand side of the crankcase and is driven by a short shaft from a skew gear on the camshaft.

3 The pump is of the non-draining variety to allow rapid pressure build up when starting from cold. Oil is drawn from the sump through a gauze screen in the oil strainer and is sucked up the pick-up and drawn into the oil pump. From the oil pump it is forced under pressure along a gallery on the right-hand side of the engine, and through drillings to the big-end, main and camshaft bearings. A small hole in each connecting rod allows a jet of oil to lubricate the cylinder wall with each revolution.

4 From the camshaft front bearing, oil is fed through drilled passages in the cylinder block and head to the front rocker pedestal where it enters the hollow rocker shaft. Holes drilled in the shaft allow for the lubrication of the rocker arms and the valve stems and pushrod ends. This oil is at a lower pressure than the oil delivered to the crankshaft bearings. Oil from the front camshaft bearing lubricates the timing gears and the timing chain. Oil returns to the sump by various passages, the tappets being lubricated by oil returning via the pushrod drillings in the block.

5 A full flow cartridge type filter is fitted and oil passes through this filter before it reaches the main oil gallery. The oil is passed directly from the oil pump to the filter.

24 Lubrication system – oil change and oil filter renewal

1 The oil filter and the engine oil must be renewed at the intervals specified in Routine Maintenance, or more frequently if the vehicle is operating under adverse conditions.

2 The oil is best drained immediately after a run, when most of the sludge and other impurities which build up will be in suspension in the oil. Place a container of sufficient capacity under the sump, remove the drain plug and allow the oil to drain into the container for at least ten minutes.

3 Whilst the oil is draining, remove the disposable cartridge type filter. Place an absorbent cloth around the base of the filter to catch the oil released from the cartridge when it is unscrewed. Unscrew the old cartridge from the filter head and discard it.

4 If the filter cartridge proves stubborn, use a strap wrench or chain wrench to undo it. As a last resort a screwdriver or similar instrument may be driven through the filter and used as a lever to undo it, but do this towards the blind end of the cartridge to minimise the danger of fragments of casing or filter material entering the engine oilways.

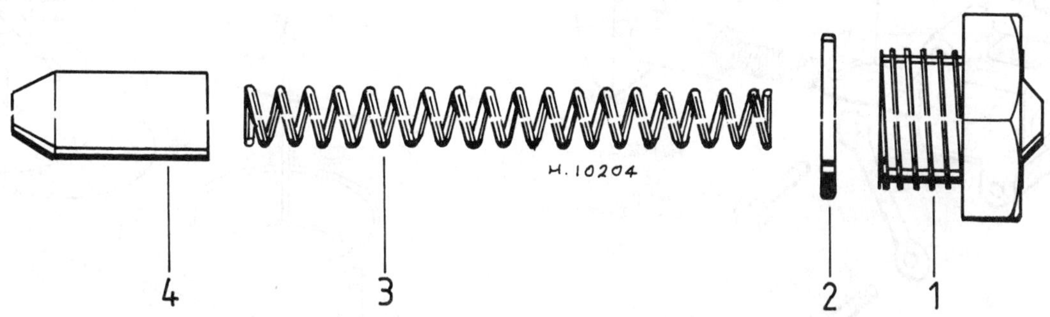

Fig. 1.18 Oil pressure relief valve (Sec 25)

1 Domed nut
2 Sealing washer
3 Valve spring
4 Valve

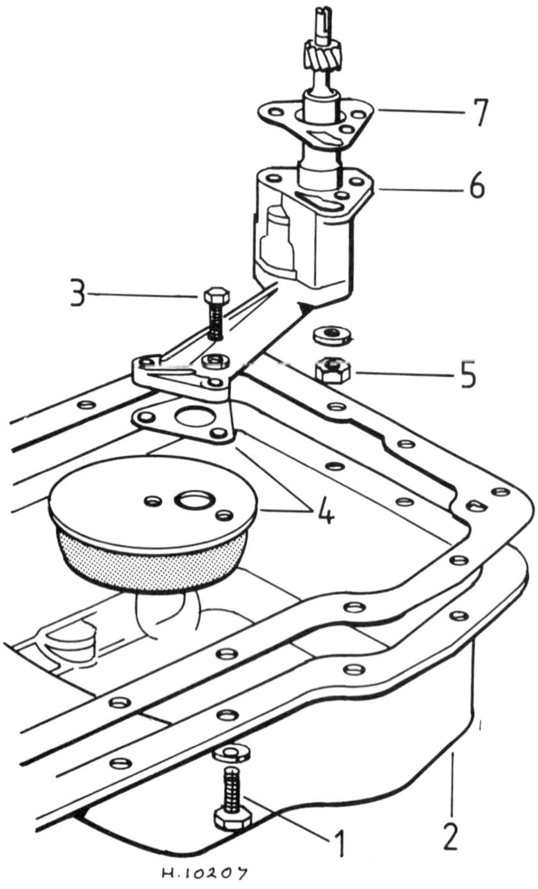

Fig. 1.19 Oil pump removal (Sec 26)

1 Sump securing bolt and spring washer
2 Sump
3 Oil strainer securing bolt and spring washer
4 Oil strainer and gasket
5 Oil pump securing nut and spring washer
6 Oil pump
7 Gasket

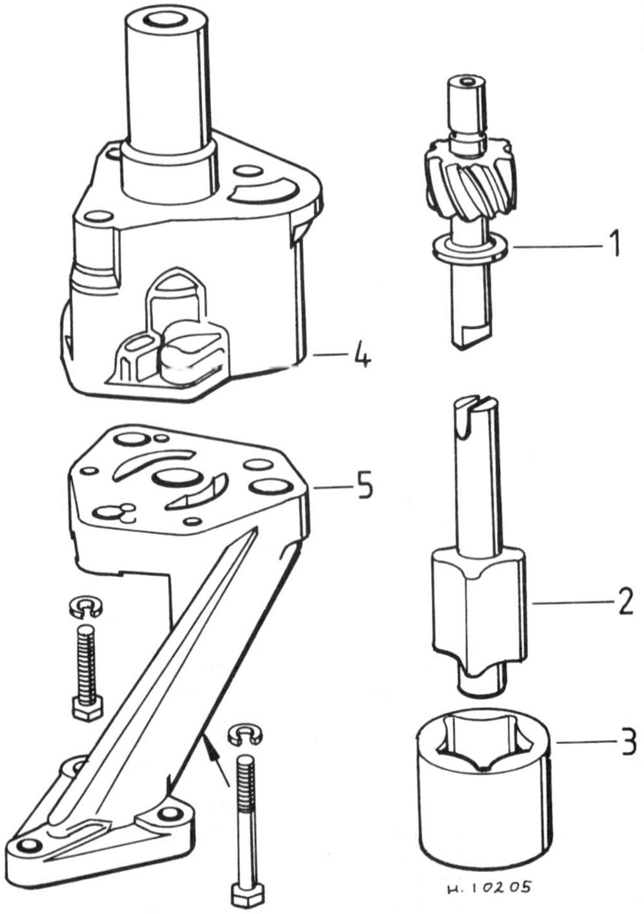

Fig. 1.20 Oil pump components (Sec 26)

1 Pump shaft
2 Inner rotor
3 Outer rotor
4 Pump body
5 Pump cover

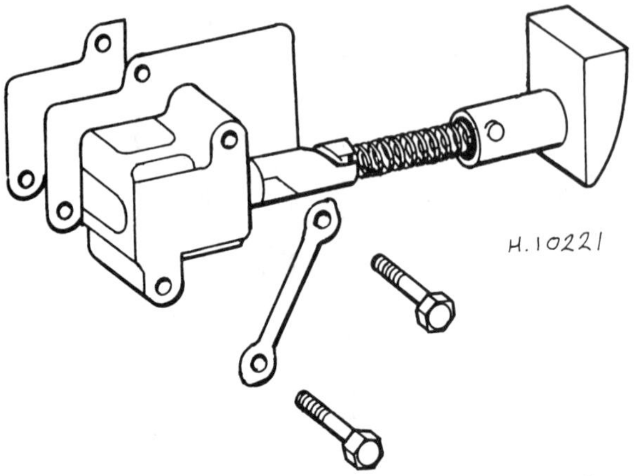

Fig. 1.21 Exploded view of timing chain tensioner (Sec 27)

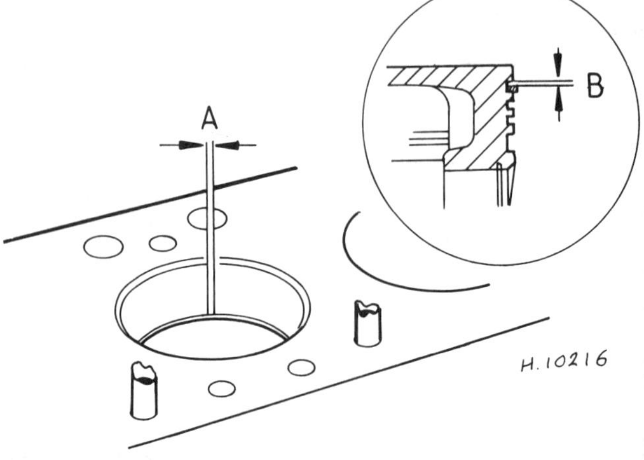

Fig. 1.22 Piston ring measurement (Sec 32)

A Ring gap *B Ring clearance*

25.1 The domed nut covers the oil pressure relief valve

5 Smear the seal on the new filter with a little clean oil and screw it onto the filter head. Tighten by hand only; **do not** use a spanner or strap wrench.
6 Make sure that the sealing washer on the sump drain plug is in good condition. Refit the plug and tighten it.
7 Fill the engine with the specified quantity and grade of oil. Start it up and allow to run for a few minutes, checking that the oil pressure warning light goes out – this may take a few seconds whilst the new filter fills up. Make sure that there are no leaks from the filter cartridge or the drain plug. Stop the engine, wait a few minutes and check the oil level, topping-up if necessary.

25 Oil pressure relief valve – removal and refitting

1 To prevent excessive oil pressure - which might result when the engine oil is thick and cold - an oil pressure relief valve is built into the left hand side of the engine at the rear. The relief valve is identified externally by a large $\frac{9}{16}$ inch domed hexagon nut (photo).
2 To dismantle the valve unscrew the domed nut and remove it complete with sealing washer. The relief spring and valve can then be easily extracted.
3 In position the valve fits over the opposite end of the relief valve spring to that resting in the dome of the hexagon nut, and bears against a machined seating in the block. When the oil pressure exceeds the specified value, the valve is forced off its seat and the oil by-passes it and returns via a drilling directly into the sump.
4 Check the condition of the spring by measuring its free length. If it is shorter than the specified length it should be replaced with a new spring.
5 Examine the valve for signs of pitting. If evident, it should be carefully lapped using cutting paste. Remove all traces of paste when a good seating has been obtained.
6 Reassembly of the relief valve is the reverse sequence to removal.

26 Oil pump – removal and dismantling

1 With the sump removed undo and remove the three bolts and spring washers that secure the oil strainer to the pump housing. Lift away the oil strainer.
2 Undo and remove the nuts and spring washers from the three studs which hold the oil pump to the underside of the crankcase. Lift away the pump and its driveshaft. Recover the pump gasket.
3 Unscrew and remove the three bolts and spring washers that secure the pump cover to the pump body. Carefully pull the cover from the two dowels which hold it in its correct position on the housing.
4 Pull out from the pump body the outer rotor and the inner rotor together with pump shaft.

27 Timing chain tensioner – removal, dismantling, reassembly and refitting

1 Remove the cover from the timing gears as described in Section 16. Unlock and remove the two screws securing the chain tensioner to the engine front plate. Note that the slipper head is under spring tension and components could be lost unless care is taken when removing the tensioner assembly.
2 Allow the spring loading against the slipper head to relax and carefully withdraw the slipper head, the spring and the inner cylinder from the tensioner.
3 To assemble, refit the chain tensioner inner cylinder and spring into the cylinder of the slipper head so that the serrated slot in the inner cylinder engages the peg in the slipper cylinder.
4 Rotate the inner cylinder clockwise against the spring tension until the lower serration in the slot engages with the peg and retains the inner cylinder in the slipper cylinder.
5 Refitting the chain tensioner is the reverse of the removal procedure.

28 Engine examination and renovation – general

With the engine stripped and all parts thoroughly cleaned, every component should be examined for wear. The following items should be checked, and, where necessary, renewed or renovated as described later.

29 Crankshaft – examination and renovation

Examine the crankpin and main journal suffaces for signs of scoring or scratches and check the ovality of the crankpins at different positions with a micrometer. If more than 0.001 in (0.0254 mm) out of round, the crankshaft will have to be reground. It will also have to be reground if there are any scores or scratches present. Also check the journals in the same fashion. BL B series engine centre main bearings are prone to failure. This is not always immediately apparent, but slight vibration in an otherwise normally smooth engine and a very slight drop in oil pressure under normal conditions are clues. If the centre main bearing is suspected of failure it should be investigated immediately, by dropping the sump and removing the centre main bearing cap. Failure to do this will result in a badly scored centre main journal. If it is necessary to regrind the crankshaft and fit new bearings, an engineering works will be able to decide how much metal to grind off and be able to supply the correct undersize shells to fit.

30 Big-end and main bearings - examination and renovation

1 Big-end bearing failure is usually accompanied by a noisy knocking from the crankcase and a slight drop in oil pressure. Main bearing failure is accompanied by vibration, which can be quite severe as the engine speed rises and falls, and a drop in oil pressure.
2 Bearings which have not broken up, but are badly worn, will give rise to low oil pressure and some vibration. Inspect the big-ends, main bearings and thrust washers for signs of general wear, scoring, pitting and scratches. The bearings should be a matt grey in colour. With lead indium bearings, should a trace of copper in colour be noticed the bearings are badly worn, for the lead bearing has worn away to expose the indium underlay. Renew the bearings if they are in this condition or if there is any sign of scoring or pitting.
3 The undersizes available are designed to correspond with regrind sizes; 0.010 in bearings are correct for a crankshaft reground 0.010 in undersize. The bearings are in fact slightly more than the stated undersize as running clearances have been allowed for during their manufacture.
4 Very long engine life can sometimes be achieved by changing big end bearings at 30 000 miles (48 000 km) and main bearings at 50 000 miles (80 000 km) irrespective of bearing wear. Normally, crankshaft wear is infinitesimal and regular changes of bearings may ensure mileages of between 100 000 (160 000 km) and 120 000 (200 000 km) miles before crankshaft regrinding becomes necessary. Crankshafts normally have to be reground because of scoring due to bearing failure.

5 Once dismantled only refit new bearing shells. It is false economy to refit old bearings.

31 Cylinder bores – examination and renovation

1 The cylinder bores must be examined for taper, ovality, scoring and scratches. Start by carefully examining the top of the bores, if they are worn fractionally a very slight ridge will be found on the thrust side. This marks the top of the piston travel. You will have a good indication of the bore wear prior to dismantling the engine, or removing the cylinder head. Excessive oil consumption accompanied by blue smoke from the exhaust is a sure sign of worn cylinder bores and piston rings.
2 Measure the diameter of the bore just under the ridge with a micrometer and compare it with the diameter at the bottom of the bore, which is not subject to such wear. If the difference between the two measurements is more than 0.006 inch (0.1524mm) then it will be necessary to fit special piston rings or to have the cylinders rebored and fit oversize pistons and rings. If no micrometer is available remove the rings from one piston and place the piston in each bore in turn about $\frac{3}{4}$ inch (19mm) below the top of the bore. If an 0.010 inch (0.254mm) feeler gauge can be slid between the piston and the cylinder wall on the thrust side of the bore then remedial action must be taken. Oversize pistons are available as stated in the Specifications.
3 Oversize pistons are accurately machined to just below the specified measurements so as to provide correct running clearances in bores bored out to the exact oversize dimensions.
4 If the bores are slightly worn but not so badly worn as to justify reboring them, special oil control rings can be fitted to the existing pistons which will restore compression and stop the engine burning oil. Several different types are available and the manufacturer's instructions concerning their fitting must be followed closely.

32 Pistons and piston rings – examination and renovation

1 If the old pistons are to be refitted, carefully remove the piston rings and then thoroughly clean them. Take particular care to clean out the piston ring grooves. Do not scratch the aluminium in any way. If new rings are to be fitted to the old pistons, then the top ring should be stepped, so as to clear the ridge left above the previous top ring. If a normal but unstepped new ring is fitted it will hit the ridge and break, because the new ring will not have worn in the same way as the old.
2 Before fitting the rings on the pistons each ring should be inserted approximately 3 inches (76mm) down the cylinder bore and the gap measured with a feeler gauge as shown in Fig. 1.22. This should be between the limits given in the specifications at the beginning of this Chapter. It is essential that the gap is measured at the bottom of the ring travel, for if it is measured at the top of a worn bore and gives a perfect fit, it could easily seize at the bottom. If the ring gap is too small rub down the ends of the ring with a very fine file until the gap is correct when fitted. To keep the rings square in the bore for measurement, line each one up in turn with an old piston in the bore upside down, and use the piston to push the ring down about 3 inches (76mm). Remove the piston and measure the piston ring gap.
3 When fitting new pistons and rings to a rebored engine the ring gap can be measured at the top of the bore as the bore will now not taper. It is not necessary to measure the side clearance in the piston ring grooves with rings fitted, as the groove dimensions are accurately machined during manufacture. When fitting new oil control rings to the pistons it may be necessary to have the grooves widened by machining to accept the new rings. In this instance the manufacturer will make this quite clear and will supply the address to which the pistons must be sent for machining.
4 When new pistons are fitted, take care to fit the exact size best suited to the particular bore of your engine. BL go one stage further than merely specifying one size piston for all standard bores. Because of very slight differences in cylinder machining during production it is necessary to select just the right piston for the bore. A range of different sizes are available either from the piston manufacturer of from the local BL stores.
5 Examination of the cylinder block face will show adjacent to each bore a small diamond shaped box with a number stamped in the metal. Careful examination of the piston crown will show a matching diamond and number. These are the standard piston sizes and will be

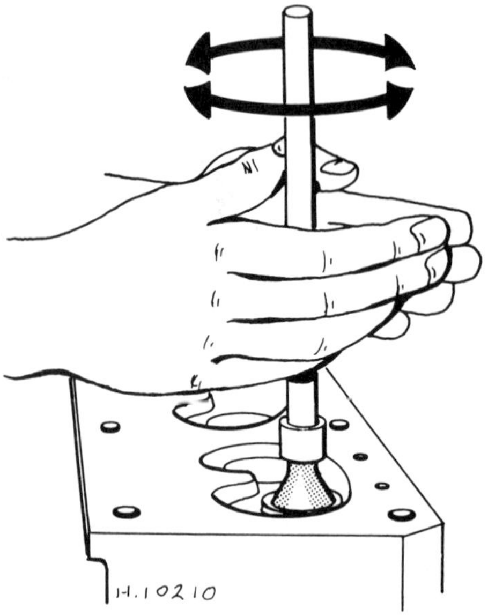

Fig. 1.23 Valve grinding using hand suction tool (Sec 34)

Lift valve off seat occasionally to spread grinding paste evenly over seat and face

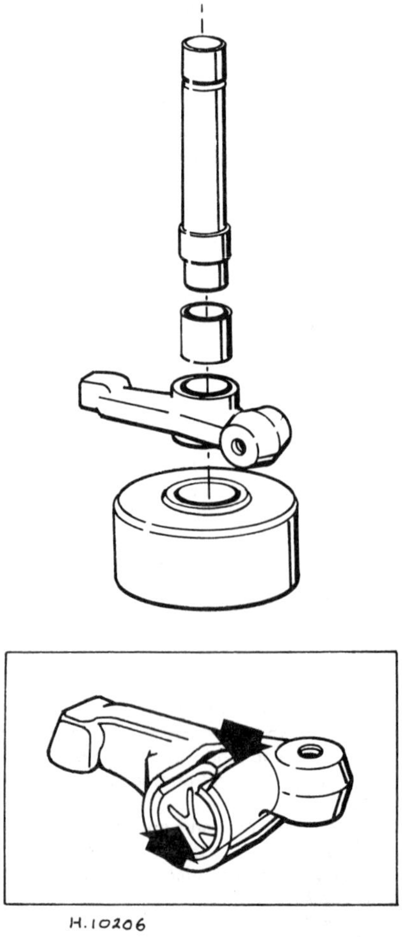

Fig. 1.24 Rocker bush removal and refitting (Sec 36)

the same for all bores. If the standard pistons are to be refitted or standard low compression pistons changed to standard high compression pistons, then it is essential that only pistons with the same number in the diamond are used. With larger pistons, the amount of oversize is stamped in an elipse on the piston crown.

6 On engines with tapered second and third compression rings, the top narrow side of the ring is marked with a T. Always fit this side uppermost and carefully examine all rings for this mark before fitting.

33 Camshaft and camshaft bearings – examination and renovation

1 Carefully examine the camshaft bearings for wear. If the bearings are obviously worn or pitted or the metal underlay is just showing through, then they must be renewed. This is an operation for your local BL garage or local engineering works as it demands the use of specialised equipment. The bearings are removed using a special drift after which the new bearings are pressed in, care being taken that the oil drillings in the bearings line up with those in the block. With another special tool the bearings are then reamed in position.

2 The camshaft itself should show no signs of wear, but if very slight, scoring marks can be removed by gently rubbing down with very fine emery cloth or an oil stone. The greatest care must be taken to keep the cam profiles smooth.

34 Valves and seats – examination and renovation

1 Examine the heads of the valves for pitting or burning, especially the heads of the exhaust valves. The valve seatings should be examined at the same time. If the pitting on the valves is very slight the marks can be removed by grinding the seats and valves together with coarse, and then fine, valve grinding paste. Where bad pitting has occurred to the valve seats it will be necessary to recut them and fit new valves. If the valve seats are so worn that they cannot be recut then it will be necessary to fit new valve seat inserts. These latter two jobs should be entrusted to a BL garage or engineering works. In practice it is very seldom that the seats are so badly worn that they require renewal. Normally it is the valve that is too badly worn, and you can easily purchase a new set of valves and match them to the seats by valve grinding.

2 Valve grinding is easily carried out. Place the cylinder head upside down on a bench with a block of wood at each end to give clearance for the valve stems. Alternatively place the head at 45° to a wall with the combustion chambers facing away from the wall.

3 Smear a trace of coarse carborundum paste on the seat face and apply a suction grinding tool to the valve head as shown in Fig. 1.23. With a semi-rotary action, grind the valve head to its seat, lifting the valve occasionally to redistribute the grinding paste. When a dull matt even surface finish is produced on both the valve seat and the valve, then wipe off the paste and repeat the process with fine carborundum paste, lifting and turning the valve to redistribute the paste as before. A light spring placed under the valve head will greatly ease this operation. When a smooth unbroken ring of light grey matt finish is produced, on both valve and valve seat faces, the grinding operation is complete.

4 Scrape away all carbon from the valve head and the valve stem. Carefully clean away every trace of grinding compound, taking great care to leave none in the ports or in the valve guides. Clean the valves and valve seats with a paraffin soaked rag then wipe with clean rag. (If an air line is available blow clean).

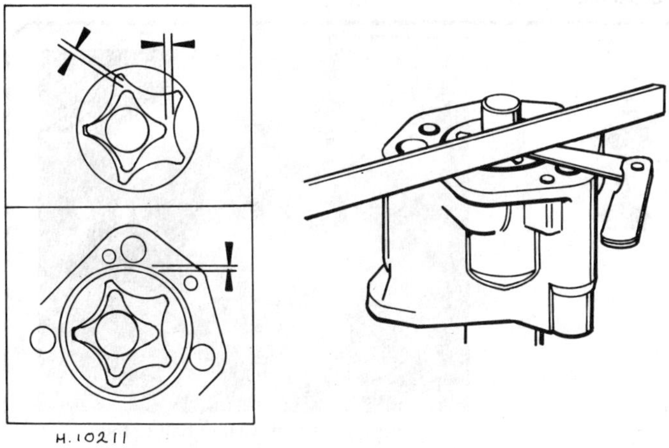

Fig. 1.25 Oil pump wear check (Sec 39)

35 Timing gears and chain – examination and renovation

1 Examine the teeth on both the crankshaft gearwheel and the camshaft gearwheel for wear. Each tooth forms an inverted V with the gearwheel periphery and if worn, the side of each tooth under tension will be slightly concave in shape when compared with the other side of the tooth, ie one side of the inverted V will be concave when compared with the other. If any signs of wear is present the gearwheels must be renewed.

2 Examine the links of the chain for side slackness and renew the chain if any slackness is noticeable when compared with a new chain. It is a sensible precaution to renew the chain every 30 000 miles (48 000 km) and at a lesser mileage if the engine is stripped down for a major overhaul. The actual rollers on a very badly worn chain may be slightly grooved.

36 Rockers and rocker shaft – examination and renovation

1 Remove the threaded plug with a screwdriver from the end of the rocker shaft and thoroughly clean out the shaft. As the shaft carries as the oil passages for the valve gear, clean out these passages and make sure they are quite clear. Check the shaft for straightness by rolling it on a flat surface. It is most unlikely that it will deviate from normal, but, if it does, you must purchase a new shaft. The surface of the shaft should be free from any wear ridges caused by the rocker arms. If any wear is present, renew the rocker shaft. Wear is likely to have occured only if the rocker shaft oil holes have become blocked.

2 Check the rocker arms for wear of the rocker bushes, at the rocker arm face which bears on the valve stem, and of the adjusting ball ended screws. Wear in the rocker arm bush can be checked by gripping the rocker arm tip and holding the rocker arm in place on the shaft, noting if there is any lateral rocker arm shake. If any shake is present, and the arm is very loose on the shaft, remedial action must be taken. It is recommended that if a forged type of rocker arm is fitted it be taken to your local BL garage or engineering works to have the old bush drawn out and a new bush fitted. The correct placement of the bush is shown in Fig. 1.24.

3 If a pressed steel rocker arm is fitted, rebushing must not be undertaken but a new rocker arm obtained.

4 Check the tip of the rocker arm where it bears on the valve head, for cracking or serious wear on the case hardening. If none is present the rocker arm may be refitted. Check the pushrods for straightness by rolling them on a flat surface. If bent they must be renewed.

37 Tappets – examination and renovation

Examine the bearing surface of the tappets which run on the camshaft. Any indentation in this surface or any cracks indicate serious wear and the tappets must be renewed. Thoroughly clean them out, removing all traces of sludge. It is most unlikely that the sides of the tappets will be worn, but if they are a loose fit in their bores and can be readily rocked, they should be discharged and new tappets fitted. It is unusual to find worn tappets, and any wear is only likely to occur after extremely high mileages.

38 Flywheel starter ring gear – examination and renovation

1 If the teeth on the flywheel starter ring gear are badly worn, or if some are missing, then it will be necessary to remove the ring. This is achieved by splitting the old ring using a cold chisel. Care must be taken not to damage the flywheel during this process.

2 To fit a new ring gear, it will be necessary to heat it gently and evenly with an oxy-acetylene flame until a temperature of approximately 662°F (350°C) is reached. (This is indicated by a light blue surface colour). With the ring gear at this temperature, fit it to the flywheel with the front of the teeth facing the clutch fitting end of the flywheel. The ring gear should be either pressed or lightly tapped onto its register and left to cool naturally when the contraction of the metal on cooling will ensure that it is a secure and permanent fit. Great care must be taken not to overheat the ring gear, for if this happens the temper of the ring gear will be lost.
3 An alternative method is to use a high temperature oven to heat the ring.
4 Because of the need for oxy-acetylene equipment or a special oven it is not practical for refitment to take place at home. Take the flywheel and new starter ring to an engineering works willing to do the job.

39 Oil pump – examination and renovation

1 Thoroughly clean all the component parts in petrol and then check the rotor endfloat and lobe clearances in the following manner.
2 Position the rotors in the pump and place the straight edge of a steel rule across the joint face of the pump. Measure the gap between the bottom of the straight edge and the top of the rotors with a feeler gauge as shown in Fig. 1.25. If the measurement exceeds 0.005 in (0.127mm) then check the lobe clearances as described in the following paragraph. If the lobe clearances are correct then remove the dowels from the joint face of the pump body and lap joint the inner face on a sheet of plate glass.
3 Measure the gaps between the peaks of the lobes and the peaks in the pump body with a feeler gauge. If the gap exceeds 0.010 in (0.254mm) then fit a new pump. This measurement is shown in Fig. 1.25.

40 Cylinder head and bore – decarbonisation

1 This operation can be carried out with the engine either in or out of the vehicle. With the cylinder head off, carefully remove with a wire brush and blunt, plastic scraper, all traces of carbon deposits from the combustion spaces and the ports. The valve stems and valve guides should also be freed from any carbon deposits. Wash the combustion spaces and ports down with petrol and scrape the cylinder head surface free of any foreign matter with an old table knife or similar article. Take care not to scratch the surfaces.
2 Clean the pistons and top of the cylinder bores. If the pistons are still in the cylinder bores then it is essnetial that great care is taken to ensure that no carbon gets into the bores for this will scratch the cylinder walls or cause damage to the piston and rings. To stop it happening first turn the crankshaft so that two of the pistons are at TDC. Place a clean non-fluffy rag into the other two bores or seal them off with paper and masking tape. The waterways and pushrod holes should be covered with a small piece of masking tape to prevent particles of carbon entering the cooling system and damaging the water pump, or entering the lubrication system and causing damage to a bearing surface.
3 There are two schools of thought as to how much carbon ought to be removed from the piston crown. One is that a ring of carbon should be left around the edge of the piston and on the cylinder bore wall as an aid to keeping oil consumption low. Although this is probably true for engines with worn bores, on unworn engines, however, the practice is to remove all traces of carbon
4 If all traces of carbon are to be removed, press a little grease into the gap between the cylinder walls and the two pistons which are to be worked upon. With a blunt scraper carefully scrape away all carbon from the piston crown, taking care not to scratch the aluminium. Also scrape away the carbon from the surrounding lip of the cylinder walls. When all carbon has been removed, scrape away the grease which will now be contaminated with carbon particles, taking care not to press any into the bores. To assist prevention of carbon build up the piston crown can be polished with a metal polish. Remove the rags or masking tape from the other two cylinders and turn the crankshaft so that the two pistons which were at the bottom are now at the top. Place non-fluffy rag into the other two bores or seal them off with paper and masking tape. Do not forget the waterways and oilways as well. Proceed as previously described.
5 If a ring or carbon is going to be left round the piston then this can be helped by inserting an old piston ring into the top of the bore to rest on the piston and ensure that carbon is not accidentally removed. Check that there are no particles of carbon in the cylinder bores. Decarbonising is now complete.

41 Valve guides – examination and renovation

1 Examine the valve guides internally for wear. If the valves are a very loose fit in the guides and there is the slightest suspicion of lateral rocking, then new guides will have to be fitted, their correct location being shown in Fig. 1.26. Try to compare them internally by visual inspection with a new guide as well.
2 Because of the need for special equipment, it is recommended that you leave the guide renewal to your BL agent or automobile engineering works.

42 Engine reassembly – general

1 To ensure maximum life with minimum trouble from a rebuilt engine, not only must every part be correctly assembled but everything must be spotlessly clean, all the oilways must be clear, locking washers and spring washers must always be fitted where needed and all bearings and other working surfaces must be thoroughly lubricated during assembly. Before assembly begins renew any bolts or studs the threads of which are in any way damaged, and whenever possible use new spring washers.
2 Apart from your normal tools, a supply of non-fluffy rag, an oil can filled with engine oil (an empty washing up liquid plastic bottle thoroughly cleaned and washed out, will do), a supply of new spring washers, a set of new gaskets and a torque wrench should be collected together.

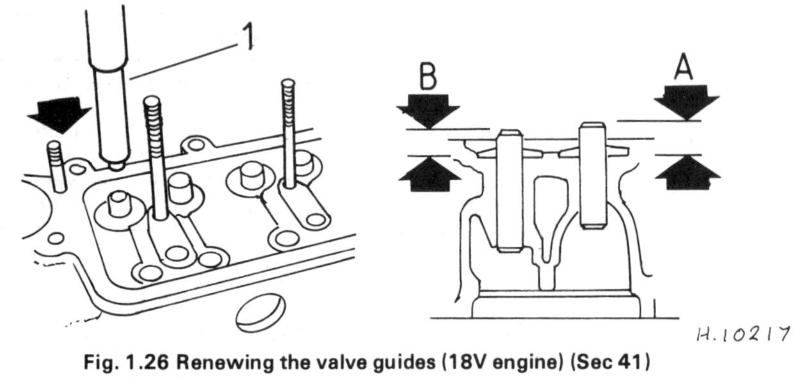

Fig. 1.26 Renewing the valve guides (18V engine) (Sec 41)

Left: Press out the old guides in direction of arrow with shaped drift (1)
Right: Press in new guides to dimension
A (inlet) 0.75 in (19.05 mm) B (exhaust) 0.625 in (15.88 mm)

43.11 Fitting a thrust washer half to the bearing cap

43 Crankshaft – refitting

1 Ensure that the crankcase is thoroughly clean and that all the oilways are clear. A thin twist drill is useful for cleaning them out. If possible blow them out with compressed air. Treat the crankshaft in the same fashion and then inject engine oil into the crankshaft oilways.
2 Commence work on rebuilding the engine by refitting the crankshaft and main bearings.
3 Fit the five upper halves of the main bearing shells to their location in the crankcase, after wiping the location clean.
4 Note that on the back of each bearing is a tab which engages in locating grooves in either the crankcase or the main bearing cap housings.
5 If new bearings are being fitted, carefully clean away all traces of protective grease with which they are coated.
6 With the five upper bearing shells securely in place, wipe the lower bearing cap housings and fit the five lower shell bearings to their caps, ensuring that the correct shell goes into the correct cap if the old bearings are being refitted. (Three upper and lower bearing shells on 16V engines).
7 Wipe the recesses either side of the centre main bearing which locate the upper halves of the thrust washers.
8 Generously lubricate the crankshaft journals and the upper and lower main bearing shells and carefully lower the crankshaft into position. Make sure that it is the right way round.
9 Introduce the upper halves of the thrust washers (the halves without tabs) into their grooves either side of the centre main bearing, rotating the crankshaft in the direction towards the main bearing tab (so that the main bearing shells do not slide out). At the same time feed the thrust washers into the locations with their oil grooves outwards away from the bearing.
10 Fit the main bearing caps in position ensuring that they locate properly. The mating surfaces must be spotlessly clean or the caps will not seat correctly.
11 When refitting the centre main bearing cap ensure that the thrust washers, generously lubricated, are fitted with their oil grooves facing outwards and the locating tab of each washer is in the slot in the bearing cap (photo).
12 Refit the main bearing cap nuts and screw them up finger tight.
13 Test the crankshaft for freedom of rotation. Should it be very stiff to turn or possess high spots a most careful inspection must be made, preferably by a skilled mechanic with a micrometer to trace the cause of the trouble. It is very seldom that any trouble of this nature will be experienced when fitting the crankshaft.
14 Tighten the main bearing nuts using a torque wrench set to the specified torque and recheck the crankshaft for freedom of rotation.
15 Using a screwdriver between one crankshaft web and a main bearing cap, lever the crankshaft forwards and check the endfloat using feeler gauges. This should be within the specified limits. If excessive, new thrust washers or oversize ones must be fitted.
16 During removal of the crankshaft gearwheel it should have been noted that there were shims behind the gearwheel. Refit the shims and then the inner Woodruff key.

44 Pistons and connecting rods – reassembly

1 Where original parts are being reassembled they must be refitted in their original positions; with new parts this does not apply. A new piston can be fitted to any connecting rod but, for example, No 2 original piston must be refitted to No 2 original connecting rod (and the right way round) and so on.
2 Assembling 16V engine pistons to their connecting rods consists of making sure that the piston and connecting rod are correctly aligned (FRONT or ▲ on the piston crown to the front of the engine; numbers on the connecting rod and cap to face the camshaft side of the engine), inserting the gudgeon pin so that the groove in it lines up with the clamp bolt hole in the rod, and fitting the clamp bolt and spring washer. Tighten the clamp bolt to the specified torque.
3 On 18V engines as the gudgeon pin in a press fit a special BL tool 18G 1150 with adaptors 18G 1150D is required to fit the gudgron pin as shown in Fig. 1.27 and should be used as follows.
4 Unscrew the large nut and withdraw the centre screw from the body a few inches. Well lubricate the screw thread and correctly locate the piston support adaptor.

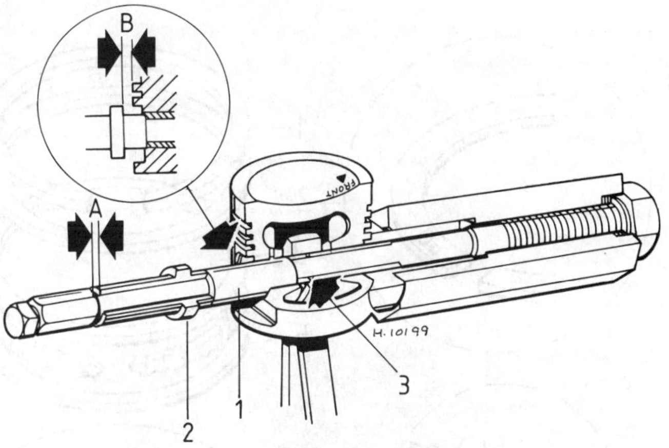

Fig. 1.27 Gudgeon pin refitting (18V engines) using BL tool 18G 1150 and adaptor 1150D (Sec 44)

1 Gudgeon pin
2 Place flange towards gudgeon pin
A Endplay 0.040 in (1 mm)
B Minimum clearance on assembly 0.080 in (2 mm)
3 Groove in parallel sleeve towards gudgeon pin

5 Carefully slide the parallel sleeve with the groove end last onto the centre screw up as far as the shoulder. Lubricate the gudgeon pin and its bores in the connecting rod and piston with a graphited oil.
6 Fit the connecting rod and piston, side marked FRONT or ▲ to the tool with the connecting rod entered on the sleeve up to the groove. Fit the gudgeon pin into the piston bore up to the connecting rod. Next fit the remover/replacer bush, flange end towards the gudgeon pin.
7 Screw the stop nut onto the centre screw and adjust the nut to give 0.040 in (1.0mm) endplay, A as shown in Fig. 1.27. Lock the nut securely with the lock screw.
8 Screw the large nut up the thrust race. Adjust the torque wrench to a setting of 12 lbf ft (1.7 kgf m) if of the 'click' type which will represent the minimum load for an acceptable fit. Use the torque wrench on the large nut, and a ring spanner on the lock screw. Pull the gudgeon pin into the piston until the flange of the remover/replacer bush is 0.080 inch (2.0mm) from the piston skirt. It is critically important that the flange is NOT allowed to contact the piston. Finally withdraw the BL service tool.
9 Should the torque wrench not 'click' or reach 12 lbf ft (1.7 kgf m) throughout the pull, the fit of the gudgeon pin in the connecting rod is not within limits; the parts must be renewed.
10 Ensure that the piston pivots freely on the gudgeon pin and is free to slide sideways. Should stiffness exist wash the assembly in paraffin,

46.5 Using a ring compressor to fit a piston

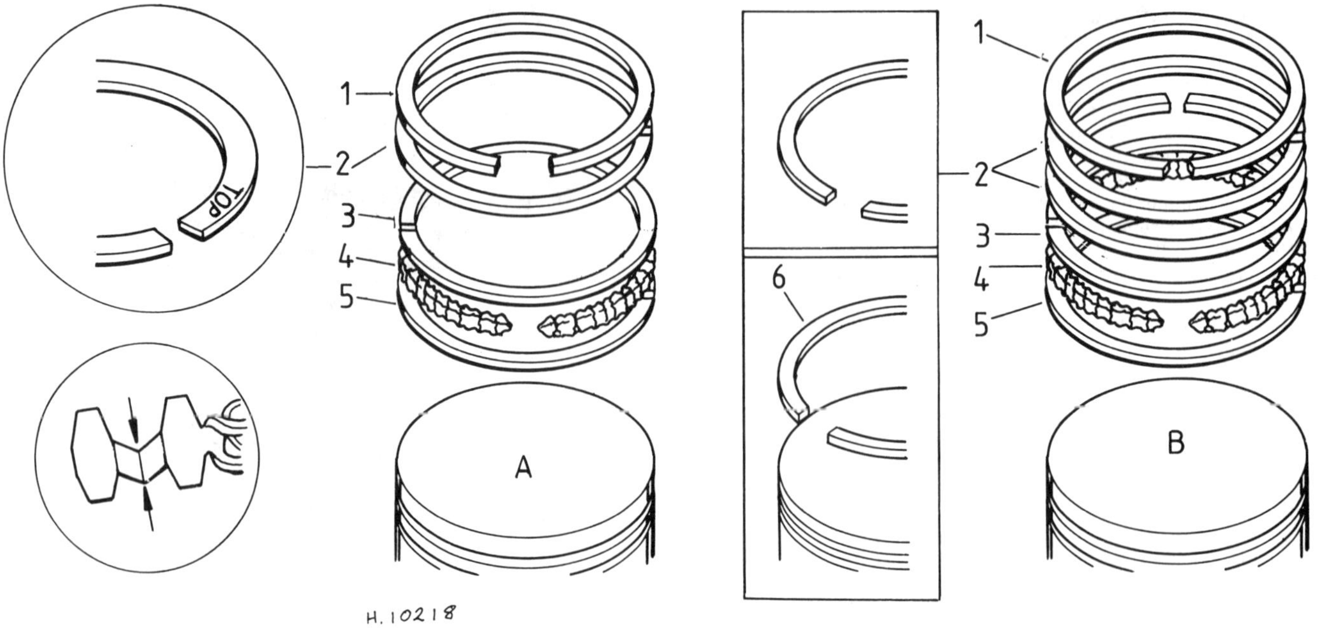

Fig. 1.28 Piston ring assemblies, A: 18V engines, B: 16V engines (Sec 45)

1 *Plain compression ring*
2 *Tapered compression ring*
3 *Top oil control ring rail*
4 *Oil control expander*
5 *Bottom oil control ring rail*
6 *(Early 16V engines) One-piece oil control ring*

lubricate the gudgeon pin with graphited oil and recheck. Again if stiffness exists dismantle the assembly and check for signs of ingrained dirt or damage. Finally check that the gudgeon pin will not foul the bores.

45 Piston rings – refitting

1 Check that the piston ring grooves and oilways are thoroughly clean and unblocked. Piston rings must always be fitted over the head of the piston and never from the bottom.
2 Refitment is the reverse procedure to removal, see Section 21.
3 Set all ring gaps 90° to each other, and away from the thrust side of the piston.
4 An alternative method is to fit the rings by holding them slightly open with the thumbs and both your index fingers. This method requires a steady hand and great care for it is easy to open the ring too much and break it.
5 The special oil control ring requires a special fitting procedure. First fit the bottom rail of the oil control ring to the piston and position it below the bottom groove. Refit the oil control expander into the bottom groove and move the bottom oil control ring rail into the bottom groove. Fit the top oil control rail into the bottom groove.
6 Ensure the ends of the expander are butting without overlapping as shown in the inset in Fig. 1.28.

46 Pistons – refitting

1 Fit pistons complete with connecting rods, to the cylinder bores as follows.
2 Wipe the cylinder bores clean with clean non fluffy rag.
3 The pistons, complete with connecting rods, must be fitted to their bores from above. As each piston is inserted into the bore ensure that it is the correct piston/connecting rod assembly for that particular bore and that the front of the piston is towards the front of the engine. Also check that the connecting rod is fitted correctly, ie with the number on the rod towards the camshaft side of the engine. Lubricate the piston well with clean engine oil.
4 Check that the piston ring gaps are 90° to each other.
5 The piston will slide into the bore only as far as the oil control ring and then it will be necessary to compress the piston rings into a clamp. The piston ring compressor should be fitted to the piston before it is inserted into the bore. If a proper piston ring clamp is not available then a suitable worm drive clip will do. Guide the piston into the bore until it reaches the ring compressor. Gently tap the piston into the cylinder bore with a wooden or plastic hammer (photo).

47 Connecting rod to crankshaft – reassembly

1 Wipe the connecting rod half of the big-end bearing location and the underside of the shell bearing clean, (as for the main bearing shells) and fit the shell bearing in position with its locating tongue engaged with the corresponding groove in the connecting rod. Always fit new shells.
2 Generously lubricate the crankpin journals with engine oil and turn the crankshaft so that the crankpin is in the most advantageous position for the connecting rod to be drawn onto it.
3 Fit the bearing shell to the connecting rod cap in the same way as with the connecting rod itself.
4 Generously lubricate the shell bearing and offer up the connecting rod bearing cap to the connecting rod. Fit the connecting rod cap retaining nuts. It will be observed that on some engines these are special twelve-sided nuts.
5 Tighten the retaining nuts to the specified torque.

48 Camshaft – refitting

1 Fit the engine front plate with a new gasket between the cylinder block and plate and tighten the retaining bolts and spring washers.
2 Wipe the camshaft bearings and generously lubricate them with engine oil.
3 Temporarily refit the camshaft gearwheel and locating plate and secure with the retaining nut. Using feeler gauges check the endfloat which should not exceed the specified value. If this maximum limit is exceeded obtain a new locating plate.
4 Remove the camshaft gearwheel retaining nut, gearwheel and locating plate.
5 Insert the camshaft into the crankcase gently, taking care not to damage the camshaft bearings with the sharp edges of the cams. Take care, the camshaft lobe edges are sharp.

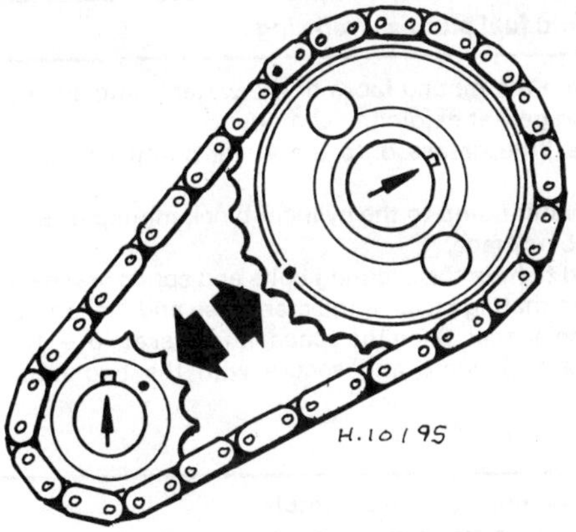

Fig. 1.29 Timing gear marks (Sec 50)

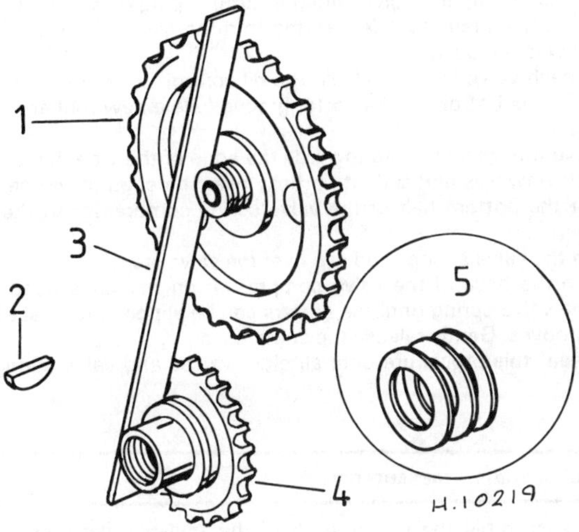

Fig. 1.30 Timing gear alignment (Sec 50)

1 Camshaft gear
2 Woodruff key
3 Straight edge
4 Crankshaft gear
5 Inset: Shims

49 Oil pump and driveshaft – refitting

Invert the cylinder block and insert the oil pump driveshaft. Fit a new oil pump to crankcase gasket over the three studs followed by the oil pump. Ensure that the shaft dog correctly engages with the drivegear dog. Fit spring washers on the studs followed by the three nuts. Tighten to the specified torque.

50 Timing gear, chain tensioner and cover – refitting

1 Before assembly begins check that the shim washers are in place on the crankshaft nose. If new gearwheels are being fitted it may be necessary to fit additional washers as detailed in paragraph 7. These washers ensure that the crankshaft gearwheel lines up correctly with the camshaft gearwheel (photo).

2 Refit the Woodruff keys in their respective slots in the crankshaft and camshaft and ensure that they are fully seated. If their edges are burred they must be cleaned with a fine file.

3 Lay the camshaft gearwheels on a clean surface so that the two timing dots are adjacent to each other. Slip the timing chain over them and pull the gearwheels back into mesh with the chain so that the timing dots, although further apart, are still adjacent to each other as shown in Fig.1.29.

4 Rotate the crankshaft so that the Woodruff key is at top dead centre. The engine should be in the upright position.

5 Rotate the camshaft so that when viewed from the front the Woodruff key is at the two o'clock position.

6 Fit the timing chain and gearwheel assembly onto the camshaft and crankshaft keeping the timing marks adjacent. If the camshaft and crankshaft have been positioned accurately, it will be found that the keyways on the gearwheels will match the position of the keys, although it may be necessary to rotate the camshaft a fraction to ensure accurate lining up of the camshaft gear wheel.

7 Press the gearwheels into position on the crankshaft and camshaft as far as they will go. If new gearwheels are being fitted they should be checked for alignment before being finally fitted to the engine. Place the gearwheels in position without the timing chain and place the straight edge of a steel rule from the side of the camshaft gear teeth to the crankshaft gearwheel and measure the gap between the steel rule and the gearwheel. Remove the gearwheels and drive keys from the crankshaft, add or remove shims as required and refit the keys. Refit the gearwheels and chain as described above.

8 Reassemble the timing chain tensioner as described in Section 27.

9 Fit the backplate and secure the tensioner assembly to the engine cylinder block with the lockwasher and two bolts. Tighten the bolts and bend up the lockwasher tabs to lock.

10 Fit a new camshaft gear lockwasher and fit the nut. Tighten to the specified torque and bend up the locking tab.

50.1 Shim washers on the crankshaft

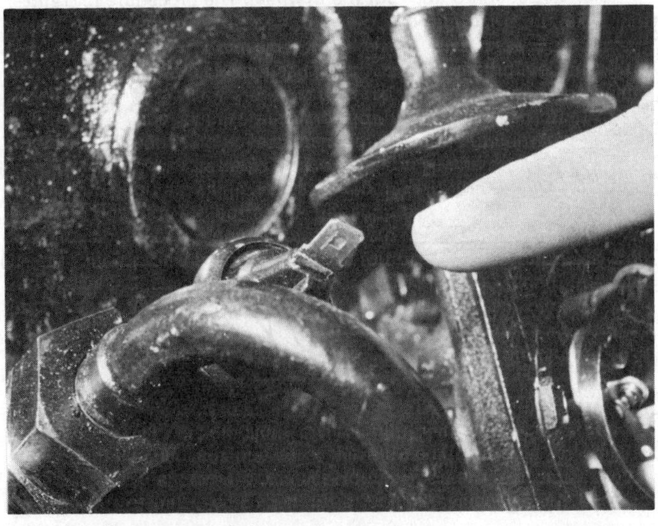

53.3 The oil pressure switch

11 Fit the oil thrower, with the face marked F towards the front of the engine, onto the crankshaft.
12 Generously oil the chain and gearwheels.
13 Ensure the interior of the timing cover and the timing cover flange is clean and generally lubricate the oil seal in the timing cover. Then with a new gasket in position, fit the timing cover to the block using the pulley to centralise the cover.
14 Screw in the timing cover retaining bolts with the flat washer next to the cover flange and under the spring washer. Tighten the bolts to the specified torques.
15 Fit the crankshaft pulley to the nose of the crankshaft ensuring that the keyway engages with the Woodruff key.
16 Fit the crankshaft retaining bolt locking washer in position and screw on the crankshaft pulley retaining nut. Tighten to the specified torque and bend over the locking washer.

51 Engine backplate – refitting

Refer to Section 8 and refit the gearbox adaptor plate (backplate).

52 Sump – refitting

1 After the sump has been thoroughly cleaned, scrape all traces of the old sump gasket from the sump flange, and fit new main bearing cap oil seal corks to the No 1 and 3 (16V engines) or 1 and 5 (18V engines) bearing caps.
2 Thoroughly clean and scrape the crankshaft to sump flange. Apply grease to the crankcase to sump flange and carefully fit a new gasket to the flange.
3 Carefully fit the sump to the underside of the crankcase.
4 Refit the sump retaining bolts, shaped washers and spring washers.
5 Tighten the sump bolts to the specified torque.

53 Oil pressure relief valve and switch – refitting

1 Assemble the valve components in the order of, valve, spring and domed nut with new washers.
2 Carefully insert the assembly into its location at the rear of the cylinder block. Tighten the domed nut fully.
3 Locate the oil pressure switch in its drilling in the side of the cylinder block just above the oil pressure relief valve and tighten firmly using an open ended spanner (photo).

54 Flywheel and clutch – refitting

Refer to Section 8 and refit the flywheel. The procedure for refitting the clutch is contained in Chapter 5.

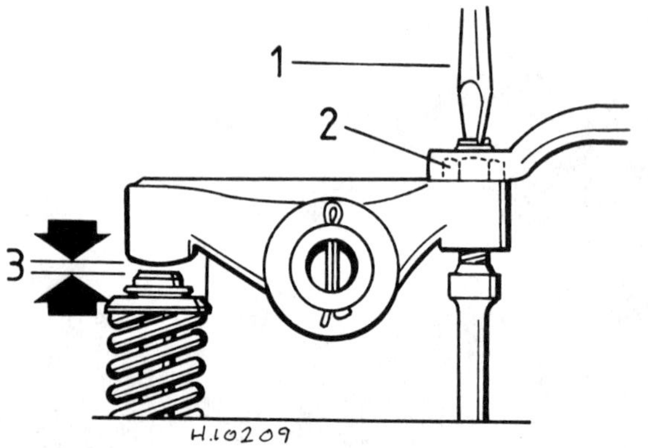

Fig. 1.31 Valve rocker clearance (Sec 60)

1 Screwdriver
2 Ring spanner on locknut
3 Clearance, see Specifications

55 Water and fuel pumps – refitting

1 Make sure the mating faces of the water pump and cylinder block are free of old gasket or jointing compound.
2 Smear a little grease onto the water pump and place on a new gasket.
3 Fit the water pump to the cylinder block mating it to the dowels in the cylinder block face.
4 Refit and tighten the securing bolts and spring washers.
5 Clean the mating faces of the crankcase and fuel pump and fit new gaskets to the spacer. Slide the spacer and gaskets over the studs.
6 Refit the fuel pump and secure with the two nuts and spring washers.

56 Valves and springs – reassembly

1 To refit the valves and the valve springs to the cylinder head, proceed as follows.
2 Rest the cylinder head on its side, or if the manifold studs are fitted, with the gasket surfaces downwards.
3 Fit each valve and valve spring in turn, wiping down and lubricating each valve stem as it is inserted into the same valve guide from which it was removed.
4 As each valve is inserted slip the oil control rubber ring into place just under the bottom of the cotter groove (use a new rubber ring if at all possible).
5 Move the cylinder head towards the edge of the work bench if it is facing downwards and slide it partially over the edge of the bench so as to fit the bottom half of the valve spring compressor to the valve head.
6 Slip the valve springs and cap over the valve stem.
7 With the base of the valve compressor on the valve head, compress the valve spring until the cotters can be slipped into place in the cotter grooves. Gently release the compressor.
8 Repeat this procedure until all eight valves and valve springs are fitted.

57 Rocker shaft – reassembly

To reassemble the rocker shaft fit the split-pin, flat washer, and spring washer at the rear end of the shaft and then slide on the rocker arms, rocker shaft pedestals, and spacing springs in the same order in which they were removed (Fig.1.9) and slip the locating plate into place. Finally, fit to the front of the shaft the spring washer, plain washer and split-pin, in that order.

58 Tappets and pushrods – refitting

1 Generously lubricate the tappets internally and externally and insert them in the bores from which they were removed through the tappet chest.
2 With the cylinder head in position, fit the pushrods in the same order in which they were removed. Ensure that they locate properly in the stems of the tappets, and lubricate the pushrod ends before fitment. Refit the two tappet covers with new cork gaskets and tighten the bolts.

59 Cylinder head – refitting

1 After checking that both the cylinder block and cylinder head mating faces are perfectly clean, generously lubricate each cylinder with engine oil.
2 Always use a new cylinder head gasket. The old gasket will be compressed and incapable of giving a good seal. Never smear grease or gasket cement either side of the gasket, for pressure leaks may blow through it.
3 The cylinder head gasket is marked FRONT and should be fitted in position according to the markings. The copper side will be uppermost (photo).
4 Carefully lower the gasket into position ensuring that the stud

59.3 The front of the cylinder head gasket is marked

59.4 Fitting the cylinder head gasket

59.5 Lowering the cylinder head into position

59.12 Using a torque wrench on the cylinder head nuts

60.5A Adjusting a rocker arm/valve clearance

60.5B Checking valve clearance with a feeler gauge

threads do not damage the side of the holes through which they pass (photo).

5 The cylinder head may now be lowered over the studs until it rests on the cylinder head gasket (photo).

6 With the cylinder head in position fit the pushrods in the same order in which they were removed. Ensure that they locate properly in the stems of the tappets.

7 Refit the two tappet covers with new cork gaskets and tighten the two securing bolts and washers.

8 The rocker shaft assembly can now be lowered over its eight locating studs. Take care that the rocker arms are the right way round. Lubricate the balljoints, and insert the rocker arm balljoints into the pushrod cups. Failure to replace the balljoints in the cups can result in the balljoints seating on the edge of a pushrod or outside it when the head and rocker assembly is pulled down tight.

9 Fit the lock plate to the second pedestal.

10 Fit the four rocker pedestal nuts and washers, and then the four cylinder head stud nuts and washers which also hold down the rocker pedestals. Pull the nuts down evenly, but without tightening them right up.

11 Fit the remaining nuts and washers to the cylinder head nuts.

12 Tighten the cylinder head nuts gradually in the sequence shown in Fig. 1.7 to half of the full-torque value listed in the Specifications, then in the same sequence, to the full torque value (photo).

13 Tighten the rocker shaft pedestal nuts to the specified torque.

14 When the engine is first run after the cylinder head has been refitted, run the engine at fast idle speed for at least 15 minutes or road test for 5 miles (8 km). Then, in the sequence shown in Fig.1.7, slacken each cylinder head nut individually a quarter turn and retighten to the full torque value.

60 Valve clearance – adjustment

1 The valve adjustments should be made with the engine cold. The importance of correct rocker/valve stem clearances cannot be overstressed as they vitally affect the performance.

2 If the clearances are set too wide, the efficiency of the engine is reduced as the valves open late and close earlier than was intended. If the clearances are set too close there is a danger that the stem and pushrods upon expansion will not allow the valves to close properly which will cause burning of the valve head and possible warping.

3 If the engine is in the vehicle, to get at the rockers, it is merely necessary to remove the two holding down dome nuts from the cover, and then to lift the rocker cover and gasket away.

4 It is important that the clearance is set when the tappet of the valve being adjusted is on the heel of the cam (ie opposite the peak). This can be done by carrying out the adjustments in the following order, which also avoids turning the crankshaft more than necessary. Valves are numbered 1 to 8 from the timing cover end.

Valve fully open	**Check and adjust**
Valve number 8 EX	*Valve number 1 EX*
Valve number 6 IN	*Valve number 3 IN*
Valve number 4 EX	*Valve number 5 EX*
Valve number 7 IN	*Valve number 2 IN*
Valve number 1 EX	*Valve number 8 EX*
Valve number 3 IN	*Valve number 6 IN*
Valve number 5 EX	*Valve number 4 EX*
Valve number 2 IN	*Valve number 7 IN*

5 The correct valve clearance is given in the Specifications at the beginning of this chapter. It is obtained by slackening the hexagonal locknut with a spanner while holding the ball pin against rotation with a screwdriver as shown in Fig. 1.31. Then, still pressing down with the screwdriver, insert a feeler gauge of the required thickness between the valve stem and head of the rocker arm and adjust the ball pin until the feeler gauge will just move in and out without nipping. Then still holding the ball pin in the correct position, tighten the locknut (photos).

6 An alternative method is to set the gaps with the engine running at idle speed. Although this method may be faster, more practice is needed and it is no more reliable.

7 Refit the rocker cover and gasket, and secure with the dome nuts and washers.

61 Distributor and distributor drive – refitting

1 It is important to set the distributor drive correctly otherwise the ignition timing will be totally incorrect. It is easy to set the distributor drive in apparently the right position, but in fact exactly 180° out, by omitting to select the correct cylinder which must not only be at TDC but must also be on its firing stroke with both valves closed. The distributor drive should therefore not be fitted until the cylinder head is in position and the valves can be observed. Alternatively, if the timing cover has not been refitted, the distributor drive can be refitted when the dots on the timing wheels are adjacent to each other.

2 The distributor driveshaft can only be fitted with the pistons half way up or down their bores. Turn the crankshaft so that the pistons are in this position (90° before or after TDC).

3 Screw a $\frac{5}{16}$ inch UNF bolt about $3\frac{1}{2}$ inches (90mm) long into the threaded hole in the top end of the shaft and fit the shaft to the engine.

4 Rotate the crankshaft so that No 1 piston is at TDC and on its firing stroke (the dots on the timing gears will be adjacent to each other). When No 1 piston is at TDC the inlet valve on No 4 cylinder is just opening and the exhaust valve closing.

5 With the pistons at TDC it will not be possible to remove the driveshaft, but it is possible to lift it sufficiently to bring it out of mesh with the drivegear on the camshaft. The driveshaft can then be turned to the correct timing position.

6 When the groove on the crankshaft pulley wheel is in line with the TDC pointer on the timing gear cover, then No 1 and No 4 piston are at TDC. Check that the No 4 cylinder valves are just rocking, ie inlet valve just opening and the exhaust valve just closing, to ensure correct stroke for No 1 cylinder.

7 Position the driveshaft so that the slot is just below the horizontal, and the larger of the two segments (one on each side of the slot) is at the top. As the gear on the shaft engages with the skew gear on the camshaft the drivegear will turn anti-clockwise until the top of the slot is adjacent to the 2 o'clock position, see Fig. 1.13.

8 Remove the bolt from the driveshaft.

9 Refit the distributor housing and lock it in position with the single bolt and spring washer. It is important that the correct bolt is used so that the head does not protrude above the face of the housing.

10 The distributor can now be refitted and the two securing bolts and spring washers which hold the distributor clamping plate to the distributor housing tightened. If the clamp bolt on the clamping plate was not previously loosened and the distributor body was not turned in the clamping plate, then the ignition timing will be as previously set. If the clamping bolt has been loosened, then it will be necessary to retime the ignition as described in Chapter 4.

62 Final assembly

1 Refit the rocker cover, using a new cork gasket, and secure in position with the two domed nuts.

2 Fit the two tappet cover plates, using new gaskets, and tighten the tappet chest bolts. Be careful not to overtighten these bolts or the covers will distort and leak oil.

3 Fit a new manifold gasket over the studs taking care not to rip it as it passes over the stud threads.

4 Refit the manifold and secure it in position with the six nuts, spring and large plain washers.

5 Insert the thermostat into its housing making sure the word FRONT marked on the flange is towards the front of the engine.

6 Fit a new gasket taking care not to rip the gasket as it passes over the threads of the three studs. Then refit the thermostat housing cover and secure with the three nuts and spring washers.

7 If the dipstick guide tube was removed this should next be inserted into its drilling in the side of the crankcase.

8 Fit a new oil filter canister (see Section 24).

9 It should be noted that in all cases it is best to reassemble the engine as far as possible before refitting it to the vehicle. This means that alternator, fanbelt and other minor attachments should be refitted at this stage.

63 Engine – refitting

1 Although the engine or engine and gearbox can be refitted by one man and a suitable hoist, it is easier if two are present. Generally refitt-

ing is the reverse sequence to removal, but note the following additional points.

2 Ensure all the loose leads, cables etc, are tucked out of the way. It is easy to trap one and cause much additional work after the engine is replaced.

3 Refit the following:

(a) Mounting nuts, bolts and washers
(b) Propeller shaft coupling
(c) Clutch slave cylinder (manual gearbox)
(d) Speedometer cable
(e) Gearbox change lever and surround (manual gearbox)
(f) Carpets
(g) Oil pressure switch cable
(h) Water temperature indicator sender unit cable
(i) Wires to coil, distributor and alternator
(j) Carburettor controls and air cleaner
(k) Exhaust manifold to down pipe
(l) Earth and starter motor cables
(m) Radiator and hoses
(n) Heater hoses
(o) Engine closed circuit breather hoses
(p) Vacuum advance and retard pipe
(q) Battery (if removed)
(r) Fuel lines to carburettor(s) and pump
(s) Bonnet

4 Finally check that the drain taps are closed and refill the cooling system with water and the engine with the specified quantity and grade of oil.

64 Engine – initial start up after overhaul and major repair

1 Make sure that the battery is fully charged and that all lubricants, coolants and fuel are replenished.

2 If the fuel system has been dismantled, it will require several revolutions of the engine on the starter motor to pump petrol to the carburettor.

3 As soon as the engine fires and runs, keep it going at a fast tickover only (no faster) and bring it up to normal working temperatures.

4 As the engine warms up, there will be odd smells and some smoke from parts getting hot and burning off oil deposits. Look for water or oil leaks which will be obvious if serious. Check also the clamp connection of the exhaust pipe to the manifold as these do not always 'find' their exact gas-tight position until the warmth and vibration have acted on them, and it is almost certain that they will need tightening further. This should be done of course, with the engine stopped.

5 When the engine running temperature has been reached, adjust the idling speed as described in Chapter 3.

6 Stop the engine and wait a few minutes to see if any lubricant or coolant drips out.

7 Retighten the cylinder head nuts as described in Section 59, if these were undone during repair.

8 Road test the car to check that the timing is correct and give the necessary smoothness and power. Do not race the engine. If new bearings and/or pistons have been fitted, it should be treated as a new engine and run in at reduced revolutions for 500 miles (800 km).

65 Fault diagnosis – engine

Symptom	Reason/s
Engine will not turn over when starter switch is operated	Flat battery Bad battery connections Bad connections at solenoid switch and/or starter motor Defective solenoid Starter motor defective Selector rod or isolator switch out of adjustment (automatic gearbox only)
Engine turns over normally but fails to fire and run	No spark at plugs No fuel reaching engine Too much fuel reaching the engine (flooding)
Engine starts but runs unevenly and misfires	Ignition and/or fuel system faults Incorrect valve clearances Burnt out valves
Lack of power	Ignition and/or fuel system faults Incorrect valve clearances Burnt out valves Worn out piston or cylinder bores
Excessive oil consumption	Oil leaks from crankshaft oil seal, rocker cover gasket, drain plug gasket sump plug washer Worn piston rings or cylinder bores resulting in oil being burnt by engine (smoky exhaust is an indication) Worn valve guides and/or defective valve stem seals
Excessive mechanical noise from engine	Wrong valve to rocker clearances Worn crankshaft bearings Worn cylinders (piston slap)
Unusual vibration	Misfiring on one or more cylinders Loose mounting bolts

Note: *When investigating starting and uneven running faults do not be tempted into snap diagnosis. Start from the beginning of the check procedure and follow it through. It will take less time in the long run. Poor performance from an engine in terms of power and economy is not normally diagnosed quickly. In any event the ignition and fuel systems must be checked first before assuming any further investigation needs to be made.*

Chapter 1 Part B 1700 cc (17V) engine

For modifications, and information applicable to later models, see Supplement at end of manual

Contents

Specifications

General

Bore	3·325 in (84·45 mm)
Type	Four-cylinder, in-line, ohc
Stroke	2·984 in (75·87 mm)
Capacity	103·74 cu in (1700 cc)
Firing order	1 – 3 – 4 – 2
Compression ratio	8·0 : 1

Crankshaft

Main journal diameter	2·1262 to 2·1270 in (54·005 to 54·026 mm)
Crankpin journal diameter	1·8754 to 1·8759 in (47·635 to 47·647 mm)
Endfloat	0·001 to 0·005 in (0·025 to 0·13 mm)
Oversize thrust washers	+ 0·003 in (0·076 mm)

Main bearings

Number and type	Five, steel-backed shell type
Running clearance	0·001 to 0·003 in (0·025 to 0·077 mm)

Big-end bearings

Type	Steel-backed shell
Running clearance	0·0015 to 0·0032 in (0·038 to 0·081 mm)

Gudgeon pins

Type	Press fit in connecting rod
Diameter	0·8125 to 0·8127 in (20·638 to 20·643 mm)

Pistons

Type	Solid skirt, combustion chamber in crown
Clearance in bore:	
Below oil control groove	0·0008 to 0·0051 in (0·02 to 0·13 mm)
Bottom of skirt	0·0009 to 0·0019 in (0·022 to 0·048 mm)

Piston rings

Compression:	
Type – top	Plain, chrome-faced
Type – second	Stepped scraper
Fitted gap	0·012 to 0·020 in (0·3 to 0·5 mm)
Ring to groove clearance	0·0015 to 0·0027 in (0·04 to 0·07 mm)
Oil control:	
Type	Twin chrome-faced rails with expander
Fitted gap	0·015 to 0·055 in (0·33 to 1·4 mm)

Camshaft

Bearings:	
Number and type	Three, direct in head and cover
Running clearance	0·0017 to 0·0037 in (0·043 to 0·094 mm)
Journal diameters	1·888 to 1·889 in (47·96 to 47·97 mm)
Endfloat	0·003 to 0·007 in (0·07 to 0·18 mm)

Timing belt tension

New belt	13 lbf (58 N)
Used belt	11 lbf (49 N)

Valves

Tappet clearance (cold – inlet and exhaust)	0·012 ± 0·001 in (0·30 ± 0·03 mm)
Adjust only if less than:	
Inlet	0·008 in (0·20 mm)
Exhaust	0·011 in (0·28 mm)
Seat angle	45°
Face angle	45° 30′
Seat diameter:	
Inlet	1·575 in (40 mm)
Exhaust	1·339 in (34 mm)
Stem diameter:	
Inlet	0·2917 to 0·2923 in (7·41 to 7·42 mm)
Exhaust	0·2909 to 0·2917 in (7·39 to 7·41 mm)
Stem to guide clearance:	
Inlet	0·001 to 0·002 in (0·03 to 0·05 mm)
Exhaust	0·0015 to 0·0027 in (0·04 to 0·073 mm)

Valve guides

Fitted height above head	0·394 in (10·0 mm)
Interference fit in head	0·0015 to 0·003 in (0·04 to 0·08 mm)

Valve springs

Free length	1·646 in (41·81 mm)
Fitted length	1·375 in (34·92 mm)
Load at fitted length	44·5 lbf (20 kgf)

Valve timing

Inlet valve:	
Opens	5° BTDC
Closes	45° ABDC
Exhaust valve:	
Opens	40° BBDC
Closes	10° ATDC

Lubrication system

System type	Wet sump, pressure fed
Oil pressure:	
Running	60 to 90 lbf/in² (4·2 to 6·3 kgf/cm²)
Idling	40 to 60 lbf/in² (2·8 to 4·2 kgf/cm²)
Relief valve spring free length	1·525 in (38·7 mm)
Warning light switch operating pressure	6 to 10 lbf/in² (0·4 to 0·7 kgf/cm²)
Oil pump type	Eccentric rotors, mounted around crankshaft

Oil pump maximum clearances (see Fig. 1.46):	
Outer rotor to housing (A)	0·004 in (0·10 mm)
Inner rotor lobe to outer rotor (B and C)	0·011 in (0·28 mm)
Outer rotor housing endfloat (D)	0·003 in (0·08 mm)
Oil filter	Full flow, disposable cartridge
Sump capacity	6·5 pints (3·75 litres)

Torque wrench settings

	lbf ft	Nm
Spark plugs – hand tight plus $\frac{1}{16}$ turn	7	10
Cylinder head bolts (oil threads):		
Pre-tighten (see text)	35	47
Fully tighten	60	81
Cam cover securing bolts	13	18
Camshaft rear end cover bolts	9	12
Petrol pump plate screws	9	12
Distributor flange nuts	18	25
Gearwheel to camshaft bolt	48	65
Manifold securing bolts	18	24
Carburettor to manifold nuts	19	25
Air filter to carburettor	19	25
Crankshaft pulley bolt	62	85
Main bearing cap bolts (oil threads)	75	100
Big-end cap nuts (oil threads)	33 ± 2	45 ± 3
Rear plate:		
8 mm screws	22	30
10 mm screws	37	50
Flywheel to crankshaft screws (oil threads)	42	58
Clutch to flywheel screws	17	23
Oil pump housing to block bolts	8	11
Oil pump backplate securing screws	2	3
Oil separator to block screws	22	30
Oil pressure switch	9	12
Oil blanking plate screws	8	11
Water pump to block bolts	8	11
Water pump pulley to hub screws	9	12
Thermostat housing securing bolts	8	11
Thermal transmitter	5	6
Cylinder block drain plug	27	37
Alternator adjusting link screw	9	12
Engine mounting bolt locknut	30	41
Other screws into alumninium:		
M6	8	11
M8	18	25
Other screws into cast iron:		
M6	8	11
M8	22	30
M10	37	50

66 General description

1 The 1700 cc (17V) O-series engine differs considerably from those dealt with in Part A of this Chapter. With a conventional four-cylinder, five main bearing layout, the engine has an overhead camshaft mounted in bearings formed in the cylinder head and the camshaft cover. As the distributor and fuel pump are driven from the camshaft, these components are mounted on the camshaft cover.

2 The camshaft is driven by an external, toothed, timing belt and operates vertically mounted valves through bucket-type tappets containing clearance adjustment shims. The belt is tensioned by an adjustable idler pulley which bears on the outside surface of the belt.

3 Oil pressure is maintained by an eccentric rotor type pump mounted, together with the filter head and a pressure relief valve, in a die casting which forms the engine front closing plate.

4 A V-belt is used to drive the coolant pump, fan and alternator from the crankshaft pulley.

5 For the procedure for changing the engine oil and oil filter cartridge, refer to Section 24, noting that the filter cartridge is on the front left of the engine.

67 Major operations possible with engine in vehicle

Refer to the information given in Section 2 of this Chapter for the 16V and 18V engines. Note in addition that the flywheel can be removed from the 17V engine while it is in the vehicle.

68 Major operations requiring engine removal

Refer to the information given in Section 3 of this Chapter for the 16V and 18V engines, noting that the flywheel can be removed from the 17V engine without having to remove the engine from the vehicle.

69 Methods of engine removal

Refer to the information given in Section 4 of this Chapter for the 16V and 18V engines.

70 Engine – removal with gearbox

1 Refer to Chapter 2 and drain the cooling system.

2 Mark the position of the bonnet hinges on the bonnet, remove the four nuts and washers and with the help of an assistant remove the bonnet. Store it safely where it cannot get damaged.

3 Disconnect the battery earth lead.

4 Remove the eight screws securing the radiator grille and remove the grille from the vehicle.

5 Spring off the retaining clip and remove the bonnet prop. Disconnect the control cable from the bonnet lock and platform. Remove the bolts securing the bonnet lock platform and remove the four outer screws which secure the platform to the radiator side blanking panels.

Lift the platform off the radiator retaining spigots. Remove the two side blanking panels.
6 Disconnect the top hose from the radiator, and the bottom hose if this was not removed during the draining. Lift out the radiator.
7 Remove the securing screws and remove the lower surround grille panel.
8 Remove the radiator top hose from the thermostat housing. Disconnect the air bleed hose from the thermostat housing and move it to one side. Disconnect the bottom cooling hose from the water pump and from the heater rail pipe; move the hose to one side.
9 Undo the two bolts securing the air cleaner to the carburettor and remove the air cleaner complete with thermostatic flap valve and flexible duct.
10 Disconnect both fuel hoses from the fuel pump and plug the hose feeding the pump.
11 Disconnect the breather hose from the carburettor. Note the restrictor in the carburettor adaptor tube.
12 Remove the nuts securing the carburettor to the manifold and remove the carburettor together with the insulator block, cable bracket and heat shield. Move it out of the way, placing it safely on the battery side of the vehicle. If an automatic gearbox is fitted, disconnect the downshift cable at the carburettor link.
13 Disconnect the brake servo vacuum hose from the adaptor on the manifold.
14 Remove the electrical wiring spade connectors from the engine ancillary components such as oil pressure switch, thermal transmitter etc and the multi-connector from the alternator. Remove the starter motor fixing bolt which retains the earth cable and remove the cable.
15 Disconnect the heater hose from the cylinder head. Remove the brake servo adaptor to release the heater rail, noting the copper washer under the rail bracket. Refit the adaptor and washer temporarily and move the rail out of the way to one side.
16 Disconnect the HT leads from the spark plugs and coil and remove the distributor cap complete with leads.
17 If not previously removed during the draining, remove the cylinder block drain plug after positioning a suitable container beneath it and drain any residual coolant. Refit the drain plug.
18 Undo the bolt securing the exhaust pipe front mounting bracket to the bracket on the gearbox bellhousing. Remove the six nuts and washers at the exhaust pipe flange to manifold joints and release the exhaust downpipes from the manifold.
19 Remove the timing belt cover by pulling it off its mountings. Undo the four bolts securing the radiator fan and remove the fan, spacer, fanbelt and pulley.
20 Position a suitable container under the sump, remove the sump drain plug and drain the oil. Refit the drain plug. Similarly, drain the gearbox oil.
21 Place a rope sling or chain around the engine and support its weight with an overhead hoist. Make certain that there is no chance of this slipping as the assembly has to be lifted out at a steep, tail down angle.
22 *Manual gearbox:* Refer to Chapter 6 and remove the gear change lever from the gearbox.
23 Mark the propeller shaft and gearbox drive flanges so that they can be reassembled in the same relative position. Remove the four securing bolts and nuts and secure the propeller shaft to one side.
24 *Manual gearbox:* Undo the finger nut securing the speedometer cable to the gearbox and remove the cable drive.
25 *Manual gearbox:* Remove the two bolts retaining the clutch slave cylinder to the gearbox, pull the cylinder off the pushrod and secure to one side out of the way. There is no need to disconnect the clutch hydraulic line.
26 *Manual gearbox:* Remove the two bolts securing each rear mounting to the gearbox crossmember and lift the engine/gearbox assembly slightly to take the weight of the gearbox off the crossmember.
27 *Automatic gearbox:* Disconnect the selector rod from both levers and remove the rod.
28 *Automatic gearbox:* Disconnect the oil cooler pipes from the gearbox. Plug the pipes and unions to prevent dirt getting in and to avoid loss of fluid.
29 *Automatic gearbox:* Disconnect the inhibitor switch leads at the harness multi-connector.
30 *Automatic gearbox:* Slacken the central nut securing the rear mounting to the crossmember and then remove the two nuts and bolts securing the rear mounting to the gearbox.
31 *Manual gearbox:* Remove the bolts and reinforcing plate securing each end of the rear crossmember to the body and remove the rear crossmember.
32 Support the rear of the gearbox with a jack.
33 Remove the nuts which secure the front engine mountings to the front crossmember brackets. Lower and remove the jack from beneath the gearbox.
34 Check all round the engine and gearbox to make sure that nothing has been left connected and then carefully raise and tilt the assembly forward and up out of the vehicle. Keep checking that nothing gets caught up until the unit is well clear and then place it where it can be cleaned before further work proceeds.

66.1 Underbonnet view of the 1700 cc (17V) OHC engine

71 Engine – removal without gearbox

1 If required, the engine can be removed on its own, leaving the gearbox in the vehicle.
2 Follow the procedure contained in the previous Section as far as fitting the lifting equipment, paragraph 20, and just take the engine weight on the hoist.
3 Remove the nuts which secure the front engine mountings to the crossmember brackets.
4 From beneath the vehicle, remove the nuts and bolts which secure the gearbox to the engine, and the exhaust support bracket to the gearbox bellhousing.
5 Remove the clutch slave cylinder securing bolts and remove the slave cylinder from the pushrod, moving it to one side out of the way. There is no need to disconnect the clutch hydraulic line.
6 Disconnect the intermediate mounting on the exhaust system by slackening the bracket clamp, rotating the clamp, and releasing the mounting from the body bracket.
7 Support the gearbox on a jack.
8 Remove the front mounting to engine retaining nuts, raise the engine slightly and remove both mountings.
9 Check that nothing has been left connected and then raise the engine further and pull it forwards. *It is essential to keep the engine and gearbox exactly in line with each other initially.* There is a fixed stud on the top right hand side of the engine backplate which can easily be damaged. Keep the two units in line until the stud has completely cleared the bellhousing.
10 Remove the engine from the vehicle and place it where it can be cleaned before further work is done on it.

72 Engine mountings – removal and refitting

1 Position a jack under the front of the engine sump and, with a

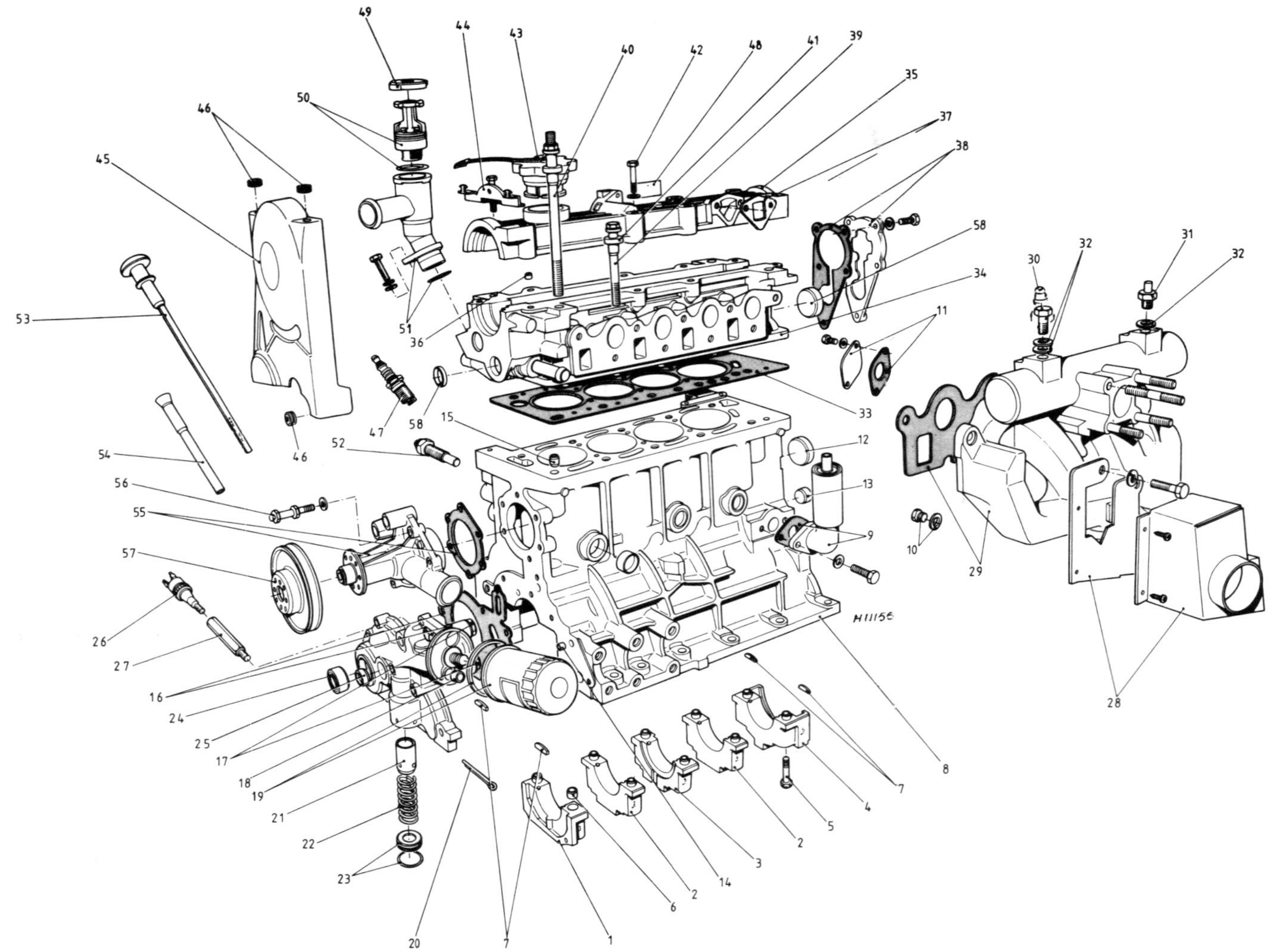

1 *Front main bearing cap*
2 *Intermediate main bearing cap*
3 *Centre main bearing cap*
4 *Rear main bearing cap*
5 *Set-screw*
6 *Ring dowel*
7 *Sealing insert*
8 *Cylinder block and crankcase*
9 *Oil separator and gasket*
10 *Coolant drain plug and washer*
11 *Blanking plate and gasket*
12 *Core plug*
13 *Plug*
14 *Dowel*
15 *Ring dowel*
16 *Oil pump and gasket*
17 *Plugs (oil cooler not fitted)*
18 *Plug (oil cooler fitted)*
19 *Oil filter cartridge and seal*
20 *Split-pin*
21 *Oil pressure relief valve plunger*
22 *Oil pressure relief valve spring*
23 *Plug and O-ring*
24 *Oil seal*
25 *Plug*
26 *Oil pressure switch*
27 *Switch extension*
28 *Hot box*
29 *Inlet/exhaust manifold and gasket*
30 *Brake servo vacuum adaptor*
31 *Distributor vacuum adaptor*
32 *Sealing washer*
33 *Cylinder head gasket*
34 *Cylinder head*
35 *Camshaft bearing cover*
36 *Ring dowel*
37 *Blanking plate and gasket*
38 *Camshaft end cover and gasket*
39 *Cylinder head bolt – short*
40 *Cylinder head bolt – long*
41 *Washer*
42 *Bolt*
43 *Oil filter and breather cap*
44 *Cover bracket*
45 *Timing belt cover*
46 *Grommet*
47 *Spark plug*
48 *Warning label*
49 *Coolant filler cap*
50 *Thermostat and O-ring*
51 *Coolant outlet elbow and O-ring*
52 *Thermal transmitter*
53 *Dipstick*
54 *Dipstick tube*
55 *Water pump and gasket*
56 *Pillar*
57 *Water pump pulley*
58 *Core plug*

Fig. 1.32 Engine external components (Sec 66)

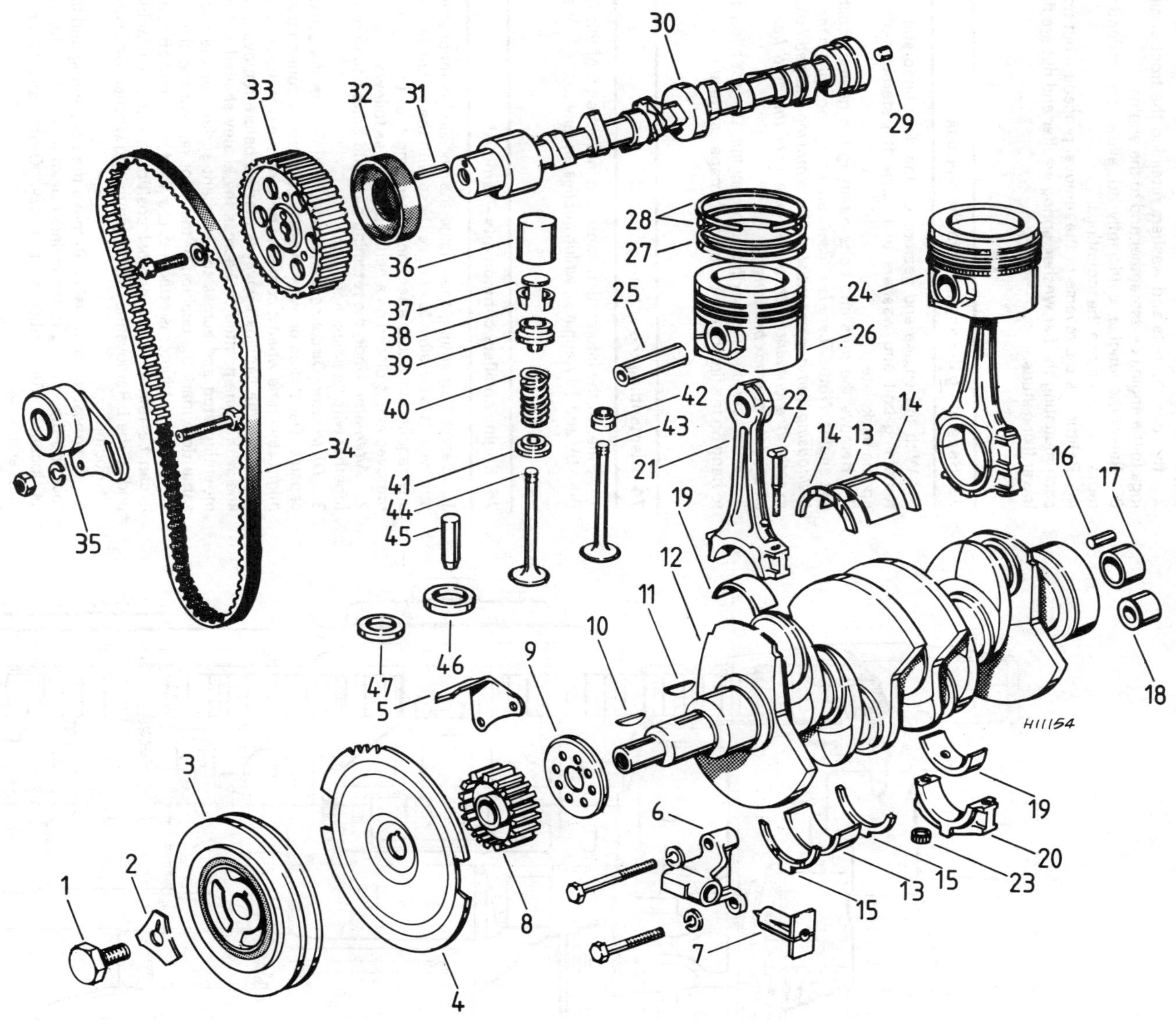

1 Pulley bolt
2 Lockwasher
3 Pulley/vibration damper
4 Timing disc
5 Timing pointer
6 Transducer bracket
7 Plug
8 Crankshaft gear
9 Gear flange
10 Pulley and gear key
11 Oil pump key
12 Crankshaft
13 Main bearings – half
14 Crankshaft upper thrust washers
15 Crankshaft lower thrust washers
16 Flywheel dowel
17 Gearbox input shaft bush
18 Converter bush
19 Big-end bearings – half
20 Connecting rod cap
21 Connecting rod
22 Connecting rod bolt
23 Connecting rod nut
24 Connecting rod and piston
25 Gudgeon pin
26 Piston
27 Oil control ring
28 Compression rings
29 Camshaft plug
30 Camshaft
31 Roll pin
32 Oil seal
33 Camshaft gear
34 Timing belt
35 Timing belt tensioner
36 Tappet
37 Shim
38 Valve collets
39 Valve spring cup
40 Valve spring
41 Valve spring seat
42 Valve stem seal
43 Inlet valve
44 Exhaust valve
45 Valve guide
46 Inlet valve seat insert
47 Exhaust valve seat insert

Fig. 1.33 Engine internal components (Sec 66)

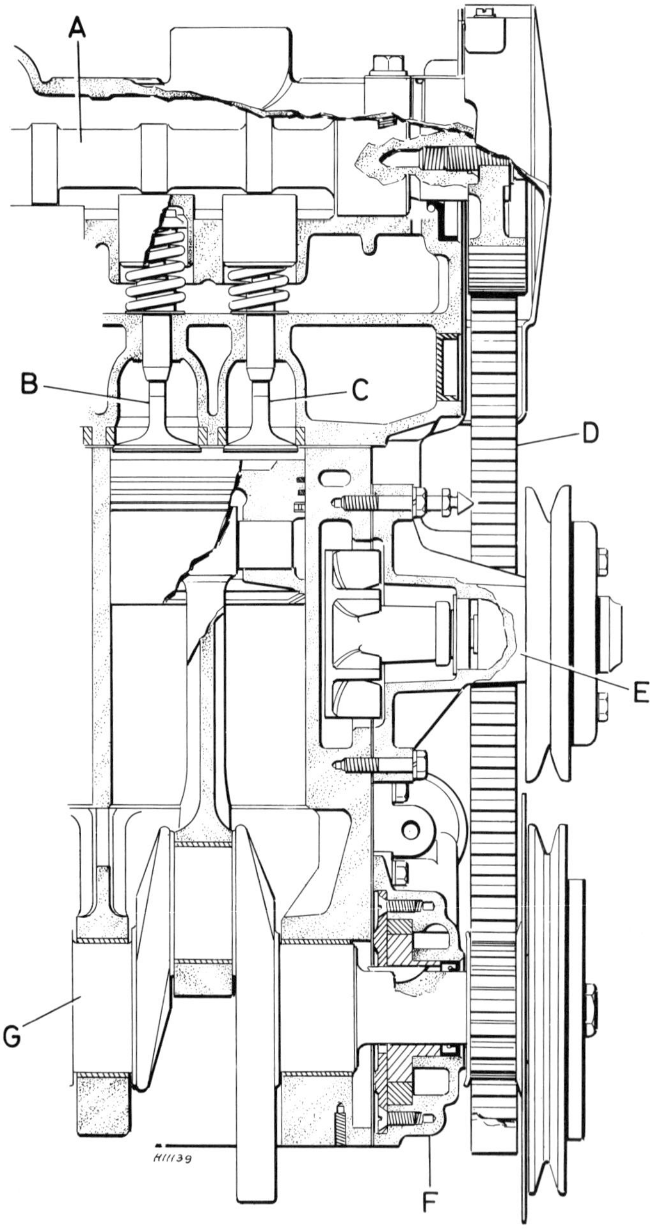

Fig. 1.34 Longitudinal section of front end of engine (Sec 66)

A *Camshaft*
B *Inlet valve*
C *Exhaust valve*
D *Timing belt*
E *Water pump*
F *Oil pump*
G *Crankshaft*

piece of wood between the sump and the jack, take the weight of the engine.
2 Remove the nuts and washers retaining the front engine mountings to the engine crossmember and to the engine.
3 Raise the engine sufficiently to allow the mountings to be removed and remove the mountings.
4 Refitting is the reverse of the removal procedure but note that on each mounting the flat washer, spring washer and nut are fitted to the lower fixing studs.

73 Separating the engine from the gearbox

1 With the engine and gearbox on the floor undo and remove the remaining bolt and washers to release the starter motor from the engine backplate.
2 Remove the nuts and bolts retaining the gearbox bellhousing to the engine. **Note**: *There is a fixed stud at the top right-hand side of the bellhousing and it is essential, when removing the gearbox, to keep it exactly in alignment with the engine until the stud has completely cleared the gearbox bellhousing.*
3 Withdraw the gearbox from the engine. **Do not** allow the weight to hang on the first motion shaft or damage can result.

74 Dismantling the engine – general

1 Refer to the information contained in Section 9 of this Chapter for the 16V and 18V engines, which applies equally to the 17V engine.

75 Engine ancillary components – removal

1 Before dismantling the engine the ancillary components must be removed. It is possible to remove these with the engine in the vehicle if any require individual attention. Presuming that the engine has been removed and is placed on a bench proceed as follows.
2 *Alternator.* Undo the retaining nuts and bolts and remove the unit together with its adjustment link.
3 *Distributor.* Disconnect the vacuum pipe from the advance/retard capsule. Depending on what work is going to be done to the engine, it might save time when it comes to refitting the distributor, if its position relative to the camshaft cover is marked before removing it. Scribe a line with a fine pointed tool across the kidney-shaped slot at the rear mounting stud and across the adjacent surface of the cover. At the same time mark the position of the rotor relative the distributor case. When the distributor is removed, provided the camshaft is not turned, it can be refitted in its original position leaving the ignition timing undisturbed. Remove the retaining nuts and washers and withdraw the distributor.
4 *Thermostat and housing.* Remove the bolt, spring and flat washer securing the housing to the cylinder head and pull the housing out of its location. Remember that a new O-ring should be fitted on reassembly.
5 *Oil filter cartridge.* Place a container under the cartridge and unscrew the cartridge from its location on the front left of the engine, catching any spilt oil in the container.
6 *Exhaust and induction manifold.* Remove the eight retaining bolts and remove the manifold complete with hot box from the cylinder head – take care as it is heavy. Retrieve the gasket from the head.
7 *Fuel pump.* Remove the two nuts retaining the pump to the camshaft cover. Remove the pump and insulator block.
8 *Coil and starter.* Remove the securing bolts and remove the coil and starter motor.
9 Remove any other minor fittings which remain on the engine.
10 The engine is now stripped of ancillary components and dismantling can proceed.

76 Flywheel and gearbox adaptor plate – removal and refitting

1 Removal or refitting the flywheel and gearbox adaptor plate is achieved in the same way as for the 18V engine. Refer to Section 8 of this Chapter for details, ignoring all references to 16V engines.

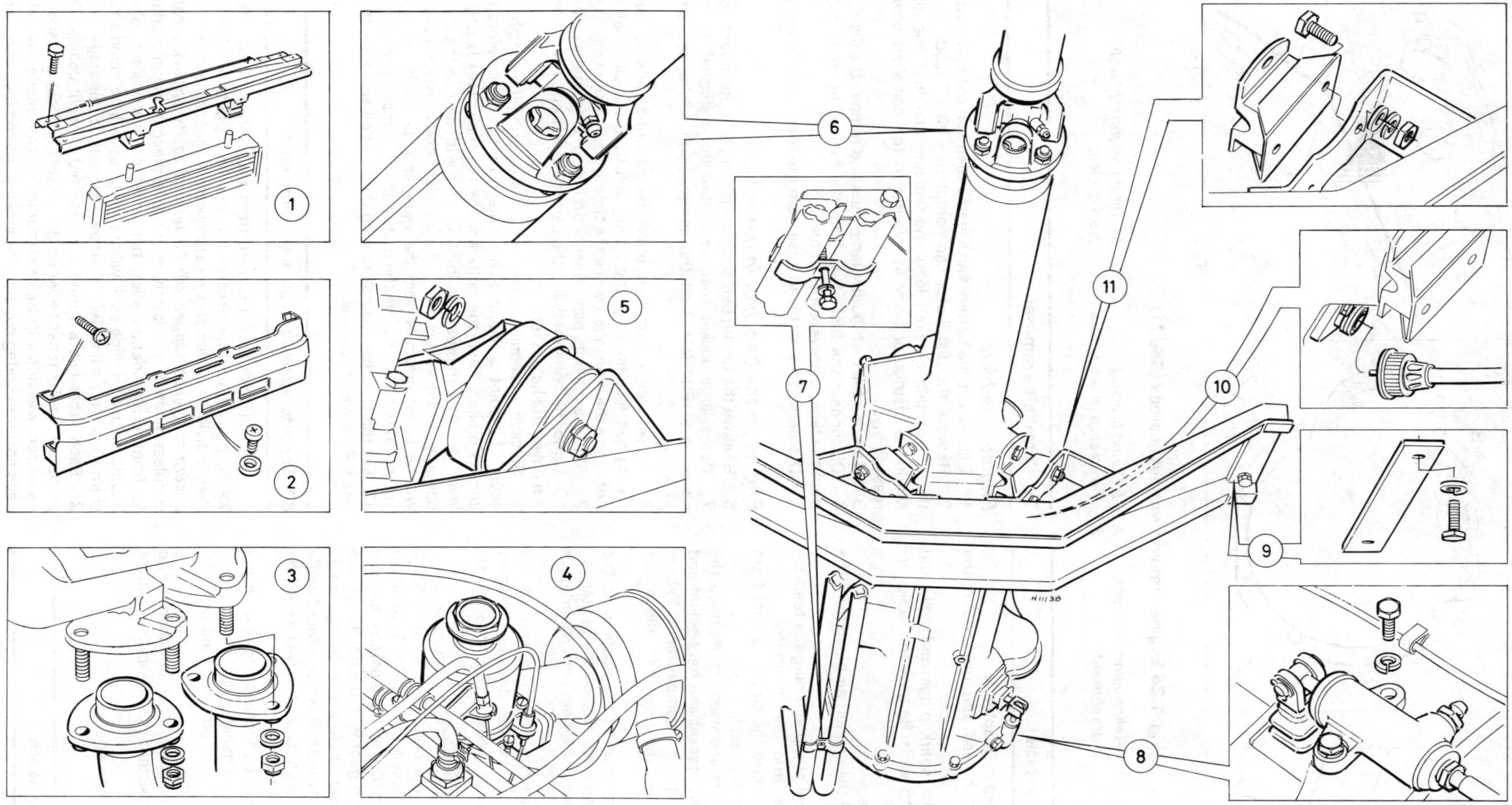

Fig. 1.35 Engine removal with gearbox (Sec 70)

1 *Bonnet lock platform*
2 *Lower surround grille panel*
3 *Exhaust manifold and down pipes*
4 *Carburettor*
5 *Engine mountings*
6 *Gearbox/propeller shaft drive flanges*
7 *Exhaust pipe front mounting bracket and bellhousing bracket*
8 *Clutch slave cylinder*
9 *Rear crossmember and reinforcing plate*
10 *Speedometer drive cable*
11 *Gearbox rear mounting*

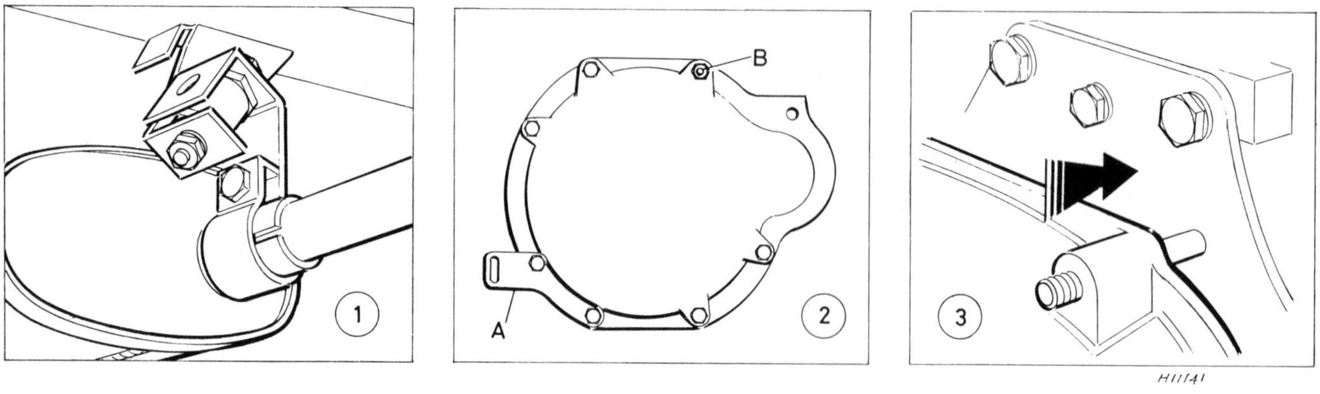

Fig. 1.36 Engine removal without gearbox (Sec 71)

1 Intermediate exhaust mounting *2 Bellhousing to engine attachments* *A Exhaust bracket* *B Fixed stud in backplate* *3 Maintain alignment until stud is clear*

77 Cylinder head removal – engine in vehicle

1 Drain the cooling system as described in Chapter 2.
2 Disconnect the battery negative (earth) lead, and tuck it out of the way.
3 Disconnect the HT leads from the spark plugs and coil, and remove the distributor cap complete with leads. Remove the cable terminal from the temperature transmitter.
4 Remove the distributor vacuum pipe.
5 Disconnect the top water hose and the coolant air bleed from the thermostat housing.
6 Disconnect both hoses from the fuel pump and plug the feed-to-pump hose. Disconnect the breather hose from the carburettor.
7 Remove the air cleaner wing nut, release the air intake duct from the manifold hot box and place the air cleaner aside.
8 Remove the carburettor securing nuts and washers. Remove the carburettor, the insulation block, support bracket and heat shield and place aside on the battery side of the engine compartment. Where automatic transmission is fitted, disconnect the downshift cable at the carburettor link.
9 Disconnect the brake servo vacuum pipe at the adaptor on the manifold. Unscrew and remove the adaptor to release the heater rail support bracket. Retrieve the washers and refit them with the adaptor on the manifold to prevent loss.
10 Disconnect the heater hose from the front of the cylinder head.
11 Remove the nut and bolt securing the exhaust downpipe bracket to the support bracket on the gearbox bellhousing. Remove the six nuts and washers to release the exhaust pipes from the manifold.
12 Remove the timing belt cover by pulling it up off its supports.
13 With a large socket spanner turn the crankshaft so that No 1 piston is at 90° BTDC and on its compression stroke. This is achieved by aligning the V in the crankshaft timing disc with the pointer, and the single dimple on the back of the camshaft gear with the pointer on the camshaft cover. This setting is necessary before the timing belt is removed. *It is most important that the crankshaft is not rotated while the camshaft is disconnected, as the pistons and valves could be damaged by coming into contact with each other.*
14 Loosen the nuts securing the timing belt tensioner to release the tension and ease the timing belt from the camshaft gear.
15 In the sequence shown in Fig. 1.38, progressively slacken and remove the cylinder head bolts and washers.
16 Lever under the two lugs marked A in Fig. 1.37 to separate the head joint and, with the help of an assistant, lift the cylinder head assembly from the block.
17 Recover the old gasket and place to one side.

78 Cylinder head removal – engine on bench

1 The sequence for removal of the cylinder head with the engine on the bench is basically identical to the later operations for removal with the engine in the vehicle. Refer to Section 77 and follow the instructions in paragraphs 12 to 17 inclusive.

79 Camshaft – removal

Engine in vehicle

1 Disconnect the battery earth lead and tuck it out of the way.
2 Remove the distributor cap complete with HT leads. Disconnect the LT (contact breaker) lead and vacuum line from the distributor. Mark the distributor body in relation to the camshaft cover to aid reassembly.
3 Unscrew the distributor clamp nuts and remove the distributor.
4 Disconnect the fuel pipes from the fuel pump and plug the line from the tank to prevent loss of fuel.
5 Disconnect the air bleed hose from the thermostat housing.

Engine on the bench or in the vehicle

6 Remove the timing belt cover by pulling it off its mountings.
7 Use a large socket spanner to turn the crankshaft so that No 1 piston is at 90° BTDC on the compression stroke, see Section 77 paragraph 13.
8 Loosen the timing belt tensioner securing nuts to release the timing belt tension. Ease the timing belt from the camshaft gear. **Do not** rotate the crankshaft while the timing belt is removed.
9 Remove the bolt, plain washer and the gear from the camshaft.
10 Undo the bolts retaining the camshaft rear end cover and remove the cover and its gasket.
11 Slacken, in sequence, the nine bolts securing the camshaft bearing cover. If a point is reached where each bolt is slack before the cover joint is broken, keep the bolts engaged in their threads and carefully lever the cover to break the joint. Do this with care as the cover can be damaged easily as it is made of aluminium and must be restrained as it will be under pressure from the valve springs. With the joint broken and all bolts slack, remove the bolts and the camshaft cover.
12 Lift the camshaft complete with the front oil seal from the cylinder head. Remove the oil seal.

80 Tappets, valves and guides – removal

1 With the camshaft removed from the cylinder head (see Section 79 for procedure if necessary). remove the tappets by lifting them off their individual valves. *It is most important that some means is used to ensure that each tappet will be refitted to its original location.* The simplest method is to lay them out on a piece of paper marked 1 to 8 for front to rear identification, but a method less prone to disturbance should be used. Inside each valve spring cup will be found an adjusting shim – keep the shims with their own tappets at this stage.
2 Each valve has a spring cup retained by split collets. There is a tendency for the taper between the cotters and cups to stick, making valve removal difficult. Before attempting to remove the valves, free each cotter to spring cup joint by striking the top face of the spring cup using a soft metal drift or short piece of tubing.
3 Again it is essential that, when removed, each valve and its spring, spring cup, spring seat and seal (inlet valve) are kept together and identified so that they can be refitted in their origjnal locations.

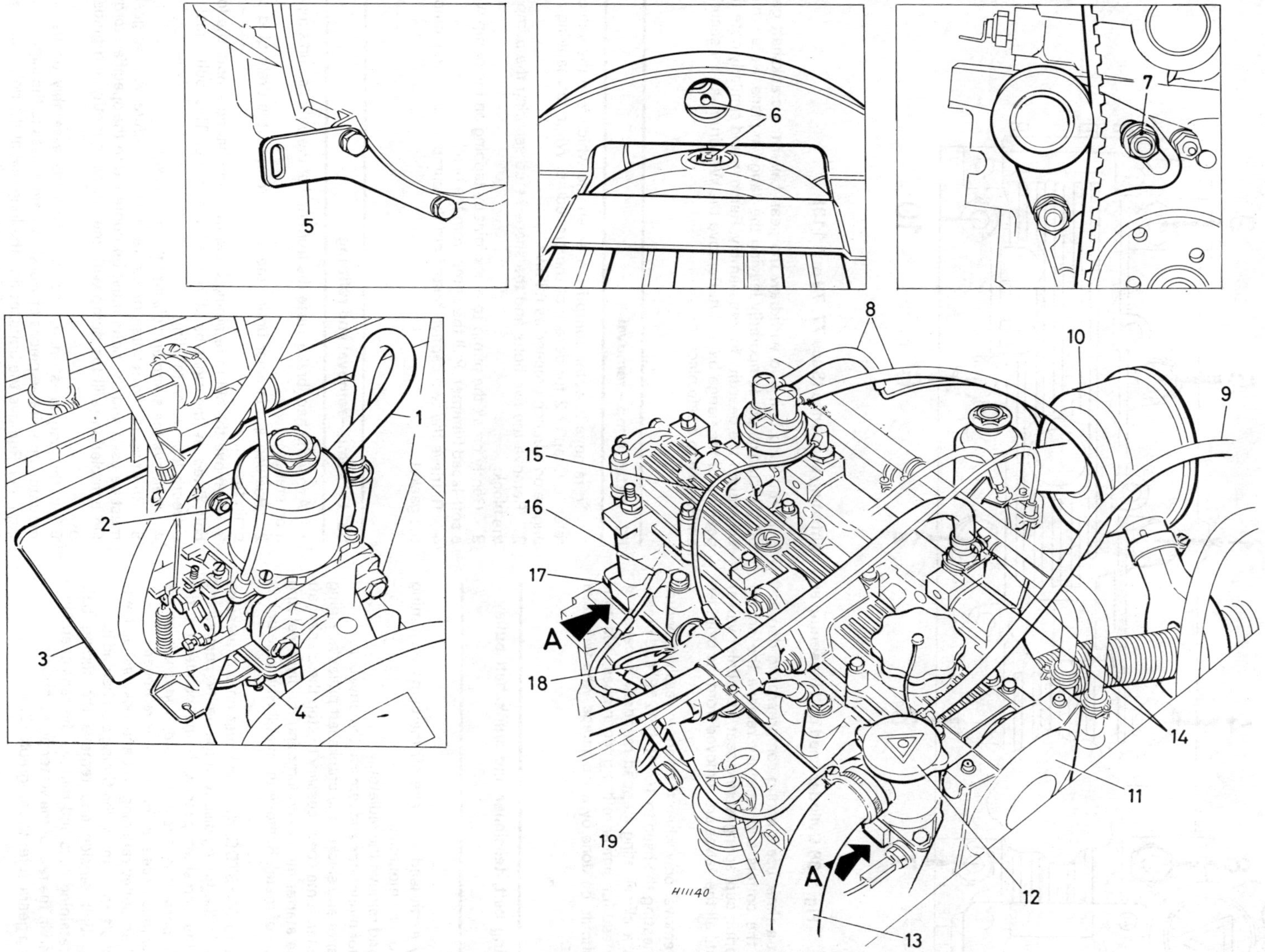

Fig. 1.37 Removing the cylinder head with engine in vehicle (Sec 77)

1 Breather hose
2 Carburettor attachment nuts
3 Heat shield
4 Exhaust joint nuts
5 Exhaust downpipe bracket
6 Camshaft gear dimple and pointer
7 Belt tensioner adjustment nut
8 Fuel hoses
9 Coolant air bleed pipe
10 Air cleaner and duct
11 Timing belt cover
12 Coolant filler cap
13 Top water hose
14 Brake servo vacuum pipe and adaptor
15 Distributor vacuum pipe
16 Spark plug HT leads
17 Cylinder head joint
18 Distributor cap
19 Cylinder block drain plug
A Cylinder head levering lugs

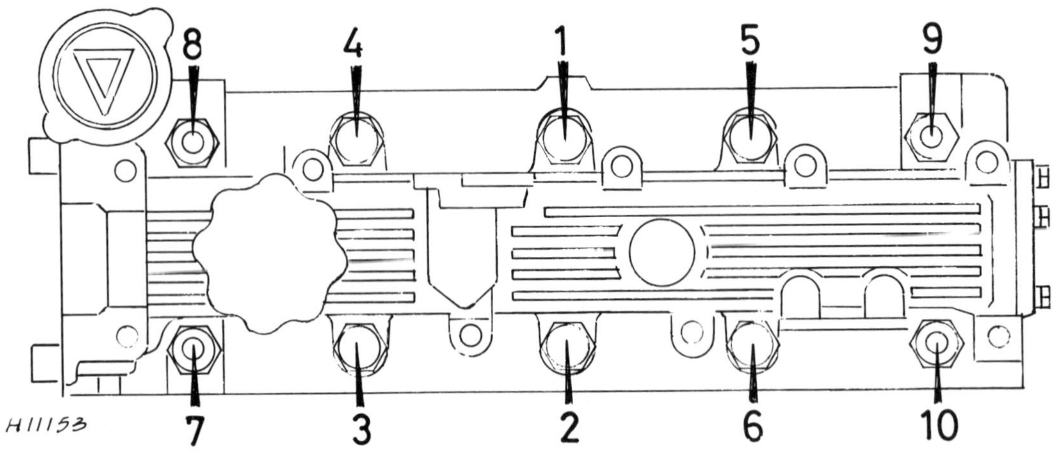

Fig. 1.38 Cylinder head bolt loosening and tightening sequence (Secs 77, 78 and 113)

Working on one valve at a time, use a spring compressor to depress the spring cup. Extract the cotters and carefully release the spring pressure. Remove the spring cup, spring, spring seat, seal (inlet valve) and valve, and repeat until all the valves are removed from the cylinder head.

4 Valve guides need removal only when they are excessively worn. The procedure involves heating the head to 100°C (212°F) and pressing the guides out with a close-fitting stepped mandrel. Because of this procedure and the need for similar refitting techniques, it is considered that this work should be done by a BL garage who will have the necessary equipment.

81 Timing gears, timing belt, tensioner and crankshaft pulley – removal

Engine in vehicle

1 Remove the battery earth lead and take off the camshaft timing bolt cover by pulling it from its mountings.

2 Refer to Chapter 2 and remove the radiator.

3 Slacken the alternator mountings and remove the drivebelt.

4 Release the tab washer and slacken the crankshaft pulley retaining bolt using a socket spanner. It may be necessary to stop the crankshaft rotating by removing the starter motor and jamming the flywheel ring gear with a suitable piece of metal. Remove the metal when the bolt is freed.

5 Turn the crankshaft to 90°BTDC, No 1 cylinder on compression stroke, by aligning the V in the timing disc with the timing pointer and the single dimple on the camshaft gear with the pointer on the camshaft cover. *It is important not to move the crankshaft or the camshaft from these positions once the timing belt is removed.*

6 Remove the crankshaft pulley retaining bolt and lockwasher. Lever the pulley off the crankshaft and remove the timing disc behind it.

7 Release the timing belt tension and remove the timing belt. Remove the tensioner retaining nuts and remove the tensioner. Note that the bolts are retained in the rear of the water pump housing.

8 To remove the timing gears, refer to paragraph 12.

Engine on bench

9 Release the tab washer and slacken the crankshaft pulley retaining bolt. Align the timing marks as described in paragraph 5.

10 Remove the pulley retaining bolt and lockwasher. Lever the pulley off the crankshaft and remove the timing disc behind it.

11 Remove the belt tensioner retaining nuts and remove the tensioner. Two bolts will be retained by the water pump housing. Remove the timing belt.

12 If it is wished to remove the timing gears, remove the camshaft gear retaining bolt and withdraw the gear. Lever the crankshaft gear off the crankshaft, temporarily leaving the flange in place. Note which way round the gear fits for reassembly. Remove and similarly note the position of the flange behind it. Remove the Woodruff key for security and keep it in a safe place.

82 Water pump – removal

1 To remove a water pump from an engine which is in the vehicle, refer to Chapter 2 for details of the procedure. When dismantling the engine on a bench proceed as follows.

2 Remove the five bolts and the shaped stud securing the pump to the block.

3 Gently break the pump to block joint by tapping on the pump with a soft-faced hammer. Pull the pump from the block.

4 Retrieve the two captive bolts from the pump housing and remove the gasket.

83 Oil sump – removal and refitting

1 Apply the handbrake, raise the front end of the vehicle and support it on stands.

2 Position a container under the sump, remove the drain plug and drain the oil.

3 Remove the sump retaining bolts from the flange – note the long bolt fitted at the right-hand rear of the sump. This bolt must be correctly located on refitting.

4 Remove the sump and discard the old gasket.

5 Refitting is the reverse of the removal procedure. A new gasket must be used and the bolts must be tightened to the specified torque. On completion refill the sump with the correct quantity and grade of oil.

6 If the engine is on the bench, place the assembly on its side, mounted on suitable pieces of wood, to gain access to the sump securing bolts. Remove the bolts, noting the long one at the right-hand rear. Remove the sump and discard the gasket.

84 Oil pick-up strainer – removal

1 Refer to Section 83 and remove the sump.

2 Remove the three retaining bolts and washers and remove the pick-up strainer assembly. Discard the gasket or O-ring seal, whichever is fitted, if it is damaged.

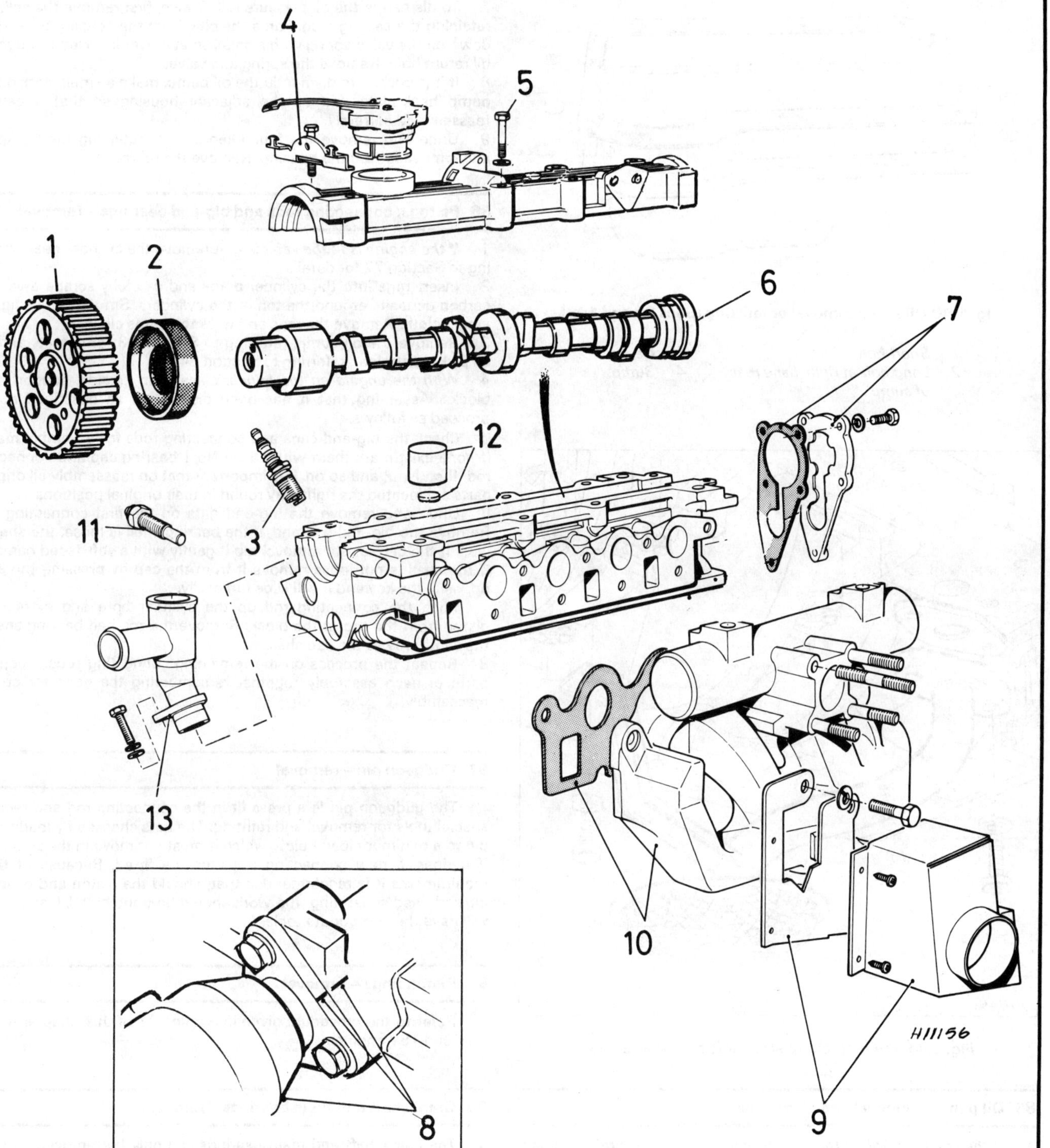

Fig. 1.39 Exploded view of cylinder head external components (Sec 79)

1	*Camshaft gear*	8	*Distributor and alignment marks*
2	*Camshaft front oil seal*	9	*Hot air box*
3	*Spark plugs*	10	*Manifold and gasket*
4	*Timing belt cover bracket*	11	*Temperature transmitter*
5	*Cover securing bolts*	12	*Tappet and shim*
6	*Camshaft*	13	*Thermostat housing*
7	*End cover and gasket*		

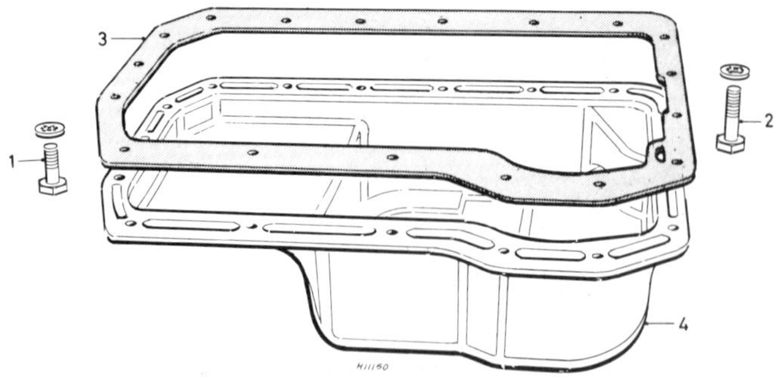

Fig. 1.40 Oil sump removal or refitting (Secs 83 and 108)

1 Short bolt
2 Long bolt at right-hand rear of sump
3 Gasket
4 Sump

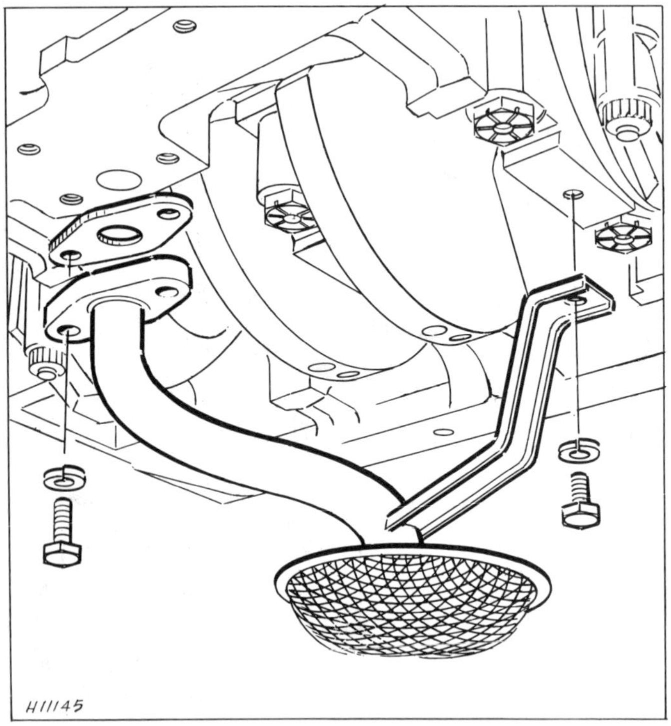

Fig. 1.41 The oil pick-up strainer (Secs 84 and 108)

85 Oil pump – removal and dismantling

1 *If the engine is still in the vehicle,* follow the procedure in Section 81, paragraphs 1 to 8, noting additionally that the fan blades must be removed, but there is no need to remove the camshaft gear. Disconnect the lead from the oil pressure switch.
2 Slacken the nine screws securing the oil pump housing to the cylinder block.
3 Remove the two screws and the electronic timing bracket noting that the long screw is at the top.
4 Remove the remaining screws and timing pointer from the oil pump housing assembly, noting the long screw adjacent to the core plug. Remove the oil pump assembly and be prepared for the release of a quantity of oil. Remove the old gasket.
5 If it has not already been removed, remove the oil filter canister and its sealing ring and discard them.
6 Undo and remove the oil pressure switch extension together with the switch.
7 To dismantle the oil pressure relief valve, first remove the split-pin retaining the sealing plug. Force the plug from the housing by pushing down on the valve spring with a small screwdriver inserted through the oil return hole. Remove the spring and valve.
8 If it is wished to dismantle the oil pump, make a small mark on the pump back-plate and on the adjacent housing so that it can be reassembled correctly.
9 Undo and remove the four Allen screws retaining the backplate and remove it from the housing. Remove the rotors.

86 Pistons, connecting rods and big-end bearings – removal

1 *If the engine is in the vehicle* first remove the cylinder head, referring to Section 77 for details.
2 Insert rags into the cylinder bores and carefully scrape away the carbon deposits around the top of the cylinders. Smooth off using fine emery cloth. Remove the rags and wipe the bores clean.
3 Remove the oil sump, referring to Section 83 for details, and the oil pick-up strainer, referring to Section 84.
4 *With the engine on the bench,* place it on its side, supported on blocks. Assuming that it has been dismantled as described above, proceed as follows.
5 Check the big-end caps and connecting rods for assembly marks. If none exist mark them with a I on No 1 bearing cap and connecting rod, II on No 2 and so on. It is important that on reassembly all original parts are located the right way round in their original positions.
6 Undo and remove the big-end nuts on the first connecting rod. Remove the big-end cap and, if the bearing shell is loose, the shell as well. If the cap resists removal tap it gently with a soft-faced hammer. If the shell is not loose remove it from the cap by pressing the edge opposite the key and it will slide out easily.
7 Push the connecting rod up the cylinder bore and extract the piston from the top of the block. Remove the top half bearing shell in the same way as the cap shell.
8 Repeat the process on the remaining connecting rods. Keep the parts of each assembly together remembering the need for correct reassembly.

87 Gudgeon pin – removal

The gudgeon pin is a press fit in the connecting rod and requires special tools for removal and refitting. The fit is checked by loading the pin to a minimum load below which it must not move in the small end. If it does, a new connecting rod must be fitted. Because of these requirements it is recommended that, should the piston and connecting rod need separating, the work should be done by a BL agent who will have the appropriate tools.

88 Piston rings – removal

Refer to the procedure given in Section 21 of this Chapter for the 16V and 18V engines.

89 Crankshaft and main bearings – removal

1 The crankshaft and main bearings can only be removed with the engine removed from the vehicle. Drain the oil, remove the sump, oil pick-up strainer, flywheel and engine backplate, oil pump housing and the big-end bearings as already described.
2 Check the identification marks on the main bearing caps. The numbers are cast on, together with arrows on No 2, 3 and 4 bearing caps pointing to the water pump end of the engine.
3 Slacken the main bearing bolts and remove the four joint seals, two on each end bearing cap.
4 Remove the intermediate caps, 2 and 4, then the end caps, 1 and 5, and lastly the centre cap, 3. With the centre cap remove the two lower half thrust washers. Make sure that the bearing shells are kept with their own bearing caps if they are to be re-used.
5 Remove the crankshaft. If the pistons are still fitted try to avoid pushing them down the bores.
6 Remove the top half thrust washers and each bearing shell.

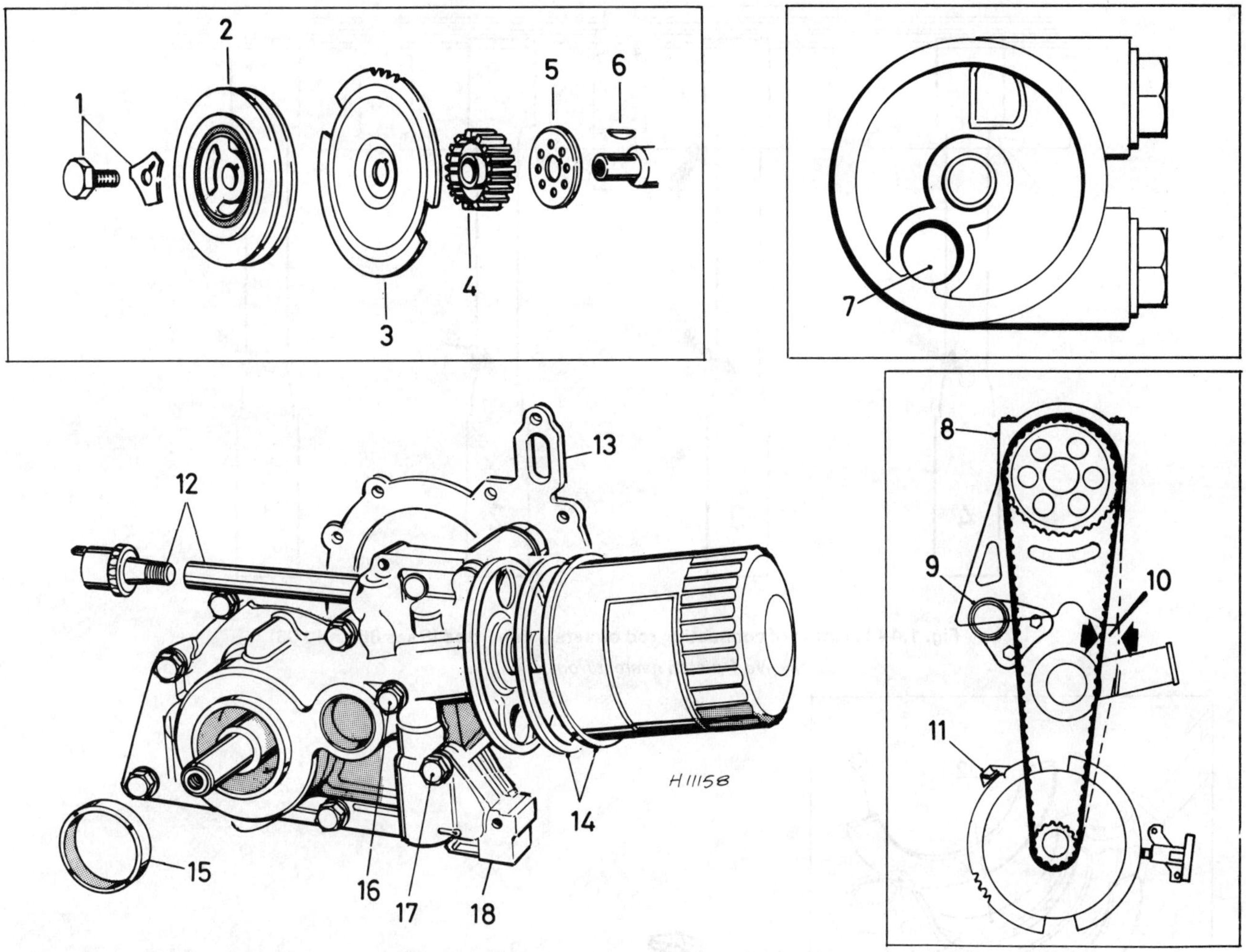

Fig. 1.42 Oil pump removal (Sec 85)

1 *Pulley retaining bolt and lockwasher*
2 *Crankshaft pulley/vibration damper*
3 *Timing disc*
4 *Crankshaft gear*
5 *Flange*
6 *Drive key*
7 *Oilway (make sure this is clear if fitting a new pump)*
8 *Timing belt cover*
9 *Timing belt tensioner*
10 *Timing belt tension measuring point*
11 *Timing pointer and disc*
12 *Oil pressure switch and extension*
13 *Oil pump gasket*
14 *Oil filter canister and seal*
15 *Crankshaft front oil seal*
16 *Long bolt near core plug*
17 *Long bolt at top of timing bracket*
18 *Oil pump housing*

90 Engine – examination and renovation – general

With the engine stripped all parts should be thoroughly cleaned and dried with non-fluffy rag. Avoid mixing parts together, remembering the need for correct reassembly of parts which have already been run in together. The items listed in the following Sections should be checked and where necessary renewed or renovated as described later. Where reference is made to earlier Sections in Part A of this Chapter, although the same principles will apply, the dimensional limits and clearances probably will not and reference should always be made to the Specifications at the beginning of Part B.

91 Crankshaft – examination and renovation

1 Examine the main bearing journals and the crankpins for scores, scratches or grooves. Check the ovality, if a micrometer of suitable size is available. If there is more than 0.001 in (0.03 mm) ovality or if the bearing surfaces are in poor condition the shaft will need regrinding.
2 A BL agent or a good motor engineering specialist will have the equipment for crankshaft regrinding and will also supply the appropriately sized bearing shells.
3 If much wear is evident it would be worthwhile making a quick check on the cylinder bores and pistons as it may be advantageous to have other parts of the engine reconditioned at the same time.

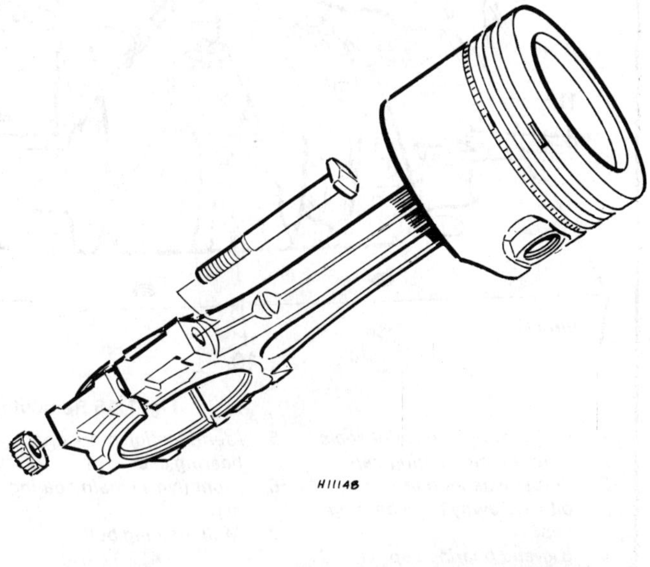

Fig. 1.43 Piston, connecting rod and big-end bearing (Sec 86)

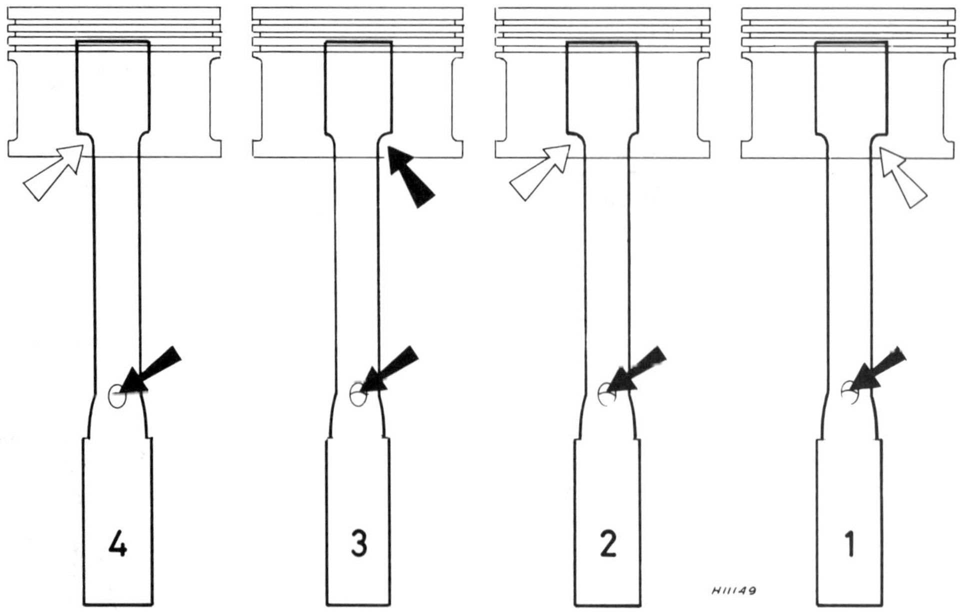

Fig. 1.44 Location of connecting rod offsets when fitted (Secs 86 and 106)
No 1 piston nearest front of engine

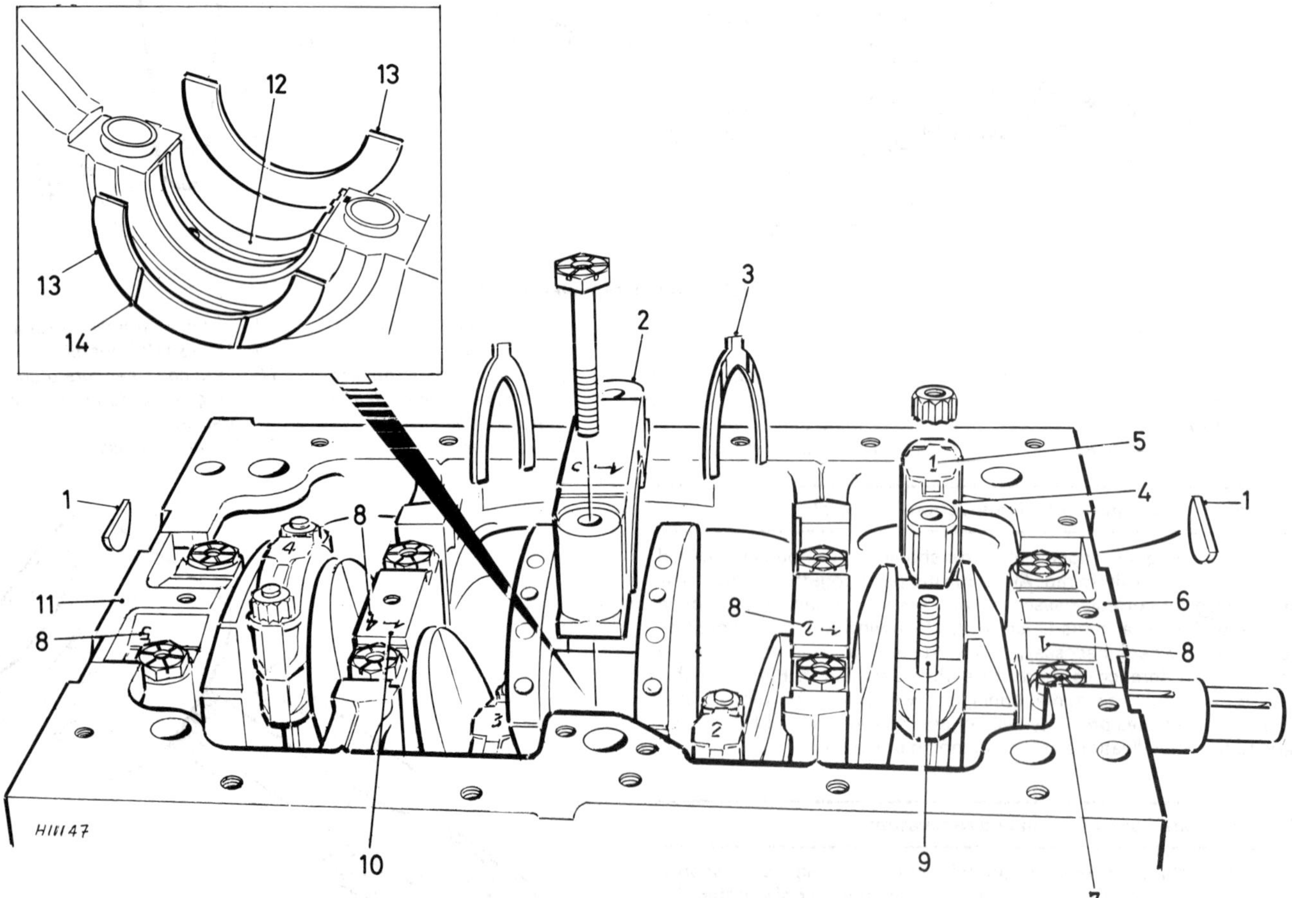

Fig. 1.45 Removing the crankshaft (Sec 89)

1 *End bearing caps joint seals*
2 *Centre main bearing cap*
3 *Thrust washer half – note oilways away from bearing cap*
4 *Big-end bearing cap*
5 *Identification mark on bearing cap*
6 *Front (No 1) main bearing cap*
7 *Main bearing bolt*
8 *Main bearing cap identification*
9 *Connecting rod identification*
10 *Main bearing intermediate caps (Nos 2 and 4)*
11 *Rear (No 5) main bearing cap*
12 *Main bearing shell*
13 *Thrust washer half*
14 *Oil groove on thrust washer*

92 Big-end and main bearings – examination and renovation

Refer to the information contained in Section 30 of this Chapter for the 16V and 18V engines.

93 Cylinder bores – examination and renovation

Refer to the information contained in Section 31 of this Chapter for the 16V and 18V engines.

94 Pistons and piston rings – examination and renovation

Refer to the information contained in Section 32 of this Chapter for the 16V and 18V engines.

95 Camshaft and camshaft bearings – examination and renovation

1 Carefully examine the camshaft journals for wear and check the diameters, see Specifications. Examine the bearings in the cylinder head and the camshaft bearing cover for scores, grooves or scratches, and check the thrust faces including the end cover.
2 Apart from carefully smoothing down any ridges on the camshaft journals or camfaces with a fine carborundum stone or fine emery cloth, and thoroughly cleaning the camshaft afterwards, no other renovation is possible for the home mechanic. In cases of advanced wear the nearest BL agent should be consulted; if any doubt exists new parts should be fitted, but this could prove very expensive.

96 Valves, valve guides and seats – examination and renovation

1 Refer to the information contained in Section 34 of this Chapter for the 16V and 18V engines.
2 Although valve guides can be renewed, as the procedure involves heating the cylinder head and pressing the guides in with a stepped mandrel it is recommended that, where worn guides need renewing, the work is done by a BL agent or a competent motor engineering specialist.

97 Timing gears and belt – examination

1 Any wear which takes place in the timing mechanism will be on the teeth of the drivebelt or will be due to stretch in the belt material. Whenever the engine is stripped for major overhaul a new belt should be fitted.
2 It is unusual for the timing gears (belt sprockets) to wear at the teeth. If the securing bolts have been inadequately tightened it is possible for the keyway or hub bore to wear. Check these two points and if wear is evident a new gear must be fitted.

101.2 Measuring the oil pump inner rotor lobe crest clearance

98 Tappets – examination and renovation

1 The shims located inside each tappet bucket must be kept with the relevant tappet and not interchanged.
2 The faces of the tappets on which the cam lobes bear should show no signs of pitting, scoring, fracturing or chipping. The tappets should not be a sloppy fit in the aluminium cylinder head. Wear is normally only encountered at very high mileages or in cases of neglected engine lubrication. Renew the tappets if they are in poor condition.

99 Flywheel starter ring gear – examination and renovation

Refer to the information contained in Section 38 of this Chapter for the 16V and 18V engines.

100 Cylinder head and bores – decarbonisation

Refer to the information contained in Section 40 of this Chapter for the 16V and 18V engines.

101 Oil pump – examination and renovation

1 After the components have been cleaned, oil them lightly and fit the outer rotor, the right way round, in the pump body. Measure the clearance between the rotor and the housing with feeler gauges and compare the measurement with that quoted in the Specifications.
2 Fit the inner rotor in the outer rotor and with feeler gauges check the clearances at the tops and bottoms of the inner rotor lobes (B and C in Fig. 1.46). Compare the measurements with those in the Specifications (photo).
3 Place a suitable straight-edge across the bed of the backplate and with feeler gauges, measure the endfloat of the outer rotor – that is the clearance between the outer rotor and the straight-edge when the outer rotor is pressed into the housing. Compare the measurement with that in the Specifications.
4 Examine the condition of the oil pressure relief valve plunger and spring. Compare the spring length with that quoted in Specifications.
5 If any of the clearances are found to be excessive the pump will have to be renewed as it is only serviced as an assembly.

102 Engine reassembly – general

Refer to the information contained in Section 42 of this Chapter for the 16V and 18V engines.

103 Crankshaft – refitting

Refer to the information contained in Section 43 of this Chapter for the 16V and 18V engines, but ignore paragraph 16 which does not apply to the 17V engine. Oil the threads of the main bearing cap bolts before fitting. When about to fit the rear main bearing cap, first lightly cover the horizontal joint with Hylomar jointing compound. Renew the four joint seals for the front and rear main bearing caps, soaking the cork seals in engine oil before fitting (photos).

104 Pistons and connecting rods – reassembly

If, because of fitting new components it is necessary to assemble a piston to a connecting rod, the work should be done by a BL agent or motor engineer specialist; see also Section 87 concerning gudgeon pin removal.

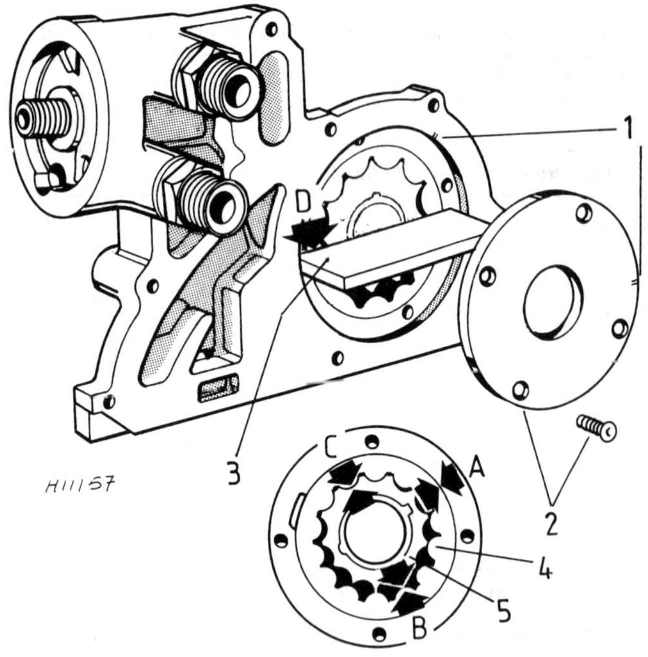

Fig. 1.46 Oil pump examination (Sec 101)

1. *Alignment marks on backplate and housing for refitting*
2. *Backplate and retaining screw*
3. *Straight edge*
4. *Outer rotor*
5. *Inner rotor*

A *Outer rotor to housing clearance*
B *Inner rotor lobe crest to outer rotor clearance*
C *Inner rotor lobe root to outer rotor clearance*
D *Outer rotor to housing endfloat (Sec 103)*

See Specifications for values of A, B, C and D

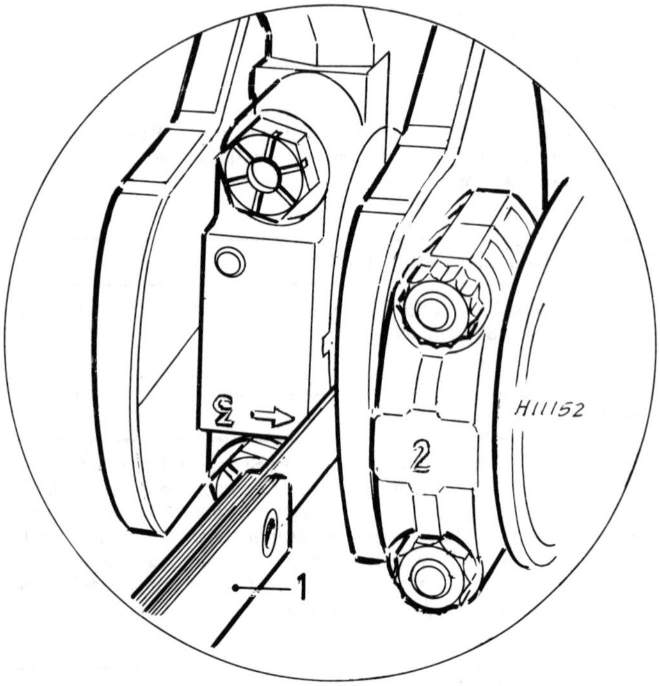

Fig. 1.47 Measuring crankshaft endfloat (Sec 103)

1. *Feeler gauges between crankshaft web and thrust washer*

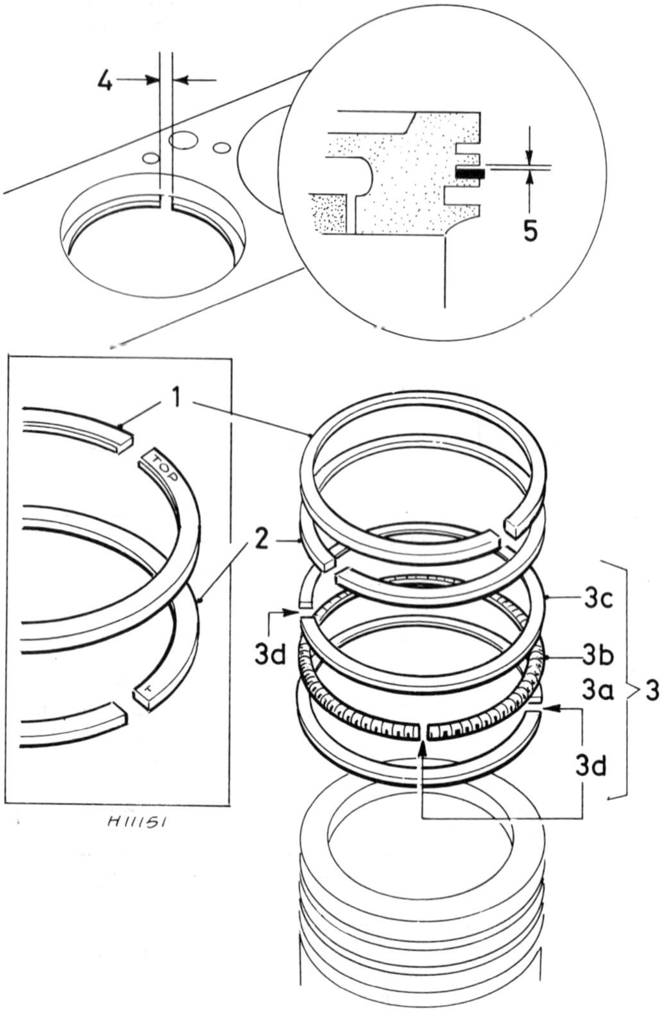

Fig. 1.48 Fitting piston rings (Sec 105)

1. *Top ring*
2. *Second ring with step towards gudgeon pin*
3. *Oil control ring*
   - a *Bottom rail*
   - b *Oil control expander*
   - c *Top rail*
   - d *Check expander ends are butting, not overlapped, and gaps at 90° to each other*
4. *Ring gap in cylinder*
5. *Ring to groove clearance*

105 Piston rings – refitting

Refer to the information contained in Section 45 of this Chapter for 16V and 18V engines, and to Fig. 1.48. The second ring on the 17V engine piston is stepped and the ring should be fitted with the step towards the gudgeon pin, the face marked TO or T towards the top of the piston.

106 Pistons and connecting rods – refitting

1 Refer to the information contained in Section 45 of this Chapter for the 16V and 18V engine and Fig. 1.44. Note that it is essential to get the correct piston/connecting rod assembly the right way round in the proper cylinder. The small end bearing in the connecting rod is offset and problems in assembly will arise if close attention is not paid to this point. The front of each piston is marked on the crown with the word FRONT, or by an arrow ▲, or by a groove. Before inserting the pistons into the cylinder block lubricate them with graphited oil; check that the gudgeon pins cannot foul the bores and that the ring gaps are properly spaced (photos).

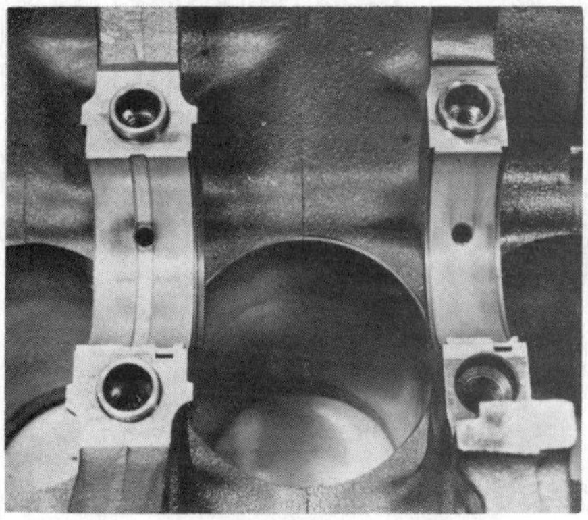

103.1a The main bearing upper shells fitted

103.1b The centre main bearing ...

103.1c ... with the upper half thrust washers being fitted

103.1d Lubricate the bearings and ...

103.1e ... lower the crankshaft into position

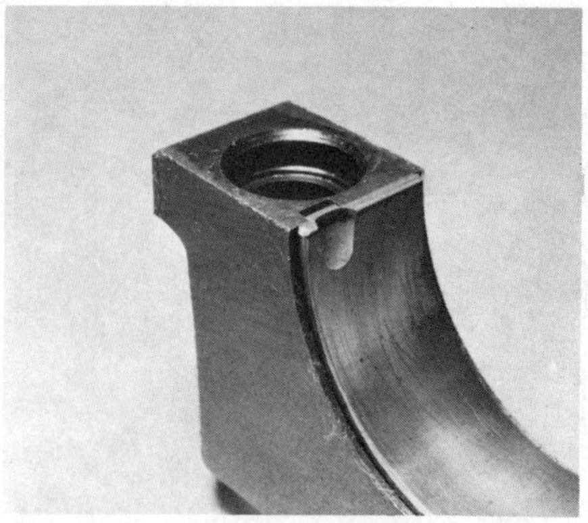

103.1f Fitting the bearing cap shells, note the locating key

103.1g The centre main bearing cap, with thrust washer halves installed,being fitted

103.1h The bearing cap bolts fitted and ...

103.1j ... being tightened to the specified torque

106.1a Markings on the piston crown

106.1b Tapping the piston into the cylinder bore with a compressor on the rings

107.1a Fitting a big-end bearing cap and ...

107.1b ... tightening the cap securing nuts

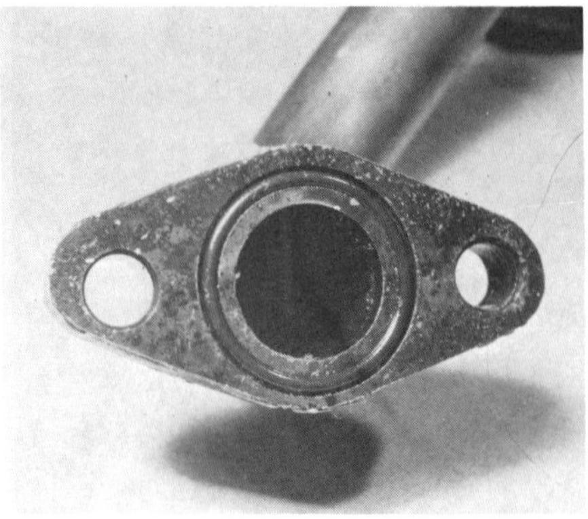
108.1a A new O-ring is fitted to the oil pick-up pipe jointflange and ...

108.1b ... the oil pick-up strainer is fitted to the cylinder block

108.2a Check that the cork seals (arrowed) are in position before ...

108.2b ... fitting the gasket and sump

109.3a Gently tap a new front seal into the oil pump housing until ...

109.3b ... it is flush with the housing face

109.5a The oil pump ...

109.5b ... backplate being fitted and ...

109.5c ... the retaining screws being tightened

109.6a After fitting the pump drive Woodruff key (note that the pulley drive key has been removed) ...

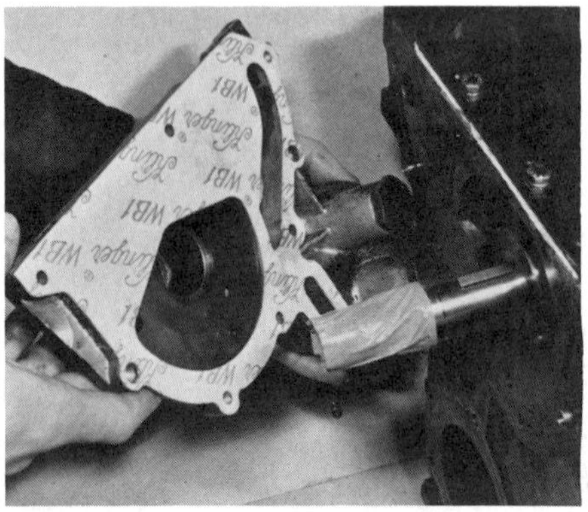
109.6b ... wrap the crankshaft with plastic tape to protect the oil seal and fit the pump using a new gasket

109.8a The timing transducer bracket fitted

109.8b The oil pump and timing pointer fitted and awaiting tightening of the retaining bolts

107 Connecting rods to crankshaft – reassembly

Refer to the information contained in Section 47 of this Chapter for the 16V and 18V engines and, in addition, lubricate the threads of the big-end studs before fitting and tightening the nuts to the specified torque (photos).

108 Oil pick-up strainer and sump – refitting

1 Fit a new O-ring seal to the pick-up pipe flange or fit a new gasket, as appropriate. Assemble the pick-up strainer to the engine and secure with the three bolts and lockwashers (photos).
2 Check that the cork seals are in position on No 1 and 5 bearing caps and fit a new gasket to the sump flange. Offer up the sump and secure with the retaining bolts and lockwashers. The long bolt fits at the right-hand rear of the sump (photos).
3 Tighten the sump bolts to the specified torque.

109 Oil pump assembly – refitting

1 Before refitting the oil pump assembly the crankshaft front oil seal should be renewed.
2 Prise out the old seal from the front of the housing, taking care not to damage the housing.
3 Dip the new seal in engine oil and insert it in the housing with the seal lip facing inwards. Carefully tap the outer edge of the seal, using a piece of flat wood and a hammer, until the seal is flush with the face of the housing (photos).
4 Fit a new O-ring to the pressure relief valve plug. Position the spring on the plug and the valve on the spring, lubricate the bore in the pump and push the assembly into the bore, valve first. Fit a new retaining split-pin.
5 Fit the backplate to the pump, checking that the marks made on dismantling are lined up. Fit the countersunk retaining screws and tighten to the specified torque (photos).
6 Fit the oil pump drive key to the crankshaft; if the pulley key is in position remove it while fitting the pump. Wrap plastic tape on the nose of the crankshaft to protect the oil seal in the pump housing (photos).
7 If a new oil pump is being fitted, check that the oilway is clear. Fit a new gasket on the engine and lubricate the crankshaft front end. Align the key slot in the oil pump with the crankshaft key and fit the pump housing to the engine, taking care not to damage the seal as it fits onto the crankshaft.
8 Fit the timing transducer bracket with the longer screw to the top.

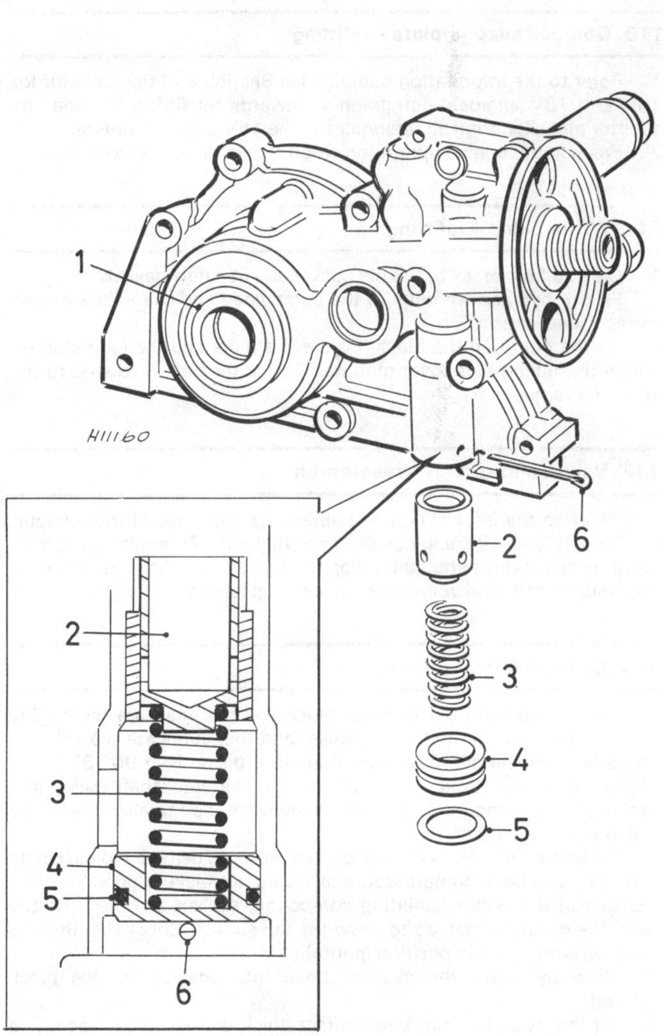

Fig. 1.49 Oil pump pressure relief valve assembly (Sec 109)

1 Oil pump assembly
2 Valve plunger
3 Valve spring
4 Plug
5 O-ring
6 Split-pin

110.1a A new oil seal fitted to the gearbox adaptor plate before ...

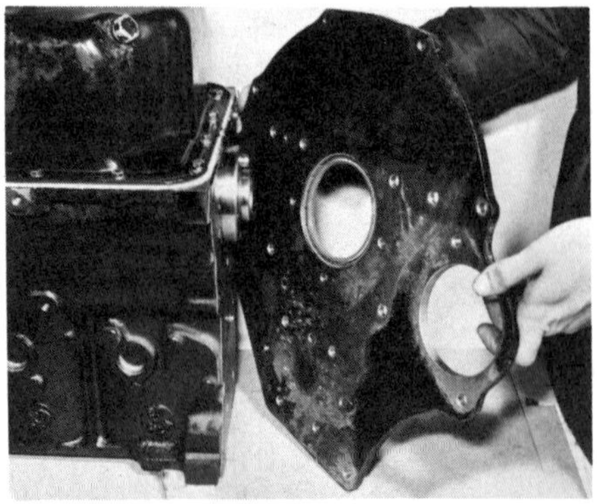

110.1b ... fitting the plate to the engine and ...

110.1c ... tightening the securing bolts

Fit the remaining screws and the timing pointer with the long screw adjacent to the core plug. Tighten all screws to the specified torque (photos).

9 Prime the pump by removing the bottom plug and injecting clean engine oil into the pump. Refit the bottom plug and remove the plastic tape from the crankshaft.

110 Gearbox adaptor plate – refitting

1 Refer to the information contained in Section 8 of this Chapter for 16V and 18V engines, paragraph 7 onwards for fitting the gearbox adaptor plate, ignoring all references to the 16V engine (photos).

2 Position the engine upright on wood blocks for further assembly.

111 Water pump – refitting

1 Clean off all traces of the old gasket and fit a new gasket.

2 Fit the two captive bolts to the pump body and use a little grease to keep them in position.

3 Fit the pump on the block. Fit the five bolts and the long stud on which the timing belt cover mounts. Tighten the bolts and stud to the specified torque.

112 Valves and springs – reassembly

Refer to the information contained in Section 56 of this Chapter for the 16V and 18V engines but note that, on 17V engines, the inlet valve oil seal, which should be dipped in oil on assembly, fits after the valve spring seat and before the valve spring (photos).

113 Cylinder head – refitting

1 Check that both the cylinder block and the head are thoroughly clean, especially the mating surfaces, and then lubricate the cylinder bores with engine oil. Make sure that No 1 piston is at 90° BTDC on the compression stroke: if necessary refit the crankshaft pulley and timing pointer temporarily and rotate the crankshaft to align the pulley notch with the pointer.

2 Fit a new cylinder head gasket – never try to use the old gasket as this will have been compressed and would not make a good joint. Fit the gasket dry without jointing compound, sealant or grease of any sort. The gasket is marked to show the top surface. Check that the two locating dowels are in position (photo).

3 Carefully lower the cylinder head into position on the block (photo).

4 Fit the cylinder head bolts with a thick washer under each. The bolts with the stud extensions on top (for fitting lifting brackets) are located at the four corner positions. The bolts should have their threads lightly oiled before fitting as the torque wrench setting for tightening is based on this condition (photo).

5 Progressively pre-tighten the bolts to the initial specified torque in the sequence shown in Fig. 1.38, and then tighten them to the full specified torque in the same sequence (photo).

6 When the engine is first run after the cylinder head has been refitted, run the engine as fast idle speed for at least 15 minutes or road test for 5 miles (8 km). Then, in the sequence shown in Fig. 1.38 slacken each cylinder head bolt by half a turn and retighten to the full torque value.

114 Valve clearance – adjustment

1 As the camshaft cover forms the upper half bearings for the camshaft, when it is removed the valve springs lift the camshaft out of the lower half bearings. To permit checking and adjusting of the valve clearances BL provide a set of three special brackets and retainer (BL tools No 18G 1301 and 1302) to hold the camshaft down with the cover removed and thereby allow access to the tappets. If it is possible, borrow or hire these brackets as they do make the job easier and quicker. However, if this is not possible, the following procedure can be used which, although it takes longer, will produce a perfectly satisfactory result.

2 *If the engine is in the vehicle,* turn the crankshaft to 90° BTDC, No 1 cylinder on compression, by aligning the V in the timing disc with the pointer and the single dimple on the back of the camshaft gear with its pointer on the camshaft cover. *Do not turn the crankshaft while the timing belt is removed.* Remove the timing belt, camshaft cover and camshaft as described in Section 79. Remove the tappet buckets, retaining the shim in each, and identify them or place them so that they can be reassembled in their original locations. Then proceed as follows.

3 With the engine on the bench and still only partly reassembled, or with the engine in the vehicle stripped as in paragraph 2, fit the first pair of tappet buckets with their shims on their valves and position the camshaft so that the cams for those valves point upwards. Work on valve pairs as follows:

1 and 4
2 and 5
6 and 7
3 and 8

Counting from the timing belt end of the engine, exhaust valves are 1-3-5-7 and inlet valves are 2-4-6-8.

4 Press the camshaft down on its bearing and measure the clearance between the cams and their tappet buckets using feeler gauges. Record the clearance. Remove the tappet buckets and shims and repeat the process with the next pair of buckets. Continue until all valve clearances have been measured (photo).

5 Compare the measured valve clearances with those specified and note that adjustment is only required if the clearance is less than the specified minimum.

6 If adjustment is necessary a new shim will be needed. The size required is calculated by adding the measured clearance to the thickness engraved on the original shim (use a micrometer to measure the shim thickness if the engraving is illegible) and subtracting the specified valve clearance from that total. New shims are available from 0.091 in (2.31 mm) to 0.117 in (2.98 mm) thickness in steps of 0.002 in (0.05 mm). Fit the selected shim into the spring cup on the valve concerned and refit the tappet bucket. It is sometimes possible to interchange two existing shims to rectify two incorrect clearances.

7 Refit all shims and tappet buckets, and lubricate with engine oil.

112.1a Top view of a valve guide

112.1b A valve being fitted followed by ...

112.1c ... the valve spring seat, valve spring and spring cap

112.1d Depressing the spring cup to fit the two collets

112.1e After fitting the adjusting shim into the spring cup the tappet is being fitted

113.2 The cleaned cylinder block fitted with a new head gasket. Note the two locating dowels

113.3 Lowering the cylinder head onto the block

113.4 The cylinder head bolts with the stud extensions are fitted at the corners

113.5 Tighten the cylinder head bolts progressively and sequentially to the specified torque

114.4 Measuring a valve clearance with feeler gauges

115.2a Lower the camshaft into the journal bearings and ...

115.2b ... check that the seal is square in the head

115.5a Lowering the camshaft cover onto the head

115.5b The timing cover mounting bracket fitted under the front bolts

115.6 With a new gasket on the head, fit the camshaft rear end cover

116.1 The crankshaft pulley Woodruff key being fitted to the crankshaft followed by ...

116.2 ... the flange and timing gear

116.3 The timing disc and the pulley/vibration damper being fitted

116.5a Notch in timing disc aligned with pointer and ...

116.5b ... the camshaft gear dimple aligned with the pointer on the cover

116.6 Tighten the pulley retaining bolt to the specified torque

116.7a The timing belt tensioner

116.7b Fitting the timing belt cover

116.8 A temporary method of assessing timing belt tension

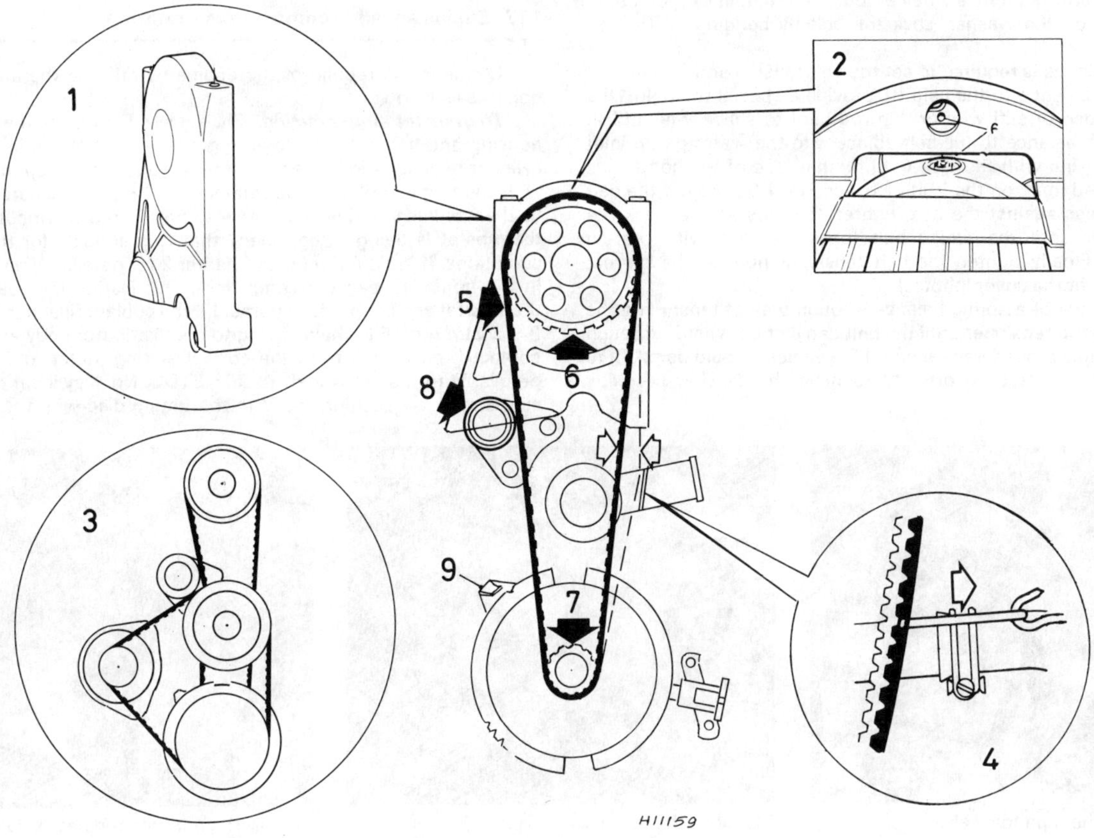

Fig. 1.50 Tensioning the timing belt (Sec 116)

1 *Timing belt cover*
2 *Camshaft gear dimple and pointer*
3 *Belt arrangement*
4 *Checking belt tension with spring balance*
5 *Timing belt*
6 *Camshaft gear*
7 *Crankshaft gear*
8 *Belt tensioner*
9 *Timing pointer*

8 If the engine is in the vehicle, reassemble by reversing the removal procedure described above. Section 115 gives details of refitting the camshaft and Section 116 includes the procedure for tensioning the timing belt. Remember to align the dimple on the camshaft gear with the pointer on the camshaft cover before fitting the timing belt, and on no account rotate the crankshaft while the timing belt is off.

115 Camshaft and gear – refitting

1 Following any necessary valve clearance adjustments, oil the camshaft journals and the tappets. Fit a new oil seal to the camshaft.
2 Lower the camshaft into the journals, making sure that the seal is square in the head and up against its register (photos).
3 Peel any old silicone rubber sealant from the joint faces of the camshaft cover and the cylinder head, and wipe the faces clean.
4 Apply a very thin bead of RTV type silicone rubber sealant – $\frac{1}{16}$ in (1.5 mm) or less in diameter – to the centre of each joint face of the camshaft bearing cover. Note that the groove around each bearing surface is to prevent sealant from spreading onto the bearing surface.
5 Immediately lower the camshaft cover onto the head and fit the securing bolts. not forgetting the bracket for the timing belt cover at the front end. Tighten the bolts progressively to the specified torque (photos).
6 With a new gasket, refit the camshaft rear end cover and secure with the six bolts and lockwashers, tightening to the specified torque (photo).
7 Fit the camshaft front seal fully into the housing, lubricating the outside diameter and the sealing lip with engine oil.
8 Locate the camshaft gear on the roll pin in the camshaft, fit the retaining bolt and washer and tighten to the specified torque.

116 Crankshaft timing gear and pulley, timing belt tensioner and timing belt – refitting

Note: *If the cylinder head is on the engine, the crankshaft should already have been set so that No 1 piston is at 90° BTDC on the compression stroke. On no account rotate the crankshaft before the timing belt has been fitted.*

1 If it is not already fitted, fit the pulley drive key to the crankshaft, then fit the timing gear flange. The flange retains the timing belt on the timing gear and should be fitted with its bevelled rim leaning away from the gear (photo).
2 Next fit the timing gear, pressing it up to the drive flange. The timing disc is next and should be fitted so that the central recess faces forwards. The centre of the disc contacts the timing gear behind it and the outer face of the disc leans away from the timing belt which runs behind (photo).
3 After the timing disc, fit the crankshaft pulley/vibration damper. The centre hub of the pulley must contact the timing disc behind it. Finally, oil the threads and fit the retaining bolt together with a new lockwasher. See Fig. 1.34 for a sectional view of the front end of the engine which will resolve any doubts on assembling the components on the crankshaft (photo).
4 Check that the wheel on the timing belt tensioner runs freely without excessive free play. If in doubt a new one must be fitted. Fit the tensioner to the two captive bolts in the water pump but only temporarily tighten the retaining nuts and washers.
5 Check that the V in the timing disc is aligned with the timing pointer. Turn the camshaft gear to align the single dimple on its rear face with the pointer on the camshaft cover. Fit the timing belt to the crankshaft and camshaft timing gears (photo).
6 Tighten the retaining bolt to the specified torque. If necessary the

bolt may be overtorqued by a small amount to bring a hexagon flat in line with a tab on the washer. Lock the bolt by bending up the tab (photo).

7 A spring balance is required to set the belt tension and a means of connecting the balance to the belt. BL provide a special tool (No 18G 1315) but a piece of stiff wire or thin rod bent to shape will suffice. Hook the spring balance to the belt adjacent to the water pump inlet pipe and pull in line with the pipe until the flat face of the belt aligns with the moulded mark on the inlet pipe, see Fig. 1.50. Adjust the belt tensioner to press against the belt, tighten the nuts and recheck the tension. Repeat until the timing belt tension agrees with that in Specifications. Finally tighten the belt tensioner nuts to the correct torque load and fit the cover (photos).

8 In the absence of a spring balance a rough way of tensioning the belt is to adjust the tensioner until the belt can just be twisted through 90° (photo). If this procedure is adopted the vehicle should be taken to a BL agent at the first opportunity to have the tension correctly adjusted.

117 Engine ancillary components – refitting

1 Continue the rebuild of the engine by refitting the ancillary components as follows.

2 *Thermostat and housing.* Fit a new O-ring to the thermostat housing and lightly lubricate with grease or petroleum jelly. Wipe the cylinder head location clean and fit the housing; line up the screw slot with the tapped hole. Fit the retaining screw and washers and tighten to the specified torque. Fit a new O-ring to the thermostat. If a new thermostat is being fitted, check that it is suitable for the operating conditions. If in doubt, refer to Chapter 2 for details. Wipe the location in the housing clean and, after lightly lubricating the seal, insert the thermostat and push it fully home. Fit the coolant filler cap.

3 *Distributor.* Fit a new O-ring to the distributor body and adjust the contact breaker points to the correct setting. Refer to Chapter 4 for details. Turn the crankshaft to 90° BTDC, No 1 cylinder on compression stroke, by aligning the V in the timing disc with the pointer and

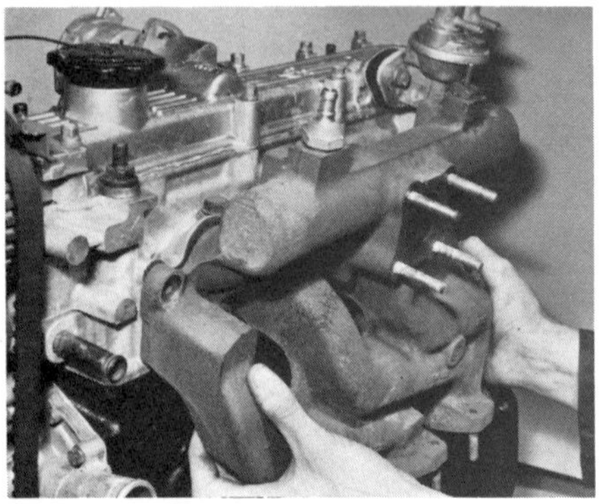
117.5a Fitting the manifolds and ...

117.5b ... the hot box

117.5c The adaptor and heat shield fitted

117.7a The flywheel being refitted and ...

117.7b ... the retaining bolts being tightened

117.8 The left-hand front engine mounting

117.9 The alternator lower adjusting bracket

117.11a The oil separator fitted and ...

117.11b ... the oil pressure transmitter on its extension

118.2 The exhaust pipe bracket fitted to the engine/bellhousing flange joint

119.1a Lowering the engine and gearbox into the vehicle

119.1b The coolant hoses connected

119.1c Heater hoses connected and carburettor fitted. Note the brake servo vacuum adaptor (arrowed)

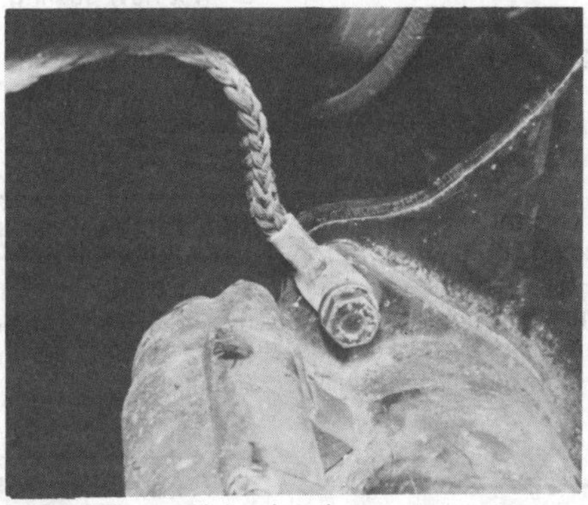

119.1d The earth lead at the starter motor connected

119.1e The rear mountings fitted and ...

119.1f ... a front mounting awaiting tightening of the securing nut

119.1g The speedometer drive cable being connected

the dimple on the camshaft gear with the pointer on the cover. Continue turning the crankshaft clockwise until the pointer lined up with the 8° BTDC mark on the timing disc. Hold the distributor with the vacuum unit 45° below the horizontal and to the left, and turn the rotor arm until it is at the 1 o'clock position. Push the distributor into its location in the camshaft cover and engage the drivegear; this will rotate the rotor to between the 2.30 and 3.30 o'clock position. Fit the securing nuts and washers finger tight. Turn the distributor in its housing until the contact breaker points are just opening and tighten the securing nuts to the specified torque. The static ignition timing is now set; refer to Chapter 4 for further details.

4 *Fuel pump.* Remove any traces of the old gaskets from the camshaft cover and the insulator block. Fit new gaskets on each side of the insulator block and fit the insulator block and fuel pump into the camshaft cover housing. Fit the retaining bolts and washers and tighten to the specified torque.

5 *Manifolds and hot box.* Scrape all traces of the old manifold gasket from the mating faces of the manifold and cylinder head. To hold the gasket in position while fitting the manifold, insert a couple of the securing bolts in the manifold and locate the gasket on these. Offer the manifold up to the engine and engage the securing bolts. Fit the remaining bolts and tighten progressively to the specified torque. If they have been removed, fit the adaptor and hot box to the manifold together with the asbestos heat shield (photos).

6 *Carburettor.* Clean the mating faces of the manifold, insulator block, cable bracket and carburettor. Using new gaskets, fit the carburettor; fit the retaining nuts and tighten to the specified torque. Where the carburettor assembly was removed before lifting the engine out of the vehicle, reassembly can be deferred until after the engine is refitted.

7 *Flywheel and clutch.* Refer to the information contained in Section 8 of this Chapter for the 16V and 18V engines for the procedure for refitting the flywheel, ignoring references to the 16V engine. Refer to Chapter 5 for the procedure for refitting the clutch (photos).

8 *Engine mountings.* If these were removed from the engine, refit the engine front mounting brackets to the engine and tighten the retaining bolts to the specified torque (photo).
9 *Alternator.* Fit the lower adjusting bracket and spacer to the engine, turning the crankshaft to line up a slot in the timing disc to provide access. Secure the alternator mounting bracket (with the alternator attached) to the cylinder block and fit the retaining bolts. Tighten to the specified torque (photo).
10 *Coil.* Fit the coil and spacer to its location behind and above the alternator and tighten the retaining bolts to the specified torque.
11 *Oil separator, oil pressure switch and thermal transmitter.* Refit these minor components to their respective locations and tighten to the specified torques (photos).
12 *Oil filter.* Fit a new oil filter cartridge and seal. Smear the seal with engine oil and hand tighten the filter only.
13 The remaining components such as the fan, fanbelt, starter motor etc can be fitted when the engine has been refitted in the vehicle.

118 Gearbox – assembly to engine

1 Examine both the flywheel/clutch assembly and the gearbox bellhousing interior for completeness, locking tabs bent up where necessary, release bearing greasing etc, to ensure that all is ready to assemble the gearbox to the engine. The clutch driven plate must be centralised as described in Chapter 5.
2 Align the engine with the gearbox. It is important that they are kept in alignment during assembly to avoid loading the gearbox first motion (input) shaft and the fixed stud at the top right side of the bellhousing. Push the gearbox and the engine together until the flanges are in contact and fit the retaining nuts and bolts. Fit the exhaust pipe bracket at the lower left location (photo). Progressively tighten the nuts and bolts.

119 Engine – refitting with gearbox

1 Engine refitting with the gearbox attached is basically the reverse of the removal procedure described in Section 70 of this Chapter. Fit new sealing rings to the exhaust manifold on assembly and fit new gaskets and hoses where necessary. When refitting the brake servo vacuum adaptor, fit the larger washer beneath the heater rail bracket. Refer to Chapter 2 for the procedure for tensioning the fanbelt (photos).

120 Engine – refitting without gearbox

1 Refitting the engine without the gearbox is the reverse of the removal procedure described in Section 71 of this Chapter.
2 When mating the engine to the gearbox, take care to engage the long stud in the top right-hand side of the backplate and keep the engine and gearbox in alignment until the joint faces mate. If necessary, slightly rotate the crankshaft to engage the first motion (input) shaft with the clutch. Fit the retaining nuts and bolts, not forgetting the exhaust bracket at the lower left section of the flange joint, and progressively tighten them. Raise the engine slightly to permit installation of the front mountings, then lower the engine and remove the hoist.
3 Fit new sealing rings to the exhaust manifold on assembly and fit new gaskets and hoses where necessary. When refitting the brake servo vacuum adaptor, fit the larger washer beneath the heater rail bracket. Refer to Chapter 2 for the procedure for tensioning the fanbelt.

121 Engine – initial start-up after overhaul and major repair

1 The procedure for initial start-up after an engine has been overhauled or had a major repair is the same as that described for the 16V and 18V engines in Part A of this Chapter. Refer to Section 64 for details.

122 Fault diagnosis – engine

1 Refer to the information contained in Section 65 for the 16V and 18V engines.
2 Excessive oil consumption can be caused by blockage of the crankcase breather.

Chapter 2 Cooling system

For modifications, and information applicable to later models, see Supplement at end of manual

Contents

Specifications

System type
Pressurised with expansion tank, thermostatic temperature control, circulation by pump and thermo-syphon

Coolant capacity (including heater)	
16V engines	11 pints (6·25 litres)
17V engines	11·15 pints (6·33 litres)
18V engines	11·75 pints (6·68 litres)

Fanbelt tension
0·25 in (6 mm) deflection when 8·0 lbf (36 N) applied between fan and alternator pulleys
OR
Belt should just slip when 11·0 lbf ft (15 Nm) torque applied in clockwise direction to alternator pulley nut

Thermostat	
Standard:	
Mark	82
Starts to open	79° to 83°C (174° to 181°F)
Fully open	93° to 96°C (199° to 205°F)
Hot climates:	
Mark	74
Starts to open	71° to 75°C (160° to 167°F)
Fully open	85° to 88°C (185° to 190°F)
Cold climates:	
Mark	88
Starts to open	85° to 89°C (185° to 192°F)
Fully open	99° to 102°C (210° to 216°F)

Expansion tank filler/relief cap
Pressure relief setting: 15 lbf/in^2 (1·03 bar)

Torque wrench settings	**lbf ft**	**Nm**
16V and 18V engines		
Water pump bolts	17	23
Water outlet elbow bolts	8	11
Water pump pulley setscrews	18	24
Thermal transmitter	16	22

17V engines	lbf ft	Nm
Water pump to block bolts	8	11
Water pump pulley to hub screws	9	12
Thermostat housing to head	8	11
Thermal transmitter	5	6
Engine block drain plug	27	37

1 General description

1 The engine cooling water is circulated by a thermo-syphon, water pump assisted system, and the coolant is pressurised. This is primarily to prevent premature boiling in adverse conditions and to allow the engine to operate at its most efficient running temperature; this being just under the boiling point of water. The overflow pipe from the radiator is connected to an expansion tank which makes topping-up virtually unnecessary. The coolant expands when hot, and instead of being forced down an overflow pipe and lost, it flows into the expansion tank. As the engine cools the coolant contracts and because of the pressure differential flows back into the radiator.
2 The cap on the expansion tank is set to a pressure of 15 lbf/in^2 (1.05 kgf/cm^2) which increases the boiling point of the coolant to 230°F (110°C). If the water temperature exceeds this figure and the water boils, the pressure in the system forces the internal valve of the cap off its seat thus exposing the expansion tank overflow pipe down which the steam from the boiling water escapes and so relieves the pressure. It is therefore important to check that the expansion tank cap is in good condition and that the spring behind the sealing washers has not weakened. Check that the rubber seal has not perished and its seating in the neck is clean to ensure a good seal. A special tool which enables a cap to be pressure tested is available at some garages.
3 The cooling system comprises the radiator, top and bottom hoses, heater hoses, the impeller water pump (mounted on the front of the engine it carries the fan blades and is driven by the fanbelt), the heater and the thermostat.
4 The system functions as follows: Cold water from the radiator circulates up the lower radiator hose to the water pump where it is pushed round the water passages in the cylinder block, helping to keep the cylinder bores and pistons cool.
5 The water then travels up into the cylinder head and circulates round the combustion spaces and valve seats absorbing more heat. Then, when the engine is at its normal operating temperature, the water travels out of the cylinder head, past the thermostat into the upper radiator hose and so into the radiator. The water passes along the radiator from one side to the other where it is rapidly cooled by the rush of cold air through the horizontal radiator core. The water, now cool, reaches the bottom hose when the cycle is repeated.
6 When the engine is cold the thermostat (a valve which opens and closes according to water temperature) maintains the circulation of the same water in the engine by returning it to the cylinder block. Only when the correct minimum operating temperature has been reached, as shown in the Specifications, does the thermostat begin to open allowing water to return to the radiator.

2 Cooling system – draining

Note: *Take care to protect the hands from escaping steam when removing the expansion tank pressure relief cap if the system is hot*
1 Before draining the cooling system, park the vehicle on level surface and remove the expansion tank pressure relief cap. Move the heater control to the hot (red) position.
2 If the antifreeze mixture is less than two years old it can be collected in a clean bowl for re-use.
3 Remove the plug, if fitted, from the radiator top hose. Undo and remove the cylinder block drain plug located on the right-hand side of the block at the rear. On 17V engines deflect the coolant from the starter motor with a piece of cardboard.
4 Slacken the radiator bottom hose clip and carefully ease the hose, from the radiator (photo).
5 When the coolant has finished running out of the cylinder block drain hole, probe the hole with a short piece of wire to dislodge any particles of rust or sediment which may be causing a blockage preventing complete draining.

3 Cooling system – flushing

1 Generally, even with proper use, the cooling system will gradually lose its efficiency as the radiator becomes choked with rust scale, deposits from water and other sediment. To clean the system out, remove the radiator filler plug, cylinder block plug and bottom hose and leave a hose running in the radiator filler plug hole for fifteen minutes.
2 Reconnect the bottom hose, refit the cylinder block plug and refill the cooling system as described in Section 4, adding a proprietary cleaning compound. Run the engine for fifteen minutes, or as instructed by the makers of the cleaning compound. All sediment and sludge should now have been loosened and may be removed by draining the system and refilling again.
3 In very bad cases the radiator should be reverse flushed. This can be done with the radiator in position, the cylinder block plug is left in position and the hose placed in the bottom hose union of the radiator. Water under pressure is forced through the orifice and out of the filler plug hole.
4 The hose is then removed and placed in the filler plug hole and the radiator washed out in the usual manner.

4 Cooling system – filling

1 Fit the cylinder block drain plug and, if the bottom hose has been removed, reconnect it. Check the tightness of all hose clips in the system. Move the heater control to the hot (red) position. On vehicles with the 17V engine remove the thermostat by removing the cap from the thermostat housing and pulling the thermostat out of the housing.
2 Fill the system through the expansion tank slowly to avoid airlocks. On 17V engines when the thermostat housing is half full, refit the thermostat, see Section 7, and continue to fill. When the expansion tank is full to the level mark, refit the expansion tank cap. On vehicles fitted with a top hose plug, continue to top-up via the plug orifice until the coolant level reaches the bottom of the plug hole. Refit the plug.
3 Run the engine until the normal operating temperature is reached, then stop the engine and allow the system to cool. Check all joints for leaks.

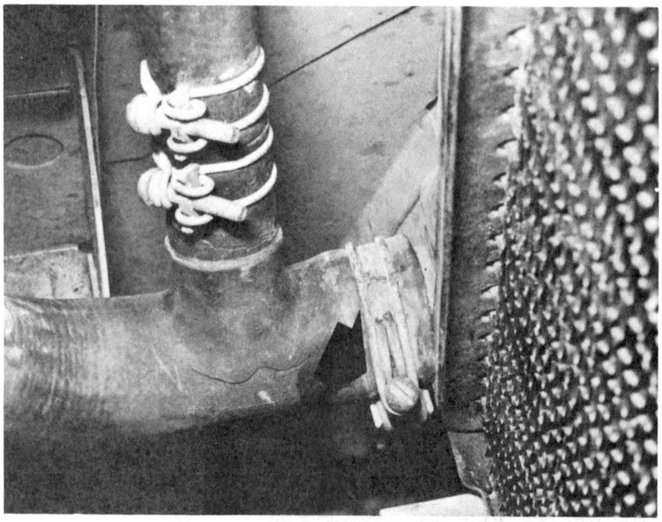

2.4 Radiator bottom hose clip (arrowed)

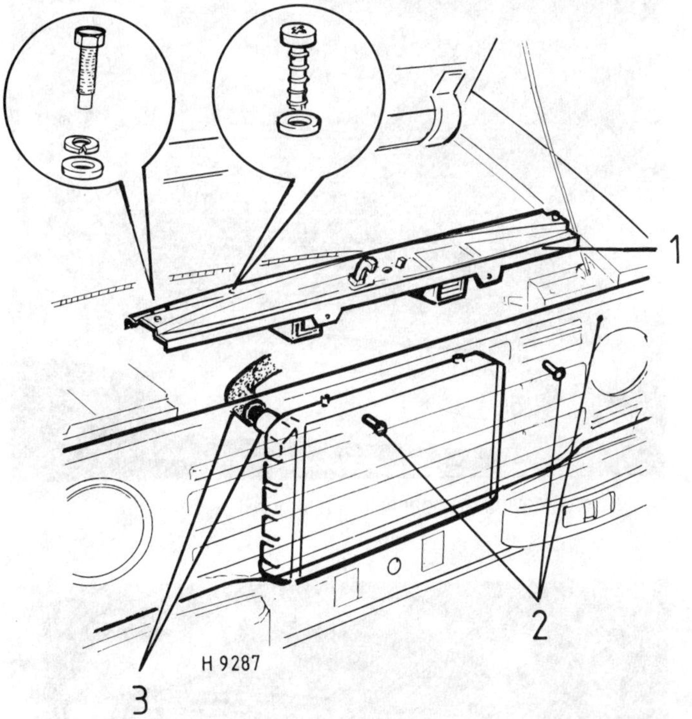

Fig. 2.1 Removal of the radiator (17V engine – others similar) (Sec 5)

1 *Bonnet lock platform*
2 *Radiator grille securing screws*
3 *Radiator and top (inlet) hose*

4 When cool, remove the expansion tank cap and top-up to the level mark. Refit the cap.
5 Remove the top hose plug, where applicable and top-up to the bottom of the orifice mark. Refit the plug.

5 Radiator – removal and refitting

Note: *Take care to protect the hands from escaping steam when removing the expansion tank pressure relief cap if the system is hot.*
1 Drain the cooling system as described in Section 2.
2 Slacken the hose clip and disconnect the top hose from the radiator.
3 Remove the radiator grille.
4 Remove the bolts retaining the bonnet lock platform (photo). Lift off the platform and move it to one side leaving the bonnet lock cable connected. On models with automatic transmission first disconnect the horn lead.
5 Remove the top and side blanking panels from around the radiator and then lift the radiator out (photos).
6 Refitting the radiator is the reverse of the removal procedure. Refill the cooling system as described in Section 4.

6 Radiator – inspection and cleaning

1 With the radiator out of the vehicle, inspect it closely for signs of leaks or damage such as cracks or impact damage to the matrix. Clean out the inside of the radiator by flushing as described in Section 3.
2 Having the radiator out of the vehicle makes it easier to turn it upside down for reverse flushing. Clean the exterior by hosing the matrix with a strong jet on the reverse face to clean out dead insects, road dirt etc.
3 It is possible to solder leaks or cracks but this should be left to a radiator repair specialist. An alternative is to use a proprietary cold setting compound but again it might be difficult to achieve a reliable repair. If extensive repairs are necessary, renewal should be considered as an easier and more reliable remedy.
4 Inspect the radiator hoses for cracks, internal and external perishing and damage caused by overtightening of the hose clips. Renew the hoses as necessary. Examine the radiator hose clips and renew them if they are rusted or distorted.

7 Thermostat – removal, testing and refitting

Note: *Take care to protect the hands from escaping steam when removing the expansion tank pressure relief cap if the system is hot*
1 Drain sufficient coolant from the system, see Section 2, to lower the level to below that of the thermostat housing.

16V and 18V engines

2 Remove the three nuts to release the housing and its joint washer from the cylinder head. Withdraw the thermostat from the cylinder head (photo).

17V engines

3 Remove the cap from the thermostat housing and pull the thermostat out (photo).

All engines

4 If the thermostat is stuck open it is defective and will have to be renewed. To test an apparently serviceable thermostat for correct functioning it should be suspended in a container of water with a thermometer and heated. Note the temperature at which the thermostat begins to open and compare this with the table in the Specifications. Discard the thermostat if it starts to open outside the required range of temperature. Continue to heat the water noting when the thermostat is fully open. Allow to cool and note when the thermostat is closed. If the thermostat does not operate correctly it must be renewed. The nominal temperature at which the thermostat should open (°C) is marked on its base.
5 Refitting a thermostat is essentially the reverse of the removal procedure, but note the follo

16V and 18V engines

6 Make sure that the mating faces of the thermostat cover and the cylinder head are clean and that the gasket is in good condition. If in doubt fit a new gasket.

17V engines

7 Wipe the inside of the housing clean and fit a new O-ring to the thermostat if necessary.

All engines

8 If the thermostat is being renewed make sure that the new one is stamped with the correct operating temperature.

8 Thermostat housing (17V engines) – removal and refitting

Note: *If the system is hot, slowly release the pressure by removing the expansion tank filler cap, taking care not to get scalded by escaping steam, before removing the thermostat cap.*
1 Remove the timing belt cover, refer to Chapter 1B.
2 Refer to Section 2 and drain the sufficient coolant from the system to lower the level below that of the thermostat housing.
3 Slacken the hose clips and disconnect the top coolant hose and the air bleed hose from the thermostat housing.
4 Remove the screw which secures the thermostat housing to the cylinder head together with its two washers (photo).
5 Pull the housing straight from its location in the head. Remove the O-ring from the base spigot.
6 Remove the thermostat from the housing by pulling it out, using a pair of pliers if necessary. Remove the O-ring.
7 Refitting is the reverse of the removal procedure. Fit new O-rings and lightly lubricate them with grease or petroleum jelly. See Section 7 for details of refitting the thermostat in its housing. Tighten the retaining screw to the specified torque and check for leaks on completion.

9 Water pump – removal and refitting

Note: *If the cooling system is hot take care to protect the hands from escaping steam when removing the expansion tank pressure relief cap.*

5.4 Removing the bonnet lock platform bolts

5.5a After removing the bonnet lock platform ...

5.5b ... remove the blanking panels and ...

5.5c ... then remove the radiator

7.2 The thermostat housing on 16V and 18V engines

7.3 Removing the thermostat (17V engines)

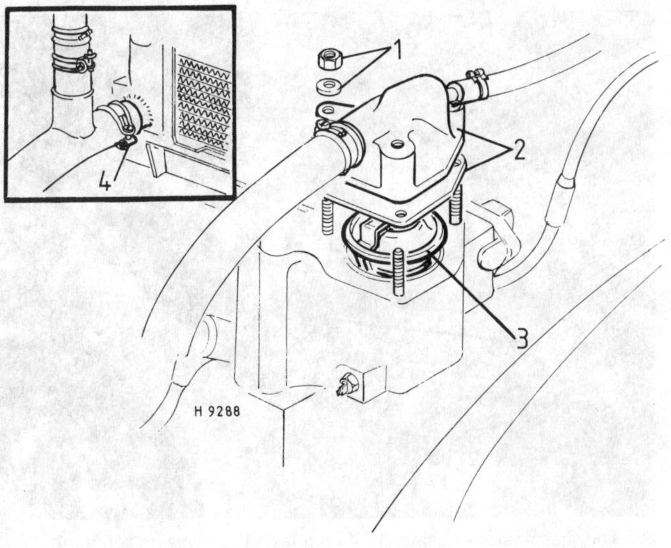

Fig. 2.2 Removal of the thermostat (16V and 18V engines) (Sec 7)

1 Thermostat housing securing nut and washer
2 Thermostat housing and gasket
3 Thermostat
4 Radiator bottom (outlet) hose connection

16V and 18V engines

1 Drain the cooling system, refer to Section 2.
2 Disconnect the battery earth cable.
3 Disconnect the cable terminal from the coolant temperature sender located under the thermostat.
4 Loosen the alternator mounting bolts and remove the fanbelt.
5 Remove the alternator pivot bolts and adjusting link, and move the alternator to one side.
6 Loosen the hose clips and disconnect the bottom radiator hose from the pump and heater pipe.
7 Remove the four setscrews and remove the pump and gasket from the cylinder block. Note that one of the four screws is different from the other three and must be refitted in its original location.
8 Refitting is the reverse sequence to removal, but the following additional points should be noted:

- *(a) Clean the mating faces of the pump and cylinder block to ensure a good leak-proof joint*
- *(b) Always use a new gasket*
- *(c) Adjust the fanbelt tension as described in Section 11*

17V engines

9 Drain the coolant, refer to Section 2, and refit the drain plug.
10 Disconnect the battery earth cable.
11 Remove the timing belt cover, refer to Chapter 1 if necessary.
12 Loosen the alternator mounting bolts and remove the fanbelt.
13 Undo and remove the fan and pulley retaining bolts and remove the fan and pulley (photo).

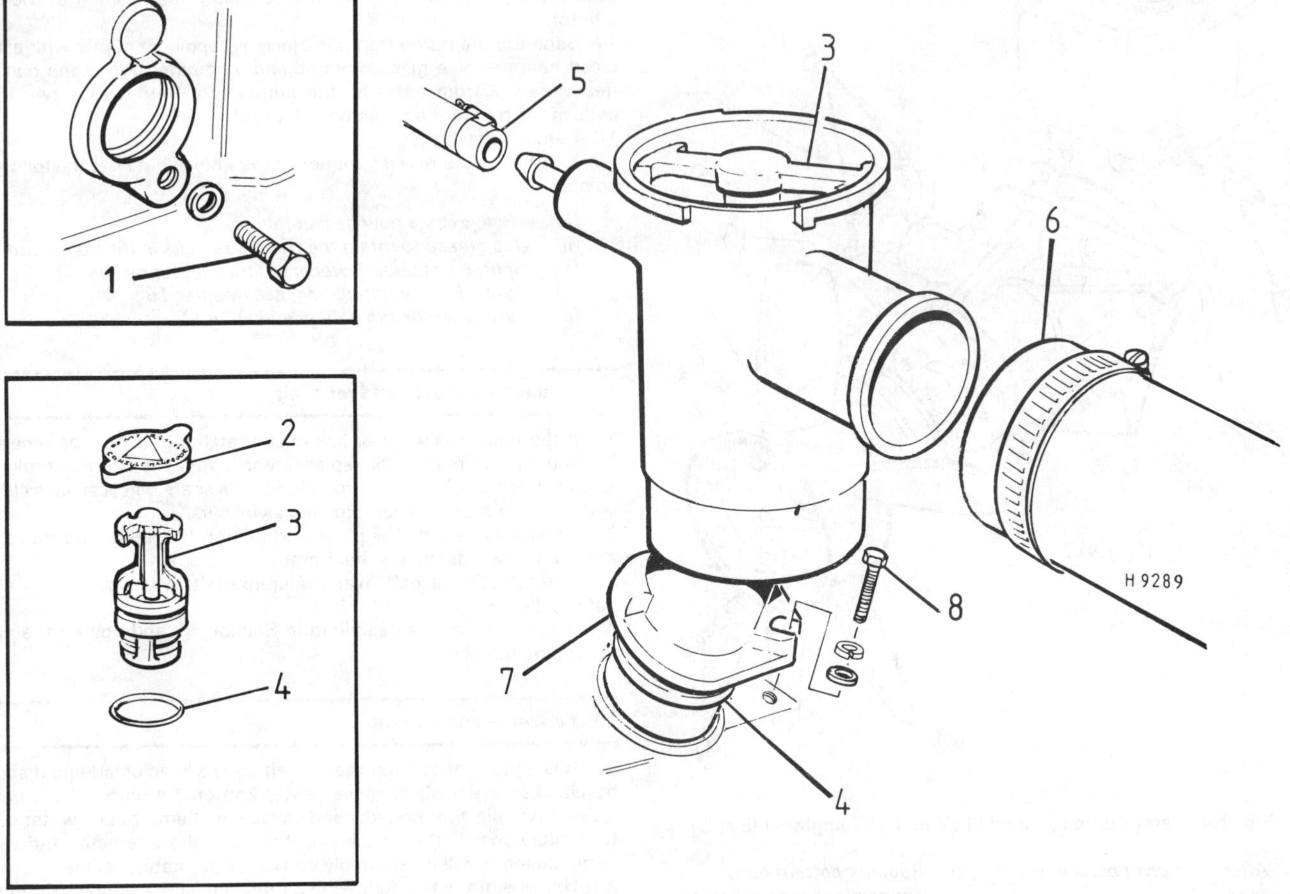

Fig. 2.3 Removal of the thermostat (17V engines) (Sec 7)

1 Cylinder block drain plug
2 Thermostat housing cap
3 Thermostat
4 O-rings
5 Air bleed pipe
6 Radiator top hose
7 Thermostat housing
8 Thermostat housing securing screw

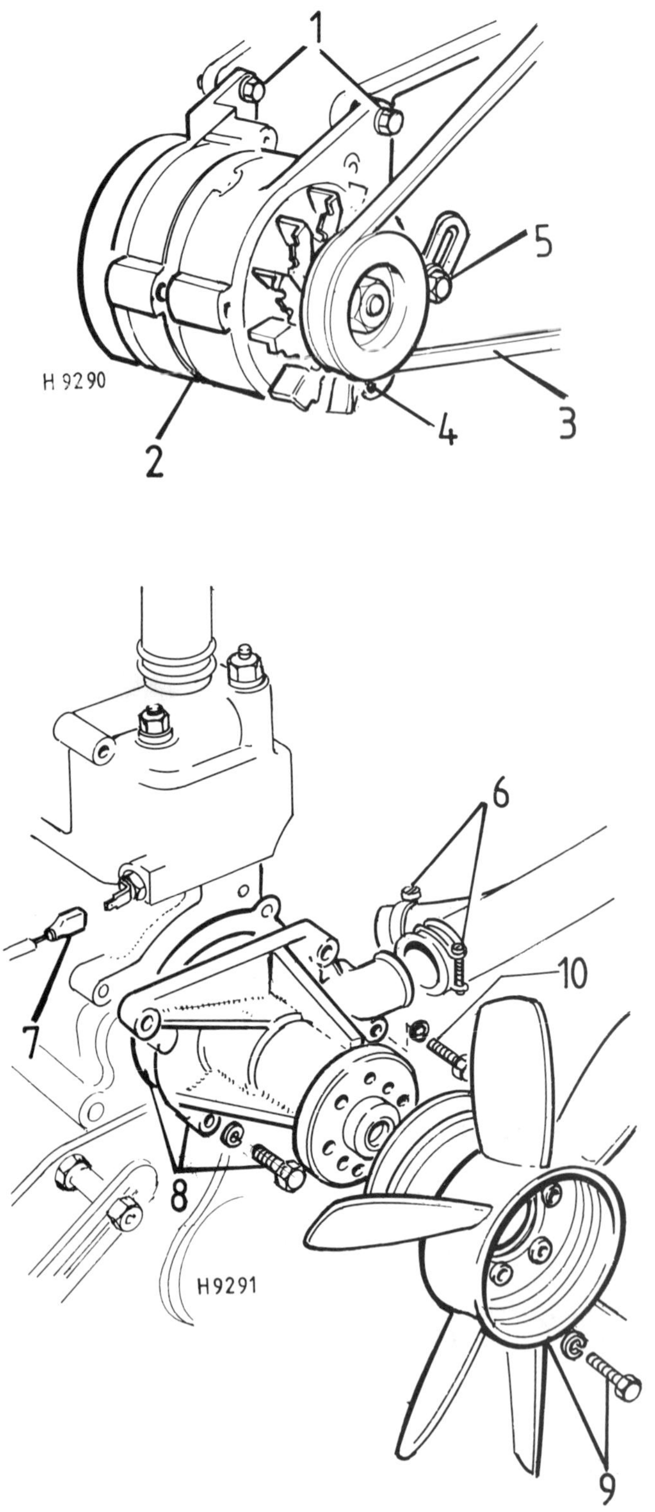

Fig. 2.4 Water pump removal (16V and 18V engines) (Sec 9)

1	*Alternator pivot bolts and nuts*	6	*Radiator bottom hose connections to pump and heater pipe*
2	*Alternator*	7	*Temperature sender terminal*
3	*Fanbelt*	8	*Water pump and gasket*
4	*Alternator adjusting link screw*	9	*Fan and securing setscrews*
5	*Alternator adjusting link nut*	10	*Water pump bolt*

8.4 The thermostat housing (17V engines): 1 Housing retaining screw 2 Temperature sender

14 Loosen the hose clip and disconnect the bottom radiator hose from the pump.

15 Remove the two nuts, spring washers and plain washers retaining the timing belt tensioner and remove the tensioner. Note that the nuts screw onto two bolts located in the rear of the water pump.

16 Remove the five bolts and the timing cover pillar stud which secure the pump to the cylinder block. Note the position of the stud (photo).

17 Separate the pump from the block by tapping it gently with a soft-faced hammer or a piece of wood and a hammer. When the pump is free extract it from between the timing belt. Retrieve the two loose bolts in the rear of the pump body (photo).

18 Remove the gasket.

19 Refitting is the reverse sequence to removal but note the following points:

- (a) *Always use a new gasket (photo).*
- (b) *Use grease to retain the two loose bolts in the pump body*
- (c) *Tighten specified fasteners to the correct torque*
- (d) *Retension the timing belt, see Chapter 1B*
- (e) *Retension the fan belt, see Section 11*

10 Fanbelt – removal and refitting

1 If the fanbelt is worn or has over stretched it should be renewed. The most usual reason for replacement is that the belt has broken in service. It is therefore recommended that a spare belt is always carried in the car. To change a belt proceed as follows.

2 Loosen the alternator pivot and slotted link bolts and move the alternator towards the engine. Remove the old belt.

3 Carefully fit the belt over the crankshaft, water pump and alternator pulleys.

4 Adjust the belt as described in Section 11 and tighten the alternator mounting bolts.

11 Fanbelt – adjustment

1 It is important to keep the fanbelt correctly adjusted and it should be checked every 6000 miles (9600 km) or 3 months. If the belt is loose it will slip, wear rapidly and cause the alternator and water pump to malfunction. If the belt is too tight the the alternator and water pump bearings will wear rapidly and cause premature failure.

2 Use one of the two following methods to check the fanbelt tension (photo):

- (a) *Using a torque spanner on the alternator pulley retaining nut apply a load of 11 lbf ft (5Nm) in a clockwise direction. If the belt tension is correct the belt will just slip at this loading*
- (b) *Apply a load of 8 lbf (36N) at right angles to the belt midway*

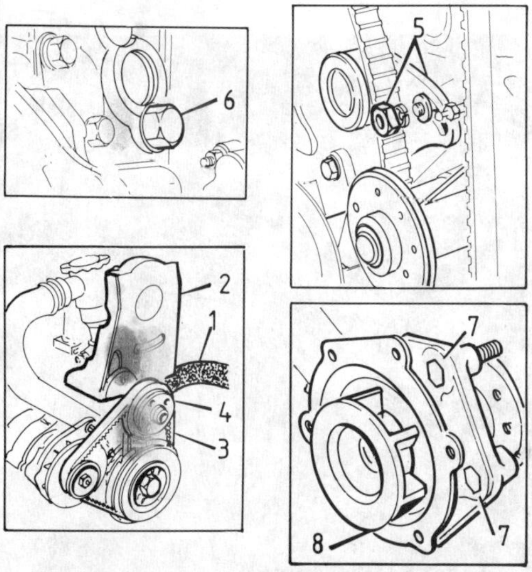

Fig. 2.5 Water pump removal (17V engines) (Sec 9)

1 Radiator bottom hose to pump
2 Timing belt cover
3 Fanbelt
4 Water pump pulley
5 Timing belt tensioner and retaining nut (one shown of two)
6 Coolant drain plug
7 Two captive bolts for tensioner
8 Water pump
9 Water pump pulley and securing bolt (one shown of four)
10 Timing cover pillar stud
11 Water pump gasket
12 Pump securing bolts
13 Radiator bottom hose connection

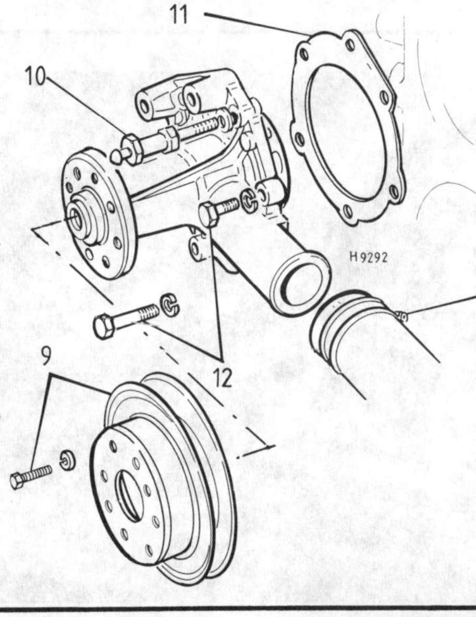

9.13 Removing the fan

9.16 Note the position of the timing cover stud

9.17 Retrieve the two loose water pump bolts (arrowed)

9.19 Refitting the water pump with a new gasket

11.2 Check the fanbelt tension between the fan and alternator pulleys

12.4 The expansion tank in its mounting bracket

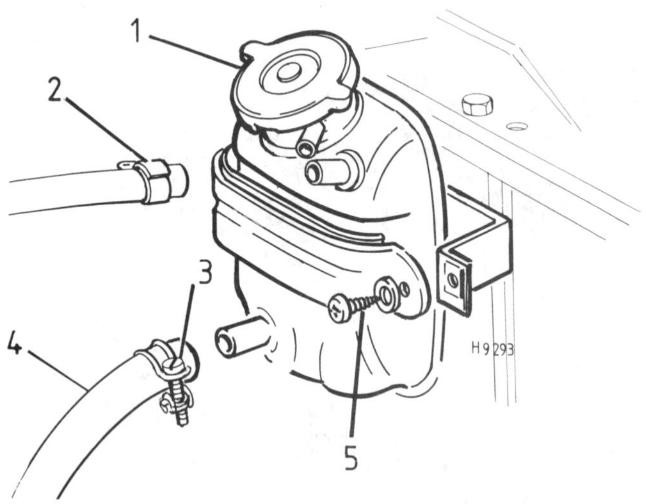

Fig. 2.6 Expansion tank fitting details (Sec 12)

1 Filler/pressure relief cap
2 Air bleed pipe connection
3 Filler hose clip
4 Filler hose (from radiator bottom hose)
5 Mounting screw (one shown of two)

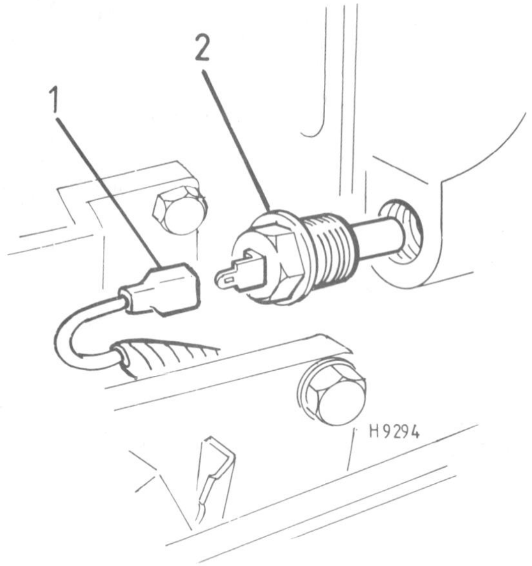

Fig. 2.7 Thermal transmitter removal (Sec 13)

1 Electrical connection *2 Thermal transmitter*

between the fan and alternator pulleys. The belt should deflect 0.25 in (6 mm)

3 To adjust the fanbelt, slacken the alternator upper and lower pivot bolts and slacken the alternator adjusting link bolt. It is easier if these bolts are only slackened just sufficiently to be able to swing the alternator stiffly with some effort. In this way the belt tension can be adjusted more quickly than by making frequent resettings.

4 Pull the alternator away from the engine to increase belt tension and tighten the adjusting link and pivot bolts. Note that leverage must only be applied to the alternator drive end bracket and it is preferable to use a wooden or soft metal lever. Check the tension and re-adjust if necessary.

5 If a new belt is being tensioned, after tightening the attachment bolts run the engine for five minutes at about 1000 rpm. Re-check the tension as described in paragraph 2 and re-adjust if necessary.

12 Expansion tank – removal and refitting

Note: *If the cooling system is hot remove the expansion tank filler cap slowly to release the pressure gradually, and protect the hands from possible scalding by escaping steam*

1 Remove the expansion tank filler cap.

2 Disconnect the air bleed hose from the expansion tank by slackening the clip and removing the hose.

3 Position a receptacle to catch spilt coolant beneath the filler hose connection, slacken the clip and disconnect the filler hose. Support the filler hose in an upright position to reduce the loss of coolant from the system.

4 Unscrew the two retaining screws and remove the expansion tank from the mounting bracket (photo).

5 The tank requires no maintenance but, while it is removed, it would be sensible to rinse it out thoroughly with clean water to remove any deposits or sediment.

6 Refitting is the reverse of removal. Top-up the system by filling to the level mark on the tank.

13 Temperature gauge and thermal transmitter

1 The thermal transmitter is placed in the cylinder head just below the thermostat and is held in position by a special gland nut to ensure

a water tight joint. It is connected to the gauge located on the instrument panel by a cable on the main ignition feed circuit and a special bi-metal voltage stabilizer.
2 To remove the thermal transmitter, partially drain the cooling system (usually 4 pints (2.27 litres) is enough). Disconnect the electrical connection and unscrew the thermal transmitter from the head. Refitting is a reverse procedure to removal.
3 For information on removing and refitting the gauge refer to Chapter 10.

14 Antifreeze mixture

1 Prior to anticipated freezing conditions, it is essential that antifreeze is added to the cooling system. Use any antifreeze conforming to BS 3151 or BS 3152. **Note**: *Do not use engine antifreeze in the windscreen washing equipment.*
2 Antifreeze solution can be left in the cooling system for up to two years, but it is advisable to have the specific gravity checked at the beginning of winter each year. This can be done at your local dealer or garage.
3 Listed below are the recommended strengths and quantities of antifreeze to use in a mixture required to give protection for varying degrees of freezing. Always use clean (preferably soft) tap water to mix the solution. The easiest way to do this is to first pour a pint or so of water into the system, then add the correct quantity of antifreeze and finally top-up with water. Run the engine to mix the solution and, when cool, top-up again with water to give the correct mixture.

Solution	Amount of Antifreeze 16V and 17V		18V		Commences freezing		Frozen solid	
%	pints	litres	pints	litres	°C	°F	°C	°F
25	*2.75*	*1.6*	*3*	*1.7*	*–13*	*–9*	*–26*	*–15*
33⅓	*3.6*	*2.1*	*4*	*2.3*	*–19*	*–2*	*–36*	*–33*
50	*5.5*	*3.1*	*5.8*	*3.3*	*–36*	*–33*	*–48*	*–53*

15 Fault diagnosis – cooling system

Symptom	Reason/s
Heat generated in engine not being successfully diposed of by radiator	Insufficient water in cooling system Fanbelt slipping (Accompanied by a shrieking noise on rapid engine acceleration) Radiator core blocked or radiator grille restricted Bottom water hose collapsed, impeding flow Thermostat not opening properly Ignition advance and retard incorrectly set (Accompanied by loss of power and perhaps, misfiring) Carburettor incorrectly adjusted (mixture too weak) Exhaust system partially blocked Oil level in sump too low Blown cylinder head gasket (Water/steam being forced down the radiator overflow pipe under pressure) Engine not yet run-in Brakes binding
Too much heat being dispersed by radiator	Thermostat jammed open Incorrect grade of thermostat fitted allowing premature opening of valve Thermostat missing
Leaks in system	Loose clips on water hoses Top or bottom water hoses perished and leaking Radiator core leaking Thermostat gasket or seal leaking Pressure cap spring worn or seal ineffective Blown cylinder head gasket (Pressure in system forcing water/steam down overflow pipe) Cylinder wall or head cracked

Chapter 3 Carburation; fuel and exhaust systems

For modifications, and information applicable to later models, see Supplement at end of manual

Contents

Specifications

Air cleaner

Type:	
Temperate and cold climates	Paper element
Hot climates	Cyclopack, cleanable element

Carburettor

16V engines

Make and type	SU HS6
Piston spring colour	Yellow
Jet size (groove at top of jet)	0·100 inch
Capstat colour	Natural
Needle:	
Temperate and cold climates	BCT
Hot climates	BDS
Idle speed	750 rpm
Fast idle speed	1300 rpm

18V engines

Make and type	SU HS6
Piston spring colour:	
Temperate and cold climates	Yellow
Hot climates	Red
Jet size (groove at top of jet)	0·100 inch
Capstat colour	Natural
Needle:	
Temperate and cold climates	BCU
Hot climates	BDT
Idle speed	750 rpm
Fast idle speed:	
Manual gearbox	1300 rpm
Automatic transmission	1500 rpm

17V engines

Make and type	SU HIF6
Specification:	
Manual gearbox	FZX 1236
Automatic transmission	FZX 1237
Piston spring colour	Red

Jet size (groove at top of jet)	0·100 inch
Needle	BEM
Idle speed	650 rpm
Fast idle speed	1100 rpm

Exhaust emission

Exhaust gas analyser reading at engine idle speed:	
16V and 18V engines	3 to 4·5% CO
17V engines	2·5% CO

Fuel pump

Make and type	SU mechanical, AUF 800 series

Fuel tank

185/200/215/230/240/250 models:	
Location	Rear, underfloor
Capacity	11 gal (50 litres)
220/250 (1979 on) 255 models:	
Location	Chassis side
Capacity	14 gal (64 litres)

Torque wrench settings

	lbf ft	Nm
16V engines		
Manifold nuts	16	22
18V engines		
Manifold nuts	15	20
Carburettor stud nuts	15	20
Float chamber securing bolt	7·5	10
17V engines		
Fuel pump plate screws	9	12
Carburettor to manifold nuts	19	25
Air cleaner to carburettor	19	25

1 General description

1 The fuel system consists of a fuel tank mounted at the rear or side of the vehicle, a mechanical fuel pump mounted on the engine, and a single SU carburettor together with the associated fuel piping.
2 A renewable paper element air cleaner is fitted to models intended for temperate climate operation; for hot climate models, the air cleaner has a cleanable element.
3 The exhaust system is a conventional arrangement of exhaust pipe and silencers.
4 The operation of individual components is described elsewhere in this Chapter.

2 Air cleaner – maintenance

1 Every 6 months or 12 000 miles (20 000 km) the air cleaner must be serviced as described below. In dusty operating conditions this servicing should be carried out more frequently (photo).

Temperate climate models

2 Unscrew and remove the wing nut and the sealing washer beneath it. Remove the air cleaner cover.
3 Withdraw and discard the used element (photo.
4 Thoroughly clean the interior of the container, including the cover.
5 Ensure that the sealing ring is in good condition.
6 Fit a new element and refit the cover. Make sure that the washer is in place under the wing nut.

Hot climate models

7 Slacken the clip securing the air intake hose to the top of the air cleaner and pull off the air intake hose.
8 Release the two straps retaining the air cleaner to its mounting bracket and remove the air cleaner from the vehicle.
9 Depending on the type of air cleaner, either unscrew the knob to detach the dust cup or release the clamp, as appropriate. Remove the baffle from the clamp type air cleaner.
10 Clean out the dust cup and, on the air cleaner with a dust valve, clean out the valve.
11 Unscrew and remove the wing nut, and remove the cleaner element. Clean the element by blowing clean air through the element from the centre, up and down the pleats.
12 Refitting is the reverse of the removal procedure. Ensure that the lips of the dust valve, where fitted, are pointing downwards.

3 Air temperature control valve – checking

1 The 17V engine has an automatic air temperature control valve controlling the air supply to the air cleaner. Check the valve as follows.
2 Rotate and disconnect the air intake tubes from the air temperature control valve.
3 Press the valve plate and release it. The plate should return to its original position. If it does not, a new valve should be fitted as a defective valve can affect engine performance and fuel consumption.
4 After checking the valve refit the air intake tubes.

4 Fuel pump – testing, removal and refitting

1 On 16V and 18V engines the fuel pump is located on the left-hand side of the crankcase. On 17V engines the fuel pump is mounted on the left-hand side of the camshaft cover over number 4 cylinder. On all engines the pump is operated by a separate lobe on the camshaft.
2 The fuel pump is a sealed unit and cannot be serviced in the event of a malfunction.
3 If a faulty pump is suspected, remove the ignition coil primary (LT) lead, remove the carburettor feed pipe at the carburettor, position a container to catch any fuel and arrange for an assistant to crank the engine on the starter. If little or no fuel flows from the pipe, the pump is defective, but make sure that the fuel pipe is not clogged or the fuel tank empty before condemning the pump.
4 To remove the fuel pump, disconnect both fuel pipes and plug the pipe from the tank to the pump to minimise fuel spillage.
5 Remove the two bolts retaining the pump to the cylinder block, or the two nuts retaining the pump to the camshaft cover, as appropriate. Remove the pump and the insulating washer or block (photo).
6 Remove all traces of the old gaskets and fit new ones. Refit the pump following the removal procedure in reverse (photo).

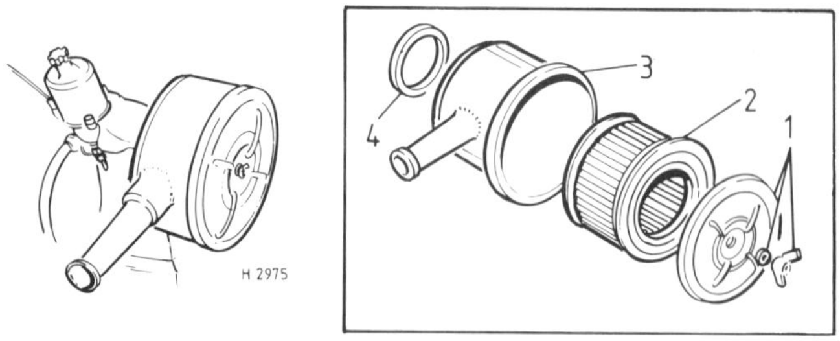

Fig. 3.1 Assembled and exploded views of the air cleaner (16V and 18V engine) (Sec 2)

1 Wing nut, fibre washer and cover
2 Element
3 Container
4 Sealing ring

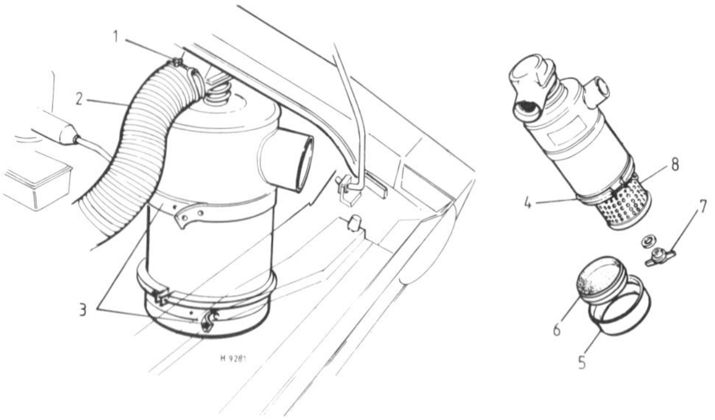

Fig. 3.2 Assembled and exploded views of the air cleaner for hot climates (Sec 2)

1 Air intake hose clip
2 Air intake hose
3 Air cleaner securing straps
4 Clamp
5 Dust cup
6 Baffle
7 Wing nut
8 Element

2.1 The air cleaner on the 17V series engine

2.3 Withdrawing the air cleaner element

4.5 Removing the fuel pump from a 17V engine

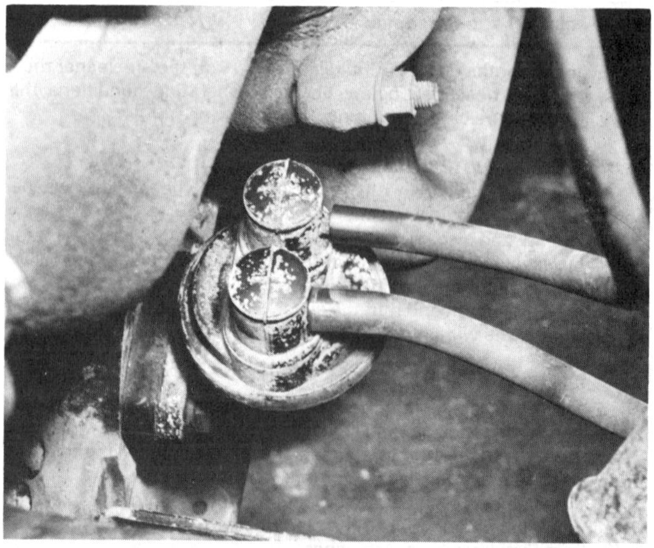

4.6 The fuel pump fitted to 16V/18V engines

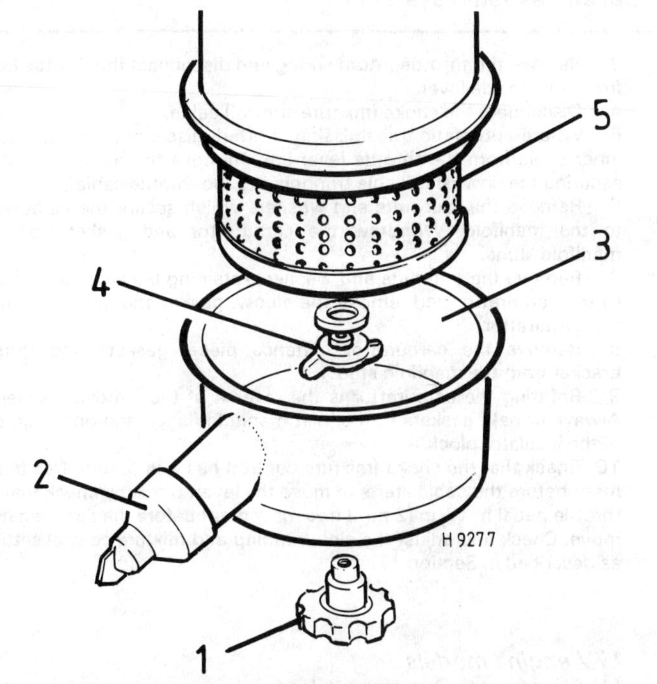

Fig. 3.3 Hot climate air cleaner with dust valve in the dust cup (Sec 2)

1 Dust cup retaining knob
2 Dust valve
3 Dust cup
4 Element retaining wing nut
5 Element

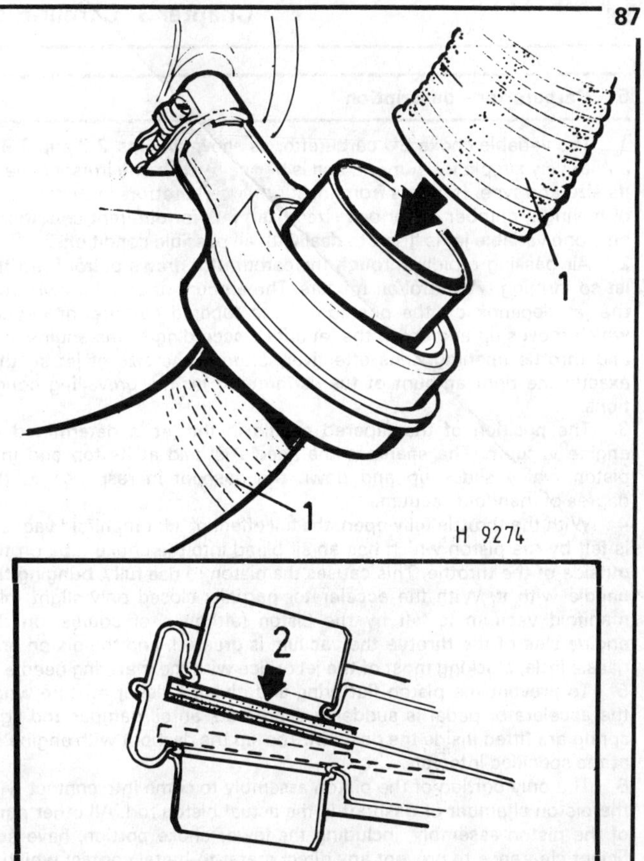

Fig. 3.4 Checking the air temperature control valve (17V engines) (Sec 3)

1 Air intake pipes
2 Press the valve and release

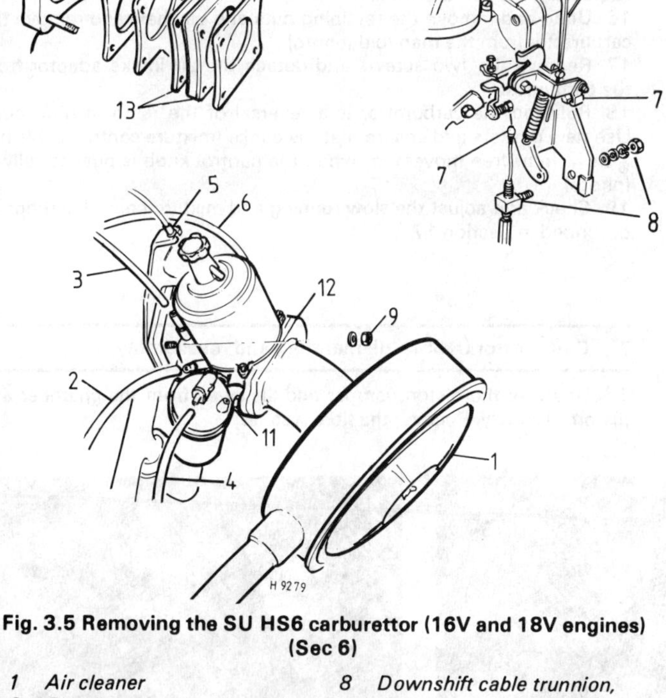

Fig. 3.5 Removing the SU HS6 carburettor (16V and 18V engines) (Sec 6)

1 Air cleaner
2 Breather pipe
3 Ignition advance vacuum pipe
4 Fuel pipe
5 Throttle cable
6 Mixture control (choke) cable
7 Downshift inner cable and throttle lever link (automatic gearbox models only)
8 Downshift cable trunnion, securing nut and washers (automatic gearbox models only)
9 Carburettor retaining nuts and washers
10 Gasket
11 Air cleaner elbow retaining nuts
12 Air cleaner elbow
13 Distance piece and gasket

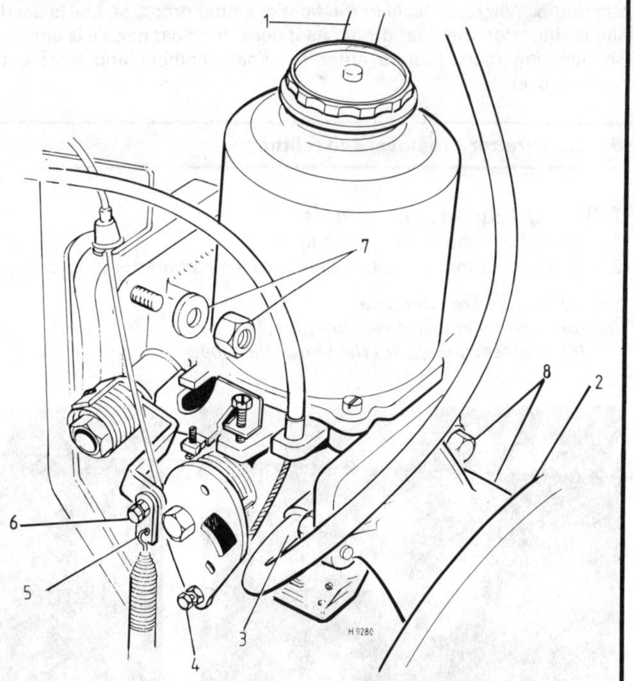

Fig. 3.6 Removing the SU HIF6 carburettor (17V engines) (Sec 6)

1 Breather hose
2 Air cleaner
3 Fuel hose
4 Mixture control (choke) connection
5 Throttle return spring
6 Throttle control connection
7 Carburettor retaining nuts and washers
8 Air intake adaptor

5 Carburettor – description

1 The variable choke SU carburettor as shown in Figs. 3.7 and 3.9 is a relatively simple instrument, and is basically the same irrespective of its size and type. It differs from most other carburettors in that instead of having a number of various sized fixed jets for different conditions, only one variable jet is fitted to deal with all possible conditions.
2 Air passing rapidly through the carburettor draws petrol from the jet so forming the petrol/air mixture. The amount of petrol drawn from the jet depends on the position of the tapered carburettor needle, which moves up and down the jet orifice according to the engine load and throttle opening, thus effectively altering the size of jet so that exactly the right amount of fuel is metered for the prevailing conditions.
3 The position of the tapered needle in the jet is determined by engine vacuum. The shank of the needle is held at its top end in a piston which slides up and down the dashpot in response to the degree of manifold vacuum.
4 With the throttle fully open, the full effect of inlet manifold vacuum is felt by the piston which has an air bleed into the choke tube on the outside of the throttle. This causes the piston to rise fully, bringing the needle with it. With the accelerator partially closed only slight inlet manifold vacuum is felt by the piston (although of course, on the engine side of the throttle the vacuum is greater), and the piston only rises a little, blocking most of the jet orifice with the metering needle.
5 To prevent the piston fluttering and giving a richer mixture when the accelerator pedal is suddenly depressed, an oil damper and light spring are fitted inside the dashpot. Top-up the dashpot with engine oil at the specified intervals.
6 The only portion of the piston assembly to come into contact with the piston chamber or dashpot is the actual piston rod. All other parts of the piston assembly, including the lower choke portion, have sufficient clearance to prevent any direct metal-to-metal contact which is essential if the carburettor is to function correctly.
7 The correct level of the petrol in the carburettor is determined by the level of the float chamber. When the level is correct the float rises and, by means of a lever resting on top of it, closes the needle valve in the cover of the float chamber. This closes off the supply of fuel from the pump. When the level in the float chamber drops, as fuel is used in the carburettor, the float drops. As it does, the float needle is unseated so allowing more fuel to enter the float chamber and restore the correct level.

6 Carburettor – removal and refitting

16V and 18V engine models

1 Disconnect the battery earth lead.
2 Disconnect the following pipes from the carburettor (photo).

(a) engine breather pipe
(b) ignition advance vacuum pipe
(c) fuel feed pipe; plug the end of the pipe

3 Release the throttle return spring and disconnect the throttle cable from the throttle lever.
4 Disconnect the choke (mixture control) cable.
5 Where automatic transmission is fitted, disconnect the downshift inner cable from the throttle lever link. Remove the nut and washers securing the downshift cable trunnnion to the throttle cable.
6 Remove the four nuts and washers which secure the carburettor to the manifold. Withdraw the carburettor and gasket from the manifold studs.
7 Remove the two nuts and washers retaining the air cleaner elbow to the carburettor and remove the elbow, gasket and air cleaner from the carburettor.
8 Remove the carburettor distance piece, gaskets and throttle bracket from the manifold studs.
9 Refitting the carburettor is the reverse of the removal sequence. Always fit new gaskets to the inlet manifold flange and one each side of the insulator block.
10 Check that the choke (mixture control) has $\frac{1}{16}$ in (2 mm) free movement before the cable starts to move the lever. Similarly check that the throttle pedal has $\frac{1}{16}$ in (2 mm) free movement before the cam begins to move. Check and adjust the slow running and mixture control settings as described in Section 17.

17V engine models

11 Disconnect the battery earth lead.
12 Remove the wing nut retaining the air cleaner and release the intake trunking from the manifold hot box. Place the air cleaner to one side.
13 Disconnect the fuel pipe and breather hose from the carburettor.
14 Disconnect the throttle and choke control cables at their connecting trunnions. Release the throttle return spring and trunnion assembly.
15 Where automatic transmission is fitted, disconnect the downshift cable from the carburettor link.
16 Undo and remove the retaining nuts and washers and remove the carburettor from the manifold (photo).
17 Remove the two screws and detach the air intake adaptor from the carburettor.
18 Refitting the carburettor is a reversal of the removal procedure. Use new gaskets and ensure that the choke (mixture control) cable has $\frac{1}{16}$ in (2 mm) free movement when the control knob is pushed fully in (photo).
19 Check and adjust the slow running and mixture control settings as described in Section 17.

7 Carburettor (HS6) – dismantling and reassembly

1 Unscrew the piston damper and lift away from the chamber and piston assembly. Recover the fibre washer.

6.2 The hose connections on the 16V/18V engine carburettor: A–Engine breather pipe, B–Ignition advance vacuum pipe, C–Fuel feed pipe

6.16 Removing the carburettor from the intake manifold studs, 17V engine

6.18 Adjust the carburettor control cables after refitting

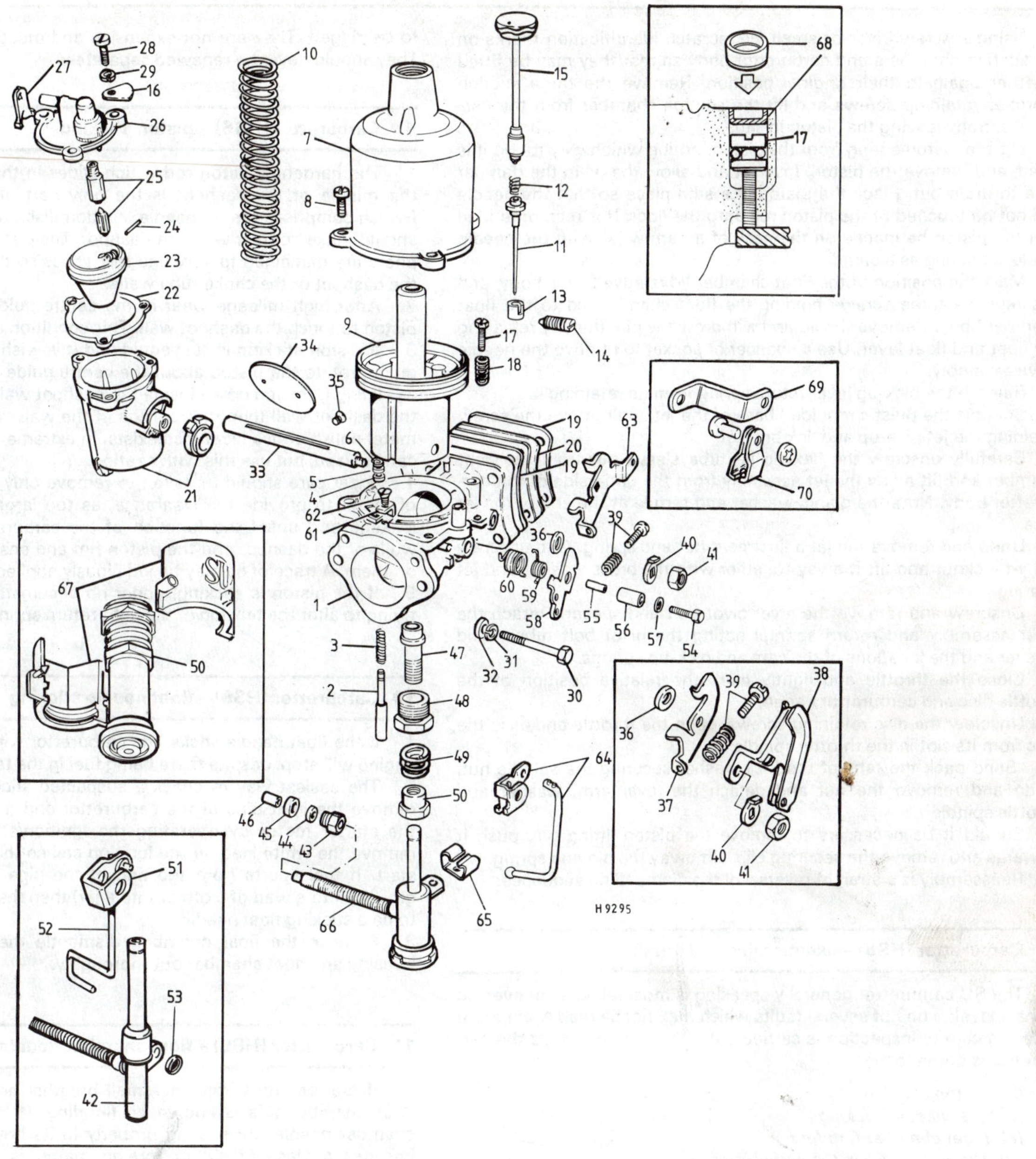

Fig. 3.7 Exploded view of the SU HS6 carburettor (16V and 18V engines) (Sec 7)

1 *Carburettor body*
2 *Piston lifting pin*
3 *Spring*
4 *Sealing washer*
5 *Plain washer*
6 *Circlip*
7 *Piston chamber*
8 *Piston chamber screw*
9 *Piston*
10 *Spring*
11 *Needle*
12 *Needle spring*
13 *Needle support guide*
14 *Needle support guide locking screw*
15 *Piston damper*
16 *Identification tag*
17 *Throttle adjusting screw*
18 *Spring*
19 *Gaskets*
20 *Distance piece*
21 *Float chamber and spacer*
22 *Float chamber gasket*
23 *Float*
24 *Float hinge pin*
25 *Needle and seat*
26 *Float chamber lid*
27 *Baffle plate*
28 *Float chamber lid screw*
29 *Spring washer*
30 *Float chamber securing bolt*
31 *Spring washer*
32 *Plain washer*
33 *Throttle spindle*
34 *Throttle disc*
35 *Throttle disc securing screw*
36 *Throttle spindle washer*
37 *Throttle return lever*
38 *Progressive throttle (snail cam)*
39 *Fast idle screw and spring*
40 *Throttle spindle nut lockwasher*
41 *Throttle spindle nut*
42 *Jet assembly*
43 *Jet flexible pipe sleeve nut*
44 *Washer*
45 *Gland*
46 *Ferrule*
47 *Jet bearing*
48 *Jet locating nut*
49 *Spring*
50 *Jet adjusting nut*
51 *Pick-up lever*
52 *Pick-up lever link*
53 *Retaining ring*
54 *Pivot bolt*
55 *Pivot bolt tube – inner*
56 *Pivot bolt tube – outer*
57 *Distance washer*
58 *Cam lever*
59 *Cam lever spring*
60 *Pick-up lever spring*
61 *Suction chamber piston guide*
62 *Guide securing screw*
63 *Throttle lever*
64 *Rod link and pick-up lever*
65 *Spring clip*
66 *Capstat jet assembly*
67 *Plastic shrouds (Carburettors with sealed adjustment)*
68 *Aluminium plug (Carburettors with sealed adjustment)*
69 *Downshift cable lever (Automatic gearbox models only)*
70 *Downshift cable link (Automatic gearbox models only)*

2 Using a screwdriver or small file, scratch identification marks on the suction chambers and carburettor body so that they may be fitted together again in their original position. Remove the three suction chamber retaining screws and lift the suction chamber from the carburettor body leaving the piston in situ.
3 Lift the piston spring from the piston, noting which way round it is fitted, and remove the piston. Invert it and allow the oil in the damper bore to drain out. Place the piston in a safe place so that the needle will not be touched or the piston roll onto the floor. It is recommended that the piston be placed on the neck of a narrow jar with the needle inside, so acting as a stand.
4 Mark the position of the float chamber lid relative to the body, and unscrew the three screws holding the float chamber lid to the float chamber body. Remove the lid and withdraw the pin, thereby releasing the float and float lever. Use a spanner or socket to remove the needle valve assembly.
5 Release the pick-up lever return spring from its retaining lug.
6 Support the plastic moulded base of the jet and remove the screw retaining the jet pick-up and link bracket.
7 Carefully unscrew the flexible jet tube sleeve nut from the float chamber and lift away the jet assembly from the underside of the carburettor body. Note the gland, washer and ferrule at the end of the jet tube.
8 Undo and remove the jet adjustment nut and spring. Also unscrew the jet locknut and lift it away together with the brass washer and jet bearing.
9 Unscrew and remove the lever pivot bolt and spacer. Detach the lever assembly and return springs noting the pivot bolt tubes, skid washer and the locations of the cam and pick-up springs.
10 Close the throttle and lightly mark the relative position of the throttle disc and carburettor flange.
11 Unscrew the disc retaining screws, open the throttle and ease the disc from its slot in the throttle spindle.
12 Bend back the tabs of the lock washer securing the spindle nut. Undo and remove the nut and detach the lever arm, washer and throttle spindle.
13 Should it be necessary to remove the piston lifting pin, push it upwards and remove the securing clip. Lift away the pin and spring.
14 Reassembly is a straight reversal of the dismantling sequence.

8 Carburettor (HS6) – examination and repair

1 The SU carburettor generally speaking is most reliable but even so it may develop one of several faults which may not be readily apparent unless a careful inspection is carried out. The common faults the carburettor is prone to are:

(a) Piston sticking
(b) Float needle sticking
(c) Float chamber flooding
(d) Water and dirt in the carburettor

2 In addition, the parts listed below are susceptible to wear after high mileages and as they vitally affect the economy of the engine they should be checked and renewed where necessary, every 24 000 miles (40 000 km).
3 *The carburettor needle:* If this has been incorrectly fitted at some time so that it is not centrally located in the jet orifice, the metering needle will have a tiny ridge worn on it. If a ridge can be seen then the needle must be renewed. SU carburettor needles are made to very fine tolerances and should a ridge be apparent, no attempt should be made to rub the needle down with fine emery paper. If it is wished to clean the needle, it can be polished lightly with metal polish.
4 *The carburettor jet:* If the needle is worn it is likely that the rim of the jet will be damaged where the needle has been striking it. It should be renewed, otherwise fuel consumption will suffer. The jet can also be badly worn or ridged on the outside from where it has been sliding up and down between the jet bearing every time the choke has been pulled out. Removal and renewal is the only answer.
5 Check the edges of the throttle and choke tube for wear. Renew if worn.
6 The washers fitted to the base of the jet and under the float chember lid may leak after a time and can cause a great deal of fuel wastage. It is wisest to renew them automatically when the carburettor is stripped down.
7 After high mileages the float chamber needle and seat are bound to be ridged. They are not expensive and must be renewed as a set. They should never be renewed separately.

9 Carburettor (HS6) – piston sticking

1 The hardened piston rod which slides in the centre guide tube in the middle of the dashpot is the only part of the piston assembly (which comprises the jet needle, suction disc and piston choke) which should make contact with the dashpot. The piston rim and choke periphery are machined to very fine tolerances so that they will not touch the dashpot or the choke tube walls.
2 After high mileage wear in the centre guide tube may allow the piston to touch the dashpot wall. This condition is known as sticking.
3 If piston sticking is suspected and it is wished to test for this condition, rotate the piston about the centre guide tube at the same time as sliding it up and down inside the dashpot wall. If the piston touches the dashpot wall then that portion of the wall must be polished with a metal polish until a clearance exists. In extreme cases fine emery cloth can be used, but use this with caution.
4 Great care should be taken to remove only the minimum amount of metal to provide the clearance, as too large a gap will cause air leakage and upset the function of the carburettor. Clean down the walls of the dashpot and the piston rim and ensure that there is no oil on them. A trace of oil may be judiciously applied to the piston rod.
5 If the piston is sticking, under no circumstances try to clear it by trying to alter the tension of the light return spring.

10 Carburettor (HS6) – float needle sticking

1 If the float needle sticks, the carburettor will soon run dry and the engine will stop, despite there being fuel in the tank.
2 The easiest way to check a suspected sticking float needle is to remove the inlet pipe at the carburettor and turn the engine over on the starter motor by operating the ignition/starter switch, but first remove the white lead on the ignition coil so that the engine does not start. If fuel spurts from the end of the pipe (direct it towards the ground, into a wad of cloth or into a jar) then the fault is almost certain to be a sticking float needle.
3 Remove the float chamber, dismantle the valve and clean the housing and float chamber out thoroughly.

11 Carburettor (HS6) – float chamber flooding

If fuel emerges from the small breather hole in the cover of the float chamber this is known as flooding. It is caused by the float chamber needle not seating properly in its housing, normally this is because a piece of dirt or foreign matter is jammed between the needle and needle housing. Alternatively the float may have developed a leak or be maladjusted so that it is not closing the float chamber needle valve even though the chamber is full of petrol. Remove the float chamber cover, clean the needle assembly, check the setting of the float as described later in this chapter and shake the float to see if any petrol has leaked into it.

12 Carburettor (HS6) – water or dirt in carburettor

1 Because of the size of the jet orifice, water or dirt in the carburettor is normally cleared.If dirt in the carburettor is suspected, lift the piston assembly and flood the float chamber. The normal level of the fuel should be about $\frac{1}{16}$ in (1·5 mm) below the top of the jet so that on flooding the carburettor the fuel should flow out of the jet hole.
2 If little or no petrol appears, start the engine (the jet is never completely blocked) and with the throttle butterfly fully open blank off the air intake. This will cause a partial vacuum in the choke tube and help suck out any foreign matter from the jet tube. Release the throttle as soon as the engine speed alters considerably. Repeat this procedure several times, stop the engine and then check the carburettor as described in paragraph 1.
3 If this fails to do the trick then there is no alternative but to remove and blow out the jet.

13 Carburettor (HS6) – jet centering

1 This operation is always necessary if the carburettor has been dismantled, but to check if this is necessary on a carburettor in service, first screw up the jet adjusting nut as far as it will go without forcing it, lift the piston and then let it fall under its own weight. It should fall onto the bridge making a soft metallic click. Now repeat the above procedure but this time with the adjusting nut screwed right down. If the soft metallic click is not audible in either of the two tests proceed as follows.
2 Disconnect the jet link from the bottom of the jet, and the nylon flexible tube from the underside of the float chamber. Gently slide the jet and the nylon tube from the underside of the carburettor body. Next unscrew the jet adjusting nut and lift away the nut and the locking spring. Refit the adjusting nut without the locking spring and screw it up as far as possible without forcing. Refit the jet and tube, but there is no need to reconnect the tube.
3 Slacken the jet locking nut so that it may be rotated with the fingers only. Unscrew the piston damper and lift away the damper. Gently press the piston down onto the bridge and tighten the locknut. Lift the piston using the lifting pin and check that it is able to fall freely under its own weight. Now lower the adjusting nut and check once again. If this time there is a difference in the two metallic clicks, repeat the centering procedure until the sound is the same for both tests.
4 Gently remove the jet and unscrew the adjusting nut. Refit the locking spring and jet adjusting nut. Top-up the damper with oil, if necessary, and refit the damper. Connect the nylon flexible tube to the underside of the float chamber and finally reconnect the jet link.

14 Carburettor (HS6) – float chamber fuel level adjustment

1 It is essential that the fuel level in the float chamber is always correct otherwise excessive fuel consumption may occur. Carburettors fitted to later models have non adjustable floats.
2 With the carburettor fitted to the engine and the float chamber full of petrol remove the piston and dashpot assembly (early models only).
3 Check that the level of fuel in the jet is about $\frac{1}{16}$ in (1·5 mm) below the top of the jet. If it is above or below this level it may be adjusted by removing the needle and seat from the underside of the float chamber lid and either adding or removing washers, so raising or lowering the relative position of the needle valve.

15 Carburettor (HS6) – needle renewal

1 Should it be necessary to fit a new needle, first remove the piston and suction chamber assembly, making the chamber for correct reassembly in its original position.
2 Slacken the needle clamping screw and withdraw the needle, guide and spring from the underside of the piston (Fig. 3.8).
3 To refit the needle assembly fit the spring and guide to the needle and insert the assembly into the piston making sure that the guide is fitted flush with the face of the piston and the flat on the guide positioned adjacent to the needle guide locking screw. Screw in the guide locking screw.

16 Carburettor (HIF6) – dismantling, inspection and reassembly

1 Assuming that you have the carburettor on the workbench, start by cleaning the exterior thoroughly with paraffin or a degreasing solvent, using a stiff brush where necessary.
2 Undo the cap at the top of the carburettor and withdraw it complete with the small damper piston. Empty the oil from the dashpot.
3 Mark the position of the float chamber cover relative to the body and remove it by unscrewing the four screws holding it down. Empty out any fuel still in the fuel chamber.
4 The float is held to the body by a pivot having a screw head on it. Unscrew and remove the pivot with its sealing washer, remove the float, unscrew the needle valve socket and remove it and the needle.
5 Dismantle the various control linkages, being sure by studying Figs. 3.6, 3.9 and 3.12 that you know how they fit together. It is an easy matter to sort this out before you take them apart but much more difficult when they are in bits.

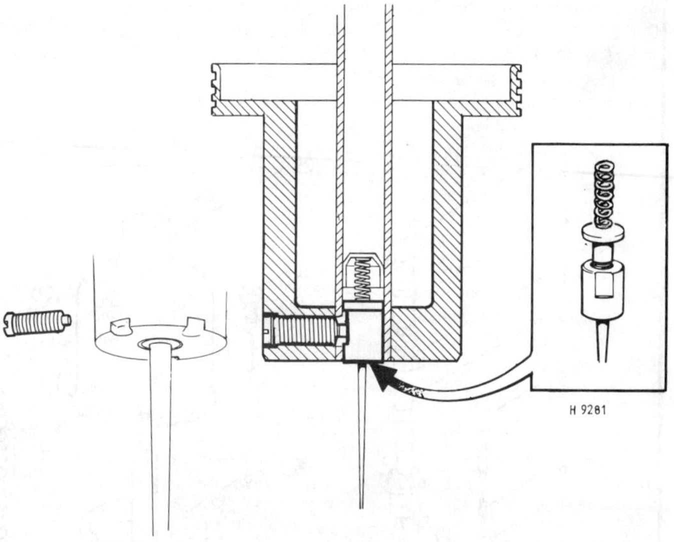

Fig. 3.8 Needle location in piston (Sec 15)

6 Unscrew the nut holding the fast idle cam, having first straightened its tab washer; take off the cam, and the spring which is contained in a small housing behind it. Undo the two screws holding down this housing and pull on the spindle which held the fast idle cam and the whole cold start assembly will come out of the body.
7 Undo the screws holding the throttle disc into its shaft, being careful not to put too much pressure on the shaft in the process (support it with the other hand). Remove the disc and withdraw the throttle shaft.
8 Mark the flanges and remove the top part of the body (suction chamber) and the piston. Be careful of the needle on the end of the piston - a good idea is to stand the piston on a narrow-necked jar with the needle hanging inside it.
9 Unscrew the jet retaining (pivot) screw and remove the bi-metal assembly holding the jet.
10 The carburettor is now sufficiently dismantled for inspection to be carried out. One or two adjusting screws and the like have been left in the body, but it is recommended that these are only removed when you are actually ensuring that the various channels are clear. Generally speaking, the SU carburettor is very reliable but even so it may develop faults which are not readily apparent unless a careful inspection is carried out, yet may nevertheless affect engine performance. So it is well worthwhile giving the carburettor a good look over when you have got it dismantled.
11 Inspect the carburettor needle for ridging. If this is apparent you will probably find corresponding wear on the inside of the jet. If the needle is ridged, it must be renewed. Do not attempt to rub it down with abrasive paper as carburettor needles are made to very fine tolerances.
12 When fitting the needle locate it carefully in the piston. The shoulder should be flush with the piston face and the engraved line should point directly away from the channel in the piston sidewall. Note that this makes the needle incline in the direction of the carburettor air cleaner flange when the piston is fitted.
13 Inspect the jet for wear. Wear inside the jet will accompany wear on the needle. If any wear is apparent on the jet, renew it. It may be unhooked from the bi-metal spring and this may be used again.
14 Inspect the piston and the carburettor body (suction chamber) carefully for signs that these have been in contact. When the carburettor is operating the main piston should not come into contact with the carburettor body. The whole assembly is supported by the rod of the piston which slides in the centre guide tube, this rod being attached to the cap in the top of the carburettor body. It is possible for wear in the centre guide to allow the piston to touch the wall of the body. Check for this by assembling the small piston in the carburettor body and sliding the large one down, rotating it about the centre guide tube at the same time. If contact occurs and the cause is worn parts, renew them. In no circumstances try to correct piston sticking by altering the tension of the return spring, although very slight contact with the body may be cured – as a temporary measure – by polishing the

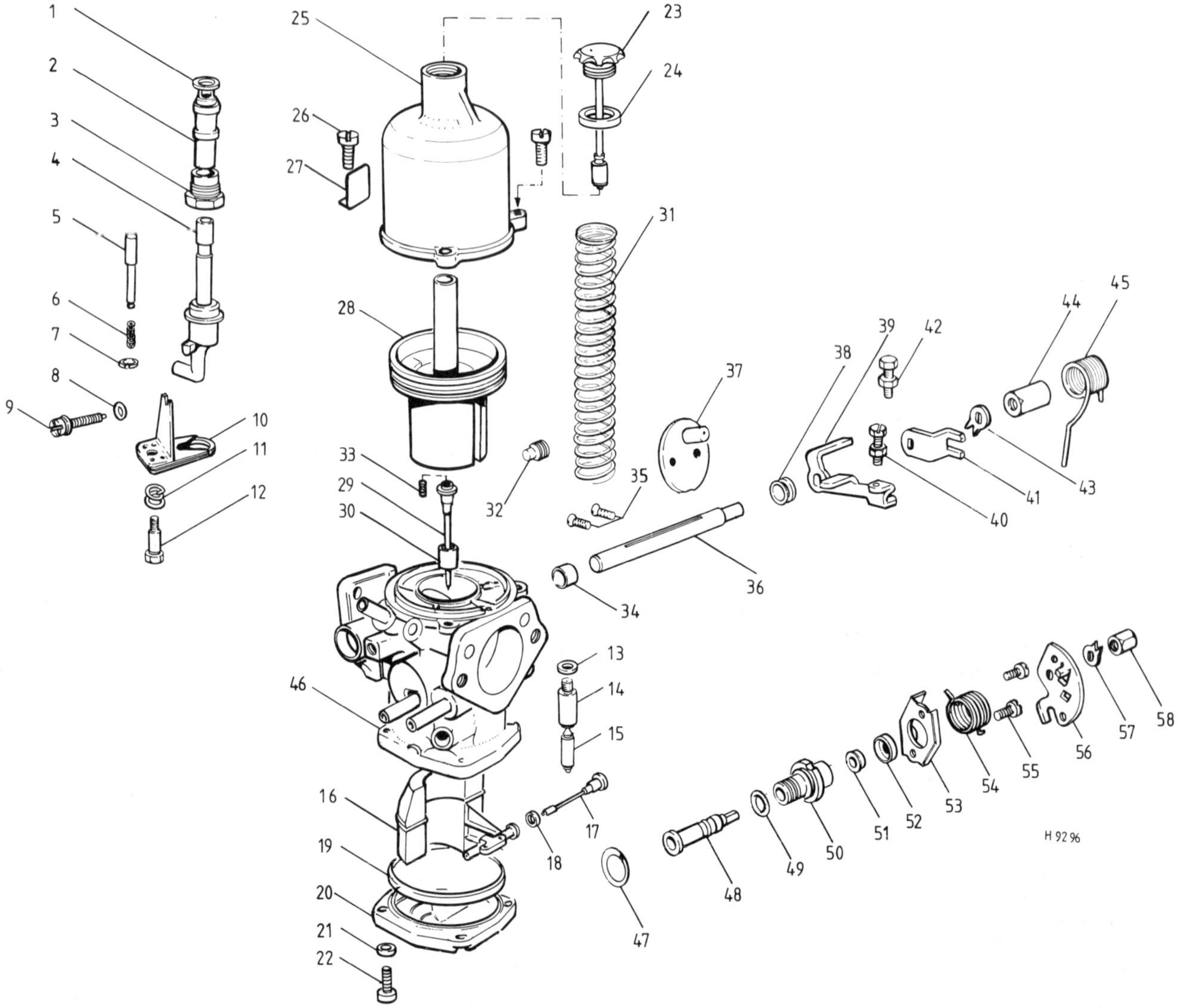

Fig. 3.9 Exploded view of the SU HIF6 carburettor (Sec 16)

1 Jet bearing washer
2 Jet bearing
3 Jet bearing nut
4 Jet assembly
5 Lifting pin
6 Lifting pin spring
7 Circlip
8 Adjusting screw seal
9 Jet adjusting screw
10 Bi-metal jet lever
11 Jet spring
12 Jet retaining (pivot) screw
13 Needle seat washer (if required)
14 Float needle seat
15 Float needle
16 Float
17 Float pivot
18 Pivot seal
19 Float chamber cover seal
20 Float chamber cover
21 Spring washer
22 Screw (4)
23 Piston damper
24 Damper washer
25 Suction chamber
26 Screw (3)
27 Identity tag
28 Piston
29 Jet needle
30 Needle guide
31 Piston spring
32 Needle retaining screw
33 Needle spring
34 Throttle spindle seal
35 Throttle disc screws
36 Throttle spindle
37 Throttle disc
38 Throttle spindle seal
39 Throttle actuating lever
40 Fast idle screw and nut
41 Throttle lever
42 Throttle adjusting screw and nut
43 Tab washer
44 Nut
45 Throttle spring
46 Body
47 Cold start seal
48 Cold start spindle
49 O-ring
50 Cold start body
51 Spindle seal
52 End cover
53 Retaining plate
54 Cold start spring
55 Retaining screw
56 Fast idle cam
57 Tab washer
58 Nut

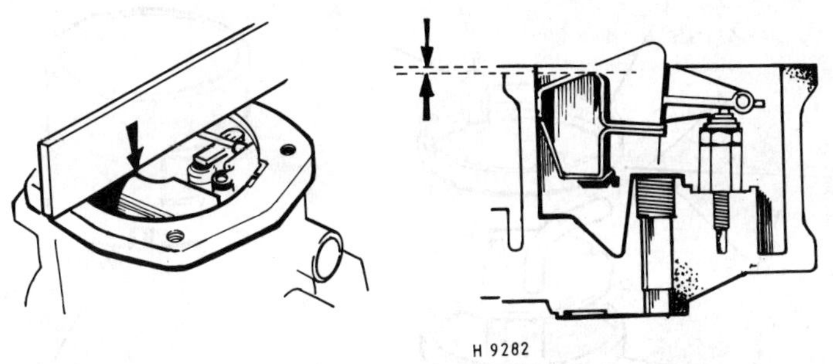

Fig. 3.10 SU HIF6 float level setting (Sec 16)
For correct value of gap (arrowed) see text

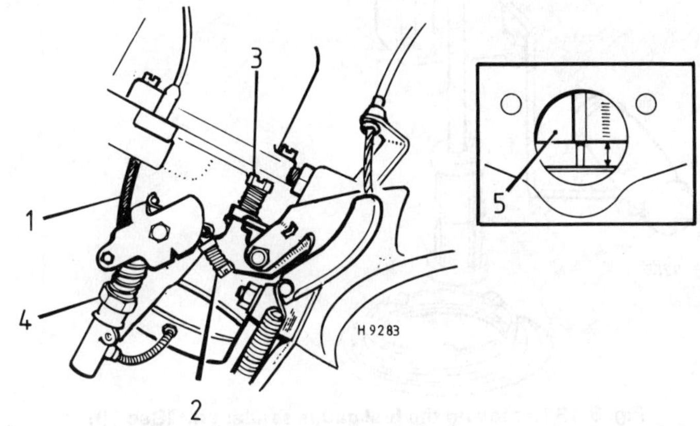

Fig. 3.11 Adjusting the HS6 carburettor (Sec 17)

1 *Mixture control (choke) cable*
2 *Fast idling adjustment screw*
3 *Throttle stop screw*
4 *Jet adjusting nut*
5 *Check free movement of the piston*

offending portion of the body wall with metal polish or extremely fine emery cloth.

15 The fit of the piston in the suction chamber can be checked by plugging the air hole in the body and assembling the piston in the chamber without its return spring, fitting the damper piston without filling the dashpot with oil. If the assembly is now turned upside down the chamber should fall to the bottom in 5-7 seconds. If the time is appreciably less than this, the piston and suction chamber should both be renewed since they are matched to each other.

16 Check for wear on the throttle shaft and bushes through which it passes. Apart from the nuisance of a sticking throttle, excessive wear here can cause air leaks in the induction system adversely affecting engine performance. Worn bushes can be extracted and new bushes fitted if necessary. The cold start device can be dismantled for cleaning, and new parts used where necessary when reassembling.

17 Reassembly is a straightforward reversal of the dismantling process. During reassembly the float level can be checked and adjusted if necessary. Invert the carburettor body and check that the needle valve is held shut by the weight of the float. Using a straightedge across the face of the float chamber measure the gap at the point arrowed (Fig. 3.10). It should be 0·04 ± 0·02 in (1·0 ± 0·5 mm). The arm can be bent carefully, if necessary, to obtain the dimensions.

18 When assembling the jet, position the adjusting screw so that the upper edge of the jet comes level with the bridge. This gives the initial position for jet adjustment.

19 When the carburettor is assembled the dashpot should be fitted with engine oil until it is $\frac{1}{2}$ in (13 mm) above the top of the hollow piston rod. Check that the piston is operating properly by lifting it with the lifting pin and letting it fall. It should hit the bridge of the carburettor with an audible metallic click. If it does not, perhaps the needle is fouling the jet (it is supposed to touch it lightly). This should not occur with careful assembly; there is no provision for centering the jet but if it is properly assembled this is not necessary.

17 Carburettor – adjustment and tuning

1 Before attempting to adjust the carburettor it is essential that the ignition timing, valve clearances, distributor contact breaker and plug gaps are all correct as these all affect engine tune. Carburettor adjust-

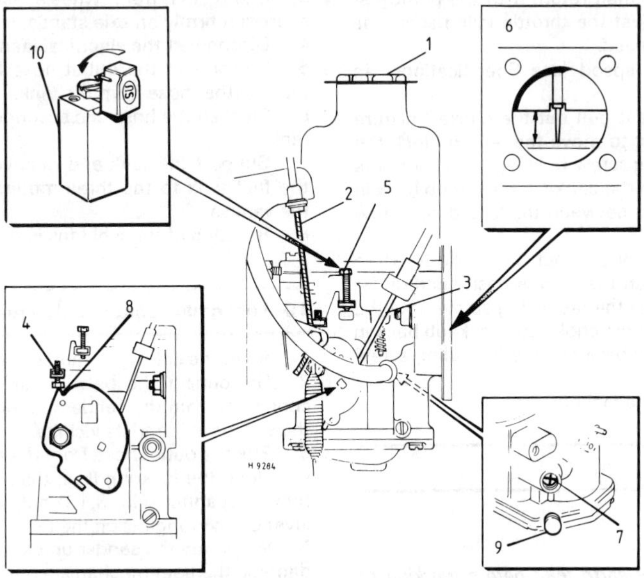

Fig. 3.12 Adjusting the HIF6 carburettor (Sec 17)

1 *Piston damper*
2 *Throttle cable*
3 *Mixture control (choke) cable*
4 *Fast idling adjustment screw*
5 *Throttle stop screw*
6 *Check free movement of the piston*
7 *Jet adjusting screw*
8 *Arrow mark on cam*
9 *Seal, mixture adjustment screw*
10 *Seal, idling adjustment screw*

ments are confined to setting the idle and fast idle speeds and the mixture strength at idle speed.
2 As a result of international legislation in respect of emission control many carburettors are sealed by the vehicle manufacturer, or by authorised service agents, after the carburettors have been adjusted. In some countries it is illegal for the home mechanic to interfere with these seals, although this is not yet the case in the UK. For accurate tuning a tachometer is indispensable and it is recommended that one be borrowed or hired if such a useful instrument is not already owned.
3 Check the oil level in the piston damper and top-up if necessary. Check that the throttle mechanical function is correct.
4 Ensure that the choke (mixture control) will return fully and that the cable has $\frac{1}{16}$ in (2 mm) free play before it starts to pull on the lever.
5 Check that a small clearance exists between the fast idling screw and the cam.
6 Remove the air cleaner as described in Section 2. Raise and lower the piston with a finger, checking that it can move smoothly without any tendency to stick or bind. If the piston sticks the carburettor should be overhauled as described in the relevant Section. Refit the air cleaner.
7 Connect up a tachometer in accordance with the maker's instructions and, on automatic transmission models, select P. Apply the handbrake and start the engine. Run it at a fast idling speed until it is properly warmed up and continue to run it for a further five minutes.
8 Before making any adjustments, increase the engine speed to 2500 rpm and maintain this speed for about half a minute to clear the inlet manifold of any excess fuel. Repeat this procedure at three-minute intervals if these adjustments cannot be completed in less time.
9 Check the engine idling speed in the Specifications and if necessary adjust by turning the throttle adjusting screw.
10 If the engine will not run smoothly at the correct idling speed, adjust the mixture setting as follows.
11 *HS6 carburettors.* Turn the jet adjusting nut one flat at a time,up to weaken or down to enrich the mixture, until the fastest idling speed is achieved, consistent with smooth running. Turn the nut up slowly (weakening) until the speed just starts to fall, then turn the nut down again very slowly until the maximum idling speed is regained. Re-adjust the throttle idling screw to obtain the specified idling speed.
12 *HIF6 carburettors.* Turn the adjusting screw clockwise to enrich or anti-clockwise to weaken the mixture strength until the fastest engine speed is obtained. Turn the screw about half a turn anti-clockwise until the speed just commences to fall and then return it to the point just before the speed began to fall. Re-adjust the throttle idling screw as required to obtain the specified idling speed.
13 Check and adjust the fast idling speed (see Specifications) as folows.
14 *HS6 carburettors.* Have an assistant pull out the choke (mixture control) knob until the linkage is about to move the jet, and lock the knob in this position. Turn the fast idling screw (2 in Fig. 3.11) until the specified idling speed is obtained. Push the choke control knob back in and check that a small clearance exists between the fast idling screw and the cam.
15 *HIF6 carburettors.* Have an assistant pull out the choke (mixture control) knob until the arrow marked on the cam is positioned under the fast idle screw (4 in Fig. 3.12). Turn the fast idling screw until the specified idling speed is obtained. Push the choke control knob back in and check that a small clearance exists between the fast idling screw and the cam.
16 Stop the engine and disconnect the tachometer.

18 Fuel tank – removal and refitting

Safety measures:

(a) Always disconnect the battery earth lead before working on the fuel tank
(b) Never allow a naked flame, smouldering cigarette or electrical apparatus within the vicinity of the fuel tank – especially while you are working on it.
(c) Be careful not to create sparks by vigorous use of metal tools
(d) Do not attempt to repair holes or faulty seams in the tank yourself – even if the tank has been empty for months. Have the tank repaired by a specialist or get a new one

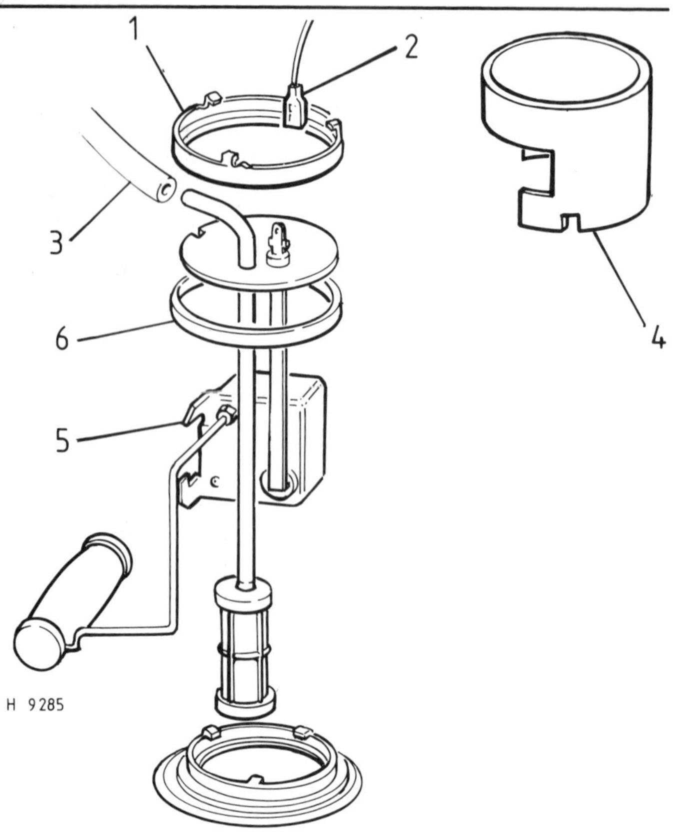

Fig. 3.13 Removing the fuel gauge sender unit (Sec 19)

1 Locking ring
2 Electrical lead
3 Outlet hose
4 Special tool for removing locking ring
5 Sender unit assembly
6 Gasket

1 Disconnect the battery earth lead.
2 Remove the drain plug and washer from the underside of the tank and drain the fuel into a sealable metal container.
3 Chock the front wheels, jack-up the rear of the vehicle, and support it firmly on axle stands.
4 Disconnect the electrical lead from the fuel gauge sender unit.
5 Disconnect the outlet hose from the fuel gauge sender unit and the breather hose from the tank.
6 Slacken the hose clip securing the filler neck connector hose to the tank.
7 Support the tank and remove the six bolts and washers securing the fuel tank to the three mounting brackets. Remove the tank from the vehicle.
8 Refitting of the fuel tank is the reverse of the removal procedure.

19 Fuel gauge sender unit – removal and refitting

1 Read the safety measures contained at the start of Section 18.
2 Disconnect the battery earth lead and disconnect the electrical connector from the sender unit on the tank. Clean off all road dirt from the sender unit and its vicinity.
3 Pull the outlet hose from the sender unit.
4 Undo the lockring from the sender unit. This really needs a special tubular spanner (Fig. 3.13) but, with care, it can be removed using a lever on the tongues on the ring.
5 Withdraw the sender unit assembly from the tank. Take care not to damage the float mechanism during removal. Remove the gasket.
6 Refitting is the reverse of the removal procedure; use a new gasket if the original is damaged.

20 Exhaust system – general

1 The exhaust system (excluding the manifold) comprises a front pipe and silencer assembly, an intermediate pipe and silencer

20.1a The exhaust manifold-to-pipe joint on the 16V and 18V engines

20.1b The exhaust manifold-to-pipe joints on the 17V engine

20.1c Two types of flexible support for the exhaust system, a strap-hanger and ...

20.1d ... a rubber cushion (arrowed)

21.5 The exhaust pipe support bracket on the 17V engine clutch bellhousing flange

assembly and a tail-pipe. The size and shape of these components varies depending on the model concerned. The manifold-to-pipe joint on 16V and 18V engine models consists of butted flanges connected by a two-piece clamp ring; on 17V engine models twin outlets are joined to twin downpipes by flange joints with sealing rings retained by nuts on studs in the manifold (photos). All other joints in the system are male and female push-fit type, secured by clamps. The whole assembly is supported by flexible mountings to the chassis (photos).

21 Exhaust system – removal and refitting

1 Even if only one component of the exhaust system is to be renewed, it is much easier, and probably quicker in the long run, to remove the complete system from the vehicle unless only the tail-pipe is being changed.
2 To remove the tail-pipe first chock the front wheels, jack-up the rear of the vehicle and support it firmly on stands.
3 Support the tail-pipe assembly and detach the support stays.
4 Slacken the clamp securing the tail-pipe to the silencer, or to the front pipe, depending on models. Remove the tail-pipe by swivelling and pulling the pipe off. Penetrating oil and a mallet may be required if the parts are heavily corroded.
5 The remainder of the system can be removed starting at the manifold end. Release the manifold clamp or undo the flange nuts,

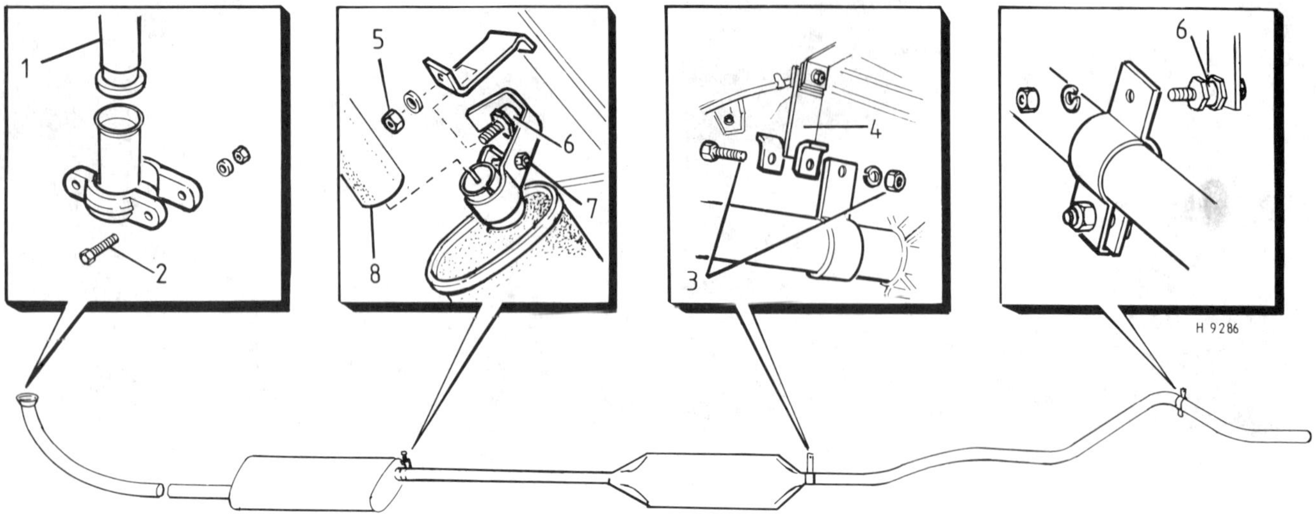

Fig. 3.14 Typical exhaust system fittings (Sec 21)

1 Manifold pipe (16V and 18V engines)
2 Manifold clamp bolt (16V and 18V engines)
3 Attachment bolt and nut
4 Flexible support hanger
5 Cushion attachment nut
6 Cushion support
7 Clamp bolt and nut
8 Exhaust pipe

depending on model, and support the front end of the exhaust pipe. Remove the supporting stays and remove the front and intermediate assemblies (photo).

6 If it is intended to repair small holes in any of the components, several proprietary materials are available. However these can only be considered a temporary repair, as, when one hole appears in an exhaust system it is certain that others will soon follow.

7 Reasembly and refitting of the exhaust system, using new components where necessary, is the reverse of the removal and dismantling sequence but note the following additional points:

(a) After refitting the system, run the engine for at least five minutes before finally tighteneing the clamps. This will relieve stresses in the system which could otherwise cause early failures
(b) Do not overtighten pipe clamps as this could cause distortion
(c) Use new sealing rings on the downpipe joints in 17V engine systems

22 Fault diagnosis

Unsatisfactory engine performance and excessive fuel consumption are not necessarily the fault of the fuel system or carburettor. In fact they more commonly occur as a result of ignition faults. Before acting on the fuel system it is necessary to check the ignition system first. Even though a fault may lie in the fuel system it will be difficult to trace unless the ignition is correct.

The table below therefore, assumes that the ignition system is in order.

Symptom	Reason/s
Smell of petrol when engine is stopped	Leaking fuel lines or unions Leaking fuel tank
Smell of petrol when engine is idling	Leaking fuel line unions between pump and carburettor Overflow of fuel from float chamber due to wrong level setting or ineffective needle valve or punctured float
Excessive fuel consumption for reasons not covered by leaks or float chamber faults	Worn needle Sticking needle
Difficult starting, uneven running, lack of power, cutting out	One or more blockages Float chamber fuel level too low or needle sticking Fuel pump not delivering sufficient fuel Intake manifold gaskets leaking, or manifold fractured
Fast idle; erratic	Air leak

Chapter 4 Ignition system

For modifications, and information applicable to later models, see Supplement at end of manual

Contents

Specifications

Spark plugs

16V and 18V engines

Make and type	Champion RN-9Y or AC Delco R42 XLS
Gap	0·035 in (0·89 mm)

17V engines

Make and type	Champion BN-9Y
Gap	0·035 in (0·89 mm)

Firing order

1, 3, 4, 2

Ignition coil

16V engines	Lucas LA12
18V engines	Lucas 11C12
17V engines	Lucas 16C6

Distributor

16V engines

Make and type	Lucas 45D4
Direction of rotation	Anti-clockwise
Dwell angle	46° to 56°
Serial number	41579
Contact breaker gap	0·014 to 0·016 in (0·35 to 0·40 mm)
Contact spring tension	18 to 24 oz (510 to 680 gm)
Centrifugal advance:	
Deceleration check *†	18° to 22° at 4000 rpm
	11° to 15° at 2800 rpm
	4° to 8° at 1600 rpm
	0° to 3° at 800 rpm
No advance below	300 rpm
Vacuum advance:	
Starts	6 in (152 mm) Hg
Finishes	14 in (356 mm) Hg
Total crankshaft degrees	14° to 18°

18V Low Compression engines

Make and type	Lucas 45D4
Direction of rotation	Anti-clockwise
Dwell angle	46° to 56°
Serial number:	
Temperate and cold climates	41566
Hot climates	41675

Contact breaker gap	0·014 to 0·016 in (0·35 to 0·40 mm)
Contact spring tension	18 to 24 oz (510 to 680 gm)
Centrifugal advance:	
Temperate and cold climates:	
Deceleration check*†	24° to 28° at 5200 rpm 20° to 24° at 3800 rpm 17° to 21° at 2800 rpm 12° to 16° at 2000 rpm 4° to 8° at 1200 rpm 0° to 4° at 800 rpm
No advance below	400 rpm
Hot climates:	
Deceleration check *†	28° to 32° at 5000 rpm 20° to 24° at 3400 rpm 14° to 18° at 2400 rpm 2° to 6° at 1200 rpm
No advance below	600 rpm
Vacuum advance:	
Temperate and cold climates:	
Starts	3 in (76 mm) Hg
Finishes	8 in (203 mm) Hg
Total crankshaft degrees	12° to 16°
Hot climates:	
Starts	3 in (76 mm) Hg
Finishes	13 in (330 mm) Hg
Total crankshaft degrees	22° to 26°

18V High compression engine

Make and type	Lucas 45D4
Direction of rotation	Anti-clockwise
Dwell angle	46° to 56°
Serial number	41415
Contact breaker gap	0·014 to 0·016 in (0·35 to 0·40 mm)
Contact spring tension	18 to 24 oz (510 to 680 gm)
Centrifugal advance:	
Deceleration check *†	28° to 32° at 5200 rpm 20° to 24° at 3100 rpm 10° to 14° at 2200 rpm 2° to 6° at 1200 rpm 0° to 1° at 600 rpm
No advance below	400 rpm
Vacuum advance:	
Starts	4 in (102 mm) Hg
Finishes	12 in (305 mm) Hg
Total crankshaft degrees	14° to 18°

17V engines

Make and type	Lucas 48D4
Direction of rotation	Anti-clockwise
Dwell angle	52° to 62°
Serial number	41740
Contact breaker gap	0·014 to 0·016 in (0·35 to 0·40 mm)
Condenser capacity	0·18 to 0·25 microfarad
Centrifugal advance:	
Deceleration check *†	32° to 36° to 5600 rpm 24° to 28° at 3600 rpm 18° to 22° at 2700 rpm 2° to 6° at 1400 rpm
No advance below	900 rpm
Vacuum advance:	
Maximum *	16° at 15 in (387 mm) Hg
Starts	7 in (178 mm) Hg

Ignition timing

Static timing for assembly:	
16V engines	11° BTDC
18V engines:	
Low compression:	
Temperate and cold climates	13° BTDC
Hot climates	12° BTDC
High compression	3° BTDC
17V engines (suggested)	7° BTDC
Stroboscopic timing:	
16V engines	16° BTDC at 1000 rpm

18V engines:	
Low compression:	
Temperate and cold climates	20° BTDC at 1000 rpm
Hot climates	17° BTDC at 1000 rpm
High compression	8° BTDC at 1000 rpm
17V engines	13° BTDC at 1500 rpm

** Crankshaft degrees and rpm*
† Vacuum pipe disconnected

Torque wrench settings

16V and 18V engines	lbf ft	Nm
Spark plugs	18	24
Distributor clamp bolt:		
Fixed nut type	5	6
Fixed bolt type	2·5	3
17V engines		
Spark plugs – hand tight plus $\frac{1}{16}$ turn	7	10
Distributor flange nuts	18	25
Screws not specified:		
Into aluminium M6	8	11
M8	18	25
Into cast iron M6	8	11
M8	22	30
M10	37	50

1 General description

1 In order that the engine may run correctly it is necessary for an electrical spark to ignite the fuel/air mixture in the combustion chamber at exactly the right moment in relation to engine speed and load. The ignition system is based on supplying low tension voltage from the battery to the ignition coil, where it is converted into high tension voltage. The high tension voltage is powerful enough to jump the spark plug gap in the cylinders many times a second under high compression pressure, providing that the ignition system is in good working order and that all adjustments are correct.
2 The ignition system comprises two individual circuits known as the low tension and high tension circuits.
3 The low tension circuit (sometimes known as the primary circuit) comprises the battery, lead to the ignition switch, to the low tension or primary coil windings (terminal +) and the lead from the low tension coil windings (terminal –) to the contact breaker points and condenser in the distributor.
4 The high tension (secondary circuit) comprises the high tension or secondary coil windings, the heavily insulated ignition lead from the centre of the coil to the centre of the distributor cap, the rotor arm, the spark plug leads and the spark plugs.
5 The complete ignition system operation is as follows. Low tension voltage from the battery is changed within the ignition coil to high tension voltage by the opening and closing of the contact breaker points in the low tension circuit. High tension voltage is then fed via the carbon brush in the centre of the distributor cap to the rotor arm of the distributor. The rotor arm revolves inside the distributor cap and each time it comes into line with one of the four metal segments in the cap, these being connected to the spark plug leads, the opening and closing of the contact breaker points causes the high tension voltage to build up, jump the gap from the rotor arm to the appropriate metal segment and so, via the spark plug lead, to the spark plug where it finally jumps the gap between the two spark plug electrodes, one being connected to earth.
6 The ignition timing is advanced and retarded automatically to ensure the spark occurs at just the right instant for the particular load at the prevailing engine speed.
7 The ignition advance is controlled by a mechanical and vacuum operated system. The mechanical governor mechanism comprises two weights which move out under centrifugal force from the central distributor shaft as the engine speed rises. As they move outwards they rotate the cam relative to the distributor shaft, and so advance the spark. The weights are held in position by two springs, and it is the tension of the springs which is largely responsible for correct spark advancement.
8 The vacuum control comprises a diaphragm, one side of which is connected via a small bore tube to the carburettor, and the other side to the contact breaker plate. Depression in the induction manifold and carburettor, which varies with engine speed and throttle opening, causes the diaphragm to move, so moving the contact breaker plate and advancing or retarding the spark. A fine degree of control is achieved by a spring in the vacuum assembly.

2 Contact breaker points – adjustment

1 To adjust the contact breaker points so that the correct gap is obtained, first release the two clips securing the distributor cap to the distributor body, or remove the two screws securing the cap to the body, and lift away the cap. Clean the inside and outside of the cap with a dry cloth. It is unlikely that the four segments will be badly burned or scored, but if they are the cap must be renewed. If only a small deposit is on the segments it may be scraped away using a small screwdriver.
2 Push in the carbon brush, located in the top of the cap, several times to ensure that it moves freely. The brush should protrude at least $\frac{1}{4}$ in (6.35 mm).
3 Gently prise the contact breaker points open to examine the condition of their faces. If they are rough, pitted or dirty it will be necessary to remove them for resurfacing, or for new points to be fitted.
4 Presuming the points are satisfactory, or that they have been

2.5 Using a feeler gauge to check the contact breaker points gap (17V engine)

cleaned and refitted, measure the gap between the points by turning the crankshaft to bring the contact breaker heel onto the peak of one of the four cam lobes as follows:

(a) Manual gearbox: engage second gear, release the handbrake and move the vehicle forwards or backwards. Reapply the handbrake
(b) Automatic transmission: Select N and with the handbrake on use a spanner on the crankshaft pulley nut to turn it clockwise

5 With the points fully open a feeler gauge corresponding to the specified contact breaker gap should be a sliding fit between the points (photo). Take care not to contaminate the points with oil.
6 If the gap varies from this amount, slacken the contact plate securing screw and adjust the contact gap by inserting a screwdriver in the notched hole at the end of the plate, turning clockwise to decrease, and anti-clockwise to increase the gap. Tighten the securing screw and recheck the gap.
7 Refit the rotor arm and distributor cap and clip the spring blade retainers into position, or refit the retaining screws.
8 On modern engines, setting the points gap with a feeler gauge must be regarded as an inferior method. For optimum engine performance, the dwell angle should be checked. The dwell angle is the number of degrees through which the distributor cam turns during the period between the instants of closure and opening of the contact breaker points. The dwell angle not only provides a more accurate setting of the contact breaker gap but this method also evens out any variations in gap caused by wear in the distributor shaft or bushes, or differences in height of any of the four cam lobes.
9 Dwell angle can only be checked on a dwell meter connected in accordance with the manufacturer's instructions.
10 The correct angle is as given in the Specifications according to engine and distributor type. Where the dwell angle is found to be too large, increase the points gap; if the angle is too small, reduce the gap.

3 Contact breaker points – removal and refitting

1 If the contact breaker points are burned, pitted, or badly worn, they must be removed and a new set fitted.
2 To remove the points, take off the distributor cap and remove the rotor arm by pulling it off the driveshaft. Unscrew the contact set securing screw and lift it from the baseplate assembly.
3 Disconnect the terminal plate (with the two wires on) from the contact breaker spring and discard the old contact set. Retain the securing screw with its spring and flat washers.
4 Wipe the new points with clean rag and petrol or methylated spirit.
5 Connect the terminal plate to the end of the contact breaker spring with the black cable uppermost.
6 Place the securing screw with its two washers in the contact set slot. Locate the fork in the contact set over the peg in the baseplate, press the pivot post into the baseplate and lightly tighten the securing screw.
7 Check that the terminal end of the contact spring is located in the insulation shoe.
8 Adjust the points gap, refer to Section 2, and lubricate the assembly, refer to Section 5. After the next 500 miles (800 km) remember to check the gap, as the cam follower will have bedded in causing the gap to decrease.

4 Condenser – removal, testing and refitting

1 The purpose of the condenser (capacitor) is to ensure that when the contact breaker points open there is no sparking across them which would waste voltage and cause wear.
2 The condenser is fitted in parallel with the contact breaker points. If it develops a short circuit, it will cause ignition failure, as the points will be prevented from interrupting the low tension circuit.
3 If the engine becomes very difficult to start, or begins to miss after several miles running, and the breaker points show signs of excessive burning, then the condition of the condenser must be suspect. A further test can be made by separating the points by hand with the ignition switched on. If this is accompanied by a strong blue flash it is indicative that the condenser has failed.
4 Without special test equipment, the only sure way to diagnose condenser trouble is to replace a suspected unit with a new one and note if there is any improvement. They are not expensive.
5 To remove the condenser from the distributor, remove the distributor cap and the rotor arm. Unscrew the condenser attachment screw, disconnect the lead from the assembly, and remove the condenser and lead.
6 Refitting of the condenser is simply a reversal of the removal process. Take particular care that the condenser lead does not short circuit against any portion of the breaker plate.

5 Distributor – lubrication

1 The distributor should be lubricated every 6000 miles (10 000 km) or 3 months, whichever occurs first. It is important not to over-lubricate as malfunctioning could easily result. Surplus lubricant must be wiped away and great care must be taken to keep the contact breaker points clean and dry.
2 Lightly smear the cam with grease or petroleum jelly but do not oil the cam wiping pad.
3 Remove the rotor arm and apply no more than two drops of engine oil to the pad in the top of the cam spindle.
4 Add one or two drops of oil through the gaps in the baseplate to lubricate the advance mechanism.
5 Every 24 000 miles (40 000 km) add a drop of oil to the moving plate bearing groove.
6 Refit the rotor arm by aligning its peg with the slot in the cam and pressing it onto the spindle. Wipe the inside of the cap clean and dry, check that no surplus lubricant is visible in the assembly, and refit the cap on the distributor.

6 Distributor – removal and refitting

16V and 18V engines

1 Disconnect the battery earth and also disconnect the lead from the

6.5 The distributor cap on 16V and 18V engines

6.14 The distributor rotor arm (17V engine) at the No 1 cylinder firing point

6.17 Refitting the distributor (17V engine)

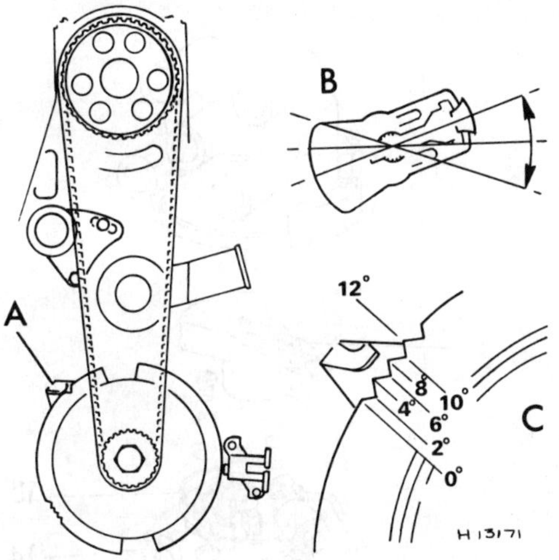

Fig. 4.1 Positioning the 17V engine distributor rotor at the No 1 cylinder firing point (Sec 6)

A Crankshaft timing pointer
B Rotor to be between 2.30 and 3.30 o'clock
C Crankshaft timing marks at specified position

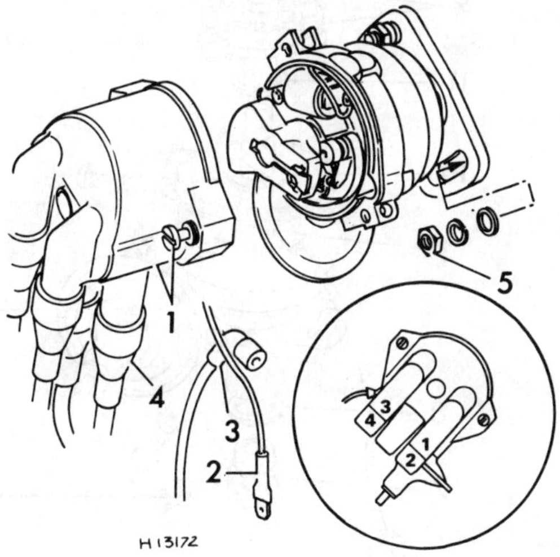

Fig. 4.2 Removing the 17V engine distributor (Sec 6)

1 Cap attachment screws
2 Low tension lead
3 Vacuum pipe
4 Spark plug leads
Inset: Identity of plug leads

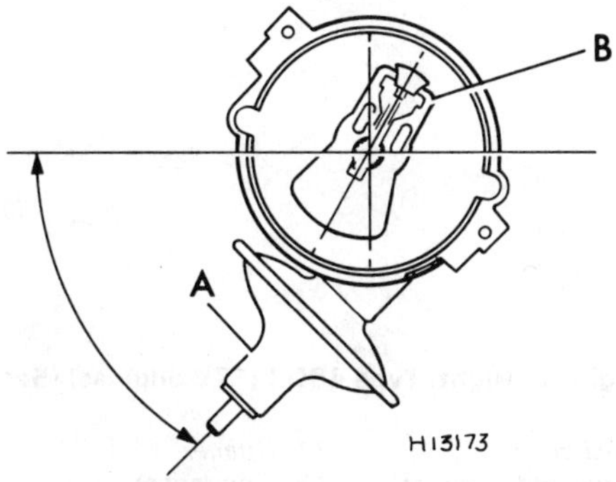

Fig. 4.3 Positioning the 17V engine distributor for refitting (Sec 6)

A Vacuum unit at 45° below horizontal and to the left
B Rotor at 1 o'clock position

oil pressure switch.
2 Disconnect the distributor low tension cable at the Lucar terminal connector.
3 Disconnect the ignition cables from the spark plugs and, if there is a risk of getting them mixed up, identify each cable with plastic tape or similar material so that each cable can be refitted in its correct position.
4 Disconnect the high tension cable from the coil.
5 Release the clips securing the distributor cap and remove the cap, complete with high tension cables, from the distributor (photo).
6 Disconnect the vacuum pipe from the distributor advance unit. Put a mark on the engine in line with the vacuum control unit to facilitate refitting of the distributor.
7 Undo and remove the two screws, spring and plain washers securing the distributor clamp plate to the cylinder block. The distributor may now be lifted up together with the clamp plate still attached.
8 If it is not wished to disturb the ignition timing, then under no circumstances should the clamp pinch-bolt, which secures the distributor in its relative position in the clamp, be loosened. Providing the distributor is removed without the clamp being loosened from the distributor, and the engine is not turned, the ignition timing will not be lost.
9 Refitting is the reverse of the removal procedure. Fit a new O-ring to the distributor body extension. When fitting the distributor, line up the vacuum unit with the mark made on the engine as described in paragraph 6, and rotate the distributor shaft until the off-centre drive dog engages the slot in the distributor driveshaft in the engine. If the engine has been turned with the distributor removed, if the clamp pinch-bolt has been loosened, or if a new distributor is being fitted, the ignition will have to be retimed, see Section 10.

17V engines

10 Disconnect the battery earth lead.
11 Unscrew the retaining screws and remove the distributor cap.
12 Disconnect the distributor low tension cable at the coil (marked –), and disconnect the vacuum pipe at the distributor vacuum unit.
13 With a spanner on the crankshaft pulley nut, turn the crankshaft to the No 1 cylinder firing point at 8° BTDC – the rotor arm in the distributor should be between the 2.30 to 3.30 o'clock position (photo) (Fig. 4.1). If it is not, rotate the crankshaft a further 360°
14 If the same distributor is going to be refitted, to avoid retiming the ignition, put a mark on the distributor mounting flange and an adjacent mark on the camshaft cover. Remove the two nuts securing the distributor to the camshaft cover and withdraw the distributor from its housing.
15 Providing the crankshaft is not turned while the distributor is off, refitting is basically the reverse of the removal procedure. Fit a new O-ring to the distributor body. When refitting the distributor, the shaft will turn due to the meshing of the skew driving gear. To allow for this the shaft must be pre-positioned as described below.
16 Hold the distributor with the vacuum unit at 45° below the horizontal to the left when looking at the rotor arm. Turn the distributor shaft so that the rotor points to the 1 o'clock position and fit the distributor in its housing. As the gears engage the rotor will turn to between the 2.30 and 3.30 o'clock position (photo).
17 Turn the distributor body to align the marks made as described in paragraph 15 and tighten the securing nuts.
18 If the crankshaft is turned while the distributor is removed, or if a new distributor is being fitted, the ignition will have to be retimed, see Section 10.

7 Distributor – dismantling

1 With the distributor removed from the engine, and placed on a bench, remove the rotor arm.
2 Extract the felt lubricating pad from the top of the cam spindle.
3 Undo the two screws retaining the vacuum unit, tilt the unit to disengage the operating arm, and remove the vacuum unit.
4 Push the low tension lead grommet into the inside of the distributor body.
5 Where the baseplate is retained by a single screw, remove the screw. Lever the slotted segment of the baseplate from its retaining groove and lift out the baseplate assembly Where the baseplate is retained by double screws, remove the screws and lift out the baseplate assembly. Note that the screw adjacent to the low tension

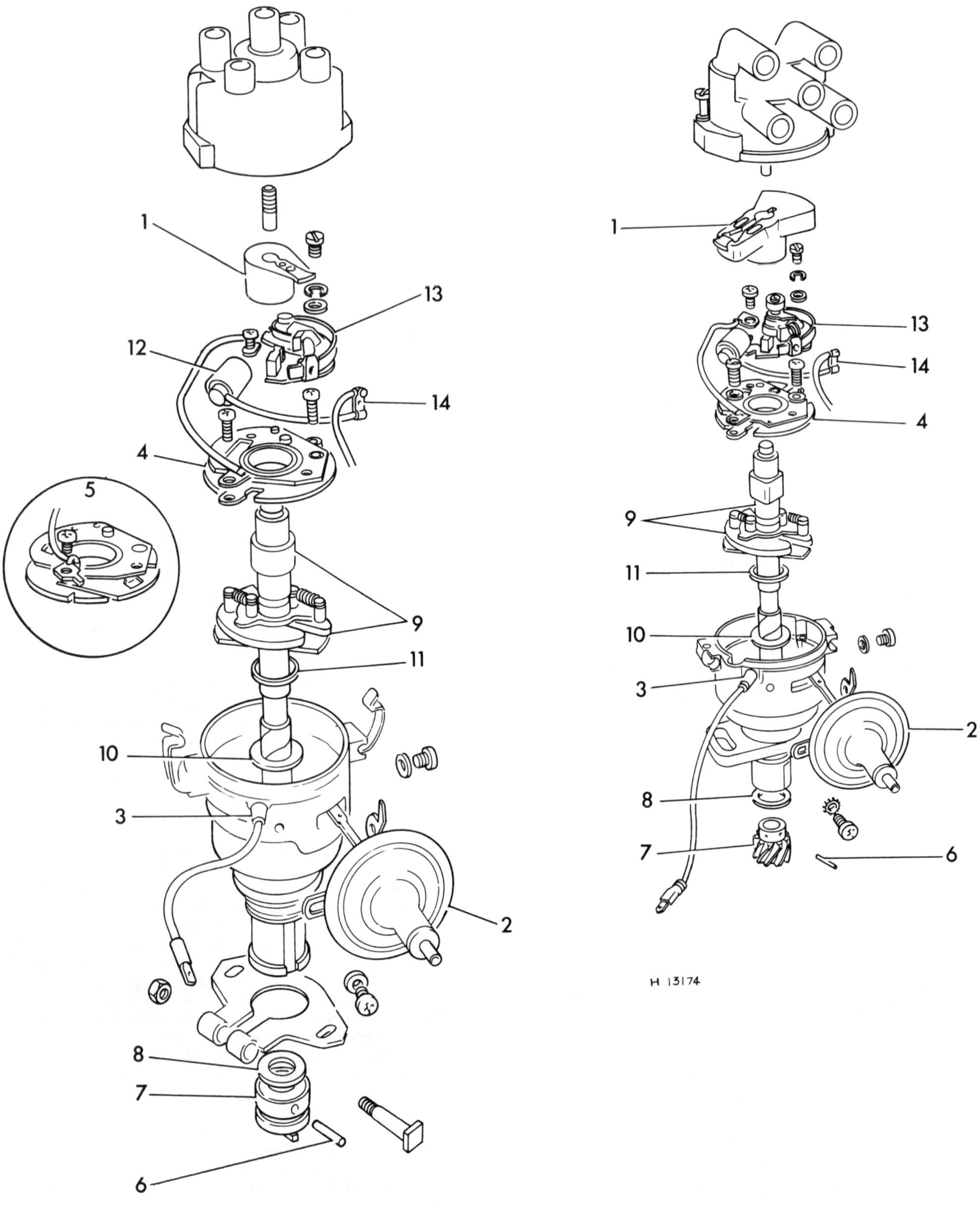

Fig. 4.4 The distributor component parts. Left: Type 45D4 (16V and 18 engines). Right: Type 48D4 (17V engines) (Sec 8)

1 *Rotor arm*
2 *Vacuum unit*
3 *Low tension lead*
4 *Baseplate, twin retaining screw type*
5 *Baseplate, single retaining screw type*
6 *Roll pin, drive dog (left), skew gear (right)*
7 *Drive dog (left), skew gear (right)*
8 *Thrust washer*
9 *Cam spindle and automatic advance weights*
10 *Steel washer*
11 *Spacer*
12 *Condenser*
13 *Contact points set*
14 *Low tension lead connector*

lead also secures the condenser earth lead.
6 Push the contact breaker spring inwards and detach the low tension connector from the spring loop.
7 Remove the earth lead and condenser, secured by one screw.
8 Remove the securing screw and lift off the contact set.
9 Remove the automatic advance control springs. Further dismantling should only be carried out if the drive dog (16V and 18V engines) or drivegear (17V engines) is to be renewed.
10 Drive out the roll pin securing the drive dog (16V and 18V engines) or the drivegear (17V engines).
11 Remove the drive dog or gear and thrust washer.
12 Remove the shaft complete with the automatic advance mechanism, distance collar and steel washer. Remove and discard the O-ring.
13 The distributor is now ready to be inspected.

8 Distributor – inspection and repair

1 Thoroughly wash all mechanical parts in petrol and wipe dry with clean, non-fluffy rag.
2 Check the contact breaker points as described in Section 3. Check the distributor cap for signs of tracking, indicated by a thin black line between the segments. Renew the cap if evident.
3 Check that the pick-up brush moves freely in the centre of the distributor cap. The tip of the brush should have a slightly bevelled edge.
4 Examine the distributor segments in the cap for excessive burning or erosion. Do not clean the segments with glasspaper or emery cloth. If they are in poor condition, renew the cap.
5 Check the rotor arm for damage, security of the metal electrode, burning and tracking. Do not clean the electrode with glasspaper or emery cloth. If the rotor arm is in poor condition, renew it.
6 Examine the baseplate assembly. If the spring between the plates is damaged, or if the plates do not move freely, renew the assembly.
7 Do not dismantle the advance mechanism beyond removing the control springs. If any of the moving parts or the cam are worn or damaged, renew the complete shaft assembly.
8 Check the fit of the shaft in its bearing. If the bearing allows more than just perceptible side movement, renew the complete distributor.

9 Distributor – reassembly

16V and 18V engines

1 Reassembly is reversal of the dismantling procedure, but note the additional points listed below.
2 Lubricate the contact points pivot post with multi-purpose grease.
3 Ensure that the spacer and steel washer are fitted on the shaft and lubricate the shaft with Rocol MP (Molypad) or equivalent.
4 Fit the thrust washer with its raised pips towards the drive dog.
5 Fit the drive dog so that the drive tongues are parallel with the rotor arm electrode and to the left of its centre line when the rotor arm points upwards and when looking at the drive dog end, see Fig. 4.5. If a new shaft is fitted it will have to be drilled through the hole in the drive dog using a $\frac{3}{16}$ in (4.75 mm) drill. Before drilling, push the shaft from the cam end and press the dog and washer against the body shank. Hold in this position while drilling. Tap the pin into position and secure it in the drive dog by ring-punching the holes. If the shaft is new, tap the drive end with a hammer to flatten the washer pips and provide the correct endfloat.
6 Where the baseplate is secured by a single screw, position the assembly so that the two downward-pointing prongs will straddle the screw hole below the cap clip. Press the baseplate into the body until it engages with the undercut. Take an accurate measurement across the spigot on which the distributor cap is located, at right angles to the slot in the baseplate. Position the earth lead and fit and tighten the baseplate securing screw. Take another measurement across the distributor cap locating a spigot. If the measurement has not increased by at least 0.006 in (0.15 mm) the baseplate assembly must be renewed. If satisfactory, check that the two downward-pointing prongs of the baseplate still straddle the screw hole and refit the vacuum unit, making sure that the operating arm is engaged with the moving plate pin (Fig. 4.6).
7 Finally set the contact breaker points as described in Section 2, and lubricate the distributor as described in Section 5. Fit a new O-ring to the distributor body extension.

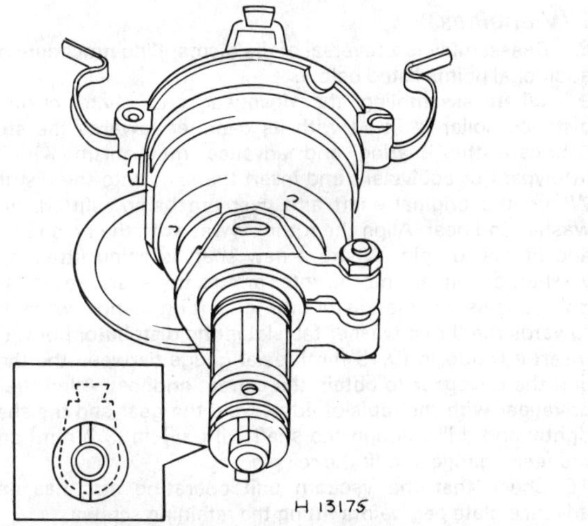

Fig. 4.5 Assembly of the drive dog to the distributor shaft, type 45D4 distributor (16V and 18V engines) (Sec 9)

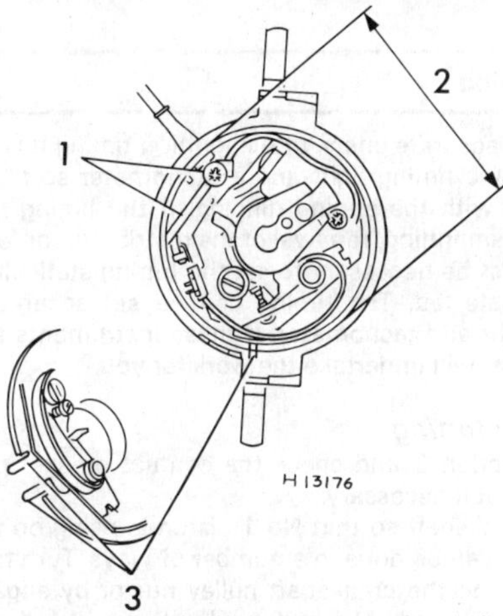

Fig. 4.6 Assembling the single-screw type baseplate to the type 45D4 distributor (16V and 18V engines) (Sec 9)

1 Positioning the earth lead
2 Measure the distributor cap locating spigot
3 Position the two downward pointing prongs

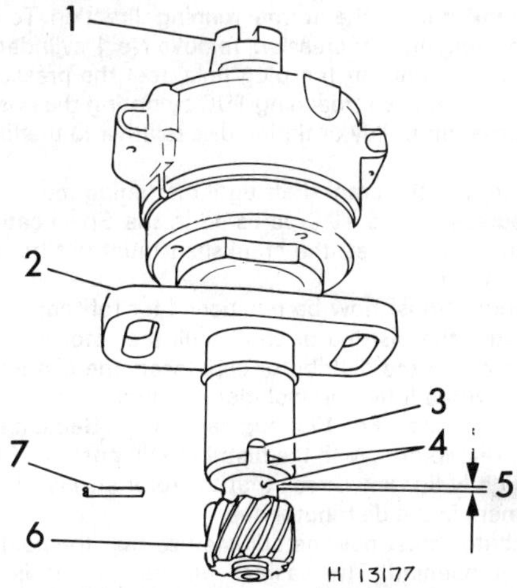

Fig. 4.7 Assembly of the drivegear to the distributor shaft, type 48D4 distributor (17V engines) (Sec 9)

1 Rear end of rotor arm towards thrust washer tab slot (3) and at right angles to mounting flange (2)
2 Mounting flange
3 Thrust washer tab slot
4 Align roll pin hole with tab slot edge
5 Endfloat, 0.005 in (0.13 mm), set by feeler gauge
6 Drivegear
7 Roll pin

17V engines

8 Reassembly is a reversal of the dismantling procedure but note the additional points listed below.

9 When assembling the drivegear and shaft, ensure that the distance collar is fitted with its chamfer towards the steel washer. Lubricate the bearing and advance mechanism with Rocol MP (Molypad) or equivalent and insert the shaft into the distributor body. Where the original shaft and gear are being refitted, fit the thrust washer and gear. Align the rotor keyway with the hole in the drivegear and fit the roll pin. Where a new shaft is being fitted, fit the thrust washer, drivegear and the rotor arm to the shaft. Set the rotor arm at right angles to the distributor mounting flange with its rear end towards the thrust washer tab slot in the distributor body, see Fig. 4.7. Insert a 0.005 in (0.13 mm) feeler gauge between the thrust washer and the drivegear to obtain the correct endfloat. Align the hole in the drivegear with the tab slot edge. Hold the gear and the shaft together tightly and drill through the shaft with a $\frac{1}{8}$ in (3.2 mm) drill. Remove the feeler gauge and fit the roll pin.

10 Check that the vacuum unit operating arm has engaged the advance plate peg before fitting the retaining screws.

11 Finally adjust the contact breaker points, see Section 2, and lubricate the distributor as described in Section 5. Fit a new O-ring.

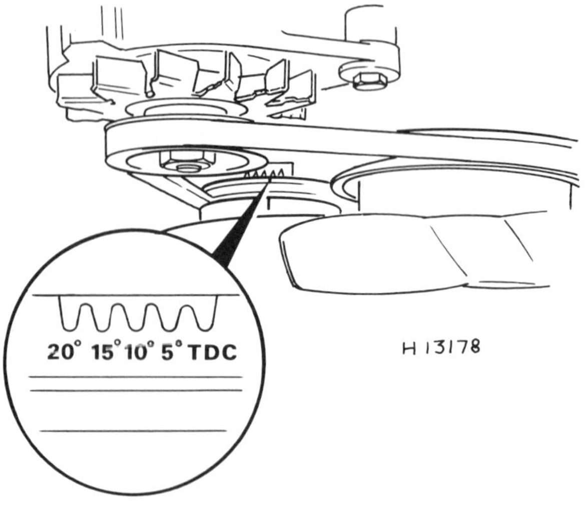

Fig. 4.8 Crankshaft timing marker (16V and 18V engines) (Sec 10)

10 Ignition timing

1 To make an accurate check of the ignition timing it is necessary to use a stroboscopic timing light and a tachometer so that the timing can be checked with the engine running. If the timing has been lost due to engine dismantling, renewal of the distributor, or for some other reason, it will first be necessary to set the timing statically so that the engine can be started. The timing can be set accurately with the stroboscopic light and tachometer. If these instruments are not available a local dealer will undertake the work for you.

Static ignition timing

2 Refer to Section 2 and check the contact breaker points gap or dwell angle. Reset if necessary.

3 Turn the crankshaft so that No 1 piston is rising on the compression stroke. This can be done in a number of ways. Turn the crankshaft using a spanner on the crankshaft pulley nut or by engaging second gear, releasing the handbrake and moving the vehicle forwards. If the vehicle cannot be moved, jack one rear wheel clear of the ground, engage top gear, releasing the handbrake and turn the raised wheel by hand. Whatever method is used remember that the crankshaft must be turned in the normal direction of rotation only, ie clockwise looking at the crankshaft pulley, so that the clearances in the drive to the distributor are taken up in the normal running direction. To ensure that No 1 piston is rising on compression, remove No 1 cylinder spark plug and, by putting a thumb on the plug hole, feel the pressure rise and check that the piston is approaching TDC by noting the position of the line on the crankshaft pulley or timing disc relative to the timing marks or pointer.

4 Continue to turn the crankshaft until the timing marks align at the number of degrees before TDC as listed in the Specifcations for the engine concerned. Once set, the crankshaft must not be moved until the distributor is fitted.

5 The distributor must now be positioned for refitting. For 16V and 18V engines turn the distributor shaft until the rotor is lined up with the No 1 segment in the distributor cap. Insert the distributor into its mounting and, with a little manipulation of the driveshaft if necessary, engage the driving dogs. For 17V engines refer to Section 6 for details of refitting the distributor. With the distributor in position, fit the retaining nuts and lightly tighten. Check that the rotor arm is still pointing to the No 1 segment in the distributor cap.

6 The distributor must now be adjusted so that the contact breaker points are just opening. The easiest way to do this is as follows. Disconnect the low tension lead from the distributor, connect a 12V low wattage bulb between the lead end and its connector, and switch on the ignition. Rotate the body of the distributor until the light goes out, indicating that the points have opened. Tighten the distributor clamp bolt and/or retaining nuts and check the timing again. When satisfactory, remove the bulb, if used, refit the distributor cap and restore any other items disturbed during the procedure.

7 Start the engine and, providing it runs smoothly, either proceed with dynamic timing adjustments if a stroboscopic light and

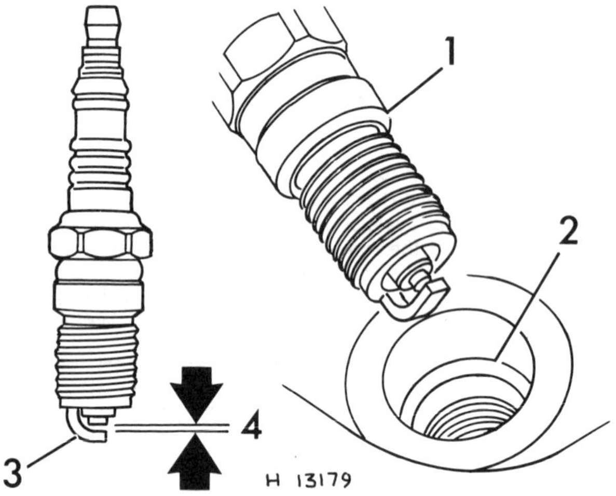

Fig. 4.9 The 17V engine spark plug (Sec 11)

1 Plug conical seat
2 Head conical seat
3 Earth electrode
4 Plug gap

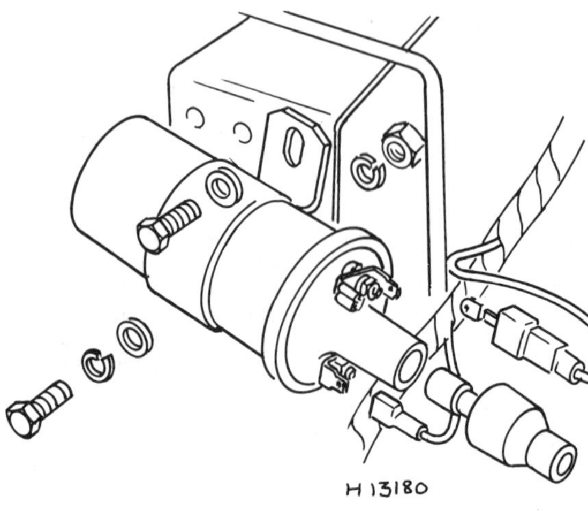

Fig. 4.10 Ignition coil fitting details (16V and 18V engines) (Sec 13)

tachometer are available or take the vehicle to a dealer to have the work done.

Dynamic ignition timing

8 Before starting the engine to make dynamic timing checks, clean the crankshaft pulley or timing disc in the vicinity of the timing mark and also the timing degrees indicator plate or pointer. White paint or chalk applied to the relevant timing marks will make them easier to see under the strobe light.

9 Disconnect the the distributor vacuum pipe and plug the end of the pipe with a suitable dowel.

10 Connect the stroboscopic timing light and tachometer to the vehicle circuits in accordance with the makers' instructions.

11 Run the engine at the speed given in the Specifications and shine the timing light onto the timing marks; the specified marks should appear in alignment. The timing may be adjusted if necessary by loosening the distributor clamp bolt (16V and 18V engines) or mounting stud nuts (17V engines) and rotating the distributor body until the timing marks are aligned. Tighten the clamp or mounting nuts to the specified torque when the adjustment is completed.

12 Check the functioning of the vacuum advance unit by running the engine at a fast idle speed and connecting the vacuum pipe to the distributor whilst watching the timing marks. The timing should immediately advance if the vacuum capsule is working correctly.

13 Stop the engine and disconnect the timing light and tachometer.

11 Spark plugs and leads

1 The correct functioning of the spark plugs is vital for the proper running and efficient operation of the engine.

2 At intervals of 6000 miles (10 000 km) the plugs should be removed, examined, cleaned, and if worn excessively, renewed. The condition of the spark plug can also tell much about the general condition of the engine.

3 If the insulator nose of the spark plug is clean and white, with no deposits, this is indicative of a weak mixture, or too hot a plug (a hot plug transfers heat away from the electrode slowly - a cold plug transfers heat away quickly).

4 If the insulator nose is covered with hard black looking deposits, then this is indicative that the mixture is too rich. Should the plug be black and oily then it is likely that the engine is fairly worn, as well as the mixture being too rich.

5 If the insulator nose is covered with light tan to greyish brown deposits, then the mixture is correct, and it is likely that the engine is in good condition.

6 If there are any traces of long brown tapering stains on the outside of the white portion of the plug, then the plug will have to be renewed, as this shows that there is a faulty joint between the plug body and the insulator, and compression is being allowed to leak away.

7 Plugs should be cleaned by a sand blasting machine, which will free them from carbon more than by cleaning by hand. The machine will also test the condition of the plugs under compression. Any plug that fails to spark at the recommended pressure should be renewed.

8 The spark plug is of considerable importance, as, if it is too large or too small the size of the spark and its efficiency will be seriously impaired. The correct plug gap is given in the Specifications.

9 To set it, measure the gap with a feeler gauge, and then bend open, or close, the outer plug electrode until the correct gap is achieved. The centre electrode should never be bent as this may crack the insulation and cause plug failure, if nothing worse.

10 *16V and 18V engines:* When refitting the plugs remember to use new washers, and tighten the plugs to the specified torque.

11 *17V engines:* When refitting plugs ensure that the seating areas in the cylinder head are perfectly clean and the plugs are clean on their conical seating face. Screw in the plug finger-tight and then gently tighten with a spanner by one sixteenth of a turn. *If this type of plug is overtightened removal will be very difficult, if not impossible.*

12 *All engines:* Wipe the insulators clean and refit the plug leads in the correct firing order which is 1, 3, 4, 2, No 1 cylinder being nearest the fan.

13 The plug leads require no maintenance other than being kept clean and wiped over regularly. At intervals of 6000 miles (10 000 km), however, pull each lead off the plug in turn and remove them from the distributor cap. Water can seep down these joints giving rise to a white corrosive deposit which must be carefully removed from the end of each cable.

12 Ignition system – fault symptoms

There are two general symptoms of ignition faults. Either the engine will not fire, or the engine is difficult to start and misfires. If it is a regular misfire, ie the engine is only running on two or three cylinders, the fault is almost sure to be in the high tension circuit. If the misfiring is intermittent, the fault could be in either the high or low tension circuits. If the engine stops suddenly, or will not start at all, it is likely that the fault is in the low tension circuit. Loss of power and overheating, apart from faulty carburettor settings, are normally due to faults in the distributor, or incorrect ignition timing.

13 Fault diagnosis – engine fails to start

1 If the engine fails to start and it was running normally when it was last used, first check that there is fuel in the petrol tank. If the engine turns over normally on the starter motor and the battery is evidently well charged, then the fault may be in either the high or low tension circuits. First check the HT circuit. **Note:** *If the battery is known to be fully charged, the ignition comes on, and the starter motor fails to turn the engine. Check the tightness of the leads on the battery terminals and also the secureness of the earth lead to the connection to the body. It is quite common for the leads to have worked loose,* even if they look and feel secure. If one of the battery terminal posts gets very hot when trying to operate the starter motor this is a sure indication of a faulty connection to that terminal.

2 One of the commonest reasons for bad starting is wet or damp spark plug leads and distributor. Remove the distributor cap. If condensation is visible internally, dry the cap with a rag and also wipe over the leads. Refit the cap.

3 If the engine still fails to start, check that current is reaching the plugs by disconnecting each plug lead in turn at the spark plug ends and holding the end of the cable about $\frac{3}{16}$ inch (4.76 mm) away from the cylinder block. Spin the engine on the starter motor.

4 Sparking between the end of the cable and the block should be fairly strong with a regular blue spark (hold the lead with rubber to avoid electric shocks). If current is reaching the spark plugs, then remove them and clean and regap them to the gap size listed in the Specifications. The engine should now start.

5 If there were no sparks from the plug leads, remove the HT lead from the coil where it enters the distributor cap and hold the end of the cable near the cylinder block. Spin the engine as before, when a rapid succession of blue sparks between the end of the lead and the block indicate that the coil is in order, and that either the distributor cap is cracked, the carbon brush is stuck or worn or the rotor arm is faulty.

6 If there are no sparks from the end of the lead from the coil, check the condition of the HT lead and its connection to the coil (photo). If all seems in order, check the low tension circuit starting with the battery.

7 Check the condition of the contact breaker points. If they are worn, pitted or dirty, clean and reset them as described in Section 2.

8 A 12V test lamp or voltmeter will be required for the next tests. Turn the crankshaft so that the contact breaker points are open, or separate them with a piece of thin card. Connect the test lamp or voltmeter leads between the coil – terminal (to which the wire to the distributor is attached) and earth. Switch on the ignition: if the lamp lights the circuit is satisfactory and you should proceed to paragraph 11. If the lamp does not light, proceed to the next paragraph.

9 Connect the test lamp or voltmeter between the coil + terminal (lead from ignition switch) and earth. Switch on the ignition: if the lamp lights, power is getting to the coil. Disconnect the lead from the coil to the distributor and recheck between the coil – terminal and earth: if the lamp lights on the + terminal but not on the – terminal the coil is faulty and must be renewed. If the lamp does not light on the + terminal then there is either a fault in the ignition switch or a break in the wiring between the battery, ignition switch and coil. A suspect ignition switch must be changed for a new one, but a quick check can be made by running a wire directly from the battery + terminal to the coil + terminal (not 17V models – see below). If the engine now runs, a fault in the ignition switch is confirmed.

10 *17V engined models* have a 6V coil and a special resistance wire between the coil and the ignition switch to absorb the other six volts

13.6a The coil on the 16V and 18V engines and ...

13.6b ... on the 17V engine

When the starter motor is operated, the resistance wire is bypassed by the starter solenoid contacts, so ensuring a good spark for starting. If the engine fires when the starter motor is operated, but will not run, a broken resistance wire should be suspected. **Do not** bypass the resistance wire with ordinary wire or the coil may overheat.

11 With a satisfactory supply of electricity to the coil assured, the final check on the LT circuit is across the contact breaker points. With the test lamp connected across the points, the points separated and the ignition turned on, no reading indicates that the condenser is short-circuiting the points and must be changed. It is possible for the condenser to fail in the open circuit mode, in which case this test will produce a reading, but there will be a strong blue flash as the points are separated and the points may well appear burnt.

14 Fault diagnosis – engine misfires

1 If the engine misfires regularly, run it at a fast idling speed, and disconnect a plug lead. Note the change in engine running, if any, and refit the lead. Repeat on all plug leads.

2 No difference in engine running will be noticed when the plug lead of the defective cylinder is removed. Removing the lead from any of the working plugs will accentuate the misfiring.

3 Remove the plug lead from the end of the defective plug and hold it about $\frac{3}{16}$ in (4.76 mm) away from the block. If sparking is fairly strong and regular the fault must lie in the spark plug.

4 The plug may be loose, the insulation may be cracked or the electrodes may have burnt away giving too wide a gap for the spark to jump across. Worse still, the earth electrode may have broken off. Either renew the plug, or clean it, reset the gap and then test it.

5 If there is no spark at the end of the plug lead, or if it is weak and intermittent, check the ignition lead from the distributor to the plug. If the insulation is cracked or damaged, renew the lead. Check the connections at the distributor cap.

6 If there is still no spark, examine the distributor cap carefully for signs of tracking. This can be recognised by a very thin black line running between two or more segments, or between a segment and some other part of the distributor. These lines are paths which now conduct electricity across the cap thus letting it run to earth. The only answer is to fit a new distributor cap.

7 Apart from the ignition timing being incorrect, other causes of misfiring due to ignition faults have already been dealt with under the section dealing with failure of the engine to start.

8 If the ignition timing is too far retarded, it should be noted that the engine will tend to overheat, and there will be quite a noticeable drop in power. If the engine is overheating and power is down, and the ignition is correct, then the carburettor should be checked, as it is likely that this is where the fault lies. See Chapter 3 for details.

Chapter 5 Clutch and actuating mechanism

For modifications, and information applicable to later models, see Supplement at end of manual

Contents

Specifications

Type	Single dry plate, hydraulically operated	
Clutch driven plate diameter		
16V engines	8 in (203·2 mm)	
17V and 18V engines	9 in (228·6 mm)	
Master cylinder bore	0·70 in (17·78 mm)	
Slave cylinder bore	1·0 in (25·4 mm)	
Torque wrench settings	**lbf ft**	**Nm**
16V and 18V engines		
Clutch to flywheel	28	37
17V engines		
Clutch to flywheel	17	23
Screws not specified:		
Into aluminium M6	8	11
M8	18	25
Into cast iron M6	8	11
M8	22	30
M10	37	50

1 General description

1 The Sherpa is fitted with an 8 inch (203·2 mm) (16V engines) or a 9 inch (228·6 mm) (17V and 18V engines) diameter coil spring clutch operated hydraulically by a master and a slave cylinder.

2 The clutch comprises a steel cover which is bolted and dowelled to the rear face of the flywheel and contains the pressure plate and clutch disc or driven plate.

3 The pressure plate, coil springs and release plate are all attached to the clutch assembly cover.

4 The clutch disc is free to slide along the splined first motion shaft and is held in position between the flywheel and pressure plate by the pressure of the springs.

5 Friction lining material is riveted to the clutch disc which has a spring cushioned hub to absorb transmission shocks and to help ensure a smooth take-off.

6 The clutch is actuated hydraulically. The clutch pedal is connected to the clutch master cylinder and hydraulic fluid reservoir by a short pushrod. The master cylinder and hydraulic reservoir are mounted on the engine side of the bulkhead in front of the driver.

7 Depressing the clutch pedal moves the piston in the master cylinder forwards so forcing hydraulic fluid through the clutch hydraulic pipe to the slave cylinder.

8 The piston in the slave cylinder moves forward on the entry of the fluid and actuates the clutch release arm by means of a short pushrod. The opposite end of the release arm is forked and is located behind the release bearing.

9 The release bearing is pushed against a central ring in the pressure plate. The ring acts on three release levers or toggles, which compress the pressure plate springs and so disengage the pressure plate from the clutch disc.

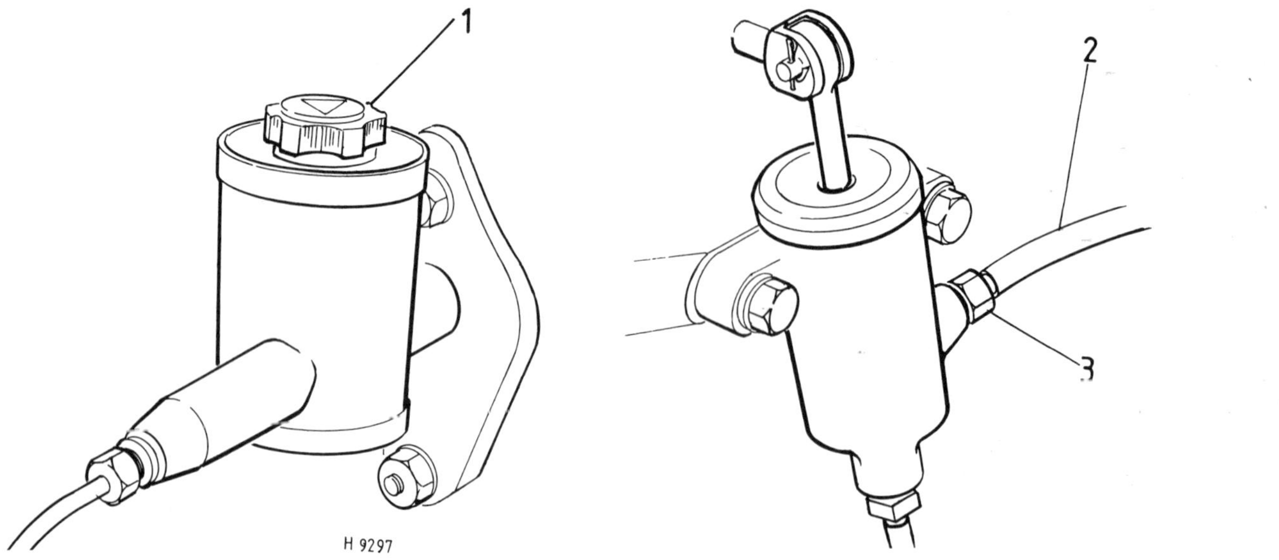

Fig. 5.1 Bleeding the clutch hydraulic system (Sec 2)

1 Master cylinder hydraulic reservoir (Sec 2)
2 Bleed tubing
3 Slave cylinder bleed nipple

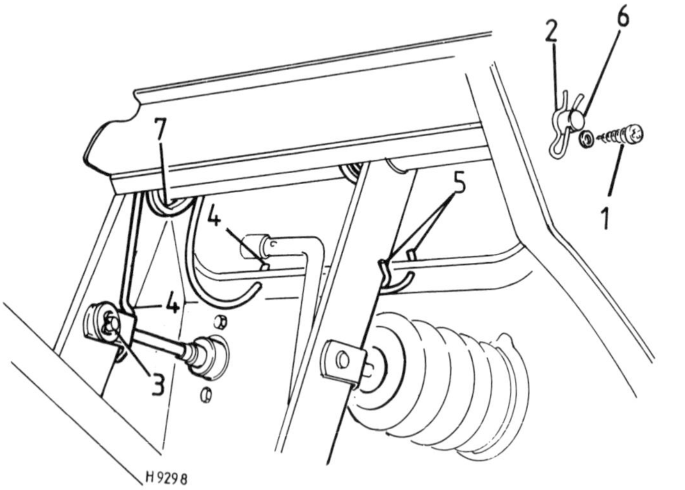

Fig. 5.2 Removing the clutch pedal (Sec 3)

1 Spring pin retaining screw and washer
2 Spring pin
3 Clevis pin
4 Clutch pedal return spring ends
5 Brake pedal return spring
6 Pedal shaft
7 Clutch pedal return spring

10 When the clutch pedal is released, the springs force the pressure plate into contact with the high friction linings on the clutch disc, and at the same time pushes the clutch disc a fraction of an inch forwards on its splines so engaging the clutch disc with the flywheel. The clutch disc is now firmly sandwiched between the pressure plate and the flywheel so the drive is taken up

11 As the friction linings on the clutch disc wear the pressure plate automatically moves closer to the disc to compensate. There is therefore no need periodically to adjust the clutch.

2 Clutch hydraulic system – bleeding

Note: *This operation will normally only be required if the clutch hydraulic system has been dismantled. Bleeding is not a routine operation.*

1 Gather together a clean jam jar, a length of rubber or plastic tubing which fits tightly over the bleed nipple on the slave cylinder, a tin of hydraulic brake fluid and someone to help.

2 Check that the master cylinder is full. If it is not, fill it. Cover the bottom two inches of the jar with hydraulic fluid.

3 Remove the rubber dust cap (if fitted) from the bleed nipple on the slave cylinder, and with a suitable spanner open the bleed nipple approximately three quarters of a turn.

4 Place one end of the tube securely over the nipple and insert the other end in the jam jar so that the tube orifice is below the surface of the fluid.

5 The assistant should now depress the pedal and hold it down at the end of its stroke. Close the bleed screw and allow the pedal to return to its normal position.

6 Continue this series of operations until clear hydraulic fluid without any traces of air bubbles emerges from the end of the tubing. Make sure that the reservoir is checked frequently to ensure that the hydraulic fluid does not drop too far thus letting air into the system.

7 When no more air bubbles appear, tighten the bleed nipple at the end of a downstroke.

8 Refit the rubber dust cap (if fitted) over the bleed nipple. Top-up the reservoir until the level is up to the bottom of the filler neck and, before refitting the cap, check that its breather hole is clear.

3 Clutch pedal – removal, repair and refitting

1 The clutch pedal is mounted with the brake pedal on a common horizontal shaft which is retained in a bracket by spring pins (photo).

2 To remove the clutch pedal, first remove the screw which retains the spring pin on the right-hand side of the bracket. Then remove the spring pins from both ends of the shaft.

3 Remove the split-pin and washer from the clevis pin which connects the clutch master cylinder pushrod to the foot pedal and remove the clevis pin.

4 Unhook both ends of the clutch pedal and brake pedal return springs.

5 Pull out the pedal shaft from the right-hand side of the bracket sufficiently to release the clutch pedal and remove the pedal.

6 Repair of the pedal is limited to pressing the bearing bushes out of the pedal tube and pressing in new bushes.

7 Refitting of the pedal is a reversal of the removal procedure, but note the following points:

(a) Engage the end of each return spring in its anchor hole in the bracket flange

(b) Connect the clutch master cylinder pushrod to the upper hole in the pedal

4 Clutch – removal and refitting

1 Remove the gearbox as described in Chapter 6.
2 With a scriber or file mark the relative position of the clutch cover and flywheel to ensure correct refitting if the original parts are to be used.
3 Remove the clutch assembly by unscrewing the six bolts holding the cover to the rear face of the flywheel. Unscrew the bolts diagonally half a turn at a time to prevent distortion of the cover flange, also to prevent an accident caused by the cover flange binding on the dowels and suddenly flying off.
4 With the bolts and spring washers removed, lift the clutch assembly off the locating dowels. The driven plate or clutch disc will fall out at this stage, as it is not attached to either the clutch cover assembly or the flywheel. Carefully make a note of which way round it is fitted.
5 It is important that no oil or grease gets on the clutch disc friction linings, or the pressure plate and flywheel faces. It is advisable to handle the parts with clean hands and to wipe down the pressure plate and flywheel faces with a clean dry rag before inspection or refitting commences.
6 To refit the clutch place the clutch disc against the flywheel with the clutch spring housing facing outwards away from the flywheel (photo). On no account should the clutch disc be refitted the wrong way round as it will be found quite impossible to operate the clutch with the friction disc incorrectly fitted.
7 Refit the clutch cover assembly loosely on the dowels. Refit the six bolts and spring washers and tighten them finger-tight so that the clutch disc is gripped but can still be moved.
8 The clutch disc must now be centralised so that when the engine and gearbox are mated the gearbox input shaft splines will pass through the splines in the centre of the hub.
9 Centralisation can be carried out quite easily by inserting a round bar or long screwdriver through the hole in the centre of the clutch, so that the end of the bar rests in the small hole in the end of the crankshaft containing the input shaft bearing bush. Moving the bar sideways or up and down will move the clutch disc in whichever direction is necessary to achieve centralisation. Take care not to damage the input shaft bearing bush in the crankshaft.
10 Centralisation is easily judged by removing the bar and viewing the driven plate hub in relation to the hole in the centre of the pressure plate. When the hub appears exactly in the centre of the release bearing hole all is correct. Alternatively, if an old input shaft can be borrowed this will eliminate all the guesswork as it will fit the bush and centre of the clutch hub exactly, eliminating the need for visual alignment (photo).
11 Tighten the clutch bolts to the specified torque in a diagonal sequence to ensure that the cover plate is pulled down evenly, and without distortion of the flange.

3.1 Clutch, brake and throttle pedal details

4.6 Refitting the clutch disc and cover assembly

4.10 Centralising the clutch disc using an old input shaft

4.12 Refitting the gearbox to the engine

12 Mate the engine and gearbox (photo), bleed the slave cylinder if the pipe was disconnected and check the clutch for correct operation.

5 Clutch – inspection

1 In the normal course of events clutch dismantling and reassembly is the term used for simply fitting a new clutch pressure plate and friction disc. Under no circumstances should the pressure plate unit be dismantled. If a fault develops in the pressure plate assembly an exchange unit must be fitted.
2 If a new clutch disc is being fitted it is false economy not to renew the release bearing at the same time. This will avoid having to renew it at a later date when wear on the clutch linings is very small.
3 Examine the clutch disc friction linings for wear or loose rivets and the disc for rim distortion, cracks and worn splines.
4 It is always best to renew the clutch driven plate as an assembly to preclude further trouble, but, if it is wished to merely renew the linings, the rivets should be drilled out, and not knocked out with a centre punch. The manufacturers do not advise that the linings only are renewed and personal experience dictates that it is far more satisfactory to renew the driven plate complete than to try to economise by fitting only new friction linings.
5 Check the machined faces of the flywheel and the pressure plate. If either is badly grooved it should be machined until smooth, or renewed. If the pressure plate is cracked or split it must be renewed.
6 Examine the hub splines for wear and also make sure that the centre hub is not loose.

6 Clutch pipe – removal and refitting

1 Wipe both pipe union ends of the hydraulic pipe to prevent the entry of dirt when the pipe is disconnected. Obtain a suitable container in which to catch hydraulic fluid spillage during the next operation, and a tapered dowel to plug the master cylinder outlet.
2 Unscrew the pipe union on the master cylinder. Remove the pipe and plug the master cylinder with the dowel to minimise the loss of fluid.
3 Unscrew the pipe union from the slave cylinder and remove the pipe. Do not allow hydraulic fluid to come into contact with the paintwork.
4 Refitting is the reverse of the removal procedure. On completion, bleed the clutch hydraulic system as described in Section 2.

7 Clutch master cylinder – removal and refitting

1 Drain the fluid from the clutch master reservoir by attaching a rubber tube to the slave cylinder bleed nipple. Undo the nipple by approximately three quarters of a turn and then pump the fluid out into a suitable container by operating the clutch pedal. Note that the pedal must be held in against the floor at the completion of each stroke and the bleed nipple tightened before the pedal is allowed to return. When the pedal has returned to its normal position loosen the bleed nipple and repeat the process, until the clutch master cylinder is empty.
2 Place a rag under the master cylinder to catch any hydraulic fluid that may be spilt. Unscrew the union nut from the end of the metal pipe where it enters the clutch master cylinder and gently pull the pipe clear. Seal the cylinder outlet and the end of the pipe to prevent dirt getting in.
3 Remove the split-pin and washer from the clevis pin connecting the master cylinder pushrod to the clutch pedal and remove the clevis pin.
4 Unscrew and remove the two nuts and spring washers securing the master cylinder and lift away. Take care not to allow any hydraulic fluid to come into contact with the paintwork as it acts as a solvent.
5 Refitting the master cylinder is the reverse sequence to removal. Bleed the hydraulic system as described in Section 2.

8 Clutch master cylinder – dismantling, examination and reassembly

1 Remove the reservoir cap and empty out all the hydraulic fluid.
2 Ease back the rubber dust cover from the pushrod end.

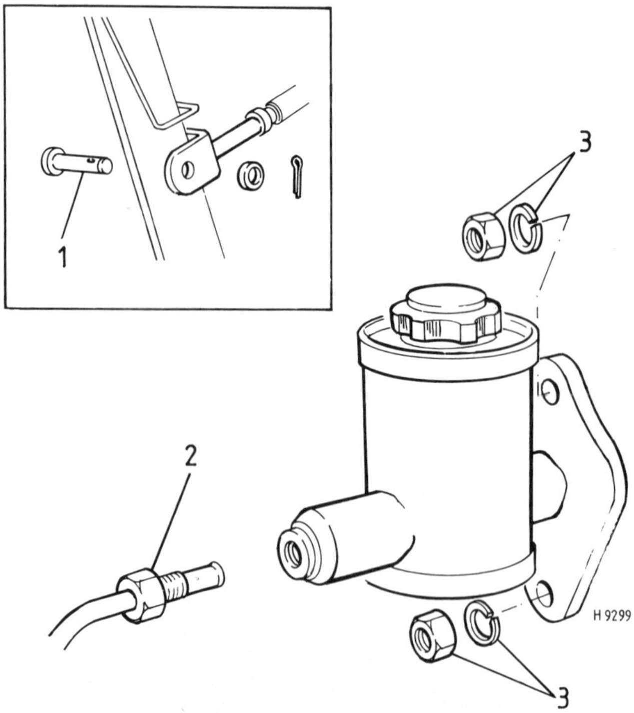

Fig. 5.3 Removing the clutch master cylinder (Sec 7)

1 Clevis pin
2 Pipe union nut
3 Master cylinder retaining nuts and spring washers

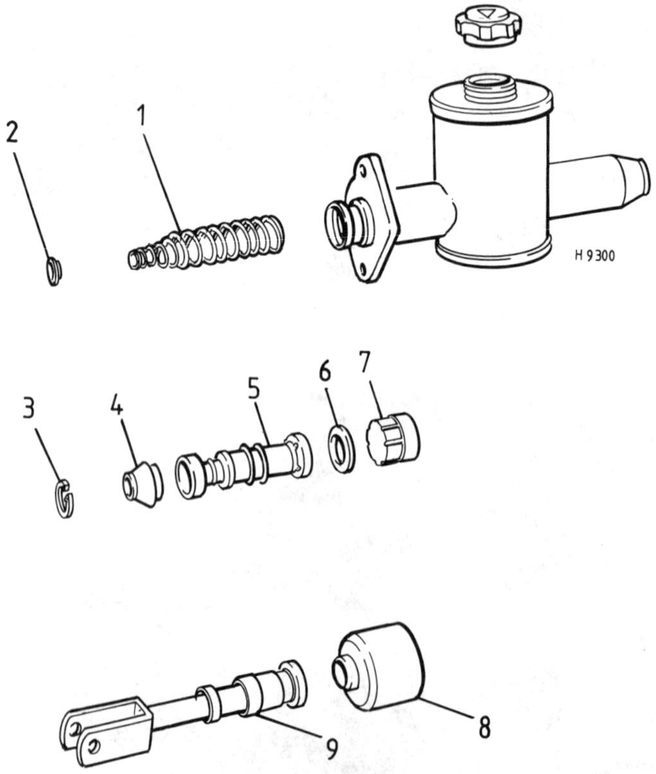

Fig. 5.4 Exploded view of clutch master cylinder (Sec 8)

1 Spring
2 Spring retainer
3 Pushrod retaining circlip
4 Secondary seal
5 Piston assembly
6 Piston washer
7 Main seal
8 Dust cover
9 Pushrod

3 Using a pair of circlip pliers release the circlip retaining the push rod assembly. Lift away the pushrod complete with rubber boot and plain washer.
4 By shaking hard, the piston with its seal, piston washer, secondary seal and spring retainer may be removed from the cylinder bore. Remove the secondary seal from the piston noting which way round it is fitted, and also that it is the same way round as the primary seal.
5 Lift away the long spring noting which way round it is fitted.
6 If the piston and seals prove stubborn, carefully use a foot-pump air jet on the hydraulic pipe connection and this should move the internal parts, but do take care as they will fly out. We recommend placing a pad over the pushrod end to catch the parts.
7 Thoroughly clean the parts in brake fluid or methylated spirit. After drying the items inspect the seals for signs of distortion, swelling, splitting or hardening, although it is recommended that new rubber parts are always fitted after dismantling as a matter of course.
8 Inspect the bore and piston for signs of deep scoring marks which, if evident, mean that a new cylinder should be fitted. Make sure the by-pass ports are clear by poking gently with a piece of thin wire.
9 As the parts are refitted to the cylinder bore make sure that they are thoroughly wetted with clean hydraulic fluid.
10 Refit the secondary cup seal onto the piston making sure it is the correct way round.
11 Insert the spring with its retainer into the master cylinder bore.
12 Next refit the main cup seal with its flat end facing towards the open end of the bore.
13 Replace the piston washer and carefully insert the piston into the bore. The small end of the piston should be towards the piston washer. Make sure that the lip of the seal does not roll over as it enters the bore.
14 Smear a little rubber grease onto the ball end of the pushrod and refit the pushrod assembly. Slide down the plain washer and secure in position with the circlip.
15 Pack the rubber dust cover with rubber grease and place over the end of the master cylinder.

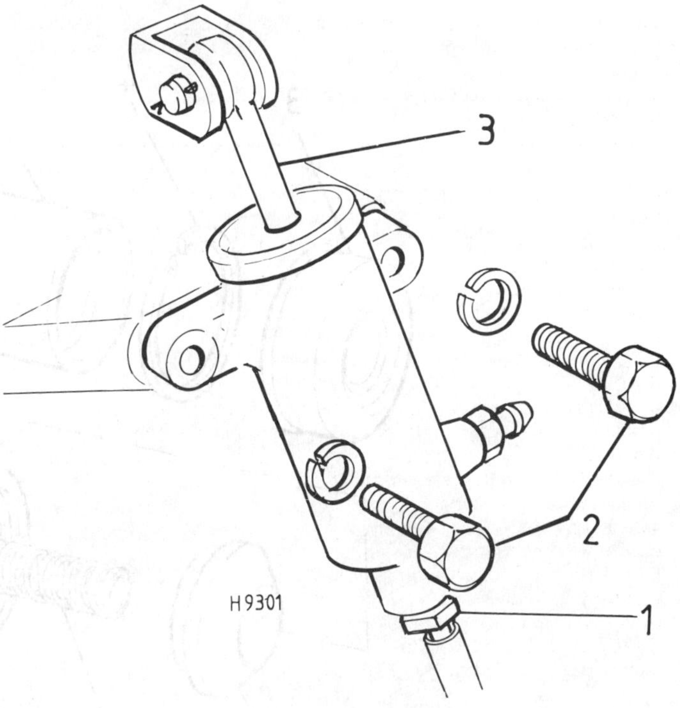

Fig. 5.5 Removing the clutch slave cylinder (Sec 9)

1 Hydraulic pipe *3 Pushrod*
2 Retaining bolts

9 Clutch slave cylinder – removal and refitting

1 Wipe the top of the master cylinder reservoir and unscrew the cap. Place a piece of polythene sheet over the top of the reservoir and refit the cap. This will stop hydraulic fluid syphoning out during subsequent operations.
2 Wipe the area around the hydraulic pipe on the slave cylinder and disconnect the metal pipe from the slave cylinder.
3 Remove the bolts and spring washers which attach the slave cylinder to the clutch housing and pull the slave cylinder away from the pushrod (photos).
4 Refitting the slave cylinder is the reverse sequence to removal.
5 After refitting, bleed the clutch hydraulic system as described in Section 2.

10 Clutch slave cylinder – dismantling, examination and reassembly

1 Clean the outside of the slave cylinder before dismantling.
2 Pull off the rubber dust cover and then remove the circlip from the cylinder. By shaking hard, the piston, cup, spring and spring retainer should come out of the cylinder bore. Note the order of assembly and which way round each part fits.
3 If the piston and seals prove stubborn, carefully use a foot-pump air jet on the hydraulic hose connection and this should remove the internal parts, but do take care as they will fly out. We recommend placing a pad over the dust cover end to catch the parts.
4 Wash all internal parts with either brake fluid or methylated spirit and dry using a non-fluffy rag.
5 Inspect the bore and piston for signs of deep scoring which, if

9.3a The clutch slave cylinder, retaining bolts removed

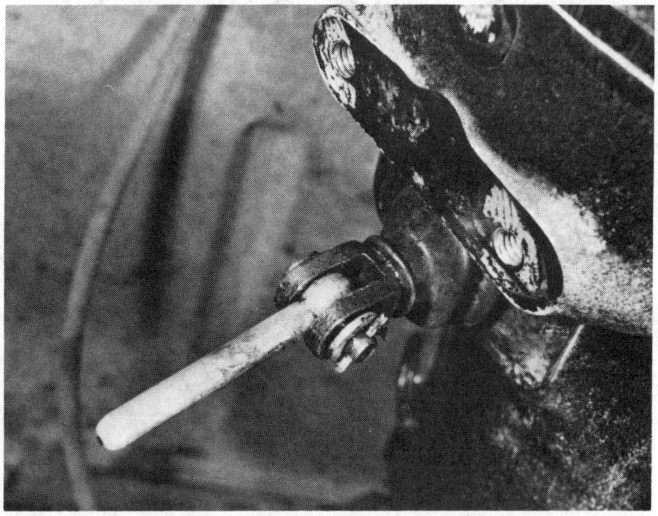

9.3b The clutch pushrod, slave cylinder removed

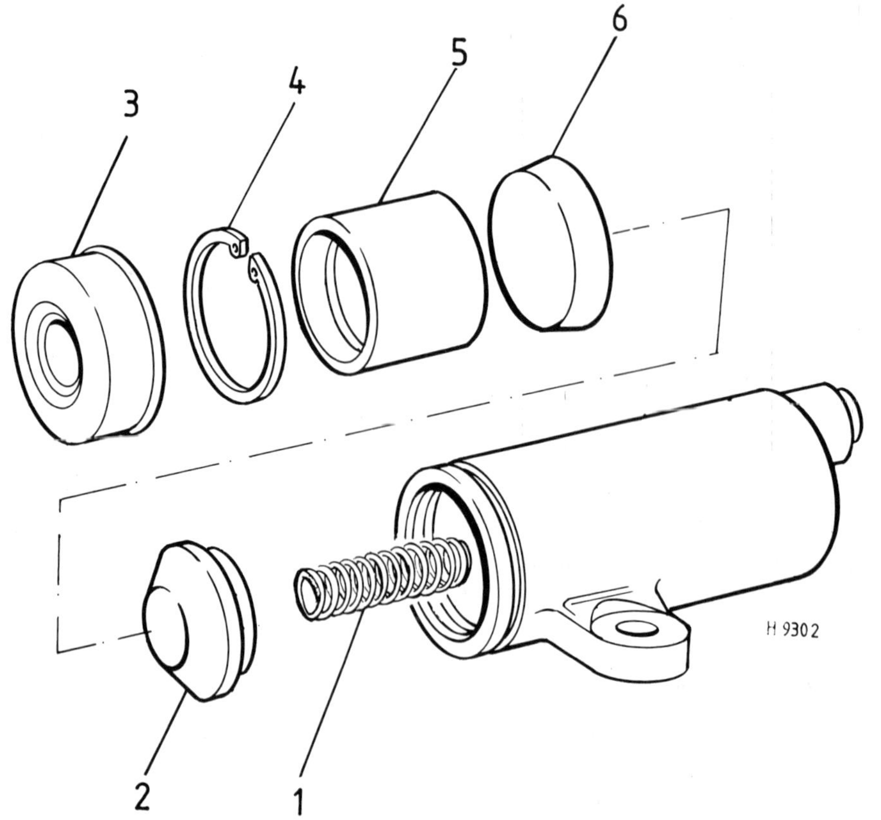

Fig. 5.6 Exploded view of clutch slave cylinder (Sec 10)

1	*Spring*	4	*Circlip*
2	*Spring retainer*	5	*Piston*
3	*Dust cover*	6	*Cup seal*

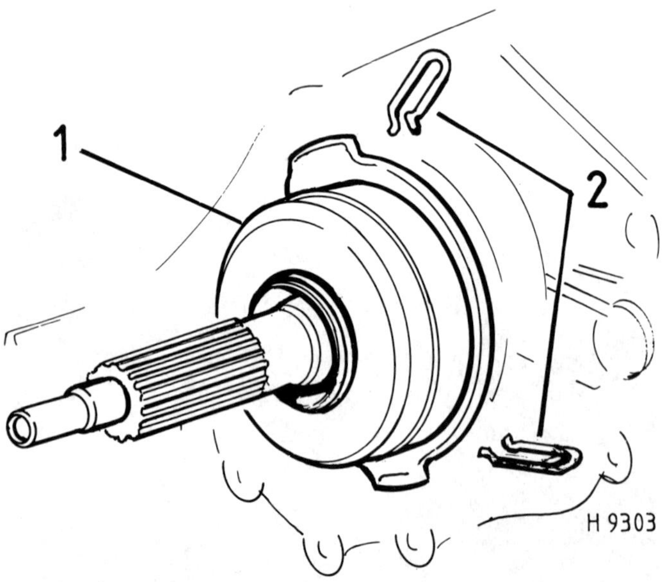

Fig. 5.7 Removing the clutch release bearing (Sec 11)

1 *Clutch release bearing* 2 *Retainer clips*

evident, means a new cylinder should be fitted.

6 Carefully examine the rubber components for signs of swelling, distortion, splitting, hardening or other wear, although it is recommended new rubber parts are always fitted after dismantling.

7 All parts should be reassembled wetted with clean hydraulic fluid.

8 Refit the spring, large end first, into the bore of the cylinder, followed by the spring retainer.

9 Fit a new cup seal and refit the piston making sure that both are fitted the correct way round.

10 Apply a little rubber grease to both ends of the pushrod and also pack the dust cover.

11 Fit the dust cover over the end of the slave cylinder engaging the lips over the groove in the body.

12 Fit the pushrod to the slave cylinder by pushing through the hole in the dust cover.

11 Clutch release bearing assembly – removal, overhaul and refitting

1 To gain access to the release bearing, it is necessary to remove the gearbox as described in Chapter 6.

2 Detach the operating lever from the release bearing and slide off the bearing assembly.

3 If the bearing is worn or shows signs of overheating it must be changed for a new bearing assembly.

4 Refitting is the reverse of the removal sequence but, on fitting the clutch release bearing, coat the rear half of the bearing bore with Duckhams Laminoid O grease or equivalent.

12 Fault diagnosis – clutch

There are four main faults to which the clutch and release mechanism are prone. They may occur by themselves, or in conjunction with any of the other faults. They are clutch squeal, slip, spin and judder.

13 Clutch squeal

1 If on taking up the drive or when changing gear, the clutch squeals, this is indicative of a badly worn clutch release bearing.
2 As well as regular wear due to normal use, wear of the clutch release bearing is much accentuated if the clutch is ridden or held down for long periods in gear, with the engine running. To minimise wear of this component the car should always be taken out of gear at traffic lights and for similar hold-ups.
3 The clutch release bearing is not an expensive item, but difficult to get at.

14 Clutch slip

1 Clutch slip is a sefl-evident condition which occurs when the clutch friction plate is badly worn, oil or grease have got onto the flywheel or pressure plate faces, or the pressure plate itself is faulty.
2 The reason for clutch slip is that due to one of the faults above, there is either insufficient pressure from the pressure plate, or insufficient friction from the friction plate to ensure solid drive.
3 If small amounts of oil get onto the clutch, they will be burnt off under the heat of the clutch engagement, and in the process, gradually darken the linings. Excessive oil on the clutch will burn off leaving a carbon deposit which can cause quite bad slip, or fierceness, spin and judder.
4 If clutch slip is suspected, and conformation of this condition is required, there are several tests which can be made.
5 With the engine in second or third gear and pulling lightly sudden depression of the accelerator pedal may cause the engine to increase its speed without any increase in road speed. Easing off on the accelerator will then give a definite drop in engine speed without the car slowing.
6 In extreme cases of clutch slip the engine will race under normal acceleration conditions. The cure is to renew the clutch driven plate, and trace and rectify any oil leak.

15 Clutch spin

1 Clutch spin is a condition which occurs when there is a leak in the clutch hydraulic actuating mechanism, there is an obstruction in the clutch either in the first motion shaft or in the operating lever itself, or oil may have partially burnt off the clutch lining and have left a resinous deposit which is causing the clutch disc to stick to the pressure plate or flywheel.
2 The reason for clutch spin is that due to any, or a combination, of the faults just listed, the clutch pressure plate is not completely freeing from the centre plate even with the clutch pedal fully depressed.
3 If clutch spin is suspected, the condition can be confirmed by extreme difficulty in engaging first gear from rest, difficulty in changing gear, and by very sudden take up of the clutch drive at the fully depressed end of the clutch pedal travel as the clutch is released.
4 Check the clutch master cylinder, the slave cylinder and the connecting hydraulic pipe for leaks. Fluid in one of the rubber dust covers fitted over the end of either the master or slave cylinder is a sure sign of a leaking piston seal. Bleed the clutch hydraulic system.
5 If these points are checked and found to be in order then the fault lies internally in the clutch, and it will be necessary to remove the clutch for examination.

16 Clutch judder

1 Clutch judder is a self-evident condition which occurs when the gearbox or engine mountings are loose or too flexible, when there is oil on the face of the clutch friction plate, or when the clutch pressure plate has been incorrectly adjusted.
2 The reason for clutch judder is that due to one of the faults just listed, the clutch pressure plate is not freeing smoothly from the friction disc, and is snatching.
3 Clutch judder normally occurs when the clutch pedal is released in first or reverse gears, and the whole car shudders as it moves backwards or forwards.

Chapter 6 Manual gearbox, overdrive and automatic transmission

For modifications, and information applicable to later models, see Supplement at end of manual

Contents

Specifications

Manual gearbox	Four forward speeds, one reverse. Synchromesh fitted to all forward speeds
Gearbox ratios	
Fourth (top)	1·000 : 1
Third	1·5754 : 1
Second	2·4706 : 1
First	3·8009 : 1
Reverse	3·5294 : 1
Endfloats	
Manual non-overdrive:	
First speed gear	0·005 to 0·007 in (0·13 to 0·18 mm)
Second speed gear	0·005 to 0·008 in (0·13 to 0·20 mm)
Third speed gear	0·005 to 0·007 in (0·13 to 0·18 mm)
Manual with overdrive:	
First speed gear	0·005 to 0·008 in (0·13 to 0·20 mm)
Second speed gear	0·005 to 0·009 in (0·13 to 0·23 mm)
Third speed gear	0·005 to 0·007 in (0·13 to 0·18 mm)
Laygear	0·002 to 0·007 in (0·05 to 0·18 mm)
Washer thicknesses available	
Thrust washer for mainshaft bearing	0·108 in (2·74 mm)
	0·110 in (2·79 mm)
	0·118 in (2·99 mm)
Laygear rear thrust washer	0·1195 to 0·1215 in (3·04 to 3·09 mm)
	0·123 to 0·125 in (3·12 to 3·17 mm)
	0·1265 to 0·1285 in (3·21 to 3·26 mm)
	0·130 to 0·132 in (3·30 to 3·35 mm)
	0·133 to 0·135 in (3·38 to 3·43 mm)
Synchronizer springs	
Free length	0·720 in (18·30 mm)
Fitted length	0·385 in (9·80 mm)
Load at fitted length	5·5 to 6·0 lbf (2·5 to 2·7 kgf)

Speedometer drivegear position	1·937 in (49·2 mm) from rear face of mainshaft bearing to front face of gear	
Overdrive		
Type	Laycock LH	
Ratio	0·82 : 1	
Relief valve spring:		
Free length	1·182 in (30 mm)	
Fitted length	0·995 in (25·3 mm)	
Load at fitted length	38·4 lbf (17·5 kgf)	
Automatic transmission		
Make	Borg Warner	
No of ratios	3 forward, 1 reverse	
Capacities		
Manual gearbox	5·25 pints (2·98 litres)	
Manual gearbox with overdrive	6·25 pints (3·5 litres)	
Automatic transmission	9 pints (5·1 litres)	
Torque wrench settings	**lbf ft**	**Nm**
Manual gearbox		
Drive flange nut	150	203
Overdrive		
Drive flange nut	110	149
Automatic transmission		
Drive flange nut	58	78
Converter to driveplate bolts	28	37
Transmission case to converter housing bolts:		
Upper	28	37
Lower	45	61
Rear extension to transmission case bolts	45	61
Oil pan to transmission case bolts	7	9
Drain plug (if fitted)	11	15
Starter inhibitor switch to transmission bolt	5	7

1 General description

1 The manual gearbox fitted to the Sherpa provides four forward gears and one reverse gear, with synchromesh fitted to all forward gears.

2 Gear selection is by means of a hand lever mounted on an extension housing bolted to the gearbox rear face. Forward or rearward movement of the gearchange lever moves the selector lever which in turn moves a selector rod to which is attached a selector fork. This moves the synchronizer coupling to engage the required gear, or in the case of the reverse gear, moves the reverse idler gear to mesh with the laygear and the reverse gear.

3 The gearbox input shaft is splined and it is onto these splines that the clutch driven plate is located. The gearbox end of the input shaft is in constant mesh with the laygear cluster, and the gears formed on the laygear are in constant mesh with the gears on the mainshaft with the exception of the reverse gear. The gears on the mainshaft are able to rotate freely which means that when the neutral position is selected the mainshaft does not rotate.

4 When the gear change lever moves the synchromesh unit outer sleeve via the selector fork, the synchromesh cup first moves and friction caused by the conical surfaces meeting takes up initial rotational movement until the mainshaft and gear are both rotating at the same speed. This condition achieved, the sleeve is able to slide over the dog teeth of the selected gear thereby giving a firm drive. The synchromesh unit inner hub is splined to the mainshaft and because the outer sleeve is splined to the inner hub engine torque is passed to the mainshaft and propeller shaft.

5 Automatic transmission is available as an optional fitting and overdrive is also available optionally on models with 17V and 18V engines. Details of this optional equipment are contained later in this Chapter.

2 Gearbox – removal and refitting

1 The gearbox can be removed as a unit with the engine as described in Chapter 1. An alternative method is to separate the gearbox bellhousing from the engine endplate, lower the gearbox and remove from under the vehicle, leaving the engine in position. Use this method if only clutch and/or gearbox repairs are to be made.

2 Disconnect the battery, raise the vehicle and put on axle stands if a ramp is not available. The higher the vehicle is off the ground the easier it will be to work underneath.

3 Undo the gearbox drain plug and drain the oil into a suitable container. When all the oil has drained out, refit the drain plug.

4 Remove the gearchange lever, referring to Section 9 for details.

5 With a scriber or file mark the gearbox and propeller shaft drive flanges to ensure correct refitting, then undo and remove the four propeller shaft flange locknuts and bolts. Using string or wire tie the propeller shaft to the exhaust pipe.

6 Using a hoist or jack, support the weight of the engine and gearbox. If a jack is being used place it under the rear of the sump with a piece of wood between the jack and the sump.

7 Unscrew the knurled nut securing the speedometer drive cable to the gearbox, remove the drive and secure it out of the way.

8 Remove the two bolts and lockwashers securing the clutch slave cylinder to the gearbox, withdraw the cylinder from the pushrod and tie it up out of the way. There is no need to disconnect the hydraulic line.

9 *16V and 18V engines.* Remove the starter motor securing bolts, withdraw the starter and move it to one side in the engine compartment.

10 *17V engines.* Disconnect the engine earth strap or cable from the starter motor retaining bolt.

11 *17V engines.* Disconnect the intermediate exhaust mounting and slacken the bracket clamp. Move the clamp in order to remove the mounting from the bracket. Undo the bolts which retain the exhaust pipe front mounting bracket to the pipe bracket and gearbox bellhousing, and remove the bracket.

12 Remove the bolts, washers and reinforcing strips which secure the ends of the gearbox crossmember to the bodyframe.

13 *Non-overdrive models.* Remove the bolts, nuts and washers which attach the mountings to the gearbox crossmember.

14 *Overdrive models.* Remove the bolts and washers which secure the mountings to the overdrive adaptor. Check that the electrical leads to the overdrive isolation switch and the overdrive solenoid have been disconnected (this should have been done when the gearchange lever

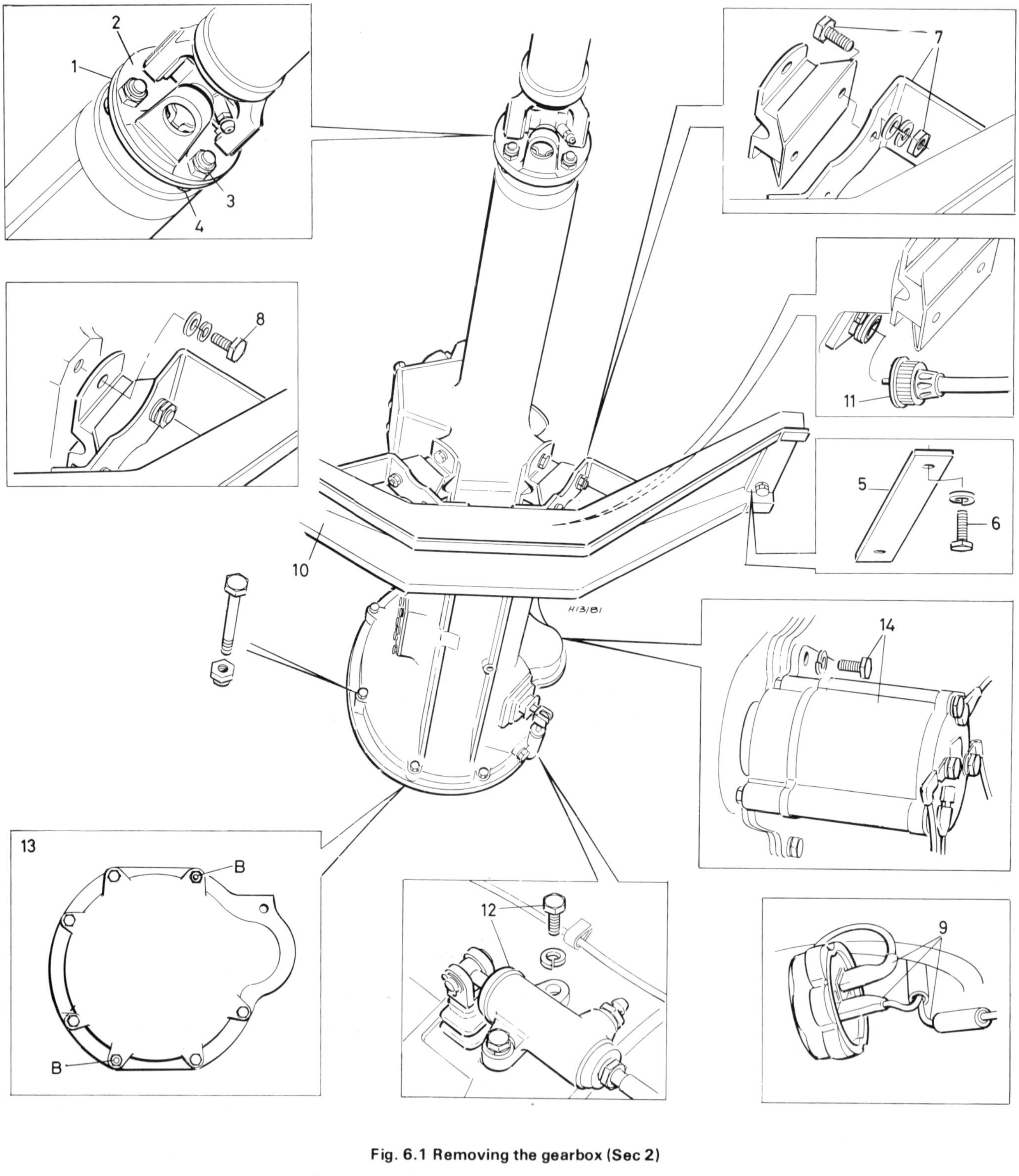

Fig. 6.1 Removing the gearbox (Sec 2)

1 Gearbox output drive flange
2 Propeller shaft drive flange
3 Locknut
4 Coupling bolt
5 Reinforcing strip
6 Crossmember securing bolt
7 Mounting to crossmember securing nut and bolt
8 Mounting to extension securing bolt
9 Overdrive inhibitor switch leads (overdrive models only)
10 Crossmember
11 Speedometer drive cable nut
12 Clutch slave cylinder and securing bolt
13 Fit the bolts B first on reassembly (16V and 18V engines)
14 Starter motor and securing bolts

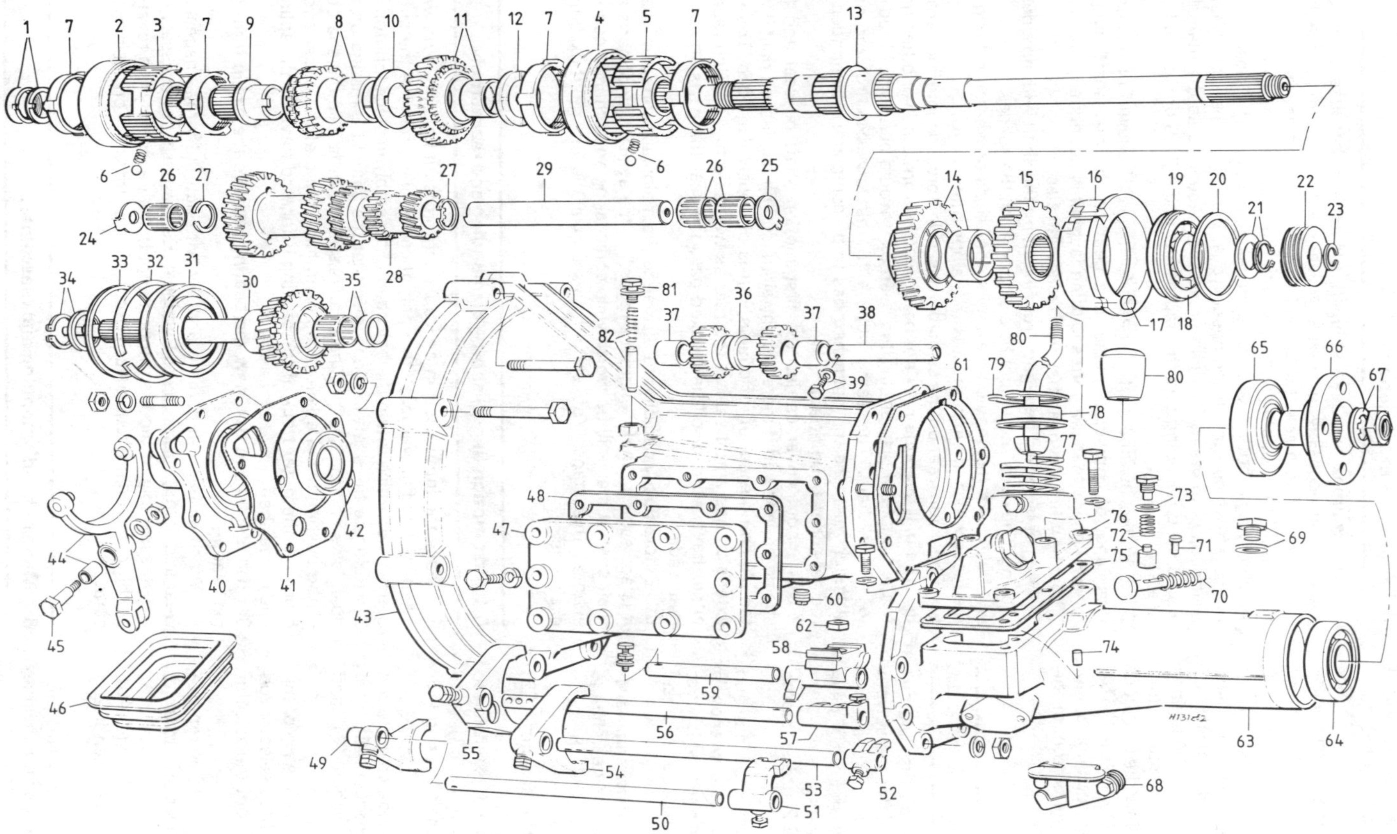

Fig. 6.2 Exploded view of the manual gearbox and extension (Sec 3)

1 *Mainshaft nut and lockwasher*
2 *Third-fourth synchro coupling*
3 *Third-fourth synchro hub*
4 *First-second synchro coupling*
5 *First-second synchro hub*
6 *Synchro ball and spring*
7 *Baulk ring*
8 *Third gear and bush*
9 *Third gear sleeve*
10 *Third gear thrust washer*
11 *Second gear and bush*
12 *Second gear thrust washer*
13 *Mainshaft (third motion shaft)*
14 *First gear and bush*
15 *Reverse gear*
16 *Bearing housing*
17 *Dowel*
18 *Bearing*
19 *Circlip*
20 *Shim*
21 *Thrust washer and circlip*
22 *Speedometer drivegear*
23 *Circlip*
24 *Thrust washer*
25 *Thrust washer*
26 *Bearings*
27 *Circlips*
28 *Laygear*
29 *Layshaft*
30 *First motion (input) shaft*
31 *Bearing*
32 *Circlip*
33 *Shim*
34 *Thrust washer and circlip*
35 *Bearing and spacer*
36 *Reverse idler gear*
37 *Bush*
38 *Reverse idler shaft*
39 *Lockscrew and lockwasher*
40 *Cover*
41 *Gasket*
42 *Oil seal*
43 *Case*
44 *Clutch operating lever and bush*
45 *Pivot bolt*
46 *Dust cover*
47 *Cover*
48 *Gasket*
49 *Reverse selector fork*
50 *Reverse selector rod*
51 *Reverse selector*
52 *Third-fourth selector*
53 *Third-fourth selector rod*
54 *Third-fourth selector fork*
55 *First-second selector fork*
56 *First-second selector rod*
57 *First-second selector*
58 *Selector lever*
59 *Lever shaft*
60 *Oil drain plug*
61 *Gasket*
62 *Bush*
63 *Rear extension*
64 *Bearing*
65 *Oil seal*
66 *Drive flange*
67 *Nut and lockwasher*
68 *Interlock arm*
69 *Plug and washer*
70 *Reverse spring and plunger*
71 *Pin*
72 *Detent spring and plunger*
73 *Plug and washer*
74 *Dowel*
75 *Gasket*
76 *Gearchange lever housing*
77 *Spring*
78 *Spring cover*
79 *Circlip*
80 *Gearchange lever*
81 *Detent plug and washer (one of three)*
82 *Detent spring and plunger (one of three)*

was removed).
15 Remove the gearbox crossmember from the vehicle.
16 Remove the nuts and bolts retaining the gearbox bellhousing to the engine.
17 If necessary, lower the engine slightly to provide the required angle to remove the gearbox, and remove the gearbox. Note that, on 17V engines, there is a long stud at the top right-hand side of the bellhousing and it is essential, when removing the gearbox, to keep it exactly in alignment with the engine until the bellhousing has completely cleared the stud.
18 If it is intended to fit a different gearbox, remove the clutch release bearing and the clutch operating pushrod from the withdrawal lever for use in the new gearbox. On non-overdrive models, also remove the rear mountings from the gearbox casing.
19 Refitting of a gearbox is the reverse of the removal procedure. Centralise the clutch driven plate, if it has been disturbed, as described in Chapter 5. When refitting the gearbox to the engine on 17V engines take care to engage the bellhousing squarely onto the fixed stud before pushing the gearbox fully into engagement. **Do not** allow the weight of the gearbox to hang on the stud or the input shaft. On 16V and 18V engines, after the gearbox is in position fit the bolts at positions B, Fig. 6.1, first. Refill the gearbox with the specified oil to the bottom of the filler/level plug.

3 Gearbox – dismantling

1 Before commencing work, clean the exterior of the gearbox thoroughly using paraffin or a water soluble solvent. After the solvent has been applied and allowed to stand for a time, a vigorous jet of water will wash off the solvent together with all oil and dirt. Finally wipe down the exterior of the unit with a dry non-fluffy rag.

Note: *All numbers in brackets refer to Fig. 6.2 unless stated otherwise.*

2 Hold the drive flange (66), and undo and remove the nut and lockwasher (67). The drive flange can be held by a mole wrench, a lever on bolts inserted in the flange or a special spanner, whichever method is most convenient. Withdraw the flange, using a universal puller if necessary.
3 Remove the six bolts and washers securing the gearchange lever housing (76) to the rear extension (63) and take off the gearchange lever extension, noting that it is located by two dowels (74). Remove the gasket (75).
4 Select fourth gear (by moving the third-fourth selector (52) forwards) and lift out the interlock arm assembly (68).
5 Remove the bolts, nuts and washers securing the rear extension to the gearbox case and remove the rear extension assembly (63). Note the location and number of any shims on the mainshaft bearing (18, 20). Remove the gasket (61).
6 Remove the ten bolts and washers securing the side cover to the gearbox and remove the side cover (47). Remove the gasket (48).
7 Undo and remove the three detent plugs and washers (81) and remove the springs and plungers or balls (82).
8 Slacken the locknuts and unscrew the locating bolts in each of the three selector forks (49, 54, and 55). Withdraw the three selector rod assemblies (50, 53 and 56) from the gearbox and lift out the selector forks.
9 Undo the nut on the clutch operating lever pivot bolt and remove the nut, washer and pivot bolt (45). Free the dust cover (46) from its location and remove the operating lever and bush (44).
10 Remove the eight nuts and washers retaining the front cover (40). Remove the front cover and retrieve any shims fitted between it and the front bearing. Remove the gasket (41).
11 Remove the lock screw (39) retaining the reverse idler shaft (38) and extract the shaft rearwards from the gearbox. Lift out the reverse idler gear (36).
12 Push the layshaft (29) out of the gearbox and lower the laygear (28) into the bottom of the gearbox.
13 Using a small drift placed on the bearing outer track, tap out the gearbox first motion (input) shaft (30) after making sure that the laygear is not in the way. Recover the spigot bearing and spacer (35) from either the input shaft or mainshaft. Remove the 3rd/4th speed synchronizer assembly (2 and 3) from the mainshaft (13).
14 Check that the laygear is out of the way and drift the mainshaft assembly rearwards out of the gearbox.
15 Remove the laygear rear thrust washer (25) noting that it has two tags. Lift out the laygear (28) and then remove the laygear front thrust washer (24) which has one tag. Remove the bearings from inside the laygear for examination if necessary.
16 This completes the dismantling of the gearbox into sub-assemblies. Further dismantling is dealt with in later Sections.

4 Gearbox – examination

1 The gearbox has been stripped, presumably because of wear or malfunction, possibly excessive noise, ineffective synchromesh or failure to stay in a selected gear. The cause of most gearbox ailments is failure of the ball bearings on the input or mainshaft and wear on the synchro rings, both the cone surfaces and dogs. The nose of the mainshaft which runs in the needle roller bearing in the input shaft is also subject to wear. This can prove very expensive as the mainshaft would need renewing and this represents about 20% of the total cost of a new gearbox.
2 Examine the teeth of all gears for signs of uneven or excessive wear and, of course, chipping. If a gear on the mainshaft requires renewal check that the corresponding laygear is not equally damaged. If it is the whole laygear may need renewing also.
3 All gears should be a good running fit on the shaft with no signs of rocking. The hubs should not be a sloppy fit on the splines.
4 Selector forks should be examined for signs of wear or ridging on the faces which are in contact with the operating sleeve.
5 Check for wear on the selector rods, selectors, and interlock arm.
6 The ball bearings may not be obviously worn but if one has gone to the trouble of dismantling the gearbox it would be shortsighted not to renew them. The same applies to the four synchronizer rings although for these the mainshaft has to be completely dismantled for the new ones to be fitted.
7 The front cover carries the first motion shaft oil seal and this should be renewed if there are any signs that oil has leaked past it into the clutch housing, or, of course, if it is obviously damaged. The rear extension has an oil seal at the rear as well as a ball bearing race. If either have worn or oil has leaked past the seal the parts should be renewed.
8 Before finally deciding to dismantle the mainshaft and renew parts it is advisable to make enquiries regarding the availability of parts and their cost. It may still be worthwhile considering an exchange gearbox even at this stage. You should reassemble the old gearbox before exchange.

5 First motion (input) shaft – dismantling and reassembly

1 Place the input shaft in a vice having protected or soft jaws, with its splined end upwards. With a pair of circlip pliers remove the circlip and washer which retain the ball bearing assembly in place (photo).
2 Relocate the shaft in the vice so that the bearing is resting on the open jaws and the shaft splined end is still upwards. Tap the shaft out of the bearing using a soft-faced hammer. Note that the offset circlip groove in the outer track of the bearing is located towards the splined end of the shaft.
3 If it is intended to renew the bearing, remove the circlip from the old bearing.
4 Wash all parts in paraffin and dry with non-fluffy rag. Inspect for wear and damage referring to Section 4. Renew parts as necessary.
5 Reassembly of the first motion (input) shaft assembly is a reversal of the dismantling sequence (photo). Use gearbox oil on all moving parts.

6 Mainshaft – dismantling and reassembly

1 Unlock the mainshaft (third motion shaft) front nut by bending up the lockwasher and remove the nut. This is designed for a C-spanner but the nut can be removed using a chain wrench. Avoid drifting it off which will damage it.
2 Remove the third speed gear, its bush and the thrust washer. The washer has two slots which interlock with the third speed synchronizer sleeve.
3 Remove the second speed gear and its thrust washer. This thrust washer has radial grooves.
4 Slide the first/second synchronizer assembly off the shaft.

5.1 The first motion shaft bearing and retaining circlip

5.5 The first motion shaft assembled

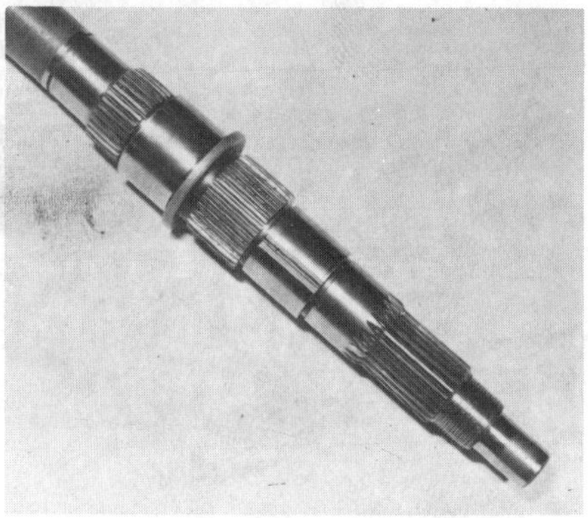

6.12a The forward end of the mainshaft

6.12b Fitting the first/second speed synchronizer and ...

6.12c ... the baulk ring

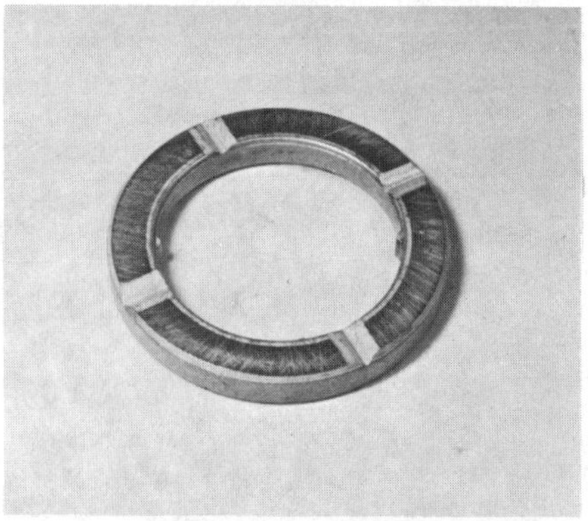

6.13a The radially grooved thrust washer ...

6.13b ... fitted to the mainshaft ...

6.13c ... followed by the second speed gear

6.14a Fitting the third gear onto its sleeve -and then ...

6.14b ... the thrust washer to engage with the dogs

6.14c Fitting the third gear assembly to the mainshaft

6.15a Fitting the baulk ring ...

6.15b ... and the third/fourth speed synchronizer assembly

6.16a Fit the lockwasher and front nut, then ...

6.16b ... tighten the nut. Note the three bearing balls located with grease in the nut peg holes to provide purchase for the chain spanner

6.17a Checking the second speed gear endfloat

6.17b The lockwasher bent into the front nut

6.18 The mainshaft bearing, with its circlip fitted, being inserted into the bearing housing

6.19a The first gear being fitted to the shaft ...

6.19b ... followed by the reverse gear

6.20a Fit the bearing assembly to the shaft

6.20b ... and drift it into position with a piece of tubing on the inner race

6.21 The thrust washer and bearing retaining circlip fitted

6.22a Fit the speedometer drivegear to the shaft ...

6.22b ... and locate it in the correct position

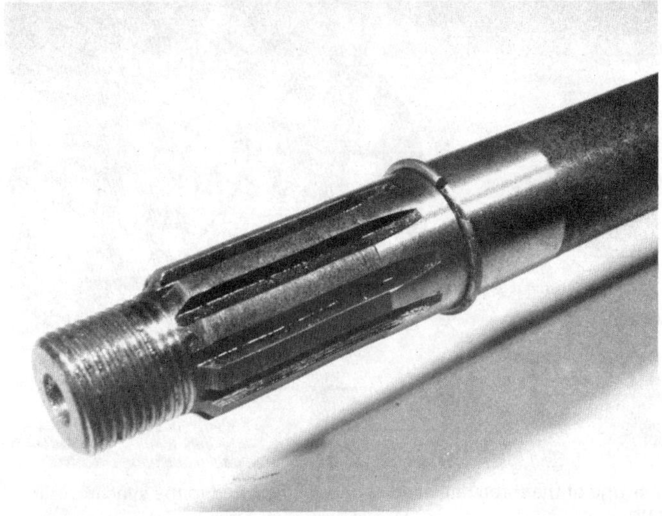
6.23 The circlip fitted to the tail of the shaft

5 Using a pair of circlip pliers, remove the circlip from the rear end of the shaft adjacent to the splines.
6 Take a measurement from one face of the speedometer drivegear to the end of its machined seating on the shaft so that it can be reassembled in the same position.
7 Lift the circlip which retains the mainshaft rear bearing on the shaft out of its groove, but leave the circlip on the shaft.
8 Support the shaft on the forward face of the first speed gear over the open, protected jaws of a vice and carefully drift the shaft downwards to release the speedometer drivegear, circlip, thrust washer, bearing housing assembly, reverse gear and the first speed gear.
9 Gently tap the rear bearing out of its housing and remove the circlip from the bearing.
10 Wash all parts in paraffin and dry with non-fluffy rag. Inspect for wear and damage, referring to Section 4. Renew parts as necessary. Overhaul of the synchro units is covered in Section 7.
11 During the reassembly of the mainshaft use gearbox oil to lubricate moving parts.
12 Fit the first/second speed synchronizer, which is the smaller of the two synchronizer assemblies, to the mainshaft. Note that the groove in the sliding coupling should be adjacent to the second speed gear which is fitted next to it (photos).
13 Fit the thrust washer (with radial grooves) and the second speed gear to the shaft (photos).
14 Fit the third gear onto its sleeve and then fit the thrust washer to engage with the sleeve dogs. Fit the assembly to the shaft with the thrust washer against the second speed gear (photos).
15 Slide the baulk ring and the third/fourth speed synchronizer assemby onto the splines of the third speed gear sleeve. (It may be convenient to remove this assembly later when assembling the mainshaft into the gearbox but for now it is best fitted on the shaft) (photos).
16 Fit the mainshaft lockwasher and front nut. Tighten the nut but do not lock it yet (photos).
17 Using feeler gauges check the endfloat of the second speed gear and the third speed gear which should be as indicated in the Specifications. If satisfactory, lock the front nut by bending the lockwasher into a slot on the nut (photos).
18 Fit the circlip to the groove in the mainshaft bearing and press the bearing into its housing with the circlip towards the flanged end of the housing. Alternatively drift the bearing home but only on the outer race (photo).
19 Fit the first speed gear to the shaft, followed by the reverse gear, with their large diameter faces adjacent to each other. Push the reverse gear towards the first speed gear and, using feeler gauges, ensure that the first gear endfloat is as quoted in the Specifications (photos).
20 Fit the bearing assembly to the shaft. If it has to be drifted on remember to drift on the inner race only, using a piece of tubing (photos).
21 Fit the thrust washer and bearing retaining circlip. This thrust washer is available in different thicknesses and the thickest washer which will allow the circlip to seat in its groove must be used (photos).
22 Fit the speedometer drivegear to the shaft using a piece of tube to tap it into position. Locate it at the dimension measured in paragraph 6 but if this was omitted, or the measurement is lost, see the Specifications where a dimension from the back of the bearing to the front of the speedometer gear is quoted (photos).
23 Finally fit the circlip to the tail end of the shaft (photo).

7 Synchronizers – dismantling, inspection and reassembly

1 Wrap a cloth around the synchronizer (to catch balls or springs) and carefully push the synchronizer hub out of the sliding coupling. Identify relevant locations so that, if the same parts are reassembled, they can all be refitted correctly.
2 Check the synchronizer springs for free length and, if a spring balance of correct range and an accurate means of measurement are available, check their load at fitted length as quoted in the Specifications. If doubt exists, renew the springs.
3 Check the fit of the baulk rings on the gears. If the rings touch the edge of the gear before the tapers are fully engaged,,or if there is almost no clearance, renew them.
4 Check for backlash between the splines in the outer sleeve and inner hub. If any play is noticeable the whole assembly should be renewed.
5 Reassembly is a fiddly business in the absence of a small ring compressor. The three balls and springs have to be fitted and compressed simultaneously whilst fitting the hub to the sliding coupling. A worm drive clip adjusted to slide easily over the hub would simplify the job (photo).
6 When fitting the hub in the coupling align the cutouts in the two parts before sliding the hub in. Note that the cutouts in the first/second speed hub are stepped as shown in Fig. 6.3. These steps must be towards the off-centre groove of the sliding coupling (photo).
7 Fit the baulk rings to the synchroniser assembly.

8 Gearbox – reassembly

1 Stand the gearbox upright on its front face and fit the laygear front thrust washer (with one tag) into its location on the layshaft front housing (photo).
2 If the laygear bearings were removed, refit them now, making sure that the two circlips are correctly fitted in their grooves inside the laygear (photo).
3 Fit the laygear into the gearbox, large gear to the front end (ie down during assembly) and fit the rear thrust washer (with two tags) into its location at the rear of the laygear. Insert the layshaft to align

7.5 One of the three balls and springs being fitted to the synchronizer hub

7.6 Assembling the hub to the sliding coupling. (Note the three pieces of split-pin and the ring compressor used to hold the balls and springs in position)

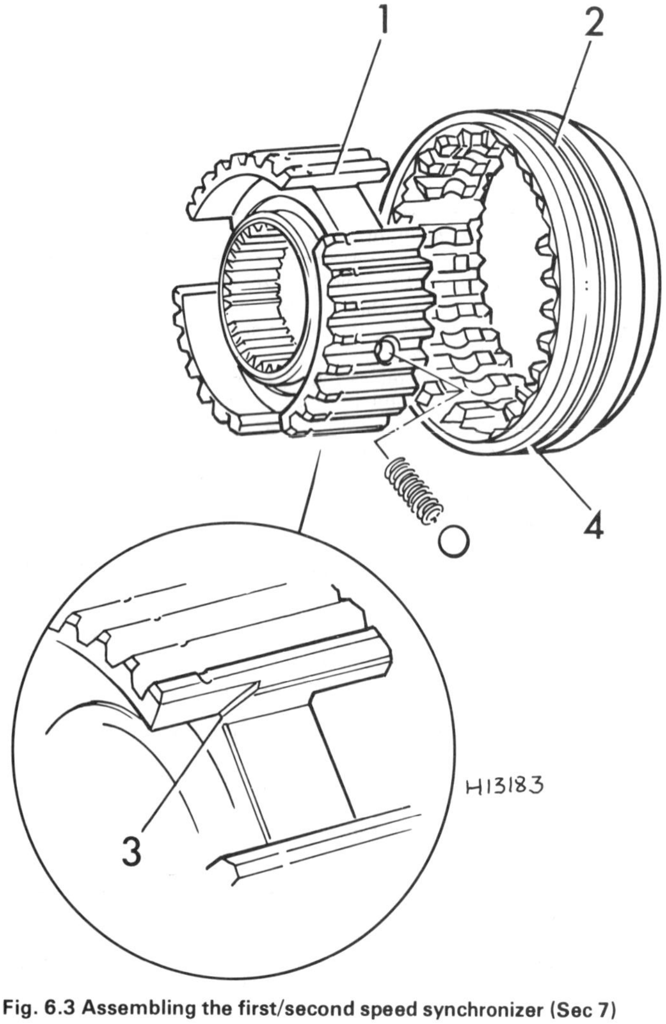

Fig. 6.3 Assembling the first/second speed synchronizer (Sec 7)

1 *Hub*
2 *Sliding coupling*
3 *Step in cut-out*
4 *Off-centre groove*

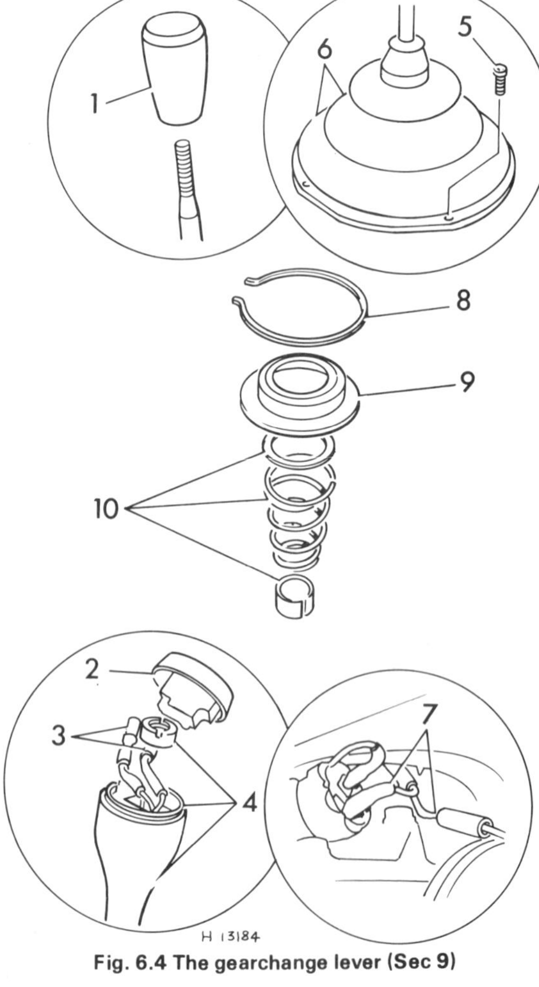

Fig. 6.4 The gearchange lever (Sec 9)

1 *Lever knob (non-overdrive models)*
2 *Lever knob cap (overdrive models)*
3 *Overdrive switch leads*
4 *Retaining nut, gear lever knob and cone (overdrive models)*
5 *Gaiter retainer securing screws*
6 *Gaiter and retainer*
7 *Overdrive isolation switch leads*
8 *Circlip*
9 *Spring cover*
10 *Washer, spring and bush*

8.1 The laygear front thrust washer located in the gearbox

8.2 Fitting a bearing to the laygear

8.3a Fit the laygear into the gearbox ...

8.3b ... and fit the rear thrust washer in its location

8.3c Leave the laygear loose in the gearbox

8.4a Insert the mainshaft assembly into the gearbox ...

8.4b ... and align the groove in the bearing housing with the locating dowel

8.7 The needle bearing and spacer fitted to the spigot on the front end of the shaft

8.8 Fitting the first motion shaft

8.9a Mesh the laygear with the mainshaft and ...

8.9b ... insert the layshaft with the stepped end at the front and positioned to ...

8.9c ... align with the recess in the front cover (arrowed)

8.11a Insert the reverse idler gear into the gearbox with the selector fork groove (arrowed) towards the front

8.11b Use a screwdriver to align the bolt hole in the shaft and ...

8.11c ... fit the locking bolt and washer

8.12a Fit the front cover shims and ...

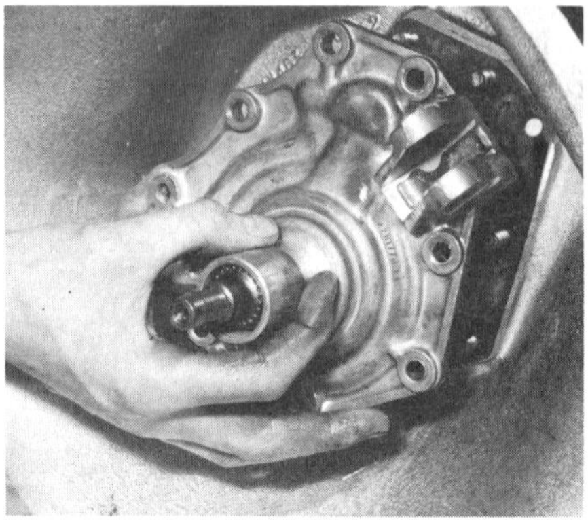

8.12b ... with a new gasket fit the front cover to the gearbox followed by ...

8.13 ... the clutch lever, pivot bolt and nut

8.14a Fit the third/fourth selector fork and ...

8.14b ... the first/second selector fork

8.14c Fit the third/fourth selector rod through the first/second selector fork

8.14d The third/fourth selector

8.14e Fit the first/second selector rod

8.14f The first/second selector

8.14g The locating screw and locknut fitted to the first/second selector before tightening of the locknut

8.15a The reverse selector fork and selector rod fitted

8.15b The reverse selector

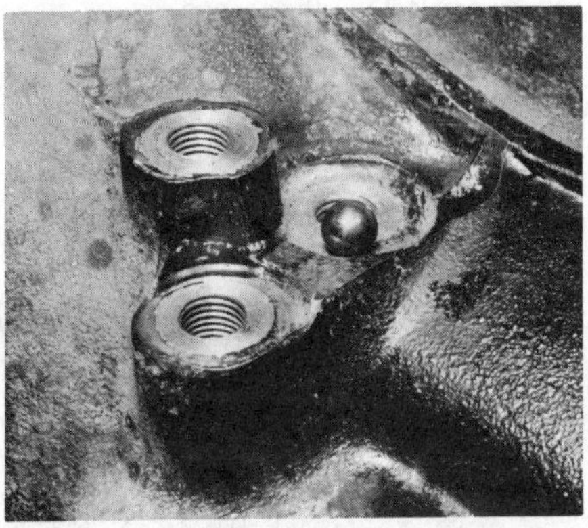

8.16a A detent ball about to be fitted ...

8.16b ... followed by the plunger, ...

8.16c ... the spring and ...

8.16d ... the detent plug

8.17 The gearbox cover being fitted with a new gasket in position

8.18 Shims being fitted on the mainshaft rear bearing

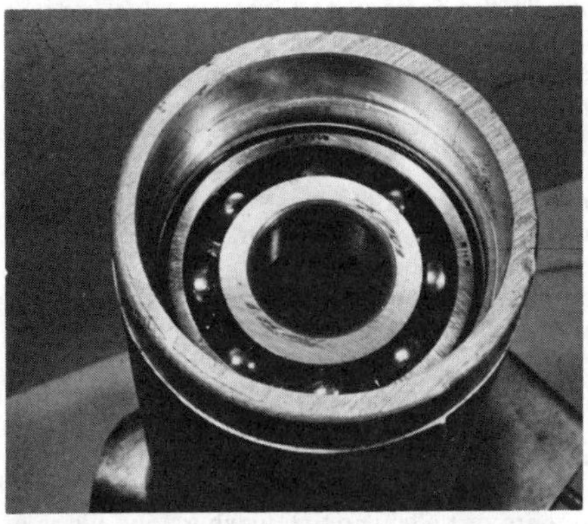

8.19a The old seal has been prised out of the rear extension and ...

8.19b ... a new seal is being fitted

8.20 The rear extension being fitted to the gearbox with a new gasket in position

8.21a Fit the drive flange to the third motion shaft in the rear extension followed by ...

8.21b ... the lockwasher and retaining nut

8.22a Engage the selector lever with the selectors, select fourth gear and ...

8.22b ... fit the interlock arm assembly

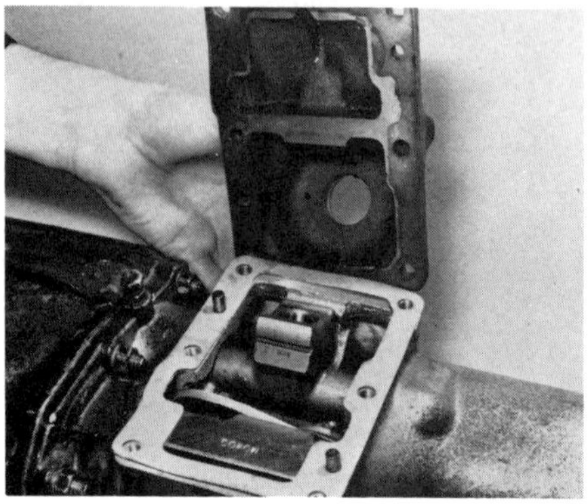

8.23 The gearchange lever housing being fitted with a new gasket in position

8.24 The speedometer driven gear being fitted

the assembly and, with feeler gauges, check that the laygear endfloat is within the specified limits. If it is not, it will be necessary to change the rear thrust washer (which is available in various thicknesses) for one which provides the correct endfloat. Once the endfloat is correct, remove the layshaft and move the laygear into the bottom of the box. Keep the thrust washers on their locations with a little grease (photos).

4 Fit the mainshaft into the gearbox from the rear end. Align the bearing housing groove with the locating dowel in the gearbox rear face and tap the bearing housing home with a soft-faced hammer (photos).

5 It is now necessary to calculate the thickness of the shim pack which fits between the rear face of the gearbox and the rear extension, and controls the nip on the mainshaft bearing outer race. Use a depth gauge to measure the depth of the bearing outer race below the rear face of the dearbox (dimension A) and the length of the spigot on the rear extension which abuts on the bearing race (dimension B). Allowing 0.015 in (0.38 mm) for the compressed thickness of the gasket, the required shim thickness is:

Dimension A
\+ *Gasket thickness (0.015 in, 0.38 mm)*
– *Dimension B*

Required shim thickness

Shims are available in thicknesses of 0.002 in (0.05 mm) and 0.004 in (0.10 mm); the shim pack should be made up to the calculated thickness or 0.001 in (0.03 mm) less.

6 If not already fitted, fit the third/fourth synchronizer to the front end of the mainshaft.

7 Fit the needle bearing and spacer to the spigot on the front end of the shaft (photo).

8 Fit the first motion (input) shaft onto the mainshaft and into the gearbox. Tap on the outer race of the bearing until it seats (photo).

9 Mesh the laygear with the associated gears and line it up with the thrust washers. Insert the layshaft with the stepped end to the front, making sure that the shaft passes through both thrust washers, and position the flat on the shaft so that it locates correctly with the front cover (photos).

10 Now calculate the shim thickness required to control the input shaft endfloat. Measure the depth of the shim register in the front cover (dimension D) and the protrusion of the input shaft bearing outer race (dimension E). The required shim thickness is:

Dimension D
\+ *0.015 in (0.38 mm) (compressed gasket thickness)*
– *Dimension E*

Required shim thickness

The shim should be the resulting thickness or 0.001 in (0.03 mm) less. This size can be made up from the two shim thicknesses available which are 0.002 in (0.05 mm) and 0.004 in (0.10 mm).

11 With the bushes in the reverse idler gear, insert it into the gearbox. The selector fork groove must be towards the front. Insert the reverse shaft from the rear, noting the alignment of the locking bolt hole in the shaft. With a screwdriver in the slot on the rear end of the shaft align the bolt hole with the hole in the gearbox and insert the locking bolt and washer. Tighten the bolt (photos).

12 Fit the appropriate shims, calculated in paragraph 10, in the shim register of the front cover and with a new gasket fit the cover to the gearbox. Fit and tighten the securing nuts and washers (photos).

13 Check the condition of the clutch lever dust cover and if necessary fit a new one to the clutch lever. Insert the lever into the bellhousing and bed the dust cover into position. Fit the clutch lever pivot bolt, washer and nut and tighten the nut (photo).

14 Fit the third/fourth and the first/second selector forks to their synchronizer assemblies and then fit the respective selector rods. In each case line up the locating screw holes in the rods with those in the forks, fit the locknuts and locating screws, tighten the screws and then the locknuts (photos).

15 Similarly fit the reverse selector fork, selector rod, locating screw and locknut. Tighten the screw and then the locknut (photos).

16 Insert the detent plungers or balls, springs, washers and plugs in each of the three locations in the gearbox, and tighten the detent plugs (photos).

9.3 Removing the gaiter retainer securing screws

9.5a Releasing the gear lever assembly retaining circlip

9.5b Withdrawing the gear lever assembly

17 Fit a new gasket on the gearbox side cover aperture, lightly smeared with a suitable gasket sealant, and then fit the cover. Fit and tighten the retaining bolts and spring washers (photo).
18 Fit the appropriate shims, as calculated in paragraph 5, on the mainshaft bearing (photo).
19 Prise out the old seal in the rear end of the extension assembly and tap a new seal into position, flush with the casting. Make sure that the seal lip faces inwards (photos).
20 Position a new gasket, lightly smeared with a suitable sealant, on the rear face of the gearbox. Fit the rear extension to the gearbox. Fit the bolts, nuts and washers and tighten (photo).
21 Smear both faces of the drive flange and the bearing face with Molygrease (Moly 500) or a suitable equivalent and fit the drive flange on the rear end of the mainshaft. Fit the lockwasher and retaining nut and tighten the nut to the specified torque (photos).
22 Engage the selector lever with the selectors and select fourth gear. Fit the interlock arm assembly (photos).
23 Lightly smear a new gasket for the gearchange lever housing with a suitable sealant and locate it on the gearbox. Fit the tower, noting the location dowels, and secure it with the four long and two short bolts and spring washers. Tighten the bolts (photo).
24 Using a new gasket, fit the speedometer driven gear and secure with its two bolts and lockwashers (photo).
25 If the gearbox was removed as a unit with the engine, it may now be re-attached. Fill the gearbox with the specified grade and quantity of oil when the engine and gearbox have been refitted and with the vehicle standing on level ground.

9 Gearchange lever – removal and replacement

1 *Non-overdrive models.* Unscrew the hand knob from the gear lever.
2 *Overdrive models.* Prise the cap from the top of the hand knob on the gear lever. Disconnect the leads from the switch and remove the cap and switch. Undo and remove the knob retaining nut, and remove the knob and the cone underneath it.
3 Remove the screws securing the gear lever gaiter retainer to the floor, and remove the retainer and gaiter from the gear lever (photo).
4 *Overdrive models.* Working underneath the vehicle, disconnect leads from the overdrive isolation switch and the snap connector on the overdrive solenoid lead.
5 Release the circlip from the gearchange lever housing and withdraw the gear lever. Take care with the spring beneath the cover (photos).
6 *Overdrive models.* Cut the plastic sleeve from the gear lever and remove the switch leads if it is intended to fit different parts.
7 Remove the spring and associated components from the gear lever.
8 Refitting is the reverse of the removal procedure, but note the following additional points:

(a) Smear the ball on the end of the gear lever with general purpose grease
(b) Overdrive models: Feed the switch leads through the groove in the gear lever knob, allowing about 2.5 in (65 mm) to connect to the switch. Use plastic tape to secure the leads to the gear lever

10 Gearbox extension rear oil seal – removal and refitting

1 Oil leaking from the gearbox extension rear oil seal indicates that the seal is defective and that renewal is necessary. This is a relatively simple job and can be done without removing the gearbox or extension.
2 Apply the handbrake and jack-up the front of the vehicle. Support it firmly on axle stands.
3 Clean off the dirt and oil residues from the vicinity of the rear oil seal. With a file or similar tool mark the propeller shaft and gearbox drive flange alignment for correct reassembly.
4 Remove the four locknuts from the flange bolts to release the propeller shaft from the gearbox and tie the shaft out of the way to the exhaust pipe.
5 Hold the gearbox drive flange with a lever on bolts inserted in the flange or with a special spanner. Undo and remove the flange retaining nut and lockwasher.
6 Withdraw the drive flange from the gearbox shaft, using a universal puller if it is tight.
7 Prise the old seal out of the gearbox extension. It doesn't matter if it gets damaged, but take care not to damage the extension or the bearing inside.
8 Wipe the seal location in the gearbox extension clean. Dip the new seal in gearbox oil and fit it to the extension. Make sure that the seal lip faces inwards and tap it carefully home until it is flush with the extension.
9 Smear both faces of the drive flange and the bearing face with Molygrease (Moly 500), or a suitable alternative, and fit the flange to the gearbox extension and shaft.
10 Refit the lockwasher and nut, and tighten the nut to the specified torque.
11 Reconnect the propeller shaft flange to the drive flange, aligning the marks made during removal. Fit and tighten the four bolts and nuts connecting the drive flanges.

11 Gearbox front oil seal – removal and refitting

1 An oil contaminated clutch can be due to oil leaking from either the engine rear seal or the gearbox front seal. In either case the gearbox will have to be removed to correct the fault. Refer to Chapter 1 if the engine seal is leaking.
2 In the case of a defective front seal in the gearbox, remove the gearbox assembly as explained in Section 2 of this Chapter.
3 Undo the nut and remove the clutch withdrawal lever pivot bolt to release the lever. Disengage the rubber dust cover in the bellhousing and remove the lever and dust cover.
4 Clean off all oil residues and clutch dust from inside the bellhousing. Remove the nuts and spring washers retaining the front cover and remove the front cover. Retrieve the shims fitted between the front cover and the first motion (input) shaft bearing, noting their number and location for correct refitting.

11.7a The clutch lever and clutch release bearing fitted with ...

11.7b ... the securing pins in position

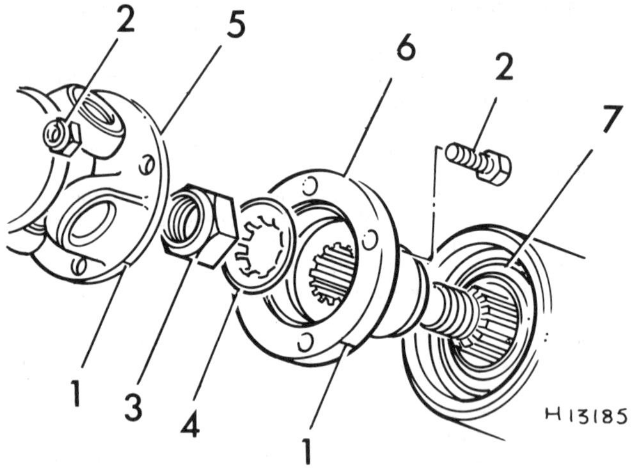

Fig. 6.5 Renewing the gearbox extension rear oil seal (Sec 10)

1 Alignment marks on the drive flanges
2 Flange coupling nut and bolt
3 Gearbox flange retaining nut
4 Nut lockwasher
5 Propeller shaft flange coupling
6 Gearbox flange coupling
7 Oil seal

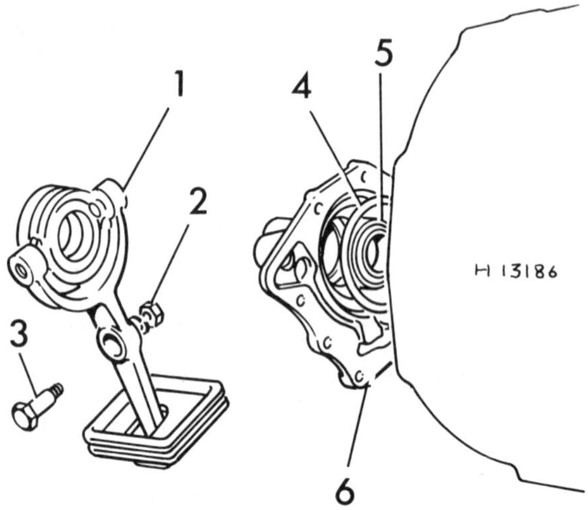

Fig. 6.6 Renewing the gearbox front oil seal (Sec 11)

1 Clutch lever
2 Pivot bolt nut
3 Pivot bolt
4 Shims
5 Oil seal in front cover
6 Front cover

5 Prise the oil seal out of the front cover, clean the aperture and fit a new seal. Make sure that the seal lip faces into the gearbox.

6 Clean off all traces of the old gasket and fit a new gasket, lightly smeared with a suitable sealant. Locate the shims in the shim register in the front cover and refit the cover to the gearbox.

7 Continue the reassembly following the removal procedure in reverse. Before refitting the clutch withdrawal lever, coat the pivot bolt and the rear half of the clutch release bearing with a good quality graphite grease (photos).

12 Overdrive – general description

Overdrive is fitted as an optional extra to the Sherpa. Only limited servicing is possible without a range of special tools and the work involved is beyond the scope of the home mechanic. Nevertheless, an understanding of the operation of the overdrive unit will enable an owner to obtain the best use of the installation and will help in identifying problems if they could arise.

The Laycock Type LH Overdrive is fitted between the gearbox and the propeller shaft. When it is engaged it provides a higher overall gear ratio than that provided by the gearbox alone. If used correctly it can reduce fuel consumption and engine wear by reducing engine rpm in top gear, but the engine must not be made to labour. An inhibitor switch in the overdrive actuation circuit prevents engagement in any but third and fourth gears. It is not necessary to use the clutch or vary the throttle opening when engaging or disengaging overdrive since the unit is designed to operate while transmitting full power. *If for any reason the overdrive fails to disengage, do not reverse the vehicle until the fault is cleared or damage may result.*

The overdrive gears are epicyclic and comprise a central sun wheel meshing with the three planet gears which in turn mesh with an internally toothed annulus gear. The planet carrier is attached to the input shaft and the annulus gear is integral with the output shaft. All gears are maintained in constant mesh.

Power is transmitted from the gearbox mainshaft to the inner member of a one-way clutch and then to the outer member when direct drive is being used. The outer member of the one-way clutch forms part of the combined annulus and output shaft, and the gear train is inoperative.

A spring loaded cone clutch engages with the friction surface on the annulus and therefore the clutch rotates with the annulus and output shaft. The sun wheel is splined to the clutch member; hence the whole gear train is locked, and overrun and reverse torque can be transmitted. Due to the helix angle of the sun wheel gearteeth additional load is imparted during overrun and reverse as the sun wheel reacts against the cone clutch.

When overdrive is engaged, hydraulic pressure causes the clutch

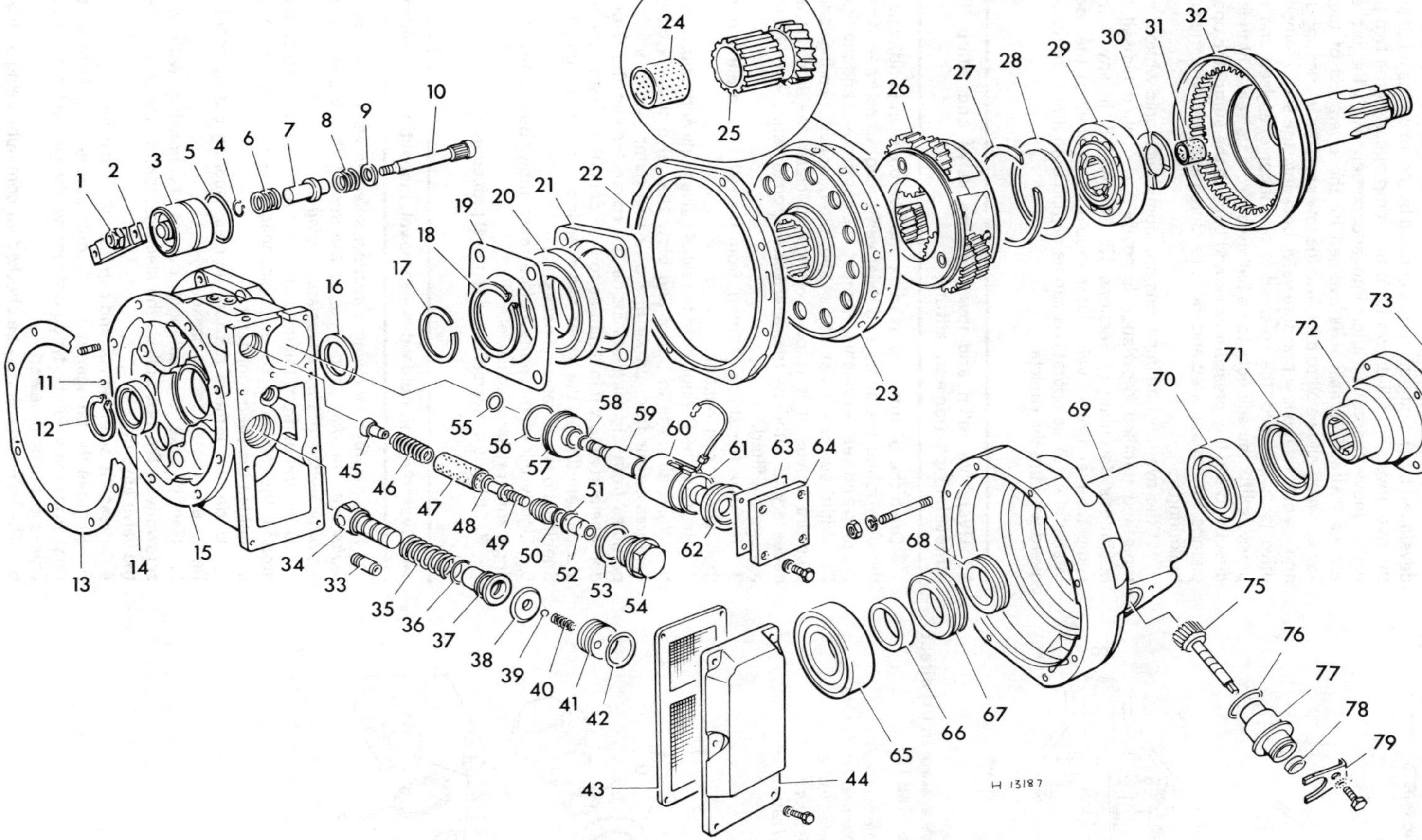

Fig. 6.7 Exploded view of the Laycock Type L.H. Overdrive (Sec 12)

1 Locknut
2 Bridge piece
3 Piston
4 Circlip
5 O-ring
6 Spring
7 Thrust rod
8 Spring
9 Spacer
10 Bolt
11 Ball
12 Circlip
13 Gasket
14 Cam
15 Case
16 Thrust washer
17 Split ring
18 Circlip
19 Bearing plate
20 Bearing
21 Bearing housing
22 Brake ring
23 Cone clutch sliding member
24 Bush
25 Sun wheel
26 Planet carrier
27 Circlip
28 Oil slinger
29 One-way clutch
30 Thrust washer
31 Bush
32 Annulus
33 Suction tube
34 Plunger
35 Spring
36 O-ring
37 Pump body
38 Valve seat
39 Ball
40 Spring
41 Plug
42 O-ring
43 Strainer
44 Sump
45 Valve
46 Spring
47 Spacer
48 Shim
49 Valve plunger
50 Valve body
51 O-ring
52 Filter
53 Washer
54 Plug
55 O-ring
56 O-ring
57 Valve base
58 Ball
59 Plunger
60 Solenoid body
61 Core
62 Cap
63 Gasket
64 Cover
65 Bearing
66 Spacer
67 Speedometer drivegear
68 Spacer
69 Casing
70 Bearing
71 Oil seal
72 Flange
73 Washer
74 Locknut
75 Speedometer pinion
76 O-ring
77 Bearing
78 Oil seal
79 Clip

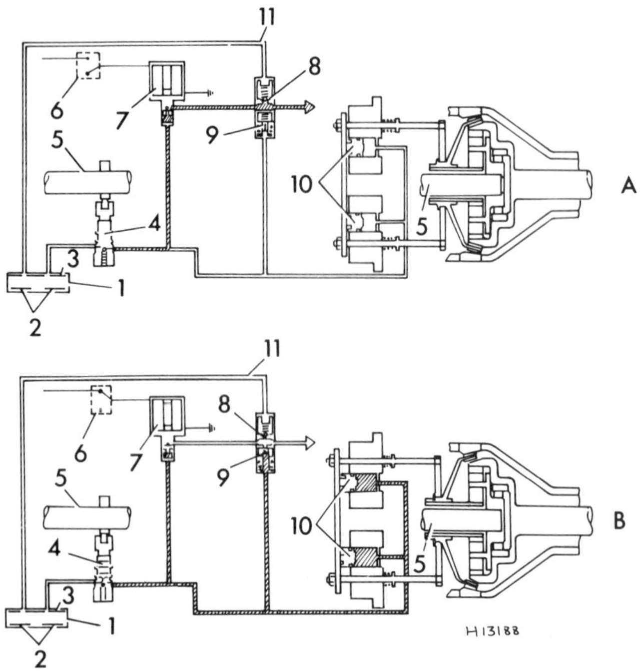

Fig. 6.8 Diagrammatic views of the overdrive system in (A) direct drive, and (B) overdrive engaged (Sec 12)

1 Sump
2 Magnets
3 Strainer
4 Pump
5 Gearbox mainshaft
6 Switch
7 Solenoid valve
8 Low pressure valve
9 Relief valve
10 Pistons
11 Oil return

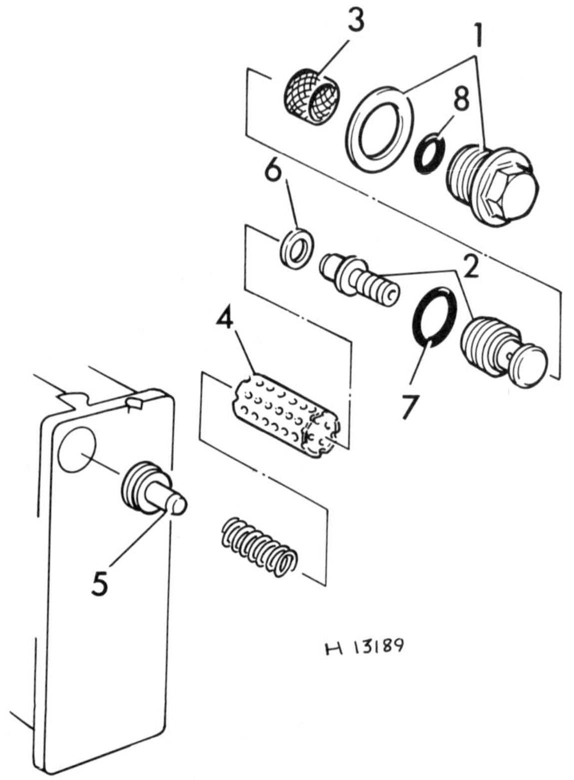

Fig. 6.9 Overdrive relief and low pressure valve (Sec 13)

1 Valve plug and washer
2 Relief valve assembly
3 Filter
4 Spacer
5 Low pressure valve
6 Shim(s)
7 Relief valve O-ring
8 Plug O-ring

to move. It disengages from the annulus and engages with the brake ring in the overdrive casing. The sun wheel, to which the clutch is attached, is therefore held stationary. The planet carrier now rotates with the input shaft and the annulus is therefore driven by the planet wheels as they rotate about their own axes. Since the output shaft speed is greater than the input shaft speed, the outer member of the clutch overruns the inner member to permit this drive.

The hydraulic pressure required to operate the cone clutch is developed by an integral pump operated by a cam on the input shaft to the overdrive unit. Oil drawn from the sump operates the two pistons which move the cone clutch into engagement with the brake ring. Excess hydraulic pressure is avoided by the operation of the relief valve, which allows oil to pass into the low pressure lubricating system and thence via the low pressure valve back into the sump. When overdrive is disengaged, the solenoid is no longer energised and the oil pressure lifts the solenoid valve ball away from its seat. Oil thus flows directly to the low pressure valve and back to the sump, the operating pistons are not actuated and the cone clutch engages with the annulus brake ring.

If the overdrive malfunctions, the home mechanic without special tools and detailed knowledge is advised to confine himself to the operations described in Sections 13, 14 and 15. In any event, the sump filter and relief valve filter should be cleaned at the specified intervals. All these operations can be carried out without removing the overdrive from the vehicle.

13 Overdrive relief and low pressure valve and sump filter – removal, inspection and refitting

1 Drain the oil from the gearbox and overdrive unit and wipe clean the area round the sump and relief valve plug. Unscrew and remove the plug and sealing washer. Remove the sump cover and the filter.

2 Withdraw the relief valve and filter, spacer, spring and low pressure valve. (If it is only wished to clean the filter, it is only necessary to withdraw the relief valve approximately $\frac{1}{2}$ in (13 mm) to remove the filter).

3 Remove the shims (if fitted) from the relief valve plunger, noting their number and location.

4 Remove the O-rings from the relief valve body and plug.

5 Wash all parts in petrol and blow the valve oilways clear with compressed air from an air line or a tyre pump. Clean any metallic particles from the magnets inside the cover. Inspect the O-rings for damage and renew if in doubt as to their condition. Check the free and loaded lengths of the relief valve spring with those given in the Specifications and renew if necessary.

6 Lightly oil all the parts before reassembly. Before fitting the O-rings, smear them with petrol.

7 Refitting is a reversal of the removal procedure.

14 Overdrive solenoid valve – removal, inspection and refitting

Note: *Before renewing a suspect solenoid valve, make sure that faults are not due to defects in the switch in the gear knob, the inhibitor switch, or their associated wiring*

1 Drain the oil from the gearbox and overdrive. Wipe clean the area around the solenoid cover. Disconnect the electrical lead to the solenoid at the snap connector.

2 Undo and remove the four screws securing the solenoid cover. Remove the cover and its gasket.

3 Gently prise the solenoid out of its housing with a small screwdriver. **Do not** pull on the electrical lead to remove it. Remove the valve assembly.

4 Press the solenoid coil and cap from their housing.

5 Remove the plunger and bal from the valve body. Remove the O-rings from the valve body; do not remove the O-ring from the plunger unless it is to be renewed.

6 The solenoid coil may be checked for continuity using an electrical multi-meter. If it is short-circuited or open-circuited it must be renewed.

7 Wash the valve parts in petrol and examine the valve ball and seat for pitting and scoring. A solenoid valve repair kit is available for the renewal of worn parts.

8 Examine the O-rings for deterioration and renew them if necessary. Clean any corrosion from the inside of the solenoid cover

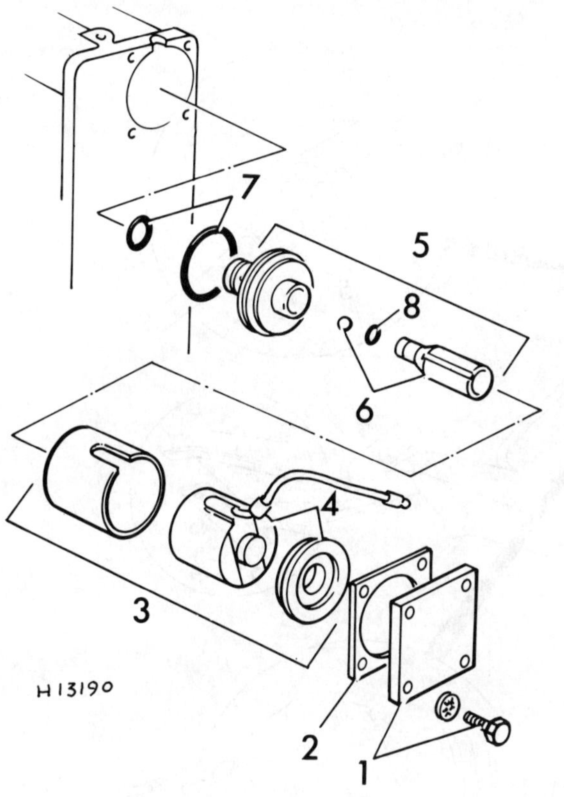

Fig. 6.10 Exploded view of overdrive solenoid valve (Sec 14)

1 Cover and retaining screw
2 Gasket
3 Solenoid assembly
4 Solenoid coil and cap
5 Solenoid valve assembly
6 Valve plunger and ball
7 Valve body O-rings
8 Valve plunger O-ring

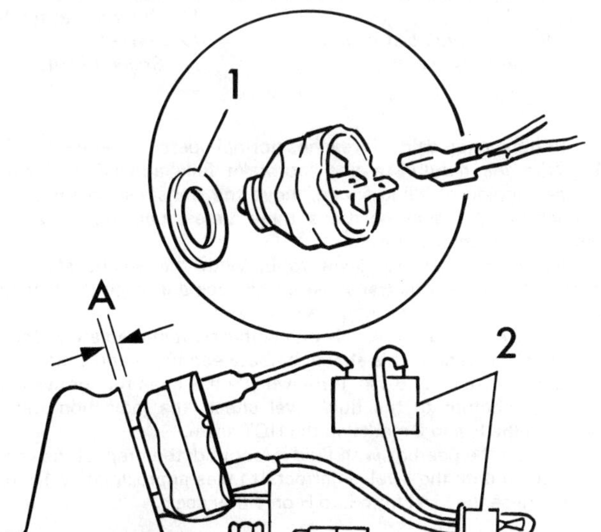

Fig. 6.11 Overdrive inhibitor switch (Sec 15)

1 Washer(s)
2 Test lamp and battery
A See text

with fine emery cloth.

9 Refitting is a reversal of the removal procedure. Smear the O-rings with gearbox oil. Make sure that the grommet on the solenoid electrical lead fits into its slot and fit a new cover gasket.

15 Overdrive inhibitor switch – removal, adjustment and refitting

1 The overdrive inhibitor switch is mounted on the gearchange lever housing; its function is to prevent the engagement of the overdrive in any but third and fourth gears.

2 The operation of the switch may be checked while it is still in place. Disconnect the leads from the switch and connect a battery and test lamp to the switch. The lamp should light only when third or fourth gear is selected. If it lights in any other position, or fails to light at all, the switch must be adjusted or renewed as described below.

3 To remove the inhibitor switch, disconnect the leads and unscrew it with a punch. Do this gently if it is hoped to re-use the switch.

4 Remove the switch, noting the position and number of the washers fitted beneath it.

5 When refitting the switch it must first be adjusted as described below.

6 Fit the switch without the washers and connect a test lamp and battery to it. Adjust the switch so that the lamp lights when first or second gear is selected.

7 Withdraw the switch until the lamp does not light with first or second gear selected, even when the gearchange lever is moved from side to side. Measure the gap (A in Fig. 6.11) between the switch and the gearchange lever housing and fit washers to a total thickness of 0.02 in (0.5 mm) more than the gap.

8 Check that the switch operates correctly when fitted with the correct thickness of washers. Reconnect the leads to the switch.

16 Overdrive – removal and refitting

1 Apply the handbrake, jack-up the front of the vehicle and support it firmly on axle stands. Disconnect the battery earth lead.

2 Undo and remove the gearbox drain plug, and drain the oil into a suitable container.

3 With a file or similar tool mark the overdrive and propeller shaft drive flanges so that they can be reassembled in the same relative position.

4 Undo and remove the nuts and washers from the four bolts in the drive flanges. Remove the bolts and separate the propeller shaft from the overdrive. Tie the propeller shaft up out of the way to the exhaust pipe.

5 Undo the knurled nut connecting the speedometer drive cable to the overdrive and disconnect the cable; tuck it out of the way.

6 Disconnect the overdrive solenoid lead at the snap connector.

7 Support the gearbox on a jack so that the weight is just taken off the overdrive mountings.

8 Remove the two bolts which secure each mounting to the overdrive adaptor.

9 Remove the bolts, washers and the reinforcing strip which retain each end of the rear crossmember to the chassis, and remove the crossmember from the vehicle.

10 Undo and remove the overdrive securing nuts and withdraw the overdrive unit. If the overdrive will not separate from the gearbox, remove the overdrive sump and filter gasket. Using a suitable lever inserted between the thrust assembly and the cone clutch, carefully free the clutch. Remove the old gasket.

11 Refitting is the reverse of the removal procedure, but note the following additional points:

(a) Use new gaskets in place of the old ones

(b) The sun wheel splines will need aligning with the planet carrier and one-way clutch. A special tool is provided by the maker for this but if an old output shaft from a gearbox can be obtained, this will do the job

(c) When fitting the overdrive to the gearbox turn the drive flange slightly to and fro to engage the shaft splines. When engaged, rotate the drive flange to locate the lowest part of the cam in line with the pump plunger

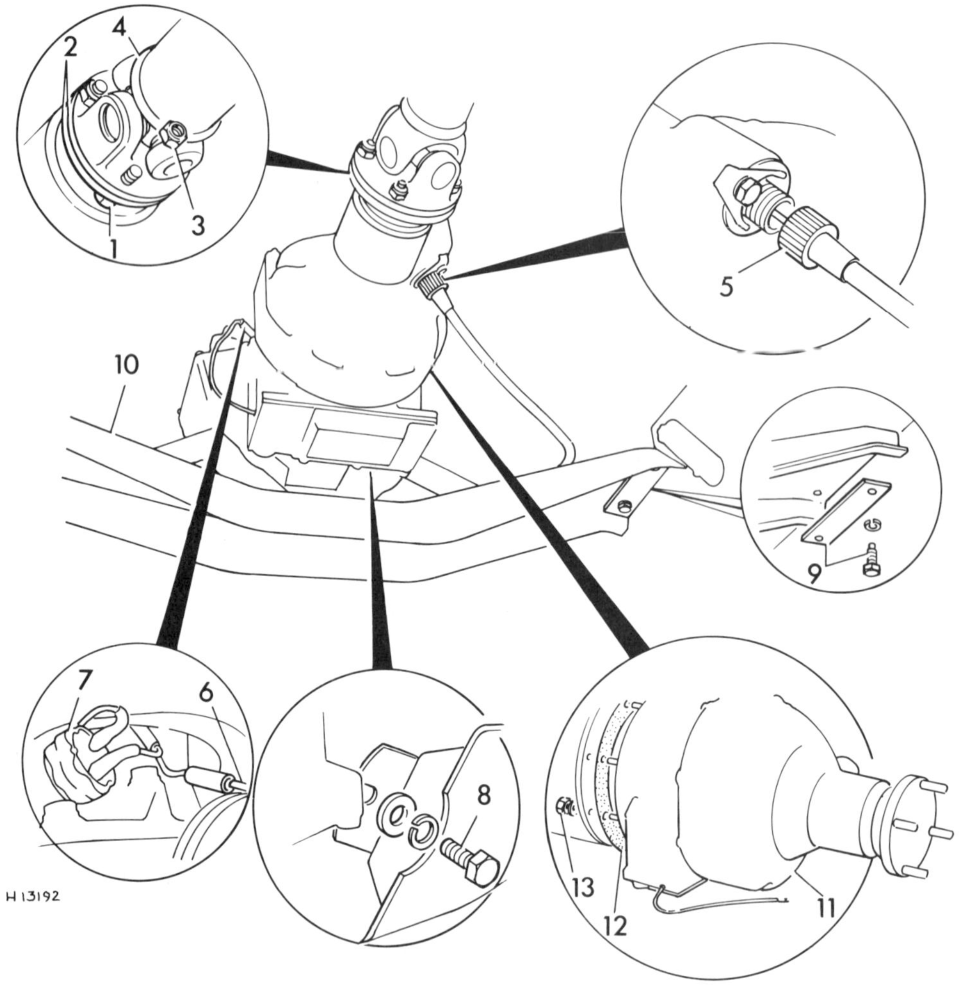

Fig. 6.12 Removing the overdrive unit (Sec 16)

1 *Flange coupling bolt*
2 *Alignment marks on coupling flanges*
3 *Locknut*
4 *Propeller shaft*
5 *Speedometer drive nut*
6 *Overdrive solenoid lead snap connector*
7 *Overdrive inhibitor switch*
8 *Overdrive adaptor mounting bolt*
9 *Reinforcing strip and retaining bolt*
10 *Rear crossmember*
11 *Overdrive assembly*
12 *Gasket*
13 *Securing nuts*

17 Automatic transmission – general

The automatic transmission takes the place of the conventional clutch and gearbox, and comprises two main elements, a torque converter and a hydraulically operated epicyclic gearbox.

The hydraulic torque converter provides additional engine torque for starting from rest and accelerating in first and second ratios. Torque multiplication is infinitely variable down to 1:1 when the stator rotates at the same speed as the impeller and turbine.

The torque/speed responsive gearbox comprises a planetary gear set providing three forward ratios and one reverse ratio.

Due to the complexity of the Borg-Warner automatic transmission unit, if performance is not up to standard, or overhaul is necessary, it is imperative that the vehicle is taken to a BL main agent who will have special equipment for accurate fault diagnosis and rectification. The content of the following Sections is therefore limited to general and servicing information.

18 Automatic transmission – fluid level checking

1 It is essential that only the specified grade of automatic transmission fluid is used for topping-up and that absolute cleanliness is observed. Check the level as follows.

2 Park the vehicle on level ground, apply the handbrake and select P.

3 Run the engine until it reaches normal operating temperature.

4 With the handbrake and footbrake fully applied, and with the engine running at idling speed, move the transmission selector lever through all the gear positions three times, pausing for about 10 seconds at each position.

5 Return the selector lever to P. With the engine still running, remove the automatic transmission dipstick and wipe the blade clean with non-fluffy rag or a clean piece of paper.

6 Re-insert the dipstick fully and withdraw it immediately. Check the fluid level indication. This should be between the H and L marks on the COLD scale (see Fig. 6.14). If the vehicle has been run for five miles or more just before to the fluid level check, the indication should be between the H and L marks on the HOT scale.

7 Top-up the gearbox with fluid if required, then repeat the checking procedures until the level is correct. It takes approximately 1 pint (0.6 litre) to raise the level from L to H on either scale.

19 Automatic transmission – removal and refitting

1 Any suspected faults must be referred to the main agent *before* removing the unit, as with this type of transmission fault diagnosis must be confirmed using special test equipment before the transmission is removed from the vehicle.

2 As the automatic transmission is heavy it is best if the vehicle is raised from the ground on ramps, but it is possible to remove the unit if

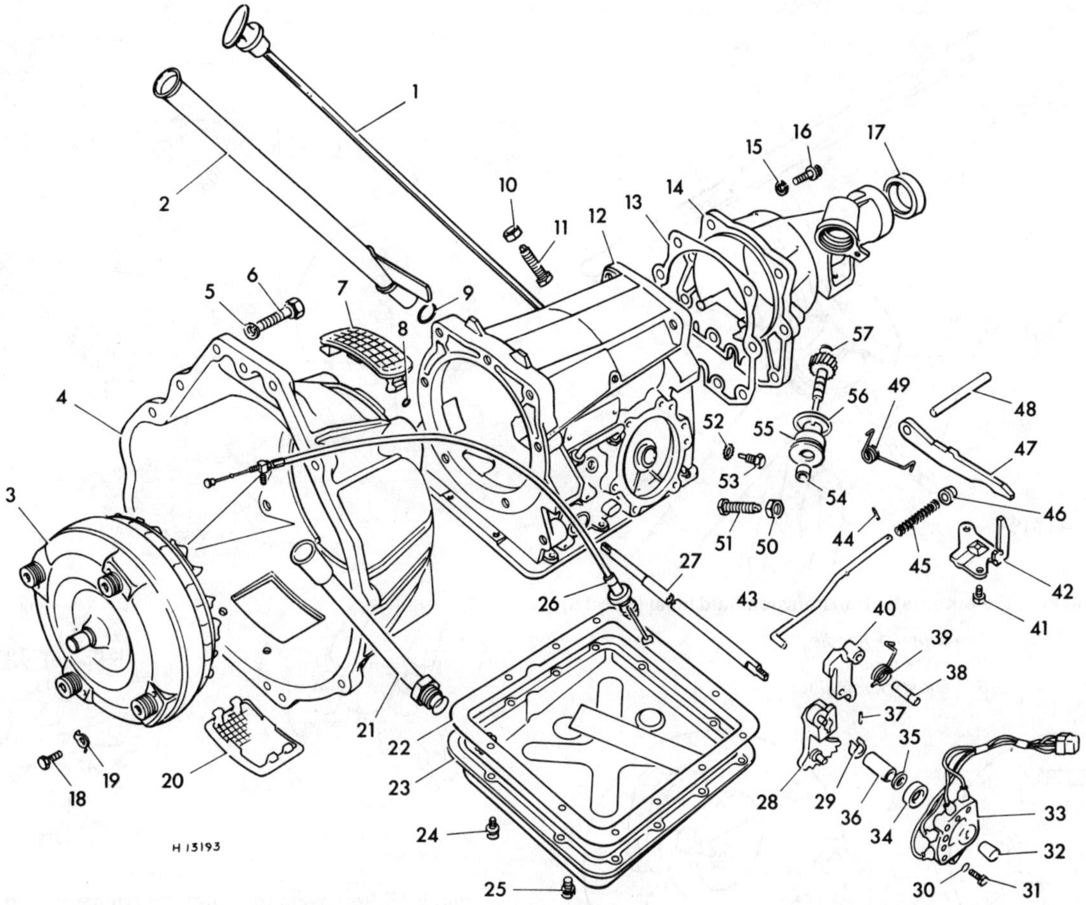

Fig. 6.13 Exploded view of the automatic transmission external components (Sec 17)

1 Dipstick
2 Filler tube
3 Torque converter
4 Torque converter housing
5 Spring washer
6 Bolt
7 Stone guard
8 O-ring
9 O-ring
10 Locknut
11 Adjusting screw for rear brake band
12 Case
13 Gasket
14 Extension housing
15 Spring washer
16 Bolt
17 Oil seal
18 Bolt
19 Lockwasher
20 Stone guard
21 Filler tube extension
22 Gasket
23 Sump
24 Bolt
25 Drain plug
26 Downshift cable
27 Control shaft
28 Detent lever
29 Spring clip
30 Washer
31 Bolt
32 Thread protector
33 Starter inhibitor/reversing lamp switch
34 Oil seal
35 Washer
36 Spacer
37 Pin
38 Pin
39 Spring
40 Parking pawl lever
41 Bolt
42 Cam plate
43 Parking rod
44 Pin
45 Spring
46 Cam
47 Parking pawl
48 Pin
49 Spring
50 Locknut
51 Adjusting screw for front brake band
52 Lockwasher
53 Bolt
54 Oil seal
55 Speedometer pinion housing
56 O-ring
57 Speedometer pinion

the vehicle is placed on high axle stands.

3 Disconnect the battery earth lead.

4 Disconnect the downshift cable from the throttle linkage.

5 Withdraw the dipstick. Unclip the breather pipe from the upper filler tube.

6 Undo the clamp or flange nuts and disconnect the exhaust pipe from the manifold.

7 Place a suitable container of about 10 pints (6 litres) capacity under the transmission sump drain plug, remove the plug and allow the fluid to drain. Refit the plug.

8 Remove the bolt, nut and washer retaining the upper filler tube bracket to the converter housing, and withdraw the upper filler tube from the lower filler tube.

9 Push the clips on the selector rod clear of the gearbox and actuator levers, and remove the selector rod.

10 Disconnect the starter inhibitor/reversing lamp switch leads at the multi-connector plug.

11 Wipe the area around the two oil cooler pipe unions, undo the unions and wrap the pipe ends in plastic to exclude dirt. Similarly seal the union adaptors on the transmission.

12 Remove the bolt, spring and plain washer retaining the speedometer cable plate and pull the speedometer cable from the drive pinion.

13 With a file or similar tool, mark the propeller shaft and transmission shaft drive flanges to permit correct reassembly. Undo and remove the bolts and locknuts from the drive flanges. Tie the propeller shaft out of the way to the exhaust pipe.

14 Support the engine with a jack (or with a sling and hoist) and take enough weight to relieve the rear mounting.

15 Disconnect each end of the rear crossmember from the chassis. Remove the nut and washers securing the crossmember to the rear transmission mounting and remove the crossmember from the vehicle.

16 Lower the engine just enough to gain access all round the transmission mounting flange. Support the transmission with a trolley jack.

17 Remove the bolts which hold the converter housing to the transmission adaptor plate and, making sure that the transmission is stable on the trolley jack, draw the transmission rearwards clear of the

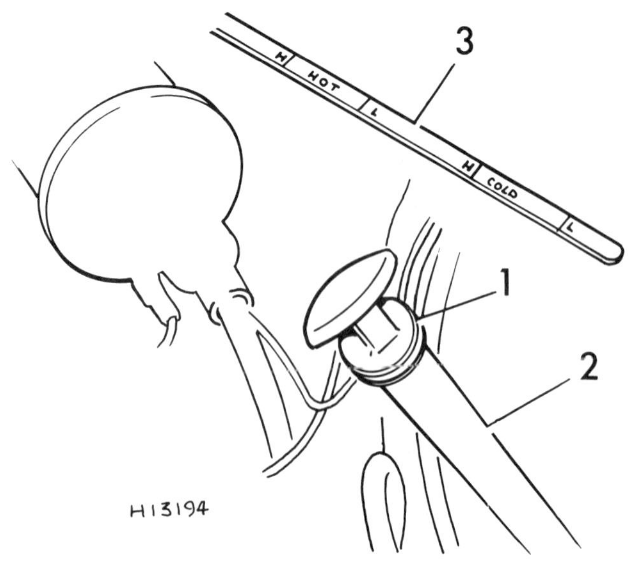

Fig. 6.14 Checking the automatic transmission fluid level (Sec 18)

1 Dipstick handle
2 Filler tube
3 Dipstick markings

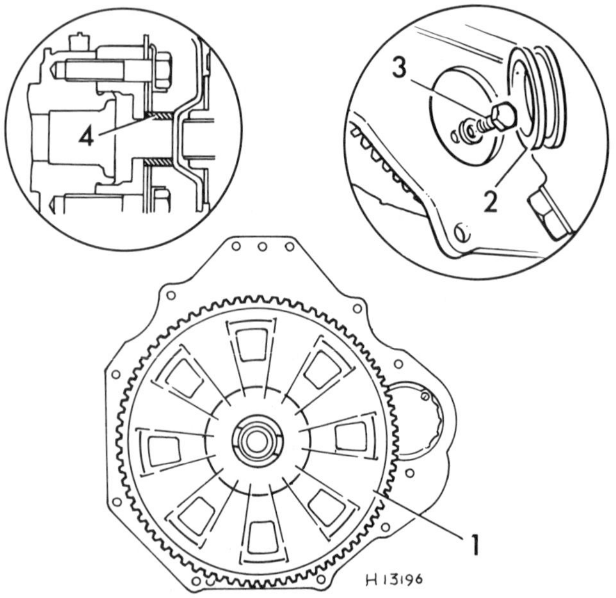

Fig. 6.16 Removing the torque converter (Sec 20)

1 Torque converter
2 Adaptor plate cover
3 Torque converter securing bolts
4 Centre spigot abutment ring

torque converter. It is important not to allow the input shaft to the transmission to take any weight.

18 Refitting the transmission is essentially the reverse of the removal procedure, but the following points should be noted:

(a) Align the torque converter and pump driving dogs and slots horizontally
(b) Carefully align the transmission and locate the input shaft and driving dogs
(c) Line up the mating marks previously made on the propeller shaft and gearbox drive flanges
(d) Refill the transmission with new fluid of the specified quantity and grade and check the level as explained in Section 18

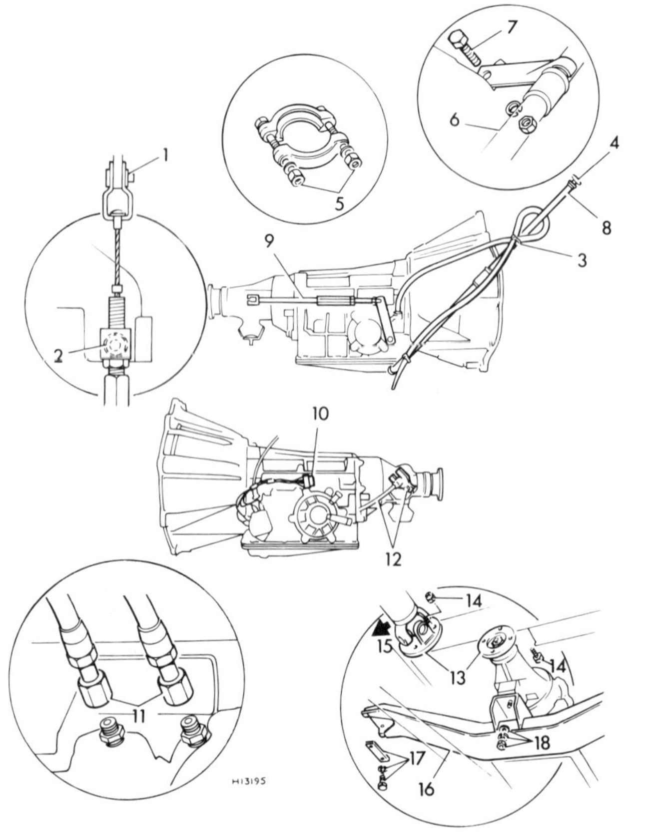

Fig. 6.15 Removing the automatic transmission (Sec 19)

1 Throttle linkage
2 Downshift cable securing nut and washers
3 Breather pipe clipped to upper filler tube
4 Dipstick
5 Exhaust pipe clamp (18V engine shown)
6 Upper filler tube
7 Upper filler tube bracket securing bolt
8 Upper filler tube
9 Selector rod
10 Starter inhibitor/reversing lamp switch lead connector plug
11 Oil cooler pipes connectors
12 Speedometer drive cable and securing bolt
13 Coupling drive flanges
14 Coupling bolts and nuts
15 Propeller shaft
16 Rear crossmember
17 Reinforcing plate and securing bolts
18 Rear mounting securing nut and washers

20 Automatic transmission torque converter – removal and refitting

1 Refer to the previous Section and remove the automatic transmission.

2 Take off the cover from the transmission adaptor plate.

3 Mark the relative positions of the driveplate and the torque converter if it is intended to refit them.

4 Rotate the crankshaft with a spanner on the crankshaft pulley nut to gain access to the four bolts securing the torque converter to the driveplate. Progressively slacken the bolts. Support the torque converter, remove the bolts and withdraw the torque converter from the engine.

5 Remove the abutment ring from the centre spigot of the torque converter.

6 **Do not** flush out the torque converter with fuel or any other petroleum based liquid. If flushing is thought to be necessary, use only the specified grade of automatic transmission fluid.

7 Refitting of the torque converter is the reverse of the removal procedure. Tighten the four bolts securing the converter to the driveplate to the specified torque.

21 Automatic transmission rear extension – removal and refitting

1 Apply the brakes, jack-up the front of the vehicle and support it firmly on axle stands. Place the automatic transmission gear selector in N.
2 Place a suitable container of about 10 pints (6 litres) capacity under the transmission sump drain plug, remove the plug and allow the fluid to drain. Refit the plug.
3 With a file or a similar tool mark the propeller shaft and transmission shaft drive flanges to ensure correct alignment on reassembly. Undo and remove the four locknuts and bolts from the drive flanges. Move the propeller shaft to one side and tie it to the exhaust pipe.
4 Hold the transmission flange and undo and remove the shaft locknut and plain washer. The flange can be held with a lever and bolts in the bolt holes. Remove the drive flange from the shaft, using a universal puller if it proves tight.
5 Remove the bolt, spring and plain washers retaining the speedometer cable plate and pull the speedometer drive cable from the drive pinion.
6 Support the gearbox with a jack and take just enough weight to relieve the extension rear mounting.
7 Remove the bolts, spring washers and the reinforcing plate securing each end of the rear crossmember to the chassis. Disconnect the rear crossmember from the rear mounting and remove the crossmember from the vehicle.
8 Remove the bolts, spring and plain washers securing the rear extension to the gearbox. Be prepared for some fluid spillage. Withdraw the rear extension from the transmission and remove the gasket.
9 If it is intended to fit a different rear extension, remove the rear mounting for transfer to the new extension.
10 Refitting is a reversal of the removal procedure, but note the following points:

(a) Use a new gasket between the transmission and extension, and tighten the securing bolts to the specified torque
(b) Tighten the output shaft flange securing nut to the specified torque
(c) Align the marks made during dismantling on the propeller shaft and output shaft drive flanges
(d) Refill the gearbox with new automatic transmission fluid, and check the level as described in Section 18

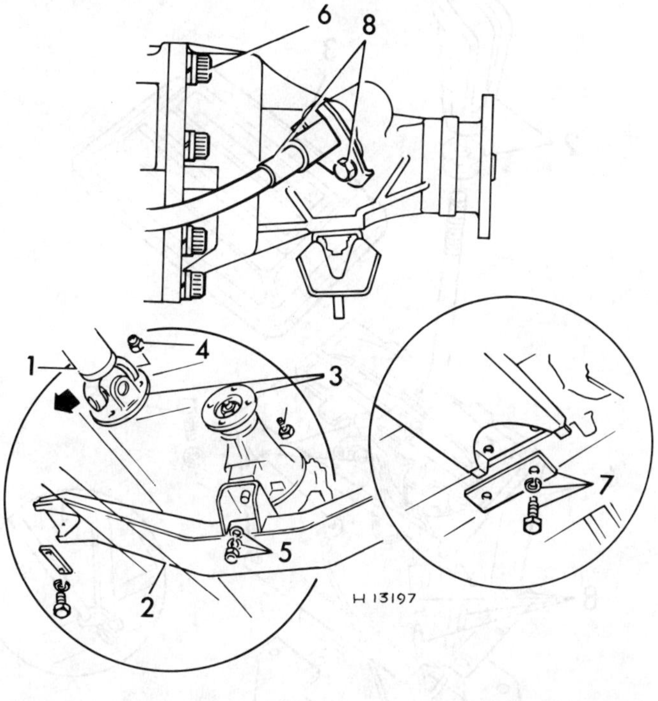

Fig. 6.17 Automatic transmission rear extension removal (Sec 21)

1 Propeller shaft
2 Rear crossmember
3 Coupling drive flanges
4 Coupling bolts and nuts
5 Rear mounting securing nut and washers
6 Rear extension securing bolts
7 Reinforcing plate and securing bolts
8 Speedometer drive and securing bolt

22 Automatic transmission rear oil seal – removal and refitting

1 Refer to Section 21 and proceed as far as removing the gearbox drive flange at the end of paragraph 4.
2 Prise out the oil seal from the extension. It doesn't matter if the seal gets damaged, but take care not to damage the rear extension.
3 Dip the new seal in clean automatic transmission fluid and gently tap it into the extension. Make sure that the seal lip faces into the gearbox and that the seal is flush with the end of the extension.
4 Reverse the removal procedure to complete the job. Tighten the output shaft flange securing nut to the specified torque and align the two flange marks when reassembling.

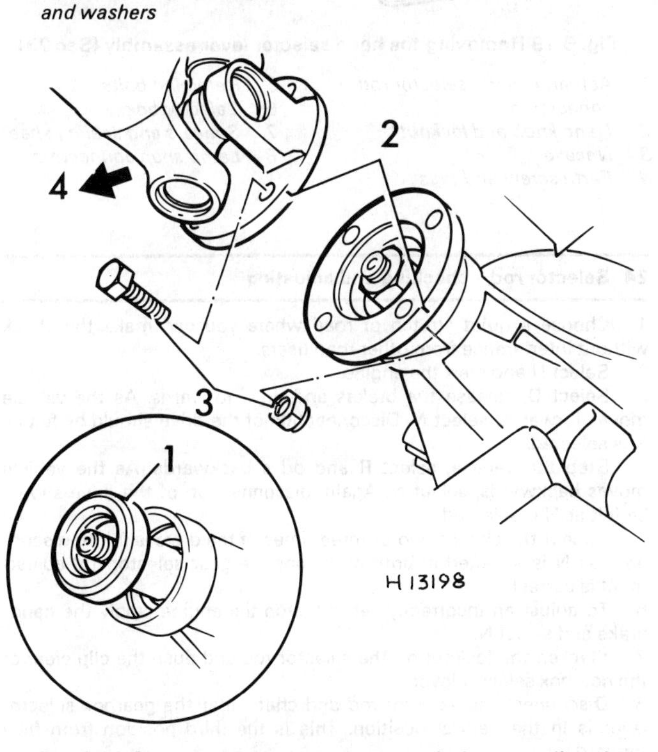

Fig. 6.18 Changing the automatic transmission rear oil seal (Sec 22)

1 Rear oil seal
2 Alignment marks on drive flanges
3 Coupling bolts and nuts
4 Propeller shaft

23 Hand selector lever assembly – removal and refitting

1 Push the selector rod clips clear of the actuator lever and disconnect the selector rod from the lever.
2 Loosen the locknut and unscrew the hand knob and locknut from the selector lever.
3 Pull the cover up from the selector box assembly.
4 Disconnect the lamp wiring at the snap connectors. Remove the earthing screw to release the earth wires and refit the screw.
5 Push aside the edge of the rubber mat surrounding the selector box to expose the retaining bolts.
6 Undo and remove the bolts and spring washers and lift out the selector box assembly.
7 If required, remove the spacers and sealing sheet from under the selector box.
8 Refitting is the reverse of the removal procedure. Check and, if necessary, adjust the selector rod, referring to Section 24.

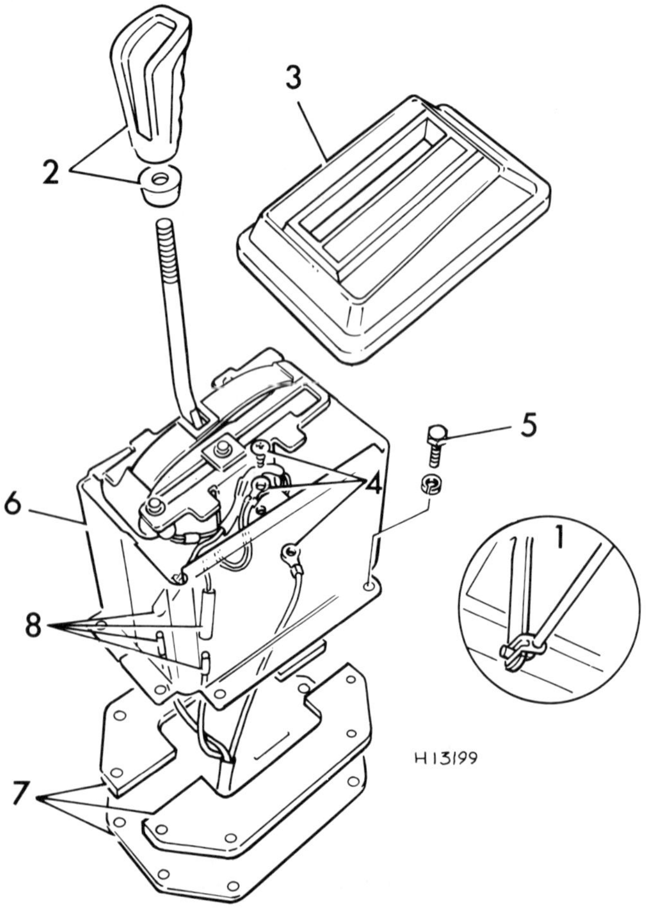

Fig. 6.19 Removing the hand selector lever assembly (Sec 23)

1 *Actuating lever/selector rod connection*
2 *Hand knob and locknut*
3 *Nacelle*
4 *Earth screw and tags*
5 *Retaining bolts*
6 *Selector box*
7 *Spacers and sealing sheet*
8 *Lamp snap connectors*

24 Selector rod – checking and adjusting

1 Choose a quiet stretch of road where you can make the check without interference from other road users.
2 Select N and start the engine.
3 Select D, release the brakes and drive forwards. As the vehicle moves forwards, select N. Disconnection of the drive should be felt as N is selected.
4 Stop the vehicle, select R and drive backwards. As the vehicle moves backwards, select N. Again, disconnection of the drive should be felt as N is selected.
5 Repeat the checks two or three times. If the drive is felt to disconnect as N is selected in both directions the gear selector rod adjustment is correct.
6 To adjust an incorrectly set rod, stop the engine, apply the handbrake and select N.
7 Slacken the locknut on the selector rod and push the clip clear of the gearbox selector lever.
8 Disconnect the selector rod and check that the gearbox selector lever is in the neutral position. This is the third position from fully forward.
9 Screw the short rod of the selector rod assembly in or out of the adjuster until the end of the rod can be located in the gearbox selector lever hole. Tighten the locknut and push the clip onto the lever to secure the selector rod.
10 Ensure that the engine will only start with N or P selected and check the rod adjustment by repeating paragraphs 1 to 5.

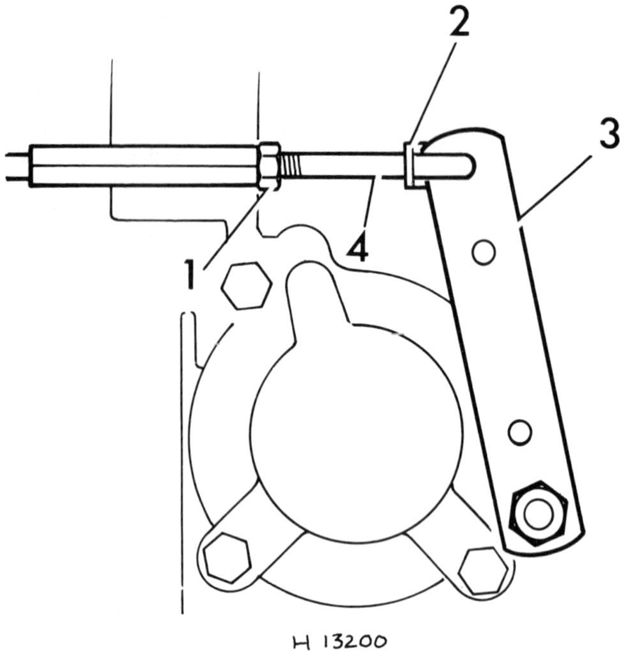

Fig. 6.20 Adjusting the selector rod (Sec 24)

1 *Selector rod locknut*
2 *Retaining clip*
3 *Gearbox selector lever*
4 *Selector rod (screw in or out to adjust)*

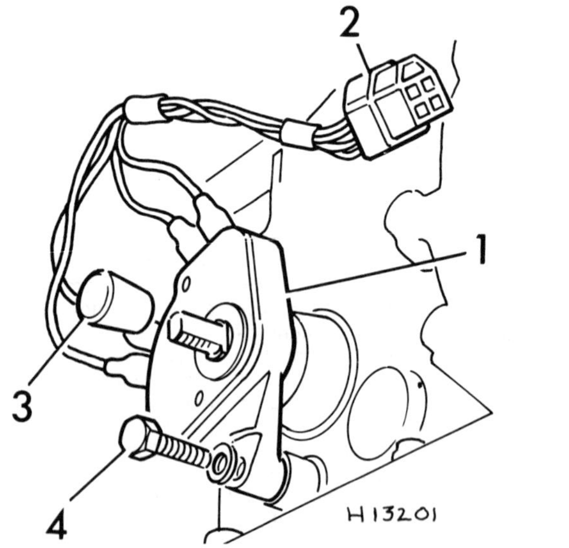

Fig. 6.21 Removing the starter inhibitor/reversing lamp switch (Sec 25)

1 *Starter inhibitor/reversing lamp switch*
2 *Multi-connector plug*
3 *Thread protector*
4 *Retaining bolt*

25 Starter inhibitor/reversing lamp switch – removal and refitting

1 Apply the handbrake, chock the wheels and disconnect the battery earth lead. **Note**: *With the battery connected the vehicle may move off if the starter is operated while the inhibitor switch is disconnected.*
2 Disconnect the switch leads at the multi-connector plug. The leads are soldered to the switch terminals.
3 Remove the thread protector from the switch spindle.
4 Undo and remove the bolt and washer retaining the switch to the gearbox case. Remove the switch.
5 Refitting is the reverse of the removal procedure. A defective switch must be renewed as no repair is possible.

26 Fault diagnosis – transmission

Symptom	Reason/s
Manual gearbox	
Ineffective synchromesh	Synchro cones worn or damaged Synchro dogs worn or damaged
Jumps out of gear	Broken selector rod detent springs or rod grooves worn Synchro sleeve groove worn Excessive gear shaft or gear endfloat Worn synchro components
Excessive noise	Low oil level or incorrect grade Worn gear teeth Worn laygear thrust washers Worn gear bushes or needle bearings
Difficulty in engaging gear	Fault in clutch system Severe wear in gear shafts or selector shafts causing ridges on shafts
Overdrive	
Overdrive does not engage	Insufficient oil in gearbox Solenoid or electrical system fault Low hydraulic pressure Damaged parts within unit
Overdrive does not disengage **(CAUTION: DO NOT REVERSE VEHICLE)**	Electrical control system fault Sticking clutch * Damaged parts within unit
Overdrive slips when engaged	Insufficient oil in gearbox Solenoid valve faulty Low hydraulic pressure Worn or glazed clutch lining
Slip in reverse or free-wheel on over-run	Worn or glazed clutch lining Broken circlip on sun wheel

** On a new unit this may be due to insufficient bedding-in. Free the clutch by giving the brake ring several sharp taps with a soft-faced mallet.*

Automatic transmission

Faults in these units are sometimes due to low fluid level which should be checked and corrected if necessary. Internal faults should be diagnosed by a BL agent or an automatic transmission specialist who will have the necessary equipment for diagnosing and rectifying the faults.

Chapter 7 Propeller shaft

For modifications, and information applicable to later models, see Supplement at end of manual

Contents

Specifications

Type	Front and rear tubular with centre bearing	
Centre joint		
Early and all automatic models	Centre flange joint	
Later models	Centre yoke joint	
Diameter of shaft		
Front	2 in (50·8 mm)	
Rear	2·25 in (57·15 mm)	
Universal joints	Hardy Spicer with roller bearings	
Torque wrench settings	**lbf ft**	**Nm**
Flange bolts/nuts	35	47
Centre bearing to frame screws	15	20

1 General description

1 Drive is transmitted from the gearbox to the rear axle by means of a finely balanced tubular propeller shaft split into two halves and supported at the centre by a rubber mounted bearing.

2 Fitted to the front and rear of the propeller shaft assembly are universal joints which allow for vertical movement of the rear axle and slight movement of the complete power unit on its rubber mountings. Each universal joint comprises a four legged centre spider, four needle roller bearings and two yokes.

3 Fore and aft movement of the rear axle is absorbed by a sliding spline in the rear half of the propeller shaft assembly. The splines mate with a sleeve and yoke assembly. When assembled a dust cap, steel washer, and cork washer seal the end of the sleeve and sliding joint.

4 The yoke flange of the front universal joint is fitted to the gearbox mainshaft flange with four bolts, spring washers and nuts, and the yoke flange on the rear universal joint is secured to the pinion flange on the rear axle in the same way.

5 Early Sherpa models, and all models fitted with automatic transmission, are fitted with a centre flange type joint connecting the front and rear halves of the propeller shaft. Later models are equipped with a centre yoke joint.

6 Two standards of propeller shaft are used, one for normal duty on low compression engine and the other for heavy duty on low compression engines and normal or heavy duty on high compression engines.

2 Propeller shaft – removal and refitting

Complete shaft removal (centre flange and centre yoke type shafts)

1 Jack-up the rear of the vehicle and support on firmly based axle stands. Alternatively position the rear of the vehicle on a ramp. Chock the front wheels.

2 The propeller shaft is carefully balanced to fine limits and it is important that it is refitted in exactly the same position after its removal. Scratch marks on the gearbox, differential pinion and propeller shaft drive flanges for correct re-alignment when refitting.

3 Support the weight of the front propeller shaft. Undo and remove the four gearbox end flange nuts and bolts (photo).

4 Support the weight of the rear propeller shaft. Undo and remove the four axle end flange nuts and bolts (photo).

5 Undo and remove the two bolts, spring and plain washers that retain the centre bearing mounting to the body brackets (photo).

6 Lift away the propeller shaft assembly from the underside of the car.

7 Refitting the propeller shaft is the reverse sequence to removal. Make sure that the alignment marks scratched before removal, paragraph 2, are lined up, and tighten the connecting bolts to the specified torque.

2.3 The gearbox end flange nuts and bolts loosened before removal

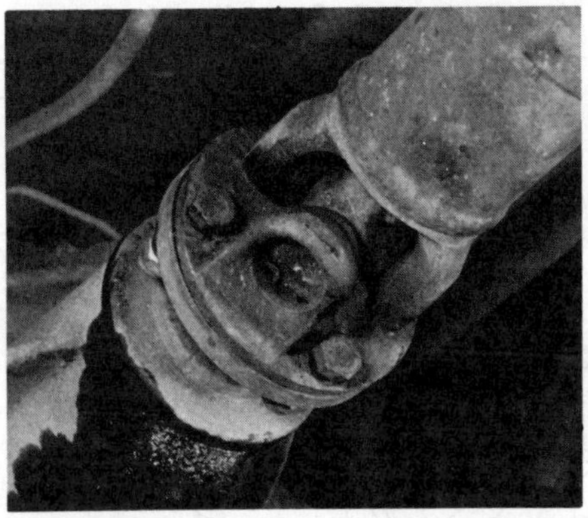

2.4 The axle end flange nuts and bolts loosened before removal

2.5 The centre bearing mounting bolts being removed

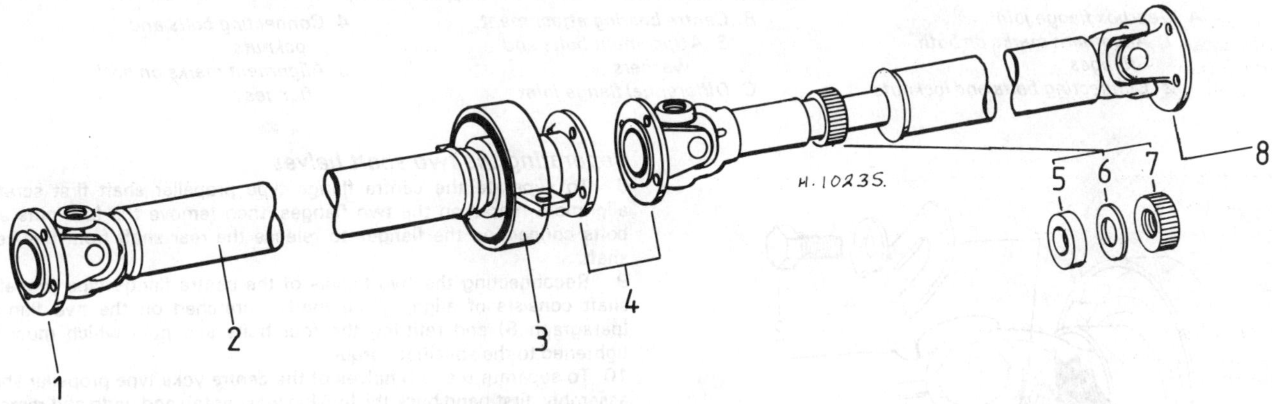

Fig. 7.1 The propeller shaft assembly, centre flange type (Sec 1)

1 Gearbox flange joint	*5 Cork washer*
2 Front propeller shaft	*6 Steel washer*
3 Centre bearing assembly	*7 Dust cap*
4 Centre flange joint	*8 Differential flange joint*

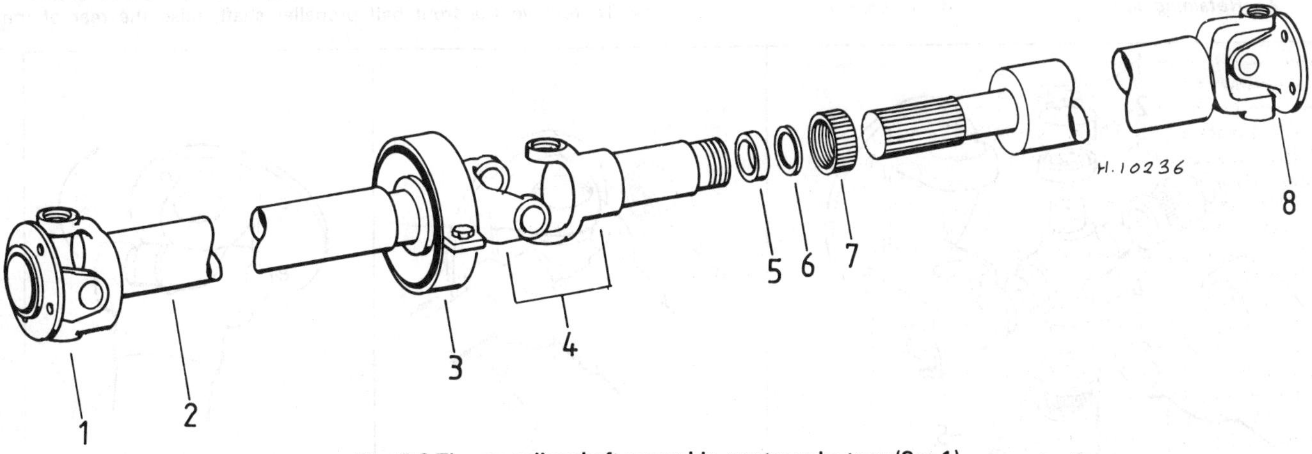

Fig. 7.2 The propeller shaft assembly, centre yoke type (Sec 1)

1 Gearbox flange joint	*5 Cork washer*
2 Front propeller shaft	*6 Steel washer*
3 Centre bearing assembly	*7 Dust cap*
4 Centre yoke joint	*8 Differential flange joint*

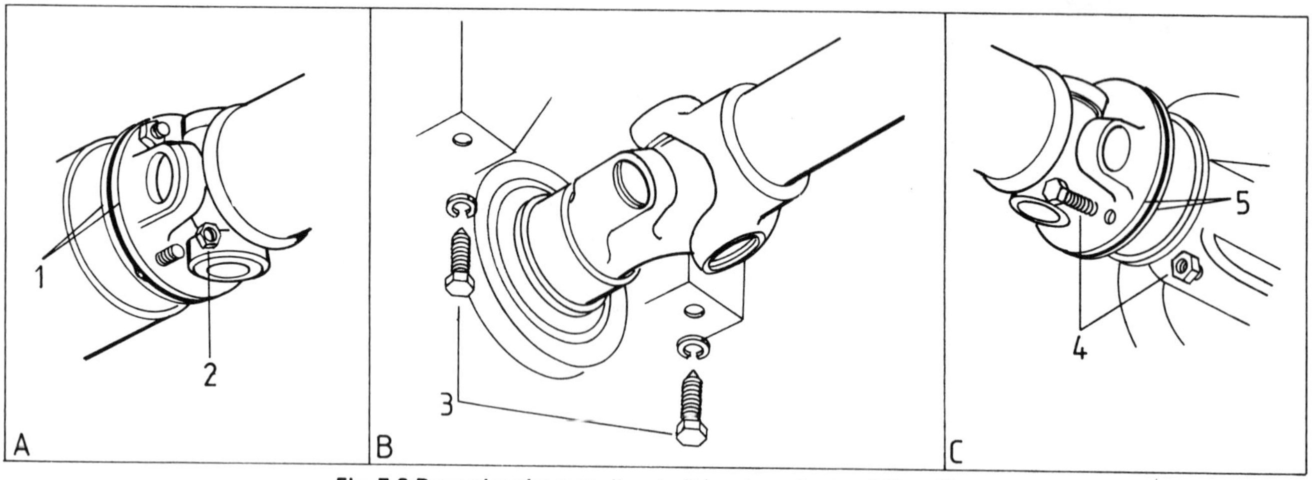

Fig. 7.3 Removing the propeller shaft (centre yoke type) (Sec 2)

A Gearbox flange joint
1 Alignment marks on both flanges
2 Connecting bolts and locknuts
B Centre bearing attachment
3 Attachment bolts and washers
C Differential flange joint
4 Connecting bolts and locknuts
5 Alignment marks on both flanges

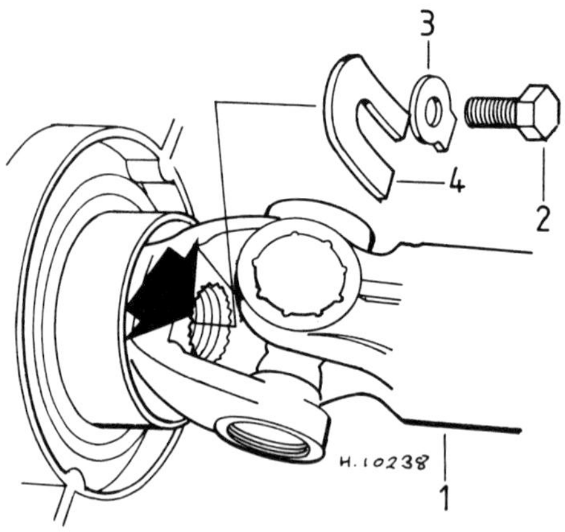

Fig. 7.4 Disconnecting the centre yoke (Sec 2)

1 Rear propeller shaft
2 Retaining bolt
3 Locking washer
4 C-washer

Separating the two shaft halves

8 To separate the centre flange type propeller shaft first scratch alignment marks on the two flanges, then remove the four nuts and bolts connecting the flanges to release the rear shaft from the front shaft.

9 Reconnecting the two halves of the centre flange type propeller shaft consists of aligning the marks scratched on the two flanges (paragraph 8) and refitting the four bolts and nuts which must be tightened to the specified torque.

10 To separate the two halves of the centre yoke type propeller shaft assembly, first bend back the locking washer tab and undo and remove the retaining bolt. Lift away the C-washer and tab washer.

11 Draw the front propeller shaft away from the rear propeller shaft universal joint splines.

12 Reconnection and refitting the two propeller shaft halves is the reverse sequence to removal.

Front or rear half propeller shaft removal (centre flange type shaft only)

13 With the centre flange type of propeller shaft either the front or rear half of the shaft can be removed independently of the other half.

14 To remove the front half propeller shaft, raise the rear of the

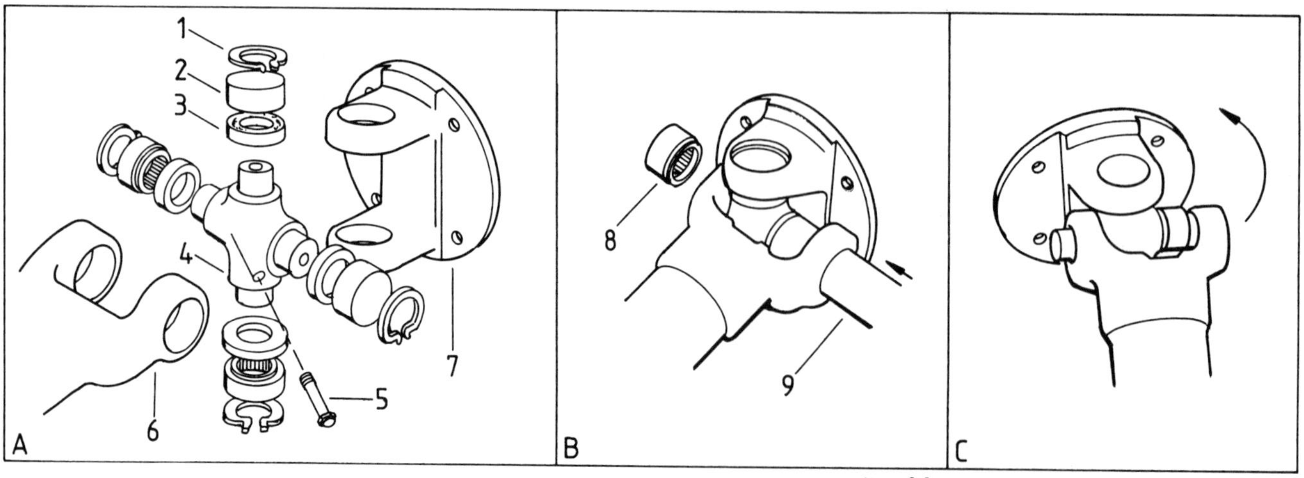

Fig. 7.5 The universal joints (Secs 3 and 4)

A Exploded view of a universal joint
1 Circlip
2 Bearing assembly
3 Seal
4 Spider
5 Grease nipple
6 Shaft yoke
7 Flange yoke
B Drifting out a bearing
8 Bearing
9 Soft metal drift
C Disconnecting the yoke from the spider

4.8 The grease nipple at the centre yoke universal joint

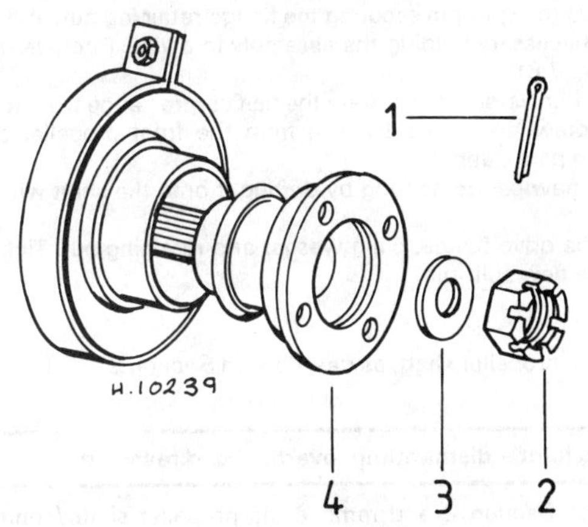

Fig. 7.6 Removing the centre flange (Sec 5)

1 *Split-pin*
2 *Retaining nut*
3 *Plain washer*
4 *Flange on front propeller shaft*

vehicle and place stands under the rear axle. Chock the front wheels.
15 Mark the propeller shaft drive flanges at the centre flange joint for correct re-alignment on refitting.
16 Remove the four nuts and bolts which connect the rear propeller shaft to the front propeller shaft and support the rear shaft.
17 Unscrew and remove the two bolts, plain and spring washers retaining the centre bearing to the body bracket.
18 Mark the gearbox and front propeller shaft drive flanges for correct re-alignment on refitting.
19 Remove the four nuts and bolts which connect the front propeller shaft to the gearbox drive flange.
20 Rotate the centre bearing through 90° and remove the front propeller shaft from the vehicle.
21 To refit the front propeller shaft reverse the removal sequence, making sure that the alignment marks made on removal are lined up. Tighten the flange connecting nuts and bolts and the centre bearing to frame bolts to the specified torque.
22 To remove the rear propeller shaft, raise the rear of the vehicle and place stands under the rear axle. Chock the front wheels.
23 Mark the propeller shaft drive flanges at the centre flange joint for correct realignment and similarly mark the differential and rear propeller shaft flanges.
24 Remove the four nuts and bolts from the centre flange joint to release the rear shaft from the front shaft. Remove the four nuts and bolts from the differential flange joint and remove the rear propeller shaft from the vehicle.
25 Refitting the rear propeller shaft consists of reversing the removal sequence. Note that the sliding yoke flange must be fitted to the front propeller shaft flange. Make sure that the alignment marks made during removal are lined up and tighten the flange connecting nuts and bolts to the specified torque.

3 Universal joints – inspection and repair

1 Wear in the needle roller bearings is characterised by vibration in the transmission, 'clonks' on taking up the drive, and in extreme cases of lack of lubrication, metallic squeaking, and ultimately grating and shrieking sounds as the bearings break up.
2 It is easy to check if the needle roller bearings are worn with the propeller shaft in position, by trying to turn the shaft with one hand, and the other hand holding the rear axle flange when the rear universal is being checked, and the front half coupling when the front universal joint is being checked. Any movement between the propeller shaft and the front, centre or rear half couplings is indicative of considerable wear. If worn, the old bearings and spider will have to be discarded and a repair kit, comprising new universal joint spiders, bearings, oil seals, and retainers purchased. Make sure this kit is available to you before dismantling.

4 Universal joints – renewal

1 Remove the propeller shaft, or the appropriate half if it is a centre flange type shaft, from the vehicle as described in Section 2.
2 Remove the grease nipple (if fitted) from the universal joint to be renewed and mark the two yokes in some way so that they can be reassembled in the same relative positions.
3 Remove the bearing retaining circlips using long-nosed pliers or a suitable screwdriver. If they are difficult to remove, apply some penetrating oil and tap the bearing housing in away from the circlip to relieve the pressure.
4 It should be possible to drive out the bearings using a soft metal drift but if this is difficult, mount the assembly in a vice and use a socket spanner which will pass through the hole in the yoke to press on the bearing, and a larger socket spanner on the other side in which to catch the opposite bearing housing as it is pressed out.
5 Repeat this operation as necessary for the other bearings and remove the spider from the yokes. It may be necessary to use some leverage to dismantle the joint.
6 Thoroughly clean the yokes and fit the new oil seals to the new spider. Pack each bearing housing with a general purpose lithium based grease. Place the spider in the yoke and pressin the bearings. Make sure that the needle rollers are not dislodged and also that the yokes are correctly aligned together.
7 Ensure that each oil seal rgisters over the lip of its bearing housing and then refit the circlips.
8 Refit the grease nipple and, with a gun containing the same general purpose lithium based grease as used in the bearing housings, grase the joint by giving it three or four strokes of the gun (photo).
9 Refit the propeller shaft as described in Section 2.

5 Centre bearing – removal and refitting

1 Refer to Section 2 and remove the propeller shaft assembly (centre yoke type shaft) or the front half of the assembly (centre flange type shaft).

Centre yoke type shaft

2 To remove the rear half of the propeller shaft proceed as described in Section 2.
3 Using a universal puller and a suitable thrust block (such as a suitable sized bolt), draw the centre bearing from the end of the front propeller shaft.
4 To fit a new bearing simply drift it into position using a piece of suitable diameter tube.
5 Reconnect the two halves of the propeller shaft as described in Section 2.

Centre flange type shaft

6 Remove the split-pin securing the flange retaining nut, and remove the nut, if necessary holding the assembly in a vice. Remove the plain washer (Fig. 7.6).

7 Using a universal puller, draw the half centre flange from the shaft. Then withdraw the centre bearing from the front propeller shaft as described in paragraph 3.

8 Fit the new centre bearing by drifting it onto the shaft with a suitable drift.

9 Refit the drive flange, plain washer and retaining nut. Tighten the nut and fit a new split-pin.

All shafts

10 Refit the propeller shaft, as described in Section 2.

6 Sliding joint – dismantling, overhaul and reassembly

1 Refer to Section 2 and remove the propeller shaft (centre yoke type shaft) or the rear half propeller shaft (centre flange type shaft) as appropriate.

2 If the splined shaft and sleeve are not marked with alignment arrows, scribe alignment marks on both so that they can be reassembled in the same relative position. Unscrew the dust cap from the sleeve and then slide the sleeve from the shaft. Take off the steel washer and the cork washer.

3 With the sleeve separated from the shaft assembly the splines can be inspected. If worn it will be necessary to purchase a new sleeve assembly, a new splined shaft assembly, or both as necessary.

4 To reassemble, fit the dust cap, steel washer, and a new cork washer over the splined part of the propeller shaft.

5 Grease the splines and then line up the arrows, or marks, on the sleeve assembly and the splined shaft, and push the sleeve over the splines. Fit the washers to the sleeve and screw up the dust cap.

6 Refit the propeller shaft (or rear half) as described in Section 2.

7 Fault diagnosis – propeller shaft

Symptom	Reason/s
Vibration when vehicle running on road	Out of balance or distorted propeller shaft Backlash in spline shaft Loose flange securing bolts Worn universal joint bearings Worn centre bearing

Chapter 8 Rear axle

For modifications, and information applicable to later models, see Supplement at end of manual

Contents

Specifications

Type	Hypoid gear, three-quarter floating	
Ratios		
185 model	4·875 : 1	
215/220/200	4·8 : 1	
240/250 LC	4·8 : 1 or 5·125 : 1	
240/250 HC	4·8 : 1	
230/250/255:		
With manual gearbox	4·8 : 1	
With automatic transmission	5·125 : 1	
Hub bearing preload	0·001 to 0·004 in (0·025 to 0·102 mm)	
Oil capacity		
185 model	2·0 pints (1·14 litres)	
All other models	3·25 pints (1·84 litres)	
Torque wrench settings	**lbf ft**	**Nm**
Pinion nut	138	187
Differential assembly retaining nuts:		
5/16 in	20	27
3/8 in	35	47
Hub nut	155	210
Spring U-bolt nuts:		
185 model	50	68
All other models	60	81
Shock absorber to axle bolts	40	54
Propeller shaft flange nuts and bolts	35	47

1 General description

The rear axle is of the three-quarter floating type and is held in place by semi-elliptic springs. These springs provide the necessary lateral and longitudinal location of the axle. The axle incorporates a hypoid crownwheel and pinion, and a two pinion differential. All repairs can be carried out to the component parts of the rear axle without removing the axle casing from the car.

The crownwheel and pinion together with the differential gears are mounted on the differential unit which is bolted to the front face of the banjo type axle casing.

Adjustments are provided for the crownwheel and pinion backlash, pinion depth of mesh, pinion shaft bearing preload, and backlash between the differential gears. All these adjustments may be made by varying the thickness of the various shims and thrust washers.

The axle half shafts are easily withdrawn and are splined at their inner ends into the splines in the differential wheels. The wheel bearing races are mounted on the outer ends of the axle casing and are secured by nuts and lockwashers.

2 Rear axle – removal and refitting

1 Chock the front wheels and slacken the rear wheel securing nuts. Jack-up the rear of the vehicle and support it on firmly based axle stands located under the body and forward of the rear axle.

2 With a scriber or file mark the pinion and propeller shaft drive flanges so that they may be refitted in their original positions.

3 Undo and remove the four nuts and bolts that secure the rear propeller shaft flange to the pinion flange. Lower the propeller shaft.

4 Undo and remove the wheel nuts and lift away the roadwheels.

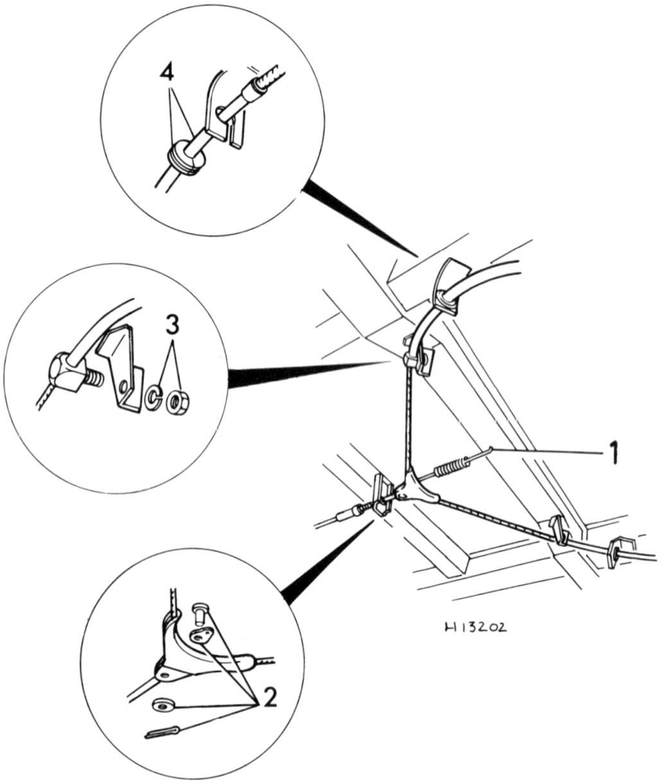

Fig. 8.1 Removing the rear axle – removing the handbrake cable (Sec 2)

1 Unhook the return spring
2 Remove the compensator clevis
3 Disconnect the outer cable end fittings
4 Withdraw the grommets and cable assembly

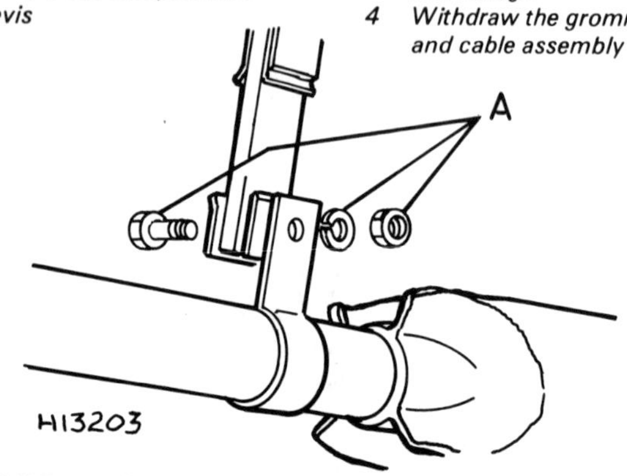

Fig. 8.2 Removing the rear axle – disconnecting the exhaust pipe (Sec 2)

A Remove the retaining nut and bolt

5 Wipe the top of the brake master cylinder reservoir, unscrew the cap and place a piece of thin polythene sheet over the top of the reservoir. Alternatively tape over the reservoir cap vent hole. Refit the cap. This will prevent hydraulic fluid syphoning out during subsequent operations.
6 Wipe the area around the union nut on the brake feed pipe at the axle bracket. Unscrew the union nut and plug the pipe to prevent the entry of dirt.
7 Undo and remove the locknut and washer from the flexible hose.
8 Detach the flexible hose from its support bracket.
9 Unhook the handbrake return spring from the chassis crossmember.
10 Remove the clevis pin which joins the handbrake compensator to the adjustment rod.
11 Remove the nuts and spring washers which secure the handbrake outer cable end fittings to the brackets on the chassis crossmember.
12 Withdraw the handbrake rear cable assembly complete with grommets from the brackets on the chassis side members.
13 Release the exhaust pipe from the rear support strap.
14 Pull the handbrake rear cable clear of the propeller shaft and exhaust pipe.
15 Remove the bolt, locknut and washers to release each shock absorber from the rear axle brackets.
16 Support the rear axle and remove the eight locknuts from the suspension U-bolts.
17 Remove the U-bolts and clamp plates. If two clamp plates are fitted to each pair of U-bolts they must both be retained for reassembly.
18 Remove the rear axle from the vehicle by lowering it to the ground and sliding it out.
19 Refitting of the rear axle is a reversal of the removal procedure. Tighten the U-bolt locknuts to the specified torque with the vehicle half laden after it has been lowered onto its roadwheels. Tighten the shock absorber bolts to the specified torque. On completion bleed the brake hydraulic system as described in Chapter 9. Check the axle oil level, which should be up to the bottom of the filler/level plug.

3 Half shaft – removal and refitting

1 Chock the front wheels and slacken the wheel nuts on the rear wheel which is to be removed. Jack-up the rear of the vehicle and support it on axle stands.
2 Remove the roadwheel and release the handbrake.
3 Remove the two countersunk screws which retain the brake drum and remove the brake drum. If the brake drum will not come off, turn the brake adjuster anti-clockwise (looking outwards) to retract the brake shoes and allow the drum to be removed. **Do not** hit the brake drum directly with a metal hammer as it is made of cast iron and may fracture.
4 Remove the countersunk screw retaining the halfshaft flange and withdraw the half shaft from the axle assembly. The bearing spacer and O-ring may come free with the half shaft. If not it is advisable to remove them and to discard the O-ring. Remove the gasket from between the half shaft flange and the wheel hub flange.
5 Carefully inspect the differential engagement splines for wear. If a new half shaft is to be fitted, apply axle oil to the splines before fitting. A coating of oil on the halfshaft itself will help to prevent corrosion.
6 Refitting the halfshaft is a reversal of the removal procedure. Use a new flange gasket and a new O-ring on assembly. Check that the bearing spacer is located properly in the wheel hub. On completion adjust the brakes as described in Chapter 9.

4 Rear hub assembly – removal and refitting

1 To remove a rear hub assembly it is first necessary to remove the half shaft as described in Section 3.
2 Remove the bearing spacer and O-ring if these were not withdrawn when the half shaft was removed.
3 Bend back the tab on the hub nut lockwasher. Using a large socket or box spanner remove the hub nut and lockwasher. *Note that the left wheel hub nut has a left-hand thread and the right hub nut a right-hand thread.*
4 Draw off the wheel hub with a universal puller or, if one is not available, temporarily bolt a wheel to the hub and, by tapping on the wheel at diametrically opposite points and working round the wheel, drift the hub off the axle assembly.
5 Examine the bearing before cleaning the hub assembly. Rotate the bearing and feel for rough running, grittiness or any slackness. Where there is any sign of wear or of overheating, indicated by bluish discolouration, the bearing must be renewed. At the same time a new oil seal should be fitted.
6 Remove the bearing, if necessary, by drifting it out of the hub with a suitable metal drift working on the oil seal side. With the bearing removed, prise out the old oil seal.
7 Thoroughly clean the hub and dry with rag. Fit a new oil seal with the lip facing towards the bearing and press it in to butt against the hub shoulder.
8 Pack the new bearing with multi-purpose grease, working it well in with the fingers. Press the bearing into the hub. This can be done in a vice using suitable packing or, alternatively, it can be drifted into the hub, but do not drift on the inner race or the bearing could be

3.4a Remove the countersunk screw retaining the half-shaft (the arrow shows the screw hole) and ...

3.4b ... withdraw the half-shaft

4.3 The rear hub retaining nut and tab washer

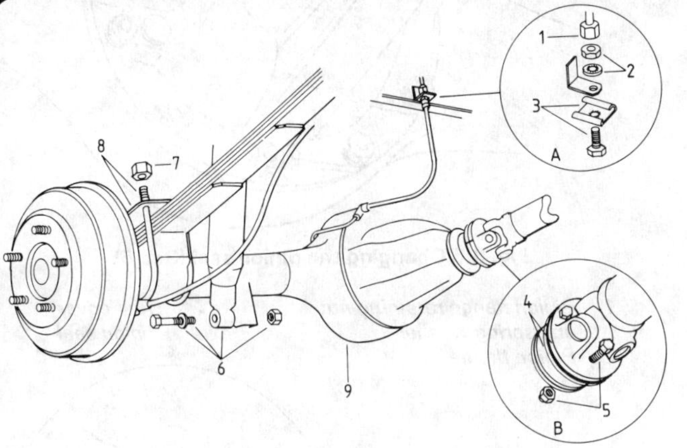

Fig. 8.3 Removing the rear axle (Sec 2)

Inset A: Disconnecting the brake hydraulic hose
1 Feed pipe union
2 Locknut and shake-proof washer
3 Hose and lockplate
Inset B: Disconnecting the propeller shaft
4 Re-alignment marks
5 Connecting bolts and locknuts
View C: Removing the axle
6 Shock absorber attachment bolt and nut
7 U-bolt locknuts
8 U-bolt and plate
9 Rear axle assembly

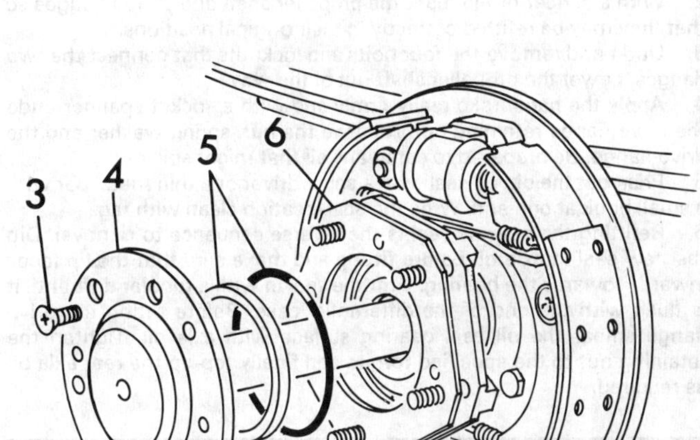

Fig. 8.4 Removing an axle half shaft (Sec 3)

1 Brake drum retaining screws
2 Brake drum
3 Half shaft retaining screw
4 Half shaft flange
5 Bearing spacer and O-ring
6 Flange gasket

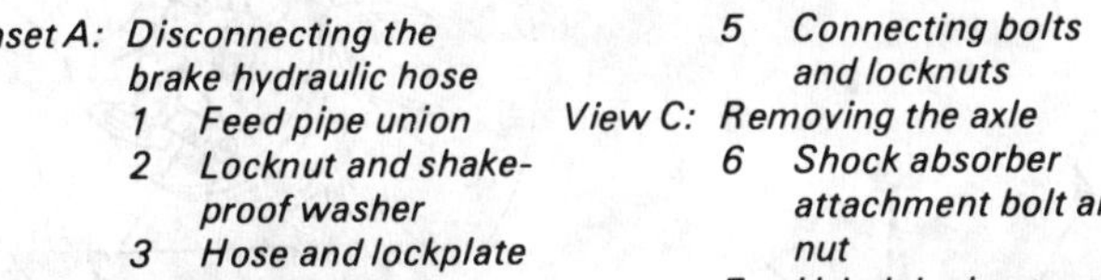

Fig. 8.5 Section through the hub bearing and seal (Sec 4)

1 Oil seal
2 Bearing

damaged. Make sure that no dirt or other foreign matter contaminates the bearing. When the bearing is fully home, butting on the housing shoulder, fill the space between the bearing and the oil seal with grease.
9 Refit the hub to the axle by drifting it into position. Again take care to keep the bearing and hub clean. The outer face of the bearing spacer should protrude from 0·001 to 0·004 in (0·025 to 0·102 mm) beyond the outer face of the hub and flange gasket when the bearing is pressed fully into position.
10 If necessary use a new lockwasher, fit and tighten the hub nut to the specified torque. If the original lockwasher is being re-used, select an unused part of the washer to bend up the locking tab.
11 Refit the half shaft as described in Section 3.

5 Pinion oil seal – removal and refitting

1 Chock the front wheels, jack-up the rear of the vehicle and support it on firmly based axle stands.
2 With a scriber or file mark the propeller shaft and pinion flanges so that they may be refitted correctly in their original positions.
3 Undo and remove the four bolts and locknuts that connect the two flanges. Lower the propeller shaft out of the way.
4 Apply the handbrake really firmly and with a socket spanner undo the drive flange retaining nut. Remove the nut, spring washer and the drive flange. Be prepared to catch any oil that might spill out.
5 Prise out the old oil seal with a screwdriver or a thin metal bar with a small hook at one end. Wipe the seal location clean with rag.
6 Refitting the new oil seal is the reverse sequence to removal. Dip the new seal in axle oil before fitting and make sure that the lip faces inwards towards the bearing. Drift the seal in with a tubular drift until it is flush with the end of the differential case. Before fitting the drive flange smear the oil seal bearing surface with axle oil. Tighten the retaining nut to the specified torque and finally top-up the rear axle oil as required.

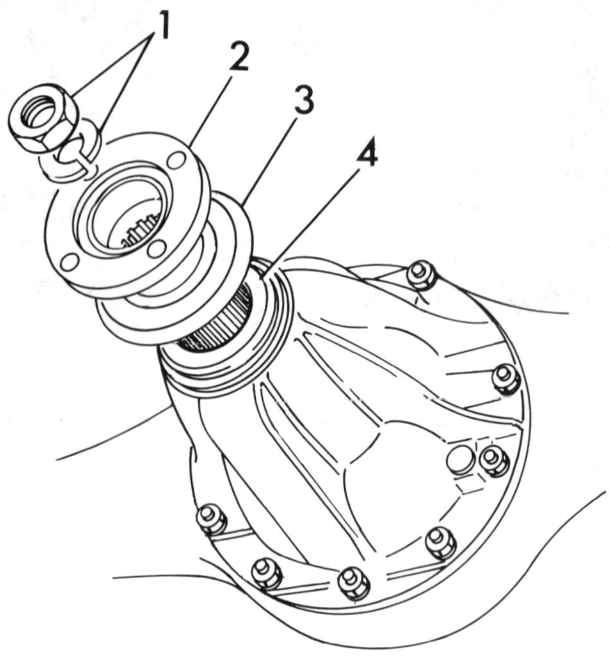

Fig. 8.6 Changing the pinion seal (Sec 5)

1 *Pinion flange retaining nut and spring washer*
2 *Pinion flange*
3 *Dust cover*
4 *Pinion seal*

6 Differential assembly – removal and refitting

1 Chock the front wheels, jack-up the rear of the vehicle and support it on firmly based axle stands.
2 Drain the oil from the rear axle into a suitable container.
3 Partially withdraw both half shafts as described in Section 3.
4 Mark the propeller shaft and pinion flanges to ensure that they can be correctly re-assembled.
5 Undo and remove the four bolts and locknuts to release the propeller shaft flange from the pinion flange and move the shaft to one side out of the way.
6 Undo and remove the nuts and the spring washers which retain the differential assembly to the axle casing.
7 Move the brake pipe three-way connector to one side off its stud.
8 Draw the assembly forwards from over the studs in the axle casing and lift it away from the vehicle. Remove the gasket from the axle casing.
9 Refitting the differential assembly is the reverse sequence to removal. The following points should be noted:

(a) Use a new gasket and make sure that the mating faces are clean
(b) Tighten the differential assembly retaining nuts to the specified torque
(c) Refill the axle with the appropriate axle oil

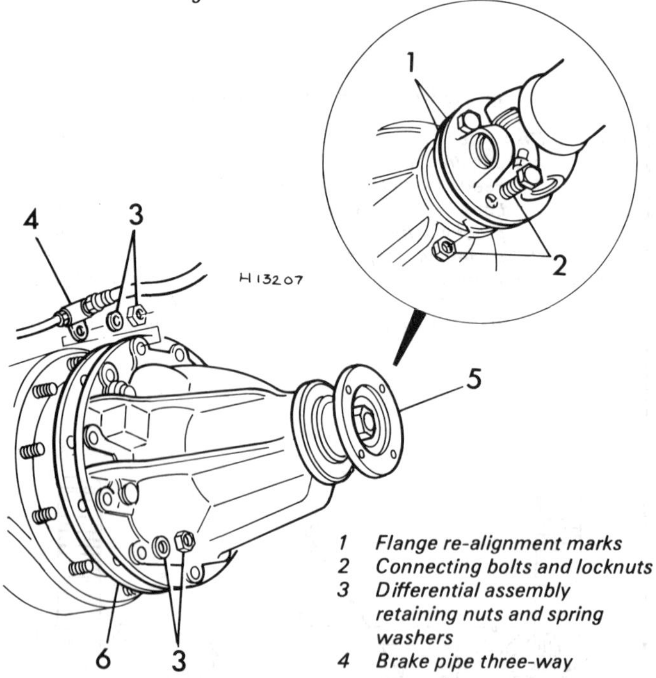

1 *Flange re-alignment marks*
2 *Connecting bolts and locknuts*
3 *Differential assembly retaining nuts and spring washers*
4 *Brake pipe three-way connector*
5 *Pinion drive flange*
6 *Axle casing gasket*

Fig. 8.7 Removing the differential assembly (Sec 6)

7 Differential assembly – overhaul

1 At some time in its life the differential assembly may require overhaul to cure severe backlash, an unacceptable whine or similar abnormal noises. Such symptoms usually indicate worn bearings coupled with worn gearing to a varying degree. Rebuilding and setting up a differential assembly is a specialised job calling for training and experience and an owner contemplating such work should first seriously consider the relative economics of obtaining parts and rebuilding (with the possibility of an unsatisfactory result) compared with obtaining a complete assembly either new, as a factory-exchanged unit or second-hand from a breaker.
2 Most garages will prefer to renew the complete differential assembly as a unit, rather than to dismantle the unit to change damaged or worn parts. To do the job correctly requires the use of a number of special and expensive tools which the majority of garages do not have.
3 The primary object of these special tools is to enable the mesh of the crown wheel to the pinion to be set very accurately and thus ensure that noise is kept to a minimum. With these factors in mind it is strongly recommended that a complete new assembly is fitted when the differential needs overhaul.

8 Fault diagnosis – rear axle

Symptom	Reason/s
Noise on drive, coasting or overrun	Shortage of oil Incorrect crownwheel to pinion mesh Worn pinion bearings Worn side bearings Loose bearing cap bolts
Noise on turn	Differential side gears worn, damaged or tight
Knock on taking up drive or during gearchange	Excessive crownwheel to pinion backlash Worn gears Worn axle-shaft splines Pinion bearing preload too low Loose drive coupling nut Loose securing bolts or nuts within unit Loose roadwheel nuts or elongated wheel nut holes
Oil leakage	Defective gaskets or oil seals possibly caused by clogged breather or oil level too high

Chapter 9 Braking system

For modifications, and information applicable to later models, see Supplement at end of manual

Contents

Specifications

System type

Drum brakes front and rear, hydraulically operated. Dual hydraulic circuit with servo assistance

Front brakes

Wheel cylinder diameter:	
185 models	0·937 in (23·80 mm)
All other models	0·875 in (22·23 mm)
Drum diameter:	
185 models	9 in (228·6 mm)
All other models	10 in (254·0 mm)
Lining width	2·75 in (69·9 mm)

Rear brakes

Wheel cylinder diameter	0·75 in (19·05 mm)
Drum diameter	9 in (228·6 mm)
Lining width	1·75 in (44·45 mm)

Minimum lining thickness

Riveted linings	Just above the rivet heads
Bonded linings	$\frac{1}{16}$ in (1·59 mm)

Master cylinder

Type	Tandem
Diameter	0·8125 in (20·64 mm)
Stroke, with servo	1·46 in (37 mm) minimum
Servo type	Lockheed type 38, mechanically operated from brake pedal

Torque wrench settings

	lbf ft	Nm
Master cylinder end plug	30	41
Master cylinder reservoir retaining screws	5	7
Front wheel cylinder retaining screws:		
Large	11	15
Small	7·5	10
Rear wheel adjuster retaining nuts	12·5 to 16·5	17 to 22

1 General description

1 The Sherpa is equipped with drum brakes on front and rear wheels. They are operated by hydraulic pressure generated in the master cylinder when the brake pedal is depressed. This pressure is transmitted to each wheel cylinder by a system of metal pipes and flexible hoses.
2 The tandem master cylinder provides pressure to the two independent hydraulic circuits. There is a separate supply to each front brake and a common supply to the back brakes. The supply to each back brake divides at a three-way connector mounted on the rear axle. The supply from the master cylinder is conducted in metal pipes. Flexible hoses are fitted at each front wheel and at the input to the rear axle three-way connector to accommodate wheel or axle movement.
3 The front brakes have two leading shoes and the rear brakes each have one leading and one trailing shoe. 185 model Sherpas have one hydraulic cylinder per wheel; all other models have two cylinders on each front wheel and one in each rear wheel. Early models had the brake linings rivetted to the shoes but all models now have bonded linings. Two snail cam adjusters are provided on each front wheel and one wedge and tappet type adjuster on each rear wheel.
4 The handbrake operates on the rear brakes only using a system of levers, rods and cables with spring-assisted return.
5 A brake servo unit is fitted as standard on all models. It is installed between the foot pedal and the master cylinder and it utilises manifold vacuum to provide additional thrust on the master cylinder pushrod, thereby reducing the amount of pedal effort required when operating the brakes. In the event of its failure the brakes will still work, but greater foot pressure will be needed.
6 Both metric and imperial (UNF) hydraulic fittings may be encountered on the Sherpa. Metric fittings are coloured gold or black and usually bear the letter M. **Do not** try to couple components with different types of thread.

2 Brakes – adjustment

1 Brake adjustment must be carried out at the specified intervals, or more frequently if pedal travel becomes excessive, in order to compensate for wear on the brake shoe linings by moving the shoes closer to the drums.
2 The brake adjusters are square-headed and easily damaged by an ill-fitting spanner. A special brake adjusting spanner is invaluable and not expensive – certainly less costly than new adjusters. Do not use a self-gripping wrench or an adjustable spanner except as a last resort.
3 If an adjuster refuses to move in either direction it may have seized up; if it cannot be moved so as to take up wear then the limit of adjustment may have been reached. In either event the brake drum will have to be removed to discover the nature of the trouble.
4 BL do not sell front brake adjusters separately from the complete backplate; however, you may be able to obtain a proprietary kit which will enable you to renew one or more adjusters after sawing or drilling the old ones off. Make sure that the kit is correct for your vehicle and follow the maker's instructions carefully. Rear brake adjusters are dealt with in Section 14.

Front brakes

5 Jack-up one of the front wheels until the tyre is clear of the ground. Make sure that the vehicle is adequately supported, preferably on an axle stand, before crawling underneath.
6 Protruding from the brake backplate will be seen two square-headed adjusters. Turn one of the adjusters in the direction of forward wheel rotation, spinning the wheel at the same time, until the wheel locks. Then back off the adjuster until the wheel is just free to spin without the linings rubbing.
7 Repeat the operation for the second adjuster on the same wheel. Then lower the wheel to the ground.
8 Repeat the complete adjustment sequence on the other front wheel.

Rear brakes

9 The rear brake adjustment procedure is as described for the front brakes, paragraphs 5 to 8, except that only one adjuster is provided in each rear wheel. Turn the adjuster clockwise, viewed from the centre of the vehicle, to lock the wheel before backing off. The adjustment can only be done with the handbrake off, so make sure that the front wheels are chocked before getting underneath (photo).
10 Finally lower the vehicle to the ground.

3 Bleeding the brake hydraulic system

1 Whenever the brake hydraulic system has been overhauled, a part renewed, or the level in the reservoir becomes too low, air will have entered the system, necessitating its bleeding. During the operation, the level of hydraulic fluid in the reservoir should not be allowed to fall below half full, otherwise air will be drawn in again.
2 Absolute cleanliness must be maintained throughout the entire bleeding operation. It is most important that no dirt or grit enters the system and all equipment that is used must be free from fuel, paraffin or any form of mineral oil.
3 Obtain a clean and dry glass jar, plastic tubing fifteen inches long and of suitable diameter to fit tightly over the bleed screw, and a supply of recommended brake fluid.
4 Check that all connections are tight and all bleed screws closed. Chock the wheels and release the handbrake. Operate the brake pedal several times to destroy the vacuum in the servo unit.
5 Fill the master cylinder reservoir and the bottom inch of the jar with hydraulic fluid. Take extreme care that no fluid is allowed to come

2.9 A rear brake adjuster (arrowed)

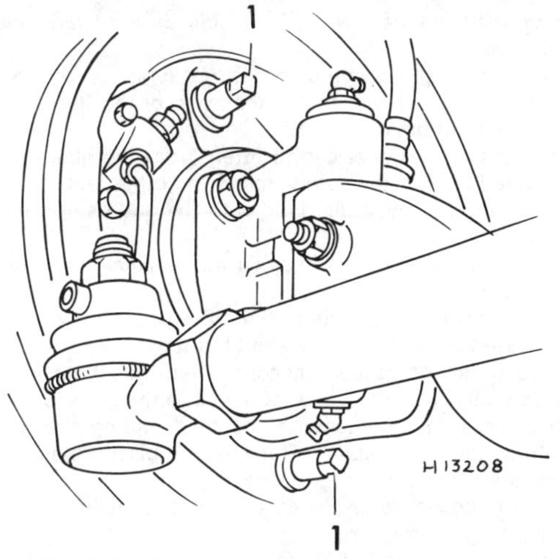

Fig. 9.1 Adjusting the brakes (Sec 2)

1 *The two adjusters on a front wheel brake*

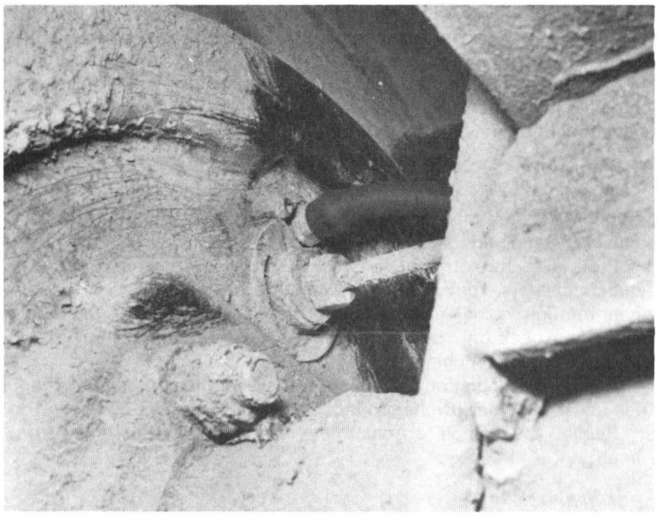
3.7 The bleed tube attached to a rear brake bleed

4.2a Remove the two screws (arrowed) retaining the front brake drum ...

4.2b ... and remove the brake drum

4.12 Fit the return springs in their correct holes and behind the shoes (front hub removed for clarity)

into contact with the paintwork as it acts as a solvent and it will damage the finish.

6 Start the bleeding procedure on one of the rear brakes. When that is done, bleed the other rear brake, then one of the front brakes and finally the other front brake.

7 Remove the rubber dust cap, if fitted, from the bleed screw and attach the bleed tube to the screw. Insert the other end of the tube in the jar containing the hydraulic fluid with the end below the level of the fluid (photo).

8 Use a suitable open ended spanner and unscrew the bleed screw about half a turn.

9 An assistant should now pump the brake pedal by first depressing it one full stroke followed by three short but rapid strokes and allowing the pedal to return of its own accord. Check the fluid level in the reservoir. Carefully watch the flow of fluid into the glass jar and, when air bubbles cease to emerge with the fluid, during the next down stroke, tighten the bleed screw. Remove the plastic bleed tube and tighten the bleed screw. Refit the rubber cap

10 Repeat the procedure on the other brakes following the sequence recommended in paragraph 6.

11 Sometimes it may be found that the bleeding operation for one or more cylinders is taking a considerable time. This is probably due to air being drawn past the bleed screw threads when the screw is loose. To counteract this condition, it is recommended that at the end of each downward stroke the bleed screw be tightened to stop air being drawn past the threads.

12 If after the bleed operation, has been completed, the brake pedal operation still feels spongy, this is an indication that there is still air in the system, or that the master cylinder is faulty.

13 When satisfactory, switch on the auxiliary circuits and apply the footbrake several times. Check that the brake failure warning light does not come on.

14 Carefully check the hydraulic system for leaks.

15 Finally, drive the vehicle and test the brakes. The pedal should feel solid, with no indication of sponginess.

16 Where only one of the hydraulic circuits has been broken it is only necessary to bleed that circuit. If the master cylinder is removed and refitted, bleed the complete system.

4 Front brake shoes – removal, inspection and refitting

1 Chock the rear wheels and apply the handbrake. Slacken the front wheel nuts, jack-up the front of the vehicle and support it on firmly based axle stands. Remove the roadwheel.

2 Undo and remove the two brake drum retaining screws and carefully pull off the brake drum. If it is tight turn back the brake adjusters, see Section 2, and, if necessary, tap the drum outwards with a soft-faced hammer (photos).

3 Using a pair of pliers release the steady springs, pins, and washers

by depressing the washers and turning them through 90°.
4 With a screwdriver, lever the trailing ends (that is, the ends without the adjuster slot) of the brake shoes from the wheel cylinders.
5 Similarly remove the leading ends (with the adjuster slot) of the brake shoes from the wheel cylinder pistons. Do not press the brake pedal while the shoes are removed.
6 Detach the brake shoes and return springs from the brake backplate.
7 Examine the brake shoe linings. The linings should be renewed if they are so worn that the rivet heads are almost flush with the surface of the lining. Where bonded linings are fitted they must be renewed when the lining material has worn down to the specified minimum thickness at its thinnest point. If the brake shoe linings are contaminated with lubricant or brake fluid, don't try to clean them: they must be renewed irrespective of the amount of wear. Brake shoes must always be renewed in axle sets, that is all four shoes on the front wheel brakes for example, and they should always be of the same, correct grade of lining material. Serious consequences could result from out of balance braking due to mixing of linings.
8 Clean the backplate and drum free from all traces of dust, preferably using methylated spirit and a stiff brush. **Do not** use compressed air as it will only disperse the dust in the air. *Brake dust contains asbestos; it is a danger to health and must not be inhaled.* Brake dust in the drums can cause brake judder or squeal if present in sufficient quantity.
9 Examine the friction surface on the inside of the drum. Normally this should be completely smooth and bright but, after some use, scores and shallow grooves will be apparent. If these are only slight they are not serious, but heavy scoring means that the drum must be reconditioned or renewed.
10 Check that each piston is free in its cylinder, that the rubber dust covers are undamaged and properly located and that there are no signs of fluid leakage.
11 Before reassembly lightly lubricate the ends of the brake shoes and other contact points of the wheel cylinders and adjusters with a good quality high melting point grease. Do not allow any grease to come into contact with the linings or rubber parts.
12 Fit the return springs to the correct holes in the brake shoes, see photo, so that they lie between the shoes and the backplate when assembled.
13 Pass the shoes and springs over the wheel hub and fit the shoe ends to the wheel cylinders in turn.
14 Refit the steady pins, springs and slotted washers.
15 Back off the brake adjusters if necessary, refit the brake drum and secure with the two retaining screws. Operate the brake several times (after making sure that the drums are fitted to the other three wheels) to centralise the shoes. Adjust the brakes as described in Section 2.
16 Refit the roadwheel and lower the vehicle to the ground. carry out a road test to ensure correct brake operation.

5 Rear brake shoes – removal, inspection and refitting

1 Chock the front wheels and slacken the rear wheel nuts. Jack-up the rear of the vehicle and support it on firmly based axle stands. Remove the roadwheel.
2 Undo and remove the two brake drum retaining screws and carefully pull off the brake drum. If it is tight turn back the brake adjusters, see Section 2, and, if necessary, tap the drum outwards with a soft-faced hammer.
3 Using a pair of pliers release the steady springs, pins and washers by depressing the washers and turning them through 90° (photo).
4 With a screwdriver, lever the bottom ends of the brake shoes from the adjuster tappets.
5 Unhook the lower return spring from the brake shoes, and remove the top ends of the brake shoes from the wheel cylinder pistons.
6 Pull the leading (front) shoe clear of the handbrake cross lever and remove the cross lever.

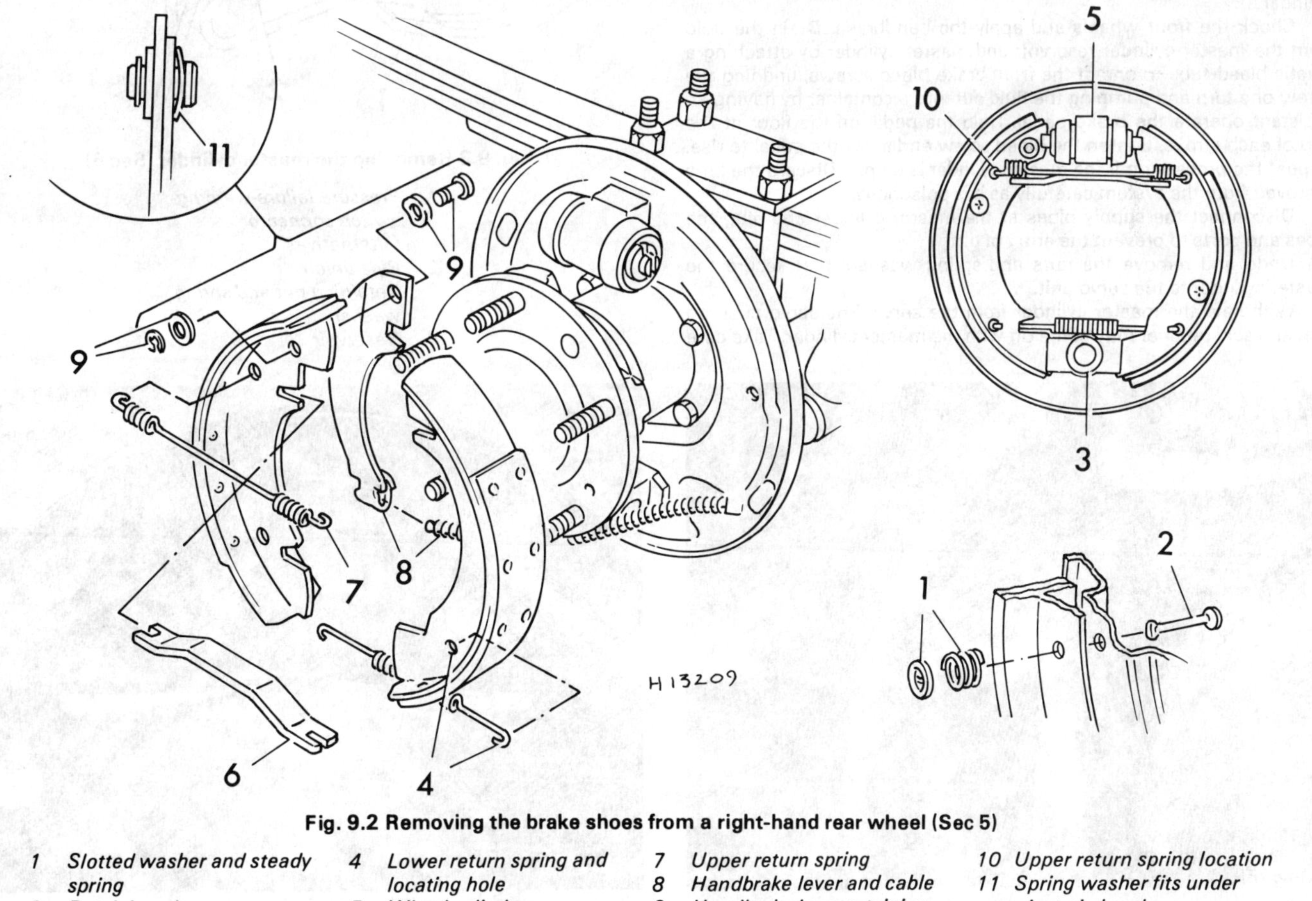

Fig. 9.2 Removing the brake shoes from a right-hand rear wheel (Sec 5)

1 Slotted washer and steady spring
2 Retaining pin
3 Brake adjuster
4 Lower return spring and locating hole
5 Wheel cylinder
6 Handbrake cross lever
7 Upper return spring
8 Handbrake lever and cable
9 Handbrake lever retaining components
10 Upper return spring location
11 Spring washer fits under pivot pin head

7 Remove the leading shoe and the upper return spring.
8 Disengage the handbrake cable from the handbrake lever on the trailing (rear) shoe. Remove the circlip, washers and pivot pin to release the handbrake lever from the trailing shoe.
9 Examining, cleaning and lubricating the rear brake assembly is a similar procedure to that for a front brake assembly as described in Section 4, paragraphs 7 to 11. Do not press the brake pedal while the shoes are removed.
10 Commence reassembly by fitting the handbrake lever to the trailing shoe. The spring washer must go under the head of the pivot pin. Fit the pivot pin with the spring washer on it to the handbrake lever and trailing shoe, and then fit the flat washer and circlip.
11 Engage the handbrake cable with the handbrake lever on the trailing shoe, and fit the upper return spring together with the leading shoe. Note that the upper spring lies on the front face of the shoes (photo).
12 Fit the cross lever between the shoes and locate the brake shoe assembly on the wheel cylinder pistons.
13 Engage the lower return spring with the brake shoes, noting that this spring lies on the rear face of the shoes. Lever the shoes into the adjuster tappets. Refit the brake shoe steady springs with their pins and slotted washers.
14 Back off the brake adjuster if necessary, refit the brake drum and secure with the two retaining screws. Operate the brake several times (after making sure that the drums are fitted to the other three wheels) to centralise the shoes. Adjust the brakes as described in Section 2.
15 Refit the roadwheel and lower the vehicle to the ground. Tighten the wheel nuts. When both rear wheels have been dealt with, carry out a road test to ensure correct brake operation.

6 Master cylinder – removal and refitting

1 Disconnect the battery earth lead and disconnect the wiring plug or lead from the pressure failure warning switch on the master cylinder.
2 Chock the front wheels and apply the handbrake. Drain the fluid from the master cylinder reservoir and master cylinder by attaching a plastic bleed tube to one of the front brake bleed screws, undoing the screw one turn and pumping the fluid out into a container by having an assistant operate the brake pedal. Hold the pedal on the floor at the end of each stroke, tighten the bleed screw and allow the pedal to rise. Repeat the process until the master cylinder is empty. Discard the fluid removed from the system carefully as it is poisonous.
3 Disconnect the supply pipes at the master cylinder and plug the pipes and ports to prevent the entry of dirt.
4 Undo and remove the nuts and spring washers that secure the master cylinder to the servo unit.
5 Withdraw the master cylinder from the servo. The servo pushrod seal and seal retainer may come off with the master cylinder. Take care not to allow any hydraulic fluid to drip onto the paintwork when removing the master cylinder from the vehicle. If necessary, drain off any residual fluid from the master cylinder after removal.
6 Refitting the master cylinder is a reversal of the removal procedure. Tighten the retaining nuts to the specified torque. Take care when connecting the union nuts to the master cylinder as it is easy to get the threads crossed.
7 Refer to Section 3 and bleed the entire brake hydraulic system.

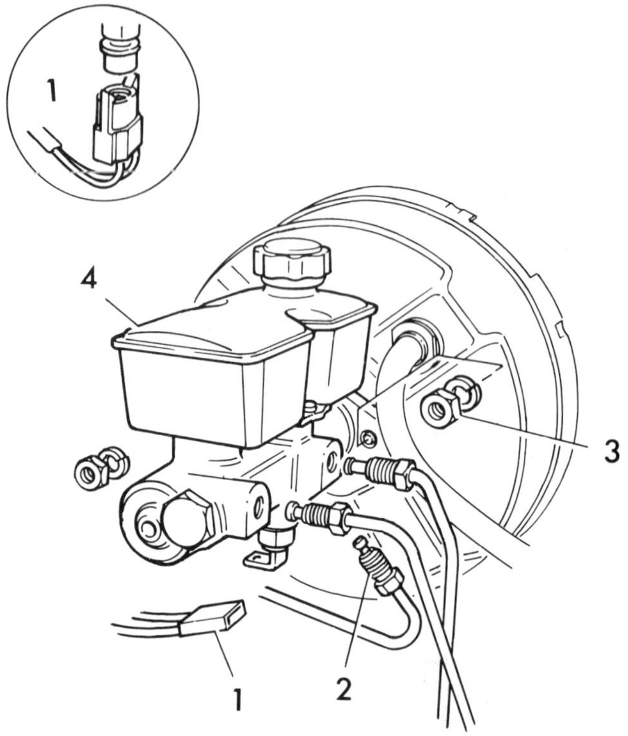

Fig. 9.3 Removing the master cylinder (Sec 6)

1 *Pressure failure warning switch connector (alternatives)*
2 *Pipe union*
3 *Retaining nut and spring washer*
4 *Reservoir*

5.3 Removing the steady spring, pin and washer from a rear brake shoe

5.11 Note that on the rear brakes the upper return spring is on the front face of the brake shoes

7 Master cylinder – dismantling, overhaul and reassembly

1 Remove the master cylinder from the vehicle as described in Section 6. Make sure that it is perfectly clean before beginning to dismantle it.
2 If the servo pushrod seal and retainer came away with the master cylinder, remove them from the cylinder now.
3 Unscrew and remove the pressure failure warning switch.
4 Mount the master cylinder in a vice with protected jaws to prevent damage to the unit.
5 Remove the two screws which secure the reservoir and lift the reservoir off the cylinder. Remove the two reservoir seals from the cylinder body.
6 With a pair of round-nosed pliers remove the circlip from the mouth (pushrod end) of the cylinder bore and withdraw the primary piston, spring retainer and return spring.
7 Insert a wooden dowel or soft metal rod into the cylinder bore and depress the secondary piston. Extract the small stop pin from the secondary fluid feed port and gently allow the secondary piston to return under control.
8 Withdraw the secondary piston, spring retainer and return spring from the master cylinder body. If these items cannot be shaken out, apply air pressure to the secondary outlet port, but be ready to prevent them flying out and possibly getting lost or damaged. Only very low air pressure is required such as from a foot pump.
9 Remove the end plug and washer retaining the pressure differential piston assembly (for operating the pressure failure warning switch). Leave the metal distance piece in place on the end plug.
10 Withdraw the pressure differential piston assembly, if necessary using low air pressure through a primary outlet port, see paragraph 8.
11 Remove the rubber seals from the pistons. These should be laid out so that their original locations can be recalled. The variety of seals provided in an overhaul kit are not individually identified but they can be sorted out by comparing them with the original seals — provided you can recall where the originals were fitted! Make sketches where necessary and label the original seals.
12 Before proceeding with the cleaning and inspection of components, examine the cylinder bore closely. If it is ridged or scored, the whole master cylinder assembly must be renewed. It is pointless, as well as dangerous, to try to service a unit with an unsatisfactory bore.
13 If the cylinder bore is perfectly smooth and with no apparent defect, clean all the components in unused brake fluid and dry with a lint-free cloth. While doing this examine each part for signs of corrosion or damage. As the components are cleaned and dried, keep them covered in a clean container until ready for assembly. *It is essential that scrupulous cleanliness is observed during reassembly.*
14 All components should be assembled wet by dipping in unused brake fluid. Seals should be manipulated with clean fingers only – do not use metal tools to fit them.
15 Fit new O-rings to the pressure differential piston.
16 Fit a piston washer to the spigot ends of the primary and

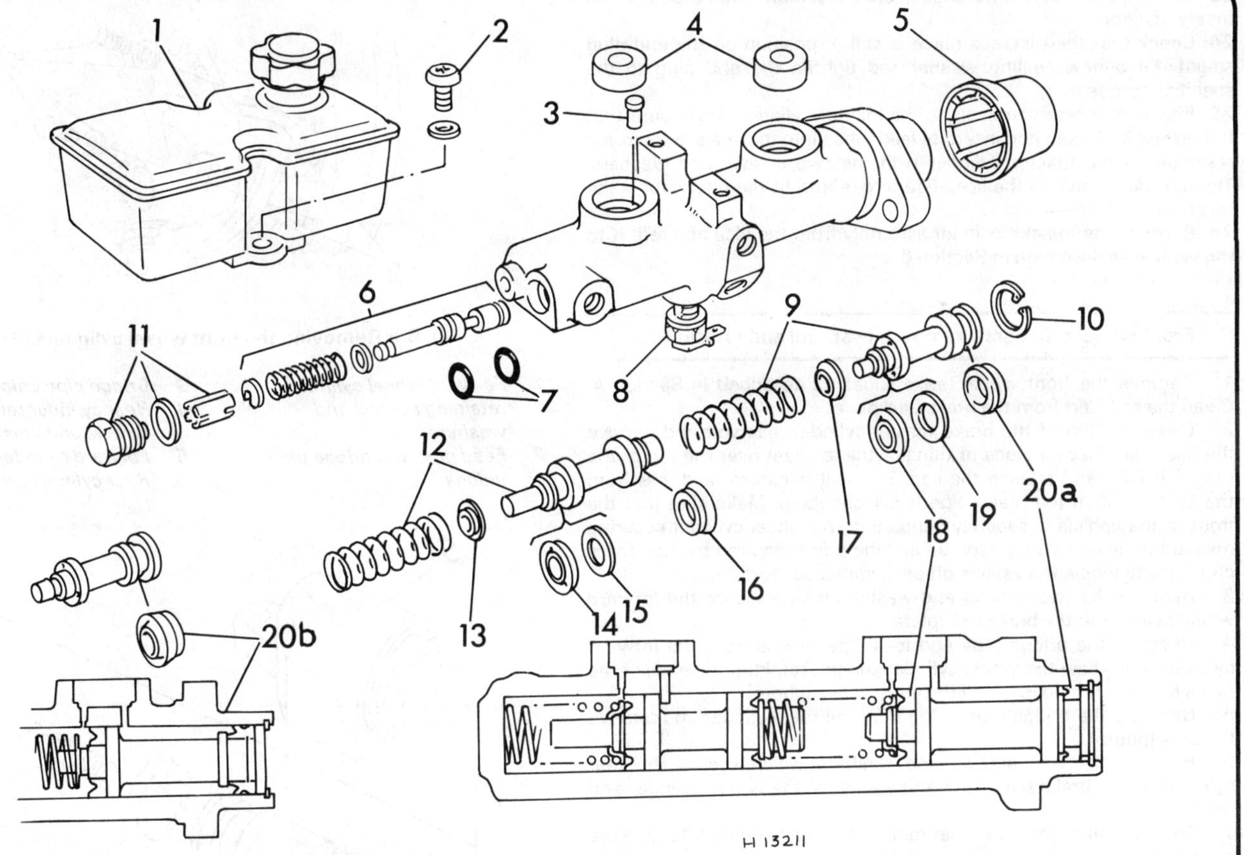

Fig. 9.4 Exploded view of the master cylinder (Sec 7)

1 *Reservoir*
2 *Retaining screw*
3 *Stop pin*
4 *Seals*
5 *Seal and retainer*
6 *Pressure differential piston assembly*
7 *O-rings*
8 *Pressure failure warning switch*
9 *Primary piston and return spring*
10 *Circlip*
11 *End plug, washer and distance piece*
12 *Secondary piston and return spring*
13 *Spring retainer*
14 *Piston seal*
15 *Piston washer*
16 *Piston seal*
17 *Spring retainer*
18 *Piston seal*
19 *Piston washer*
20a *Piston seal (thin)*
20b *Alternative piston seal (thick)*

secondary pistons.

17 Fit a piston seal following the piston washer on each piston. These two seals are identical and they must be fitted, using the fingers only, with their lip facing away from the piston washer.

18 Of the remaining three piston seals, which are different from each other, select the thinnest and fit it to the secondary piston with its lip facing towards the primary spring seat.

19 Examine the assembly identification tag. If it has a small hole this indicates that the secondary seal on the primary piston should be a thin one; no hole in the identification tag indicates that it should be a thick seal. Fit the appropriate seal to the primary piston with its lip facing towards the primary spring seat. The surplus seal provided in the kit can be discarded. It will now be appreciated why the recommendation to lay out the original seals for comparison was made in paragraph 11.

20 Compare the primary and secondary piston return springs. The secondary spring is the shorter of the two. Fit this spring and the spring retainer to the secondary piston and insert the components into the master cylinder. Make sure that the seal lips enter without being turned back.

21 Press the secondary piston into the cylinder using a wooden dowel or a soft metal rod. Insert the stop pin when the piston head has passed the feed port.

22 Fit the return spring and spring retainer to the primary piston and insert the components into the master cylinder, again making sure that the seals enter without being turned back. Refit the circlip using round-nosed pliers and check that it expands into its groove.

23 Insert the pressure differential piston assembly into its bore in the master cylinder.

24 Check that the distance piece is still in position on the end plug spigot. Fit a new sealing washer and tighten the end plug to the specified torque.

25 Fit the two reservoir seals to the master cylinder. Make sure that the reservoir is clean and insert its feed tubes into the seals. Secure the reservoir to the master cylinder with the two screws and washers. Tighten the screws to the specified torque and fit the filler cap to the reservoir.

26 Remove the master cylinder assembly from the vice and refit it to the vehicle as described in Section 6.

8 Front wheel cylinders – removal, overhaul and refitting

1 Remove the front wheel brake shoes as described in Section 4. Clean the road dirt from the brake backplate.

2 Clean the top of the brake master cylinder reservoir and remove the filler cap. Place a piece of thin polythene sheet over the top of the reservoir filler and replace the cap. This will minimise fluid loss from the system when the brake pipe is disconnected. Make sure that the front of the vehicle is securely supported: the wheel cylinder securing nuts are done up to a high torque and the effort required to undo them could easily topple the vehicle off an unstable support.

3 Remove the two screws and washers which secure the forward wheel cylinder to the brake backplate.

4 Unscrew the bridge pipe and feed pipe unions from the forward cylinder. Withdraw the wheel cylinder and gasket. Plug the ends of the pipes to prevent fluid loss and the entry of dirt (photo).

5 Unscrew the bridge pipe union from the rear cylinder and remove the pipe (photo).

6 Remove the two screws and washers which secure the rear cylinder to the brake backplate and withdraw the wheel cylinder and gasket.

7 To overhaul a wheel cylinder remove the dust cover protecting the open end of the bore.

8 Withdraw the piston and return spring from the wheel cylinder body. Use low air pressure to blow the piston out if necessary.

9 Remove the seal from the piston.

10 Before proceeding with the cleaning and inspection of the wheel cylinder components, examine the cylinder bore closely. If it is ridged or scored, the whole wheel cylinder assembly must be renewed. It is pointless, as well as dangerous, to try to service a unit with an unsatisfactory bore.

11 If the cylinder bore is smooth and with no apparent defect, clean the wheel cylinder components in unused brake fluid and dry with a lint-free cloth. While doing this, examine each part for signs of corrosion or damage. As the components are cleaned and dried, keep them

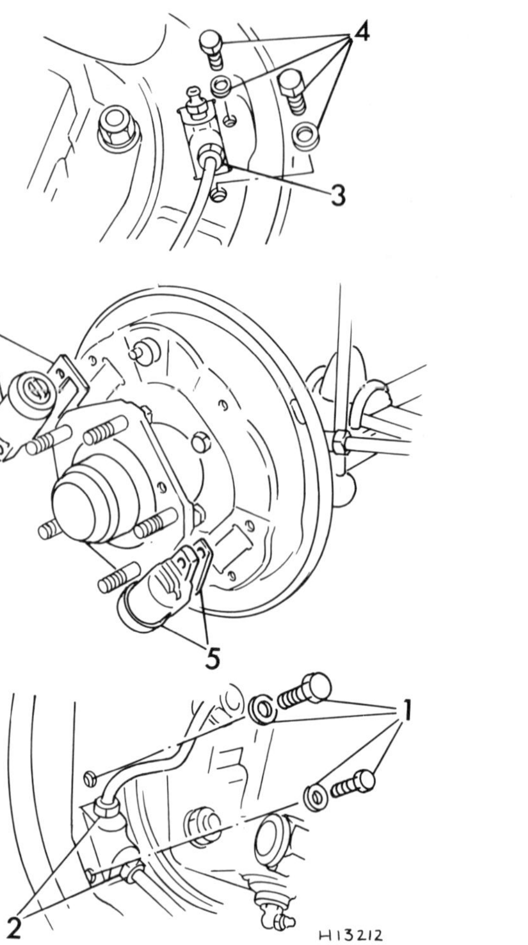

Fig. 9.5 Removing the front wheel cylinders (Sec 8)

1 Forward wheel cylinder retaining screws and washers
2 Feed pipe and bridge pipe unions
3 Bridge pipe union
4 Rear cylinder retaining screws and washers
5 Forward cylinder and gasket
6 Rear cylinder and gasket

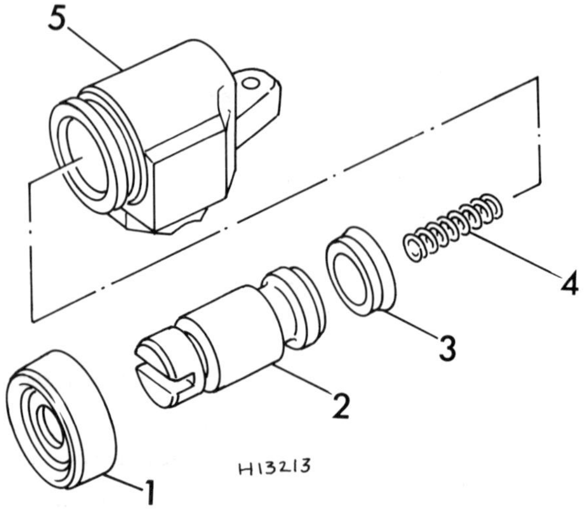

Fig. 9.6 Exploded view of a front wheel cylinder (Sec 8)

1 Dust cover
2 Piston
3 Seal
4 Return spring
5 Cylinder

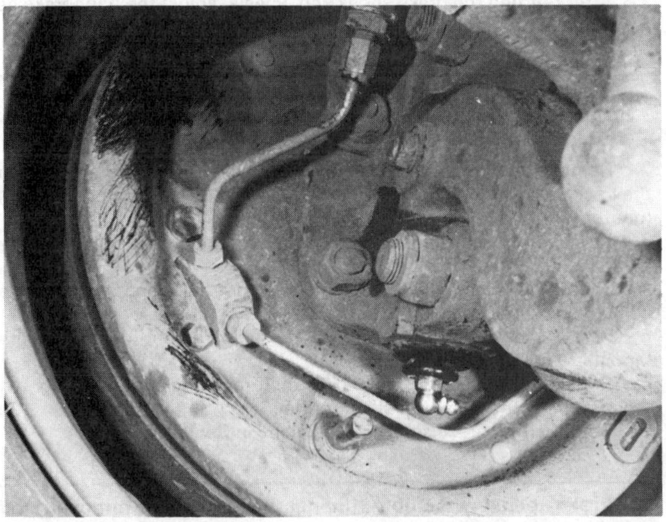

8.4 The feed pipe and bridge pipe on the front brake backplate

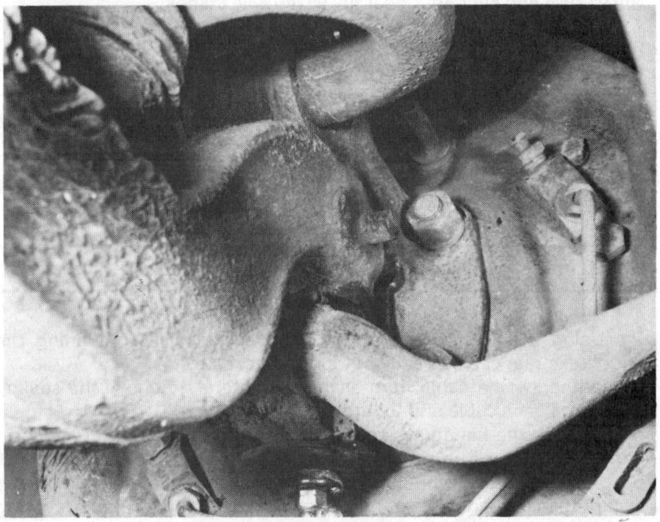

8.5 The bridge pipe between the front and rear cylinders

covered in a clean container until ready for assembly. *It is essential that scrupulous cleanliness is observed during reassembly.* Discard the old rubber seal and dust cap.

12 Immerse all components in clean, unused brake fluid and assemble whilst wet.

13 Fit the new seal, manipulating it with the fingers only, with its flat face towards the piston. Insert the spring and then the piston into the wheel cylinder body, making sure that the seal lip does not roll or become trapped.

14 Refit the dust cover, engaging the lips in the piston groove and the wheel cylinder groove.

15 Repeat the overhaul operation, paragraphs 7 to 14, on the other cylinder.

16 Refitting the wheel cylinders is a reversal of the removal procedure, paragraphs 1 to 5. When refitting the cylinders, use new gaskets, and tighten the retaining screws to the specified torque.

17 On completion of assembly adjust the brakes, Section 2; remove the piece of polythene film from the master cylinder reservoir and bleed the hydraulic system, Section 3; and make a road test to test the brakes.

9 Rear wheel cylinders – removal, overhaul and refitting

1 Remove the rear wheel brake shoes as described in Section 5. Clean the road dirt from the brake backplate.

2 Clean the top of the brake master cylinder reservoir and remove the filler cap. Place a piece of thin polythene sheet over the top of the reservoir filler and replace the cap. This will minimise fluid loss from the system when the brake pipe is disconnected. Alternatively tape over the vent hole in the reservoir cap.

3 Unscrew the union and remove the brake pipe from the wheel cylinder. Plug the end of the pipe to prevent fluid loss and the entry of dirt.

4 Using a suitable screwdriver, spring off the circlip which secures the wheel cylinder to the backplate. Remove the wheel cylinder and its gasket.

5 Overhauling a rear wheel cylinder is broadly similar to the procedure described for the front wheel cylinders in Section 8, paragraphs 7 to 14 inclusive. The main difference is that the rear wheel cylinder has two dust covers, pistons and seals. The work detailed must therefore be repeated for the duplicate parts.

6 Refitting the wheel cylinder is a reversal of the removal procedure, paragraphs 1 to 4. When refitting the cylinder, use a new gasket and a new circlip.

7 On completion of assembly adjust the brakes, Section 2; remove the piece of polythene or tape from the master cylinder reservoir and bleed the hydraulic system, Section 3; and make a road test to test the brakes.

10 Handbrake cable – adjustment

1 Chock the front wheels and completely release the handbrake lever. Pull up the lever to the second ratchet tooth (click) from the off position.

2 Jack-up the rear of the vehicle and support it on firmly based axle stands.

3 Slacken the handbrake cable adjuster screw locknut and turn the adjuster until the rear cable is slack (photo).

4 Refer to Section 2 and adjust the rear brakes.

5 Turn the cable adjuster screw until the brakes just begin to drag when the wheels are spun by hand. Test on both rear wheels.

6 Hold the adjuster screw with one spanner and tighten the locknut with another. Check the wheel drag again.

7 Fully release the handbrake lever and check that the rear wheels rotate freely with no drag. Apply the handbrake lever to the third click and check that the rear wheels are held. If necessary repeat the adjustment procedure until satisfactory.

8 Lower the vehicle to the ground.

11 Handbrake lever assembly – removal and refitting

1 Chock the wheels. Working under the cab floor, remove the split-pin, anti-rattle washer and clevis pin which attach the handbrake lever rod to the bellcrank lever (photo).

2 Inside the cab, release the handbrake rod gaiter from the floor and undo and remove the three screws and spring washers which secure the handbrake mounting bracket to the cab floor. Remove the handbrake lever assembly.

3 If it is required to remove the handbrake rod from the handbrake lever, drive out the rivet securing the two together. This can be done by supporting the assembly so that the rivet has clearance to come out, eg on the partly opened jaws of a vice or over a suitable socket spanner, and tapping the rivet out with a parallel pin punch.

4 To refit the handbrake rod to the lever it is necessary to use a clevis pin. Re-drill the rivet hole in the handbrake lever on one side with a $\frac{5}{16}$ in (7·94 mm) drill. Using clevis pin CLZ 0511, washer WA 108051 and a split-pin, or suitable alternatives, assemble the rod to the lever.

5 Refit the handbrake lever to the vehicle by reversing the removal procedure. On completion check, and if necessary adjust, the handbrake as described in Section 10.

12 Handbrake front cable – removal and refitting

1 Chock the wheels and release the handbrake. Jack-up the rear of

the vehicle and support it on irmly based axle stands.
2 Unhook the handbrake return spring from the chassis crossmember.
3 Remove the split-pin, anti-rattle washer and the clevis pin which secure the cable to the bellcrank lever under the cab (photo 11.1).
4 Undo the cable adjuster locknut and unscrew the adjuster sleeve from the adjuster rod. Remove the cable assembly.
5 Refitting is the reverse of the removal procedure. Adjust the handbrake as described in Section 10 and lower the vehicle to the ground.

13 Handbrake rear cable assembly – removal and refitting

1 Refer to Section 5 and remove the rear brake shoes.
2 Using a suitable screwdriver, remove the circlips retaining the handbrake rear cable to each rear brake backplate.
3 Withdraw the cable from each backplate and remove the sealing rings from each cable end (photo).
4 Unhook the handbrake return spring from the chassis crossmember.
5 Remove the split-pin, anti-rattle washer and the clevis pin which secure the compensator to the adjustment rod. Note the return spring anchor tag under the head of the clevis pin.
6 Remove the nuts and washers which secure the brake outer cable end fittings to the brackets on the chassis crossmember.
7 Remove the cable assembly, complete with grommets, from the brackets on the chassis side-members. Remove the grommets from the cable.
8 Refitting the handbrake rear cable is the reverse of the removal procedure. If necessary renew the sealing rings and circlips at the brake drum ends of the cable. When fitting the clevis pin at the cable compensator, fit the handbrake return spring anchor tag under the head of the clevis pin.

9 After refitting the brake shoes and wheels, adjust the handbrake as described in Section 10.

14 Rear brake shoe adjusters – removal, overhaul and refitting

1 Refer to Section 5 and remove the rear brake shoes.
2 Remove the two nuts and washers which secure the adjuster to the backplate and pull out the adjuster. Repeat on the other rear wheel (photo).
3 Remove the two tappets from the adjuster body and screw out the wedge adjuster.
4 Thoroughly clean all components and coat the tappets and wedge adjuster with a good quality corrosion resistant grease.
5 Screw the wedge adjuster into the adjuster body and refit the tappets. Wipe off surplus grease.
6 Refit the wedge adjuster unit to the brake backplate and tighten the retaining nuts to the specified torque.
7 Refit the rear brake shoes as described in Section 5.

15 Brake pedal assembly (manual gearbox) – removal and refitting

1 Remove the split-pin, washer and clevis pin which secure the servo unit pushrod to the brake pedal (photo).
2 Remove the screw and plain washer which secure the pedal shaft spring clip to the pedal bracket. Also remove the spring clips on the other end of the pedal shaft.
3 Unhook the brake pedal and clutch pedal return springs from the pedals and the bracket.
4 Unclip the main wiring harness from the dashboard side panel. Place the harness and the soundproofing material to one side.

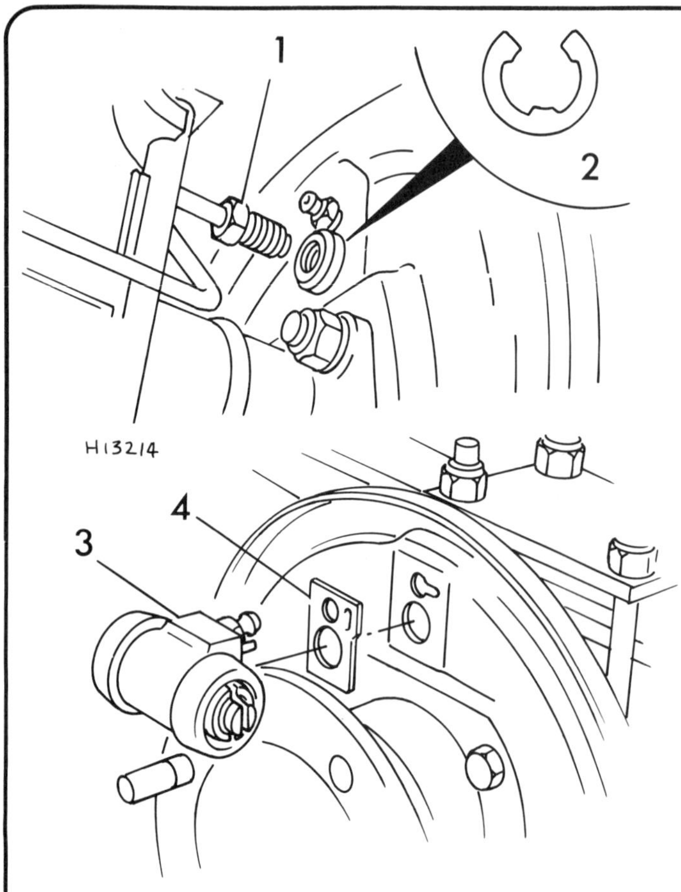

Fig. 9.7 Removing the wheel cylinder from a rear wheel (Sec 9)

1 *Brake pipe union*
2 *Circlip*
3 *Wheel cylinder*
4 *Gasket*

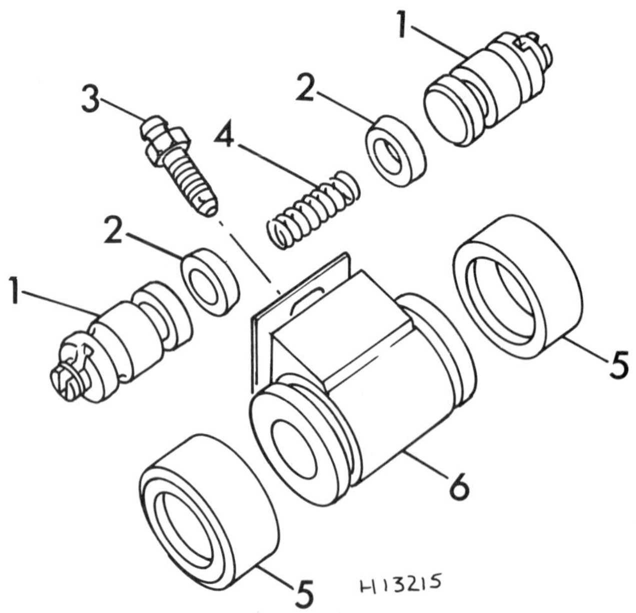

Fig. 9.8 Exploded view of a rear wheel cylinder (Sec 9)

1 *Pistons*
2 *Seals*
3 *Bleed screw*
4 *Return spring*
5 *Dust covers*
6 *Wheel cylinder*

10.3 The handbrake cable adjuster is just forward of the cable compensator

11.1 The handbrake bellcrank lever

13.3 The handbrake cable attachment at the rear brake backplate

14.2 The rear brake adjuster and shoe tappet housing

15.1 The brake pedal connection to the servo unit pushrod (manual gearbox)

17.2 The servo unit assembly (clutch fluid reservoir on right)

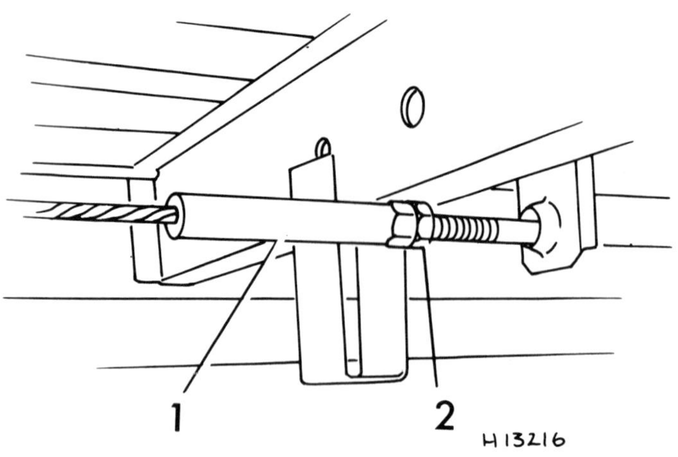

Fig. 9.9 Handbrake cable adjuster (Sec 10)

1 Adjuster screw *2 Locknut*

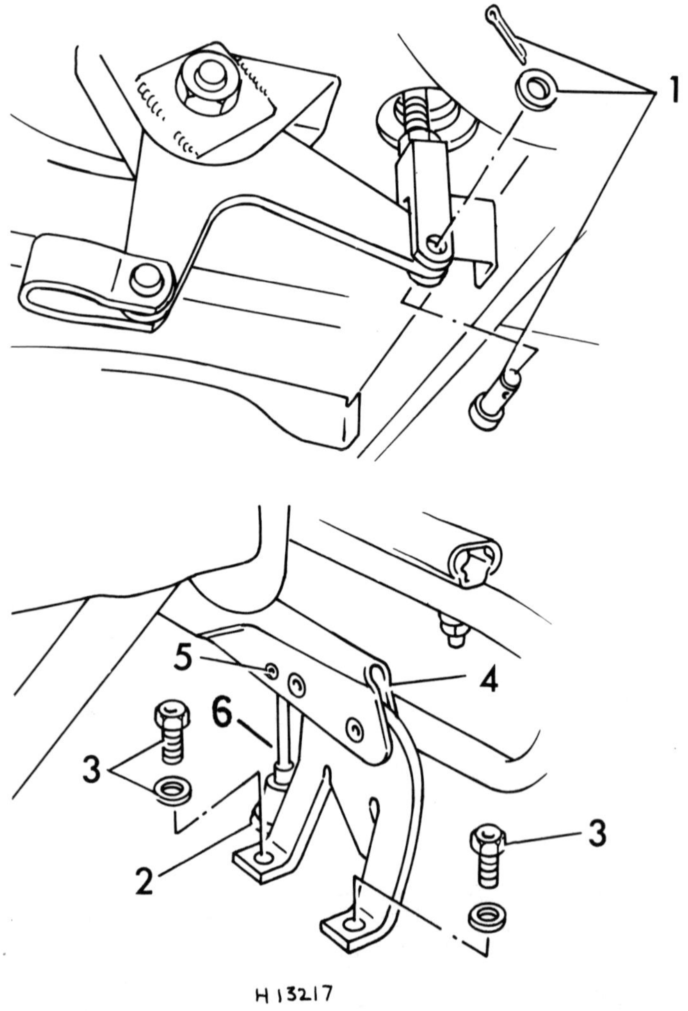

Fig. 9.10 Removing the handbrake lever assembly (Sec 11)

1 Split-pin, anti-rattle spring and clevis pin
2 Gaiter
3 Retaining screws and spring washers
4 Handbrake lever
5 Rivet
6 Handbrake rod

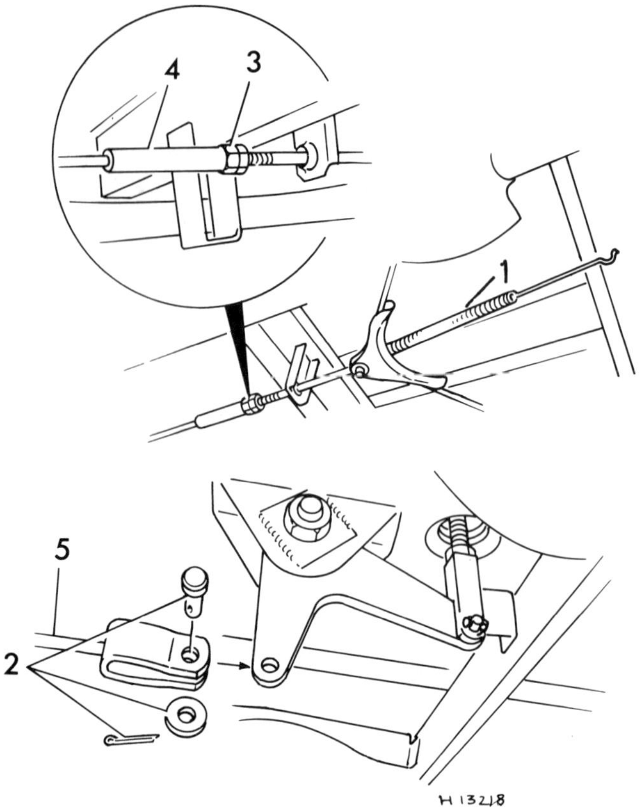

Fig. 9.11 Removing the handbrake front cable assembly (Sec 12)

1 Return spring
2 Split-pin, washer and clevis pin
3 Adjuster locknut
4 Adjuster screw
5 Front cable end

5 Remove the pedal shaft from the bracket, leaving the clutch pedal to hang on its pushrod.
6 Remove the brake pedal and remove the return spring and the rubber footpad from the pedal.
7 Before refitting the brake pedal, grease the pedal bushes and smear the pedal shaft with a general purpose grease. Omit this greasing, however, if plastic bushes are fitted.
8 Refit the brake pedal assembly by reversing the removal procedure.

16 Brake pedal assembly (automatic transmission) – removal and refitting

1 Disconnect the battery earth lead.
2 Disconnect the direction indicator switch wiring harness at the multi-connector block.
3 Disconnect the ignition/starter switch wiring harness at the three snap connectors.
4 Unclip the direction indicator switch wiring harness from the facia.
5 Working under the bonnet, make alignment marks to show the relationship of the steering inner column to the universal joint for correct reassembly.
6 Remove the nut, washer and bolt which secure the steering inner column to the universal joint.
7 Remove the two shear bolts which secure the steering column to the facia and the pedal bracket. These will have to be drilled out or removed with a stud extractor.
8 Lift out the steering column assembly.
9 Remove the split-pin, washer and clevis pin which secures the servo unit pushrod to the brake pedal.
10 Remove the screw and plain washer which secure the pedal shaft spring clip to the pedal bracket. Remove the spring clip on the other end of the pedal shaft.
11 Unhook the pedal return spring from the pedal and the bracket.

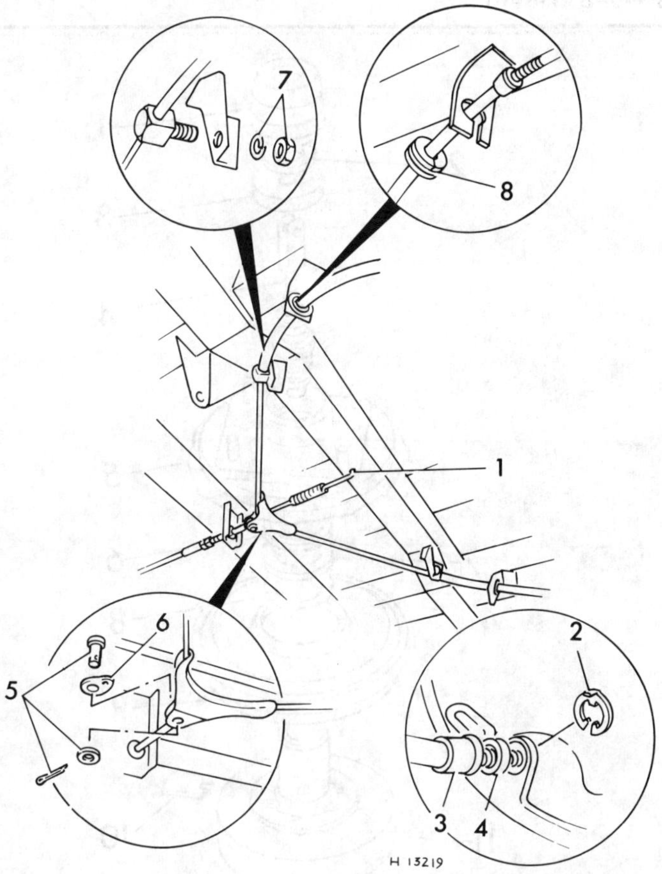

Fig. 9.12 Removing the handbrake rear cable assembly (Sec 13)

1 *Return spring*
2 *Circlip*
3 *Cable*
4 *Sealing ring*
5 *Split-pin, washer and clevis pin*
6 *Return spring anchor tag*
7 *Outer cable end fitting retaining nut and washer*
8 *Grommet*

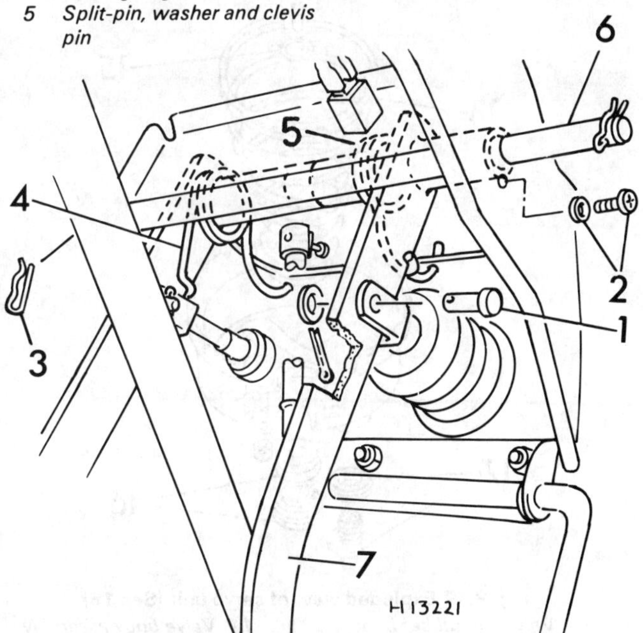

Fig. 9.14 Brake pedal (manual gearbox) removal (Sec 15)

1 *Clevis pin*
2 *Spring clip retaining screw and washer*
3 *Spring clip*
4 *Clutch pedal return spring*
5 *Brake pedal return spring*
6 *Pedal shaft*
7 *Brake pedal*

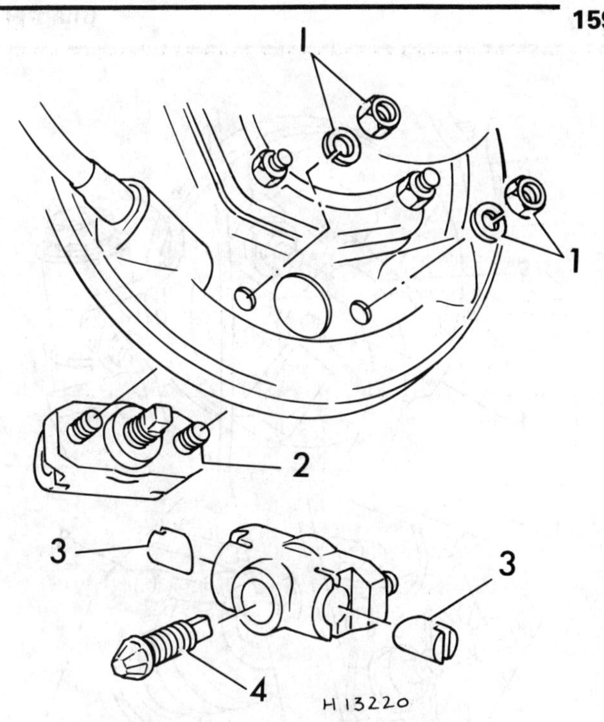

Fig. 9.13 Rear brake shoe adjuster details (Sec 14)

1 *Retaining nuts and spring washers*
2 *Adjuster assembly*
3 *Tappets*
4 *Wedge adjuster*

Fig. 9.15 Brake pedal (automatic transmission) removal (Sec 16)

1 *Spring clip*
2 *Brake pedal*
3 *Pedal shaft*
4 *Return spring*
5 *Spring clip retaining screw and washer*
6 *Clevis pin*
7 *Shear bolts*
8 *Direction indicator switch harness connector*
9 *Ignition/starter switch harness connectors*
10 *Steering column assembly*
11 *Steering column/universal joint bolt*
12 *Bush*

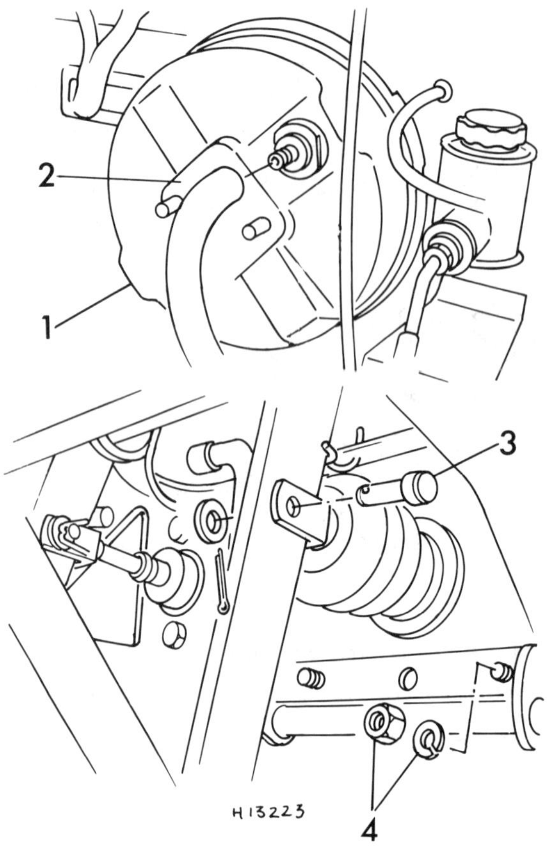

Fig. 9.16 Removal of the servo unit (Sec 17)

1 Servo unit
2 Vacuum pipe
3 Clevis pin
4 Servo retaining nuts and spring washers

12 Unclip the main wiring harness from the dashboard side panel. Place the harness and the soundproofing material to one side.
13 Support the brake pedal and pull out the pedal shaft.
14 The brake pedal assembly can now be removed. Remove the plastic bushes at each end of the pedal tube and remove the rubber footpad from the pedal.
15 Refitting the brake pedal is a reversal of the removal procedure. Using new shear bolts to secure the steering column to the facia and pedal bracket.

17 Servo assembly – removal and refitting

1 Disconnect the wiring plug or snap connector from the pressure failure warning switch on the brake master cylinder.
2 Undo the hose clip and disconnect the vacuum hose from the servo (photo).
3 Remove the nuts and washers to release the master cylinder from the servo.
4 Release the brake hydraulic pipe from its clip on top of the mounting bracket.
5 Carefully withdraw the master cylinder to clear the servo studs and support the master cylinder. Be careful not to strain the hydraulic pipe. Retrieve the servo pushrod seal and seal retainer if these are withdrawn with master cylinder.
6 Remove the split-pin, washer and clevis pin attaching the servo unit pushrod to the brake pedal.
7 Undo and remove the four nuts and spring washers which secure the servo unit to the brake pedal bracket and remove the servo from the front of the bulkhead.
8 Refitting of the servo is the reverse of the removal procedure. After refitting, start the engine and check the servo operation.

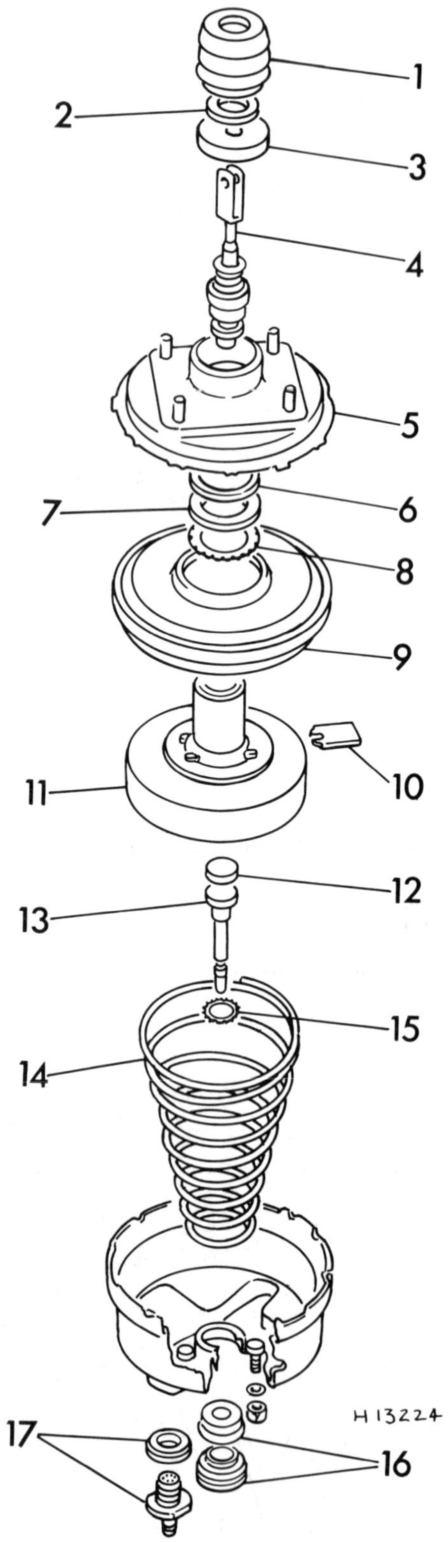

Fig. 9.17 Exploded view of servo unit (Sec 18)

1 Valve rod rubber boot
2 Filter retainer
3 Filter
4 Valve rod and plunger
5 End cover
6 Seal
7 Nylon bearing
8 Bearing retainer
9 Diaphragm
10 Key
11 Valve body assembly
12 Reaction disc
13 Pushrod
14 Return spring
15 Pushrod retainer
16 Pushrod seal and seal retainer
17 Non-return valve and grommet

18 Servo assembly – dismantling, overhaul and reassembly

1 In the normal course of events it is recommended not to attempt overhauling a servo assembly but to renew the unit if performance is unsatisfactory. However, the only requirements for overhaul are a kit of repair parts which can be obtained from a main dealer, and three simple special servicing tools which can easily be made.
2 Remove the servo assembly as described in Section 17.
3 If they were not already withdrawn with the master cylinder, remove the pushrod seal and seal retainer from the servo vacuum shell.
4 The servo must be held securely during dismantling and to overcome this difficulty a dummy flange is required so that, with the flange attached to the servo, the assembly can be secured in a vice. Make up a dummy flange from a piece of scrap metal, referring to Fig. 9.18. The only critical dimensions are the sizes and spacings of the three holes. Attach the dummy flange to the studs which normally retain the master cylinder to the servo and then turn the servo over and secure the dummy flange firmly in a vice.
5 Take off the rubber boot from the end cover and valve rod.
6 Remove the filter retainer from the valve body, withdraw the filter and slide it over the valve rod fork.
7 The next step requires another special tool which can be made from scrap material, referring to Fig. 9.19. Again the critical dimensions are the size and locations of the four stud holes and the central clearance hole. Attach this end cover removal tool to the four studs by which the servo is attached to the bulkhead and secure it with a couple of nuts.
8 Put alignment marks on the end cover and vacuum shell for use during reassembly and then turn the tool anti-clockwise as far as the stops allow, at the same time pressing downwards against the thrust of the return spring. If necessary, prise back the retaining lugs to allow the end cover to lift clear under spring pressure. Remove the end cover, the removal tool, and the valve body assembly.
9 Lift out the return spring from the vacuum shell.
10 Remove the valve body assembly from the end cover. Prise out the bearing retainer and remove the nylon bearing and rubber seal from the end cover.
11 Separate the diaphragm from the valve body. Push the valve rod and plunger into the valve body and remove the key. Withdraw the rod and plunger from the valve body.
12 Prise out the pushrod retainer from the valve body, and remove the pushrod and reaction disc.
13 Remove the non-return valve and its grommet from the vacuum shell.
14 Clean all the components in clean, unused brake fluid and dry with a lint-free cloth. Examine all parts for wear, damage or corrosion and renew any doubtful components. Mark the position of the adjuster on the pushrod and check that the torque required to turn it is not less than 5 lbf in (0·56 Nm). Renew the unit if it is found to be lower. Restore the position of the adjuster to the alignment marks. Discard the parts for which replacements are provided in the repair kit.
15 During reassembly it is most important that scrupulous cleanliness is observed at all stages.
16 Commence reassembly by fitting the non-return valve and its grommet to the vacuum shell.
17 Lubricate the reaction disc with disc brake lubricant and insert it into the valve body. Fit the pushrod into the valve body, ensuring that the large face is squarely seated on the reaction disc. Fit the pushrod retainer.
18 Coat the contact surfaces of the valve rod and plunger with a little disc brake lubricant. Refit the valve rod and plunger assembly into the valve body, ensuring that the rubber seal is seating centrally. Press gently on the valve rod and plunger, and fit the key to retain the assembly in the valve body.
19 Fit the diaphragm to the valve body, ensuring that the inner edge of the diaphragm is correctly seated in the groove in the valve body.
20 Coat the end cover seal and the nylon bearing with disc brake lubricant and fit them to the end cover with the lip of the seal facing away from the bearing, towards the bulkhead mounting end of the unit.
21 The next step requires the use of the third and last special servicing tool. It is a setting tool and can be turned out of any suitable metal to the size shown in Fig. 9.20.
22 Using the setting tool, press the bearing retainer into the end

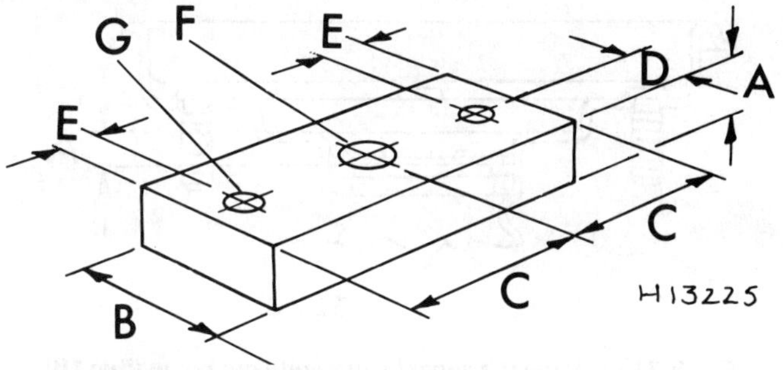

Fig. 9.18 Dummy servo mounting flange (Sec 18)

A $\frac{1}{2}$ *in (12.7 mm)*
B $1\frac{1}{2}$ *in (38.1 mm)*
C $1\frac{1}{2}$ *in (38.1 mm)*
D $\frac{3}{4}$ *in (19.05 mm)*
E $\frac{3}{8}$ *in (9.53 mm)*
F *One hole* $\frac{1}{2}$ *in (12.7 mm) diameter*
G *Two holes 0.335* $\pm$ *0.005 in (8.5* $\pm$ *0.13 mm) diameter*

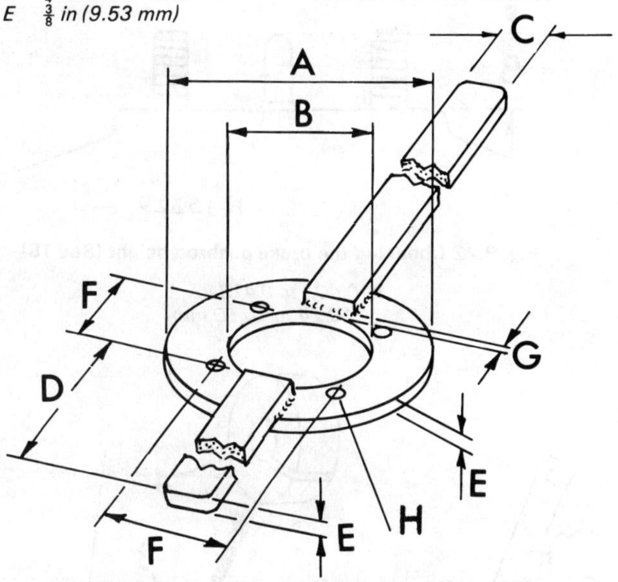

Fig. 9.19 Servo end cover removal tool (Sec 18)

A $6\frac{7}{8}$ *in (174.6 mm)*
B $3\frac{7}{8}$ *in)98.43 mm)*
C $1\frac{1}{2}$ *in (38.1 mm)*
D *12 in (304.8 mm)*
E $\frac{1}{4}$ *in (6.35 mm)*
F *3.543 in (90 mm)*
G $\frac{1}{8}$ *in (3.175 mm)*
H *Four holes* $\frac{11}{32}$ *in (8.73 mm) diameter*

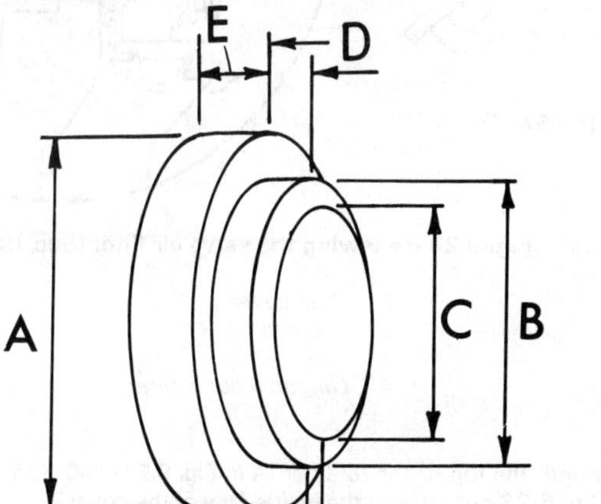

Fig. 9.20 Servo bearing retainer setting tool (Sec 18)

A $2\frac{1}{2}$ *in (63.5 mm) diameter*
B $1\frac{31}{32}$ *in (50 mm) diameter*
C $1\frac{19}{32}$ *in (40.5 mm) diameter*
D *0.270* $\pm$ *0.005 in (6.86* $\pm$ *0.13 mm)*
E $\frac{1}{2}$ *in (12.7 mm)*
F $\frac{1}{16}$ *in (1.6 mm) chamfer*

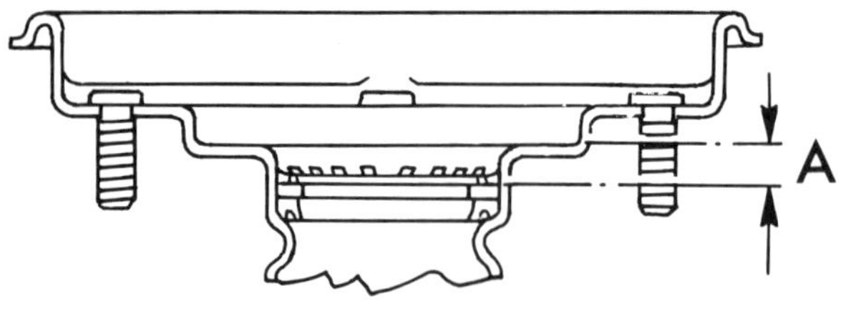

Fig. 9.21 Pressing the servo bearing retainer home (Sec 18)

A 0.265 to 0.275 in (6.73 to 6.98 mm)

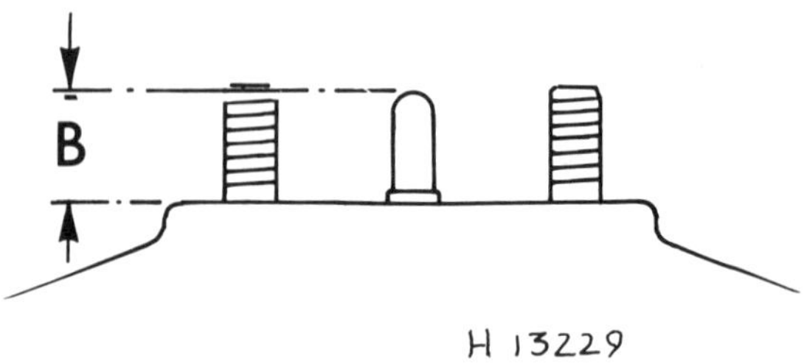

Fig. 9.22 Checking the brake pushrod height (Sec 18)

B 0.403 to 0.413 in (10.24 to 10.50 mm)

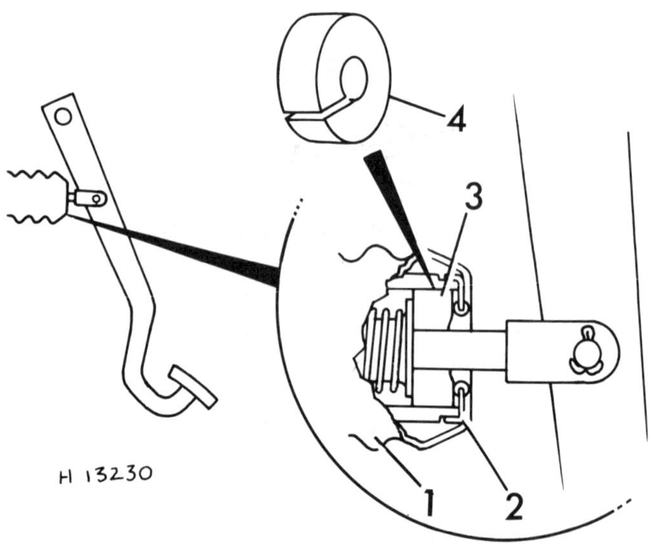

Fig. 9.23 Renewing the servo air filter (Sec 19)

1 *Dust cover*
2 *End cap*
3 *Filter*
4 *Diagonal cut in filter*

cover until the top of the retainer (A in Fig. 9.21) is 0·265 to 0·275 in (6·73 to 6·98 mm) below the inside face of the cover.
23 Fit the return spring to the vacuum shell.
24 Smear the edge of the diaphragm with disc brake lubricant where it comes into contact with the rim of the vacuum shell and the end cover.
25 Fit the valve body and the diaphragm assembly to the end cover and then fit the end cover to the vacuum shell. Using the end cover tool and pressing down against the spring pressure, turn the tool clockwise so that the alignment marks made in paragraph 8 come into line and the stops are engaged. Make sure that the diaphragm is not trapped by the edge of the cover during assembly.
26 Fit a new filter and the filter retainer to the valve body, and then fit the rubber boot to the end cover.
27 Remove the assembly from the vice and remove the dummy flange from the master cylinder fixing studs.
28 Check that the height of the pushrod (B in Fig. 9.22) is 0·403 to 0·413 in (10·24 to 10·49 mm) above the master cylinder mounting face on the vacuum shell. If necessary, an adjustment will have to be made on the pushrod adjuster; one full turn of the adjuster hexagon alters the length of the rod by about 0·035 in (0·89 mm).
29 Fit the seal retainer and seal to the vacuum shell, ensuring that the seal locates on the smooth portion of the pushrod.
30 Remove the end cover tool and refit the servo assembly to the vehicle as described in the Section 17.

19 Servo assembly – air filter renewal

1 Every 36 000 miles (60 000 km) the servo air filter should be renewed. Proceed as follows.
2 Pull back the dust cover on the brake pedal pushrod and slide the end cap up the rod. Pull out the old air filter, if necessary cutting it free from the pushrod.
3 Cut the new filter diagonally to the centre hole, fit it over the pushrod and press it into the housing.
4 Refit the end cap and the dust cover.

20 Flexible hoses – inspection, removal and refitting

1 Inspect the condition of the flexible hydraulic hoses leading from the chassis mounted metal pipe to the brake backplates. If any are swollen, damaged, cut, or chafed, they must be renewed.
2 Unscrew the metal pipe union nut from its connection to the hose, and then holding the hexagon on the hose with a spanner, unscrew the attachment nut and washer.
3 The chassis end of the hose can now be pulled from the chassis mounting bracket and will be quite free.
4 Disconnect the flexible hydraulic hose at the backplate by unscrewing it from the brake cylinder. **Note**: When releasing the hose from the backplate, the chassis end must always be freed first.
5 Refitting is a straight reversal of the above procedure. Bleed the hydraulic system on completion.

21 Hydraulic pipes and hoses – general

1 Carefully examine all brake pipes, pipe connections and unions periodically.
2 First examine for signs of leakage where the pipe unions occur. Then examine the flexible hoses for signs of chafing and fraying and, of course, leakage. This is only a preliminary part of the flexible hose inspection, as exterior condition does not necessarily indicate the interior condition, which will be considered later.
3 The steel pipes must be examined carefully and methodically. They must be cleaned off and inspected for any signs of dents, corrosion, or other damage. Corrosion should be scraped off and, if the depth of pitting in the pipes is significant, they will need renewing. This is particularly likely in those areas underneath the car body where the pipes are exposed to road and weather conditions.
4 If any section of pipe is to be taken off, first wipe and then remove the fluid reservoir cap and place a piece of polythene over the reservoir. Refit the cap. This will stop syphoning during subsequent operations. Alternatively, tape over the vent in the reservoir cap.
5 Rigid pipe removal is usually quite straightforward. The unions at each end are undone, the pipe and union pulled out and the centre sections of the pipe removed from the body clips. Where the pipe unions are exposed to full force of road and weather they can sometimes be very tight. As one can only use an open-ended spanner and the unions are not large, burring of the flats is not uncommon when attempting to undo them. For this reason a self-locking grip wrench is often the only way to remove a stubborn union.
6 Removal of flexible hoses is described in Section 20.
7 With the flexible hose removed, examine the internal bore. If it is blown through first, it should be possible to see through it. Any specks of rubber which come out, or signs of restriction in the bore, means

that the rubber lining is breaking up and the pipe must be renewed.

8 Rigid pipes which need renewing can usually be purchased at any garage where they have the pipe, unions and special tools to make them up. All they need to know is the total length of the pipe, the type of flare used at each end with the union, and the length and thread of the union.

9 Refitting of the pipe is a straightforward reversal of the removal procedure. If rigid pipes have to be made up it is best to get all the sets (bends) in them before trying to install them. Also if there are any acute bends, ask your supplier to put these in for you on a special tube bender, otherwise you may kink the pipe and thereby decrease the bore and fluid flow.

10 With the pipes replaced, remove the polythene or tape from the reservoir cap and bleed the system as described in Section 3.

22 Fault diagnosis – braking system

Symptom	Reason/s
Pedal travels almost to floor before brakes operate	Brake fluid level too low Wheel cylinder leaking Master cylinder leaking Brake flexible hose leaking Brake pipe fractured Brake system unions loose Shoe linings over 75% worn
Brake pedal feels springy	New linings not yet bedded-in Brake drums badly worn or cracked Master cylinder or servo securing nuts loose
Brake pedal feels spongy and soggy	Wheel cylinder leaking Master cylinder leaking (bubbles in fluid) Brake pipe or flexible hose leaking Unions in brake system loose and leaking
Excessive effort required to brake car	Faulty vacuum servo unit Shoe linings badly worn Harder linings fitted than standard Linings and/or brake drums contaminated with oil, grease etc
Brakes uneven and pulling to one side	Linings or brake drums contaminated with oil, grease etc Tyre pressure unequal Brake shoes incorrectly fitted Different type linings fitted to each wheel
Brakes tend to bind, drag or lock-on	Brakes overadjusted Air in hydraulic system Handbrake cable overtightened Master cylinder reservoir vent blocked Master cylinder by-pass port blocked Wheel cylinder seized Shoe pull-off springs broken, stretched or loose

Chapter 10 Electrical system

For modifications, and information applicable to later models, see Supplement at end of manual

Contents

Specifications

System type ... 12 volt, negative earth

Battery capacity and charging rate

Battery type:	
A9	40 Ah; 3·5 amp
A11	50 Ah; 5·0 amp
A13	60 Ah; 6·0 amp
CP11	66 Ah; 6·5 amp
CP13/11	68 Ah; 6·5 amp

Alternator

Types	**16ACR**	**18ACR**
Output at 14V, 6000 rpm	34 amps	43 amps
Rotor winding resistance at 20°C (68°F)	3·3 ohm ± 5%	3·2 ohm ± 5%
Maximum permissible rotor speed	15 000 rpm	15 000 rpm
Brush length, new	0·5 in (12·7 mm)	0·5 in (12·7 mm)
Brush spring tension, brush face flush with brush box	9 to 13 ozf (255 to 368 gf)	9 to 13 ozf (255 to 368 gf)

Starter motor

Type	Pre-engaged, Lucas M35J or 2M100	
	M35J	**2M100**
Light running current	65 amp at 8000 to 10 000 rpm	45 amp at 5200 to 7000 rpm
Lock torque	7·0 lbf ft (0·97 kgf m) at 390 amp	14 lbf ft (1·94 kgf m) at 463 amp
Brush spring tension	28 ozf (0·8 kgf)	36 ozf (1·02 kgf)
Minimum brush length	$\frac{5}{16}$ in (8 mm)	$\frac{3}{8}$ in (10 mm)
Armature endfloat	0·01 in (0·25 mm)	0·01 in (0·25 mm)
Bush fitting mandrel:		
Commutator end	0·438 in (11·12 mm)	0.438 in (11·12 mm)
Driving end	0·473 in (12·01 mm)	0·473 in (12·01 mm)
Commutator thickness (minimum)	0·08 in (2 mm)	0·14 in (3·5 mm)

Wiper motor

Type	Lucas 14W (two-speed)
Running current (rack disconnected):	
Low speed	1·5 amp
High speed	2·0 amp
Wiper speed (after 60 secs):	
Low speed	46 to 52 rpm
High speed	60 to 70 rpm
Armature endfloat	0·002 to 0·010 in (0·05 to 0·25 mm)
Brush spring tension	5 to 7 ozf (140 to 200 gf)
Minimum brush length	0·18 in (4·7 mm)
Maximum pull to move racking in tube	6 lbf (2·7 kgf)

Light bulbs

	Watts	Part Number
Headlamp, sealed beam unit, right-hand drive	60/45	GLU 101
Headlight, sealed beam unit, left-hand drive (except France)	60/50	GLU 114
Headlamp (France only)	45/40	BFS 411
Sidelamp	5	GLB 207
Front flasher lamp	21	GLB 382
Stop/tail lamp	21/5	GLB 380
Rear flasher lamp	21	GLB 382
Number plate lamp	5	GLB 989
Interior/load space lamp	6	GLB 254
Brake failure warning lamp	1·5	GLB 280
Warning and panel lamps:		
Printed circuit instruments	1·2	GLB 286
Standard instruments	2·2	GLB 987
Automatic transmission selector lamps	2	GLB 281

Torque wrench settings

	lbf ft	Nm
Alternator pulley nut	25	33·9
Windscreen wiper motor yoke bolts	1·17 (14 lbf in)	1·6

1 General description

1 The electrical system is of the 12 volt negative earth type. The major components comprise a 12 volt battery, a Lucas alternator which is fitted to the front right-hand side of the engine and is driven from the pulley on the front of the crankshaft, and a starter motor which is mounted on the rear right-hand side of the engine.
2 The battery supplies a steady amount of current for the starting, ignition, lighting and other electrical circuits, and provides a reserve of electricity when the current consumed by the electrical equipment exceeds that being produced by the alternator.
3 The battery is charged by a Lucas 16ACR or 18ACR alternator, details of which will be found in Section 6.
4 When fitting electrical accessories to vehicles with a negative earth system it is important, if they contain silicone diodes or transistors, that they are connected correctly, otherwise serious damage may result to the component concerned. Items such as radios, tape recorders, electronic tachometer, automatic dipping, parking lamp and anti-dazzle mirrors should all be checked for correct polarity.
5 It is important that the battery positive earth lead is always disconnected if the battery is to be boost charged or if any body repairs are to be carried out, using electric arc welding equipment. Serious damage can be caused to the more delicate instruments, especially those containing semi-conductors.

2 Battery – removal and refitting

1 The battery is in a special carrier fitted on the right-hand wing valance of the engine compartment. It should be removed once every three months for cleaning and testing. Disconnect the positive and then the negative leads from the battery terminals by slackening the clamp retaining nuts and bolts or by unscrewing the retaining screws if terminal caps are fitted instead of clamps.
2 Unscrew the clamp bar retaining nuts, and lower the clamp bar to the side of the battery. Carefully lift the battery from its carrier. Hold the battery vertical to ensure that none of the electrolyte is spilled.
3 Refitting is a direct reversal of this procedure. Refit the negative lead after the positive lead and smear the terminals with petroleum jelly to prevent corrosion, NEVER use an ordinary grease as applied to other parts of the car.

3 Battery – maintenance and inspection

1 Normal weekly battery maintenance consists of checking the electrolyte level of cells to ensure that the separators are covered by ¼ inch of electrolyte. If the level has fallen, top up the battery using distilled water only. Do not overfill. If the battery is overfilled or any electrolyte spilled, immediately wipe away excess as electrolyte attacks and corrodes most metals it comes into contact with very rapidly.
2 If the battery is of the Lucas 'Pacemaker' design a special topping up procedure is necessary as follows:

a) The electrolyte levels are visible through the translucent battery case or may be checked by fully raising the vent cover and tilting to one side. The electrolyte level in each cell must be kept such that the separator plates are just covered. To avoid flooding, the battery must not be topped up within half an hour of it having been charged from any source other than from the generating system fitted to the car

b) To top-up the levels in each cell, raise the vent cover and pour distilled water into the trough until all the rectangular filling slots are full of distilled water and the bottom of the trough is just covered. Wipe the cover seating grooves dry and press the cover firmly into position. The correct quantity of distilled water will automatically be distributed to each cell

c) The vent must be kept closed at all times except when being topped up

3 Another type of battery has a removable, one-piece filler cover which should be removed and the separator plates examined to see that they are covered in electrolyte. To top-up this type of battery:

(a) Make sure that all the filling valves are raised

(b) Pour distilled water into the filling trough until all the valves and the connecting trough are full

(c) Press the cover firmly in position – the electrolyte level should not be above the high marks, if these exist on the walls of the battery case

4 As well as keeping the terminals clean and covered with petroleum jelly, the top of the battery, and especially the top of the cells, should be kept clean and dry. This helps to prevent corrosion and ensures that the battery does not become partially discharged by leakage through dampness and dirt.
5 Inspect the battery securing nuts, battery clamp plate, tray and battery leads for corrosion (white fluffy deposits on the metal which are brittle to the touch). If any corrosion is found, clean off the deposit with ammonia and paint over the clean metal with an anti-rust, anti-acid paint.
6 At the same time inspect the battery case for cracks. Cracks are frequently caused at the top of the battery case by pouring in distilled water in the middle of winter AFTER instead of BEFORE a run. This gives the water no chance to mix with the electrolyte and so the former freezes and splits the battery case.
7 If topping up becomes excessive and the case has been inspected for cracks that could cause leakage, but none are found, the battery is being overcharged and the alternator will have to be tested by a specialist or a dealer.
8 With the battery on the bench at the three monthly interval check, measure the specific gravity with a hydrometer to determine the state of charge and condition of the electrolyte. There should be very little variation between the different cells and, if a variation in excess of 0.025 is present, it will be due to either:

(a) Loss of electrolyte from the battery at some time caused by spillage or a leak, resulting in a drop in the specific gravity of the electrolyte when the efficiency was replaced with distilled water instead of fresh electrolyte

(b) An internal short circuit caused by buckling of the plates or similar malady pointing to the likelihood of total battery failure in the near future

9 The specific gravity of the electrolyte for fully charged conditions at the electrolyte temperature indicated, is listed in Table A. The specific gravity of a fully discharged battery at different temperatures of the electrolyte is given in Table B.

Table A
Specific gravity – battery fully charged

1.268 at 100°F or 38°C electrolyte temperature
1.272 at 90°F or 32°C electrolyte temperature
1.276 at 80°F or 27°C electrolyte temperature
1.280 at 70°F or 21°C electrolyte temperature
1.284 at 60°F or 16°C electrolyte temperature
1.288 at 50°F or 10°C electrolyte temperature
1.292 at 40°F or 4°C electrolyte temperature
1.296 at 30°F or -1.5°C electrolyte temperature

Table B
Specific gravity – battery fully discharged

1.098 at 100°F or 38°C electrolyte temperature
1.102 at 90°F or 32°C electrolyte temperature
1.106 at 80°F or 27°C electrolyte temperature
1.110 at 70°F or 21°C electrolyte temperature
1.114 at 60°F or 16°C electrolyte temperature
1.118 at 50°F or 10°C electrolyte temperature
1.122 at 40°F or 4°C electrolyte temperature
1.126 at 30°F or -1.5°C electrolyte temperature

4 Battery – electrolyte replenishment

1 If the battery is in a fully charged state and one of the cells maintains a specific gravity reading which is 0.025 or lower than the others and a check of each cell has been made with a voltage meter to check for short circuits (a four to seven second test should give a steady reading of between 1.2 and 1.8 volts), then it is likely that electrolyte has been lost from the cell with the low reading at some time.
2 Top-up the cell with a solution of 1 part sulphuric acid to 2.5 parts of water. If the cell is already fully topped up draw some electrolyte out of it with a hydrometer. The total capacity of each cell is approximately 1/3 pint.
3 When mixing the sulphuric acid and water NEVER ADD WATER TO SULPHURIC ACID – always pour the acid slowly onto the water in a glass container. IF WATER IS ADDED TO SULPHURIC ACID IT WILL EXPLODE.
4 Continue to top-up the cell with the freshly made electrolyte and to recharge the battery and check the hydrometer readings.

5 Battery charging

1 In winter time when a heavy demand is placed on the battery, such as when starting from cold, and much electrical equipment is continually in use, it is a good idea to occasionally have the battery fully charged from an external source at a rate of 3.5 to 4 amps.
2 Continue to charge the battery at this rate until no further rise in specific gravity is noted over a four hour period.
3 Alternatively, a trickle charger, charging at the rate of 1.5 amps can be safely used overnight.
4 Special rapid 'boost' charges which are claimed to restore the power of the battery in 1 to 2 hours are most dangerous unless they are thermostatically controlled as they can cause serious damage to the battery plates through overheating.
5 While charging the battery note that the temperature of the electrolyte should never exceed 100°F (37.8°C).

6 Alternator – general description

1 The Lucas 16ACR series alternator is fitted as standard to models covered by this manual. As an optional extra the 18ACR alternator is available and this is a similar machine with a greater output. The main advantage of the alternator lies in its ability to provide a high charge at low revolutions. Driving slowly in heavy traffic with a dynamo invariably means no charge is reaching the battery. In similar conditions even with the wipers, heater, lights and perhaps radio switched on, the alternator will ensure a charge reaches the battery.
2 The alternator is of the rotating field ventilated design and comprises principally a laminated stator, on which is wound a star connected 3 phase output winding, and a twelve pole rotor carrying the field windings. The alternator case houses a rectifier pack for converting the AC output of the machine to DC for battery charging, and an output control regulator.

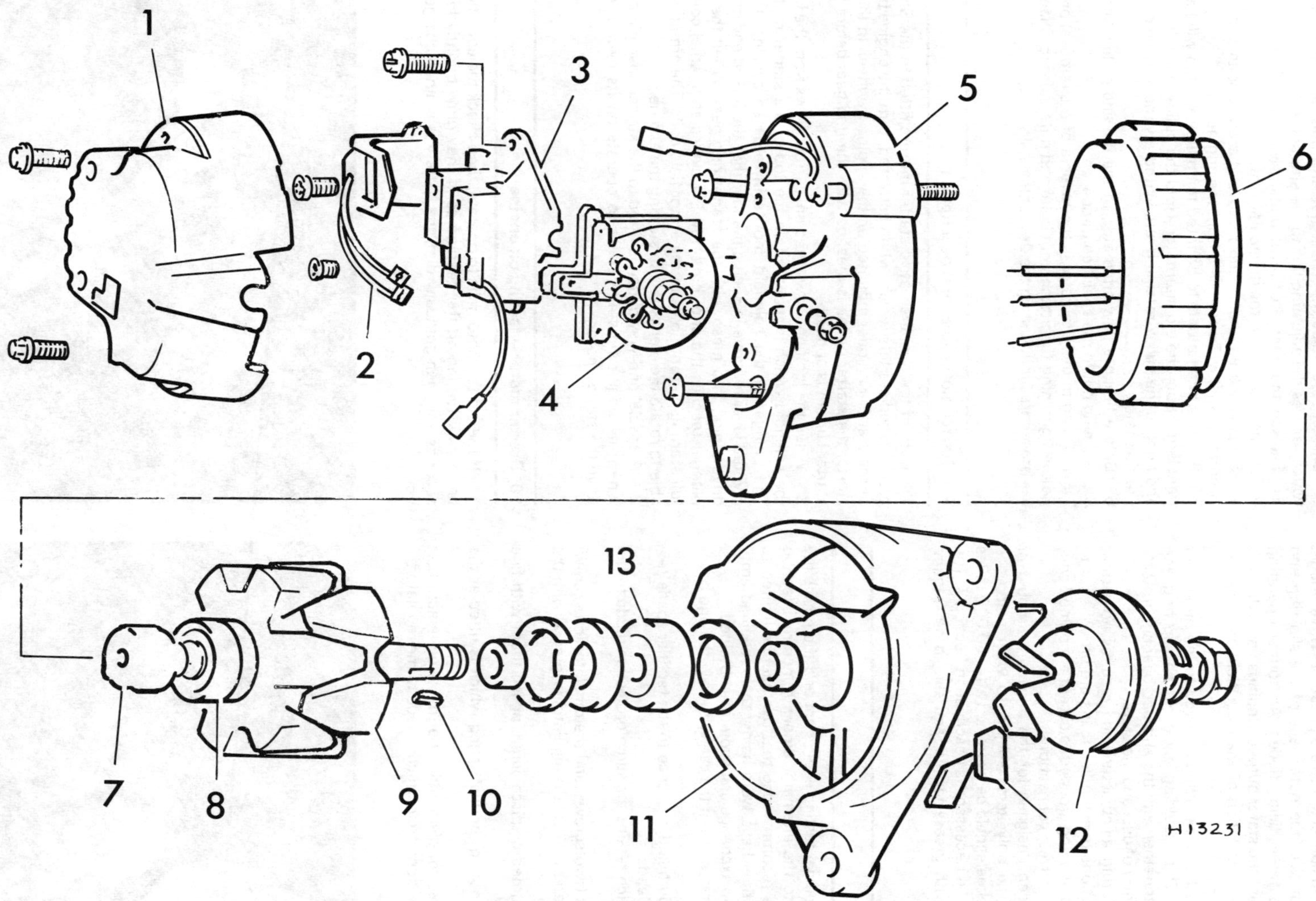

Fig. 10.1 Alternator component parts (Sec 6)

1 *Moulded end cover*
2 *Connector*
3 *Brushbox assembly*
4 *Rectifier pack*
5 *Slip ring end bracket*
6 *Stator windings*
7 *Slip ring moulding*
8 *End bearing (slip ring)*
9 *Pressure ring*
10 *Key*
11 *Drive end bracket*
12 *Fan and pulley*
13 *End bearing (drive end)*

3 The rotor is belt driven from the engine through a pulley keyed to the rotor shaft. A pressed steel fan adjacent to the pulley draws cooling air through the machine. Rotation is clockwise viewed on the drive end. Maximum permissible rotor speed is 15 000 rpm.
4 Rectification of alternator output is achieved by six silicone diodes housed in a rectifier pack and connected as a 3 phase full wave bridge. The rectifier pack is attached to the outer face of the slip ring end bracket and contains also three 'field' diodes. At normal operating speeds, rectified current from the stator output windings flows through these diodes to provide self excitation of the rotor field, via brushes bearing on face type slip rings.
5 The brushgear is housed in a moulding screwed to the outside of the slip ring end bracket. This moulding encloses the slip ring and brush gear assembly, and together with the shielded bearing, protects the assembly against the entry of dust and moisture.
6 The regulator is set during manufacture and requires no further attention. Electrical connections to external circuits are brought out to Lucar connector blades, these being grouped to accept a moulded connector socket which ensure correct connections.
7 The equipment has been designed for the minimum amount of maintenance in service , the only items subject to wear being the brushes and bearings. Brushes should be examined after 60 000 miles (100 000 km) and renewed if necessary as described in Section 7. The bearings are pre-packed with grease for life, and should not require any further attention.

7 Alternator – overhaul

1 Due to the need for special testing equipment and the possibility of damage being caused to the alternator diodes if incorrect testing methods are adopted, it is recommended that overhaul be limited to renewal of the brushes if they have worn below their specified limit. Where other faults occur, an exchange unit should be obtained.
2 With the alternator removed from the car, withdraw the end cover (two screws).
3 Record the wiring colours and sequence to the rectifier spade terminals and disconnect the leads.
4 Remove the screw which retains the surge protection diode to the end bracket.
5 Remove the brush box (two screws) and the regulator (one screw) from the end bracket. On later models the regulator is integral with the brushgear.
6 If the slip rings are burned or discoloured, polish them with fine grade glass paper (not emery).
7 Remove the brushes, noting the leaf spring which is fitted at the side of the inner brush.
8 Reassembly is a reversal of dismantling but make quite sure that the internal terminal connections are correctly made to the regulator.

8 Alternator – removal and refitting

1 To remove the alternator, first release the spring clip and remove the multi-connector from the alternator end cover (photo).
2 Slacken the adjustment link nut and bolt, and remove the bolt which secures the adjusting link to the alternator.
3 Slacken the alternator pivot bolts, swing the alternator towards the engine and remove the drivebelt (photo).
4 Support the alternator and remove the pivot bolts and nuts. The alternator can now be removed from the engine.
5 If a new alternator is to be fitted the pulley and fan will have to be transferred from the old alternator. Hold the pulley in a soft-jawed vice and remove the pulley retaining nut. Withdraw the fan and pulley from the alternator.
6 Before fitting a new alternator, fit the fan and pulley, and tighten the retaining nut to the specified torque.
7 Refitting of the alternator is the reverse of the removal procedure. Before tightening the attachment bolts and nuts, refer to Chapter 2 for the procedure for adjusting the drivebelt tension.

9 Starter motor – general description

1 Both of the two alternative starter motors fitted to the Sherpa are pre-engaged starters which, when activated, turn the engine by means of the starter ring gear mounted on the engine flywheel. In this type of starter the starter pinion is meshed with the ring gear before the main starter current is applied.
2 When the ignition is switched on, current flows from the battery to the solenoid which is mounted on the top of the starter motor body. The plunger in the solenoid moves inwards so causing a centrally pivoted lever to move in such a manner that the forked end pushes the drive pinion into mesh with the starter ring gear. When the solenoid plunger reaches the end of its travel, it closes an internal contact and full starting current flows to the stator field coils. The armature is then able to rotate the crankshaft so starting the engine.
3 A special one-way clutch is fitted to the starter drive pinion so that when the engine fires and starts to operate on its own, it does not drive the starter motor.

10 Starter motor – testing on engine

1 If the starter motor fails to operate, then check the condition of the battery by turning on the headlamps. If they glow brightly for several seconds and then gradually dim, the battery is in an uncharged condition.

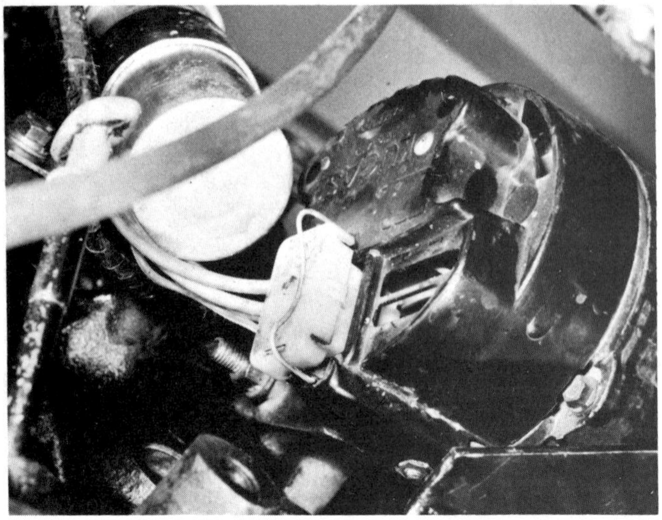
8.1 The alternator multi-connector and spring clip

8.3 The alternator with the drivebelt removed (17V engine)

2 If the headlamps glow brightly and it is obvious that the battery is in good condition then check the tightness of the battery wiring connections (and in particular the earth lead from the battery terminal to its connection on the body frame). Check the tightness of the connections at the starter motor. Check the wiring with a voltmeter for breaks or shorts due to failure of insulation.
3 If the wiring is in order then check that the starter motor solenoid switch is operating. To do this, connect a 12V test lamp between the STA terminal of the solenoid and earth while an assistant operates the ignition/starter switch. If the solenoid is operating there will be an audible click and the lamp will light. If the solenoid does not operate, remove the red and white lead from its connector on the solenoid and connect the test lamp between the end of the lead and earth. Operate the ignition/starter switch again: if the lamp lights, electricity is reaching the solenoid, which must be defective. If the lamp fails to light, the ignition/starter switch or associated wiring is at fault.
4 If the battery is fully charged, with wiring in order, and the solenoid is working but the starter motor fails to operate, then it will have to be removed from the vehicle for examination.

11 Starter motor – removal and refitting

1 Disconnect the battery earth lead. Apply the handbrake.
2 Raise and support the front of the vehicle firmly on axle stands.
3 Disconnect the thin cables from their connectors on the solenoid.
4 Pull back the rubber cover on the large cable terminal on the solenoid. Remove the retaining nut and washer, and remove the cable from the solenoid.
5 Remove the two bolts securing the starter motor and their spring washers. Withdraw the starter motor from the engine.
6 Refitting of the starter motor is the reverse of the removal procedure (photo).

12 Starter motor solenoid – removal and refitting

1 If the need arises, the starter motor solenoid can be removed from the vehicle without removing the starter motor. Proceed as follows.
2 Disconnect the battery earth lead and apply the handbrake. Raise the front of the vehicle and support it firmly on axle stands.
3 Disconnect the thin cables from their connectors on the solenoid.
4 Pull back the rubber cover on the large cable terminal on the solenoid. Remove the retaining nut and washer, and remove the cable from the solenoid.
5 Remove the nut and washer securing the cable connector to the STA terminal on the solenoid.
6 Remove the two nuts and washers (type M35J starter), or the two setscrews and washers (type 2M100 starter), securing the solenoid to the starter.
7 Remove the solenoid coil unit, lifting it to release the solenoid plunger from the engagement lever.
8 Before refitting a solenoid coil unit to a starter motor, lubricate the operating surfaces of the plunger connecting link and the engagement lever with Shell SB2628 grease (temperate and cold climates), or Shell Retinax A grease (hot climates), or a suitable equivalent alternative grease. Position and engage the plunger with the engagement lever correctly and fit the solenoid to the starter by reversing the removal procedure.

13 Starter motor (M35J pre-engaged) – dismantling and reassembly

1 Before commencing overhaul of a defective starter motor, make sure that spare parts are available and enquire as to their price. The time and cost of reconditioning an old motor may compare unfavourably with the cost of an exchange starter motor.
2 Commence dismantling by removing the solenoid (Section 12). Remove the rubber sealing block from the drive end bracket.
3 Remove the retaining ring (if fitted) from the engagement lever pivot pin and remove the pin.
4 Unscrew and remove the screws which secure the drive end bracket and remove the bracket.
5 Remove the sealing cover (if fitted) from the engagement lever.
6 Remove the armature end cap from the commutator end bracket.
7 Withdraw the split-pin and then take off the shim(s) and thrust plate from the end of the armature shaft.
8 Remove the armature complete with its internal thrust washer(s), and remove the thrust washer(s) from the shaft, noting their number and location.
9 Undo and remove the four screws securing the commutator end bracket to the starter motor body.
10 Carefully detach the end bracket from the yoke, at the same time disengaging the field brushes from the brushgear, and remove the bracket.
11 Move the thrust collar clear of the jump ring. Remove the jump ring and then withdraw the drive assembly from the armature shaft.
12 Remove the spring ring to free the engagement lever, thrust washers and spring from the one-way clutch drive.
13 With the motor stripped, the brushes and brushgear may be inspected. To check the brush spring tension, use a new brush fitted into each holder in turn and, with an accurate spring balance, load the brush until it protrudes approximately 1/16 in (1.6 mm) from the holder. Make a note of the spring balance reading which should be as given in the Specifications. Weak springs should be renewed.
14 Inspect the brushes for wear. If they are longer than the specified minimum, check their fit in the holders. Brushes shorter than the

11.6 The starter motor fitted and connected

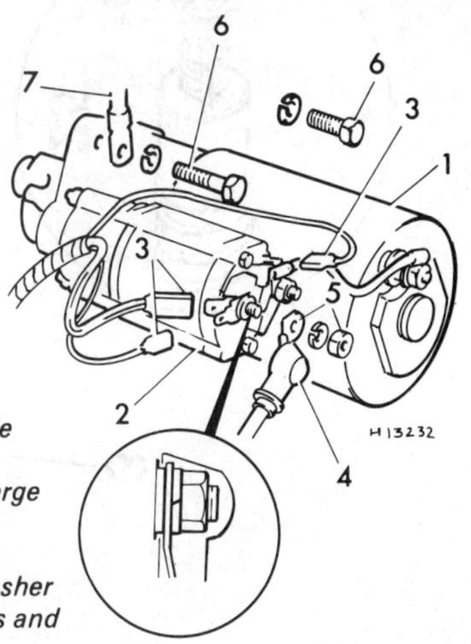

1 *Starter motor*
2 *Solenoid*
3 *Disconnect thin cable terminals*
4 *Protective boot on large cable*
5 *Large cable terminal retaining nut and washer*
6 *Motor retaining bolts and washers*
7 *Earth lead*

Fig. 10.2 Removing the starter motor (Type M35J is shown but Type 2M100 is similar) (Sec 11)

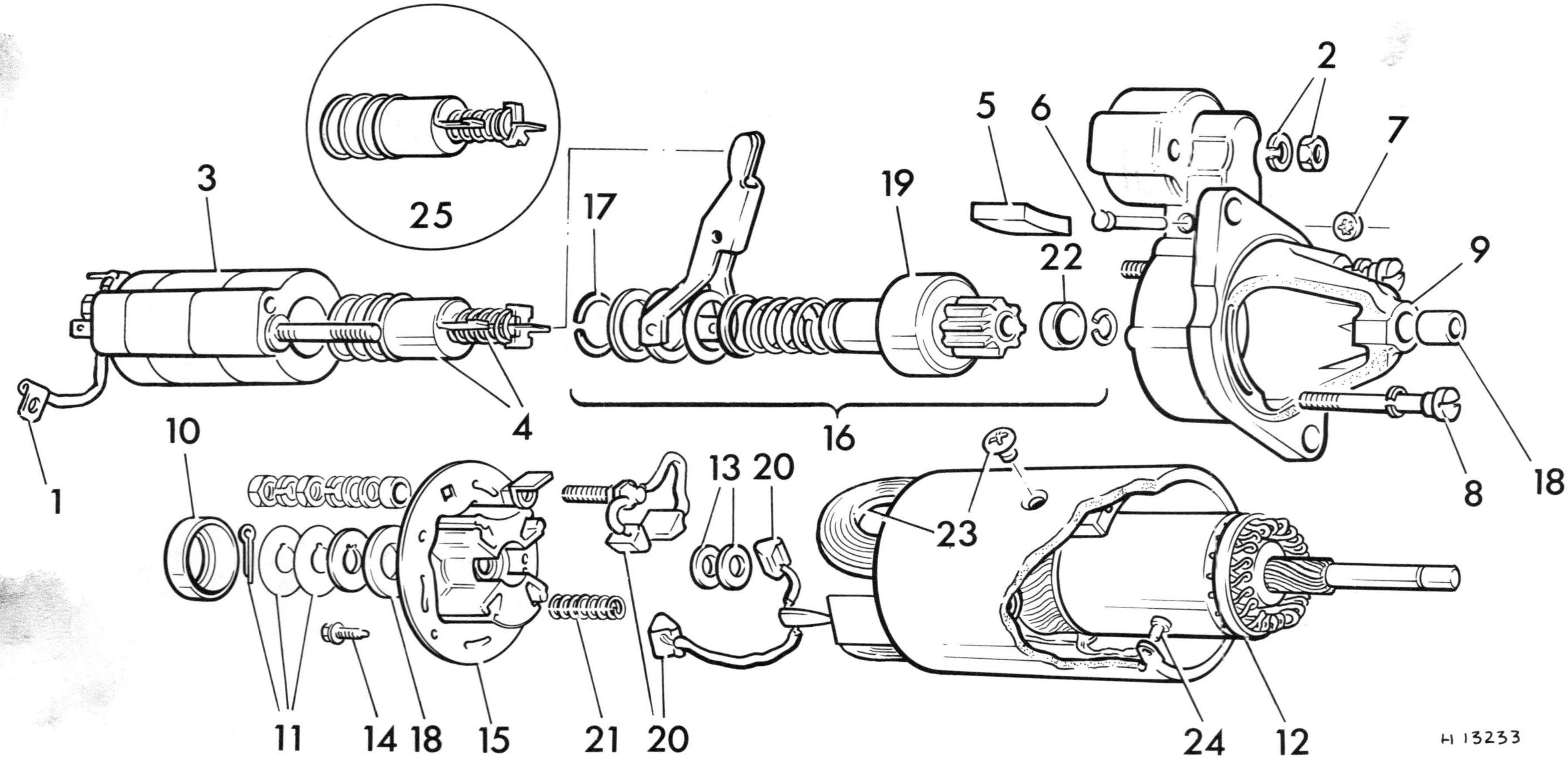

Fig. 10.3 Exploded view of Type M35J starter motor (Sec 13)

1 Link from solenoid STA terminal
2 Solenoid retaining nut and washer
3 Solenoid coil unit
4 Solenoid plunger and return spring
5 Rubber sealing block
6 Engagement lever pivot pin
7 Pivot pin retaining ring
8 Drive end bracket securing bolt and washer
9 Drive end bracket
10 Armature end caps
11 Split-pin, shims and thrust plate
12 Armature
13 Thrust washers
14 Commutator end bracket retaining screw
15 Commutator end bracket
16 Drive assembly
17 Spring ring retaining engagement lever, thrust washers and spring
18 Armature bushes
19 Roller clutch
20 Commutator brushes
21 Commutator brush spring
22 Thrust collar
23 Field winding and retaining screw
24 Field winding to yoke retaining rivet
25 Alternative type of solenoid plunger

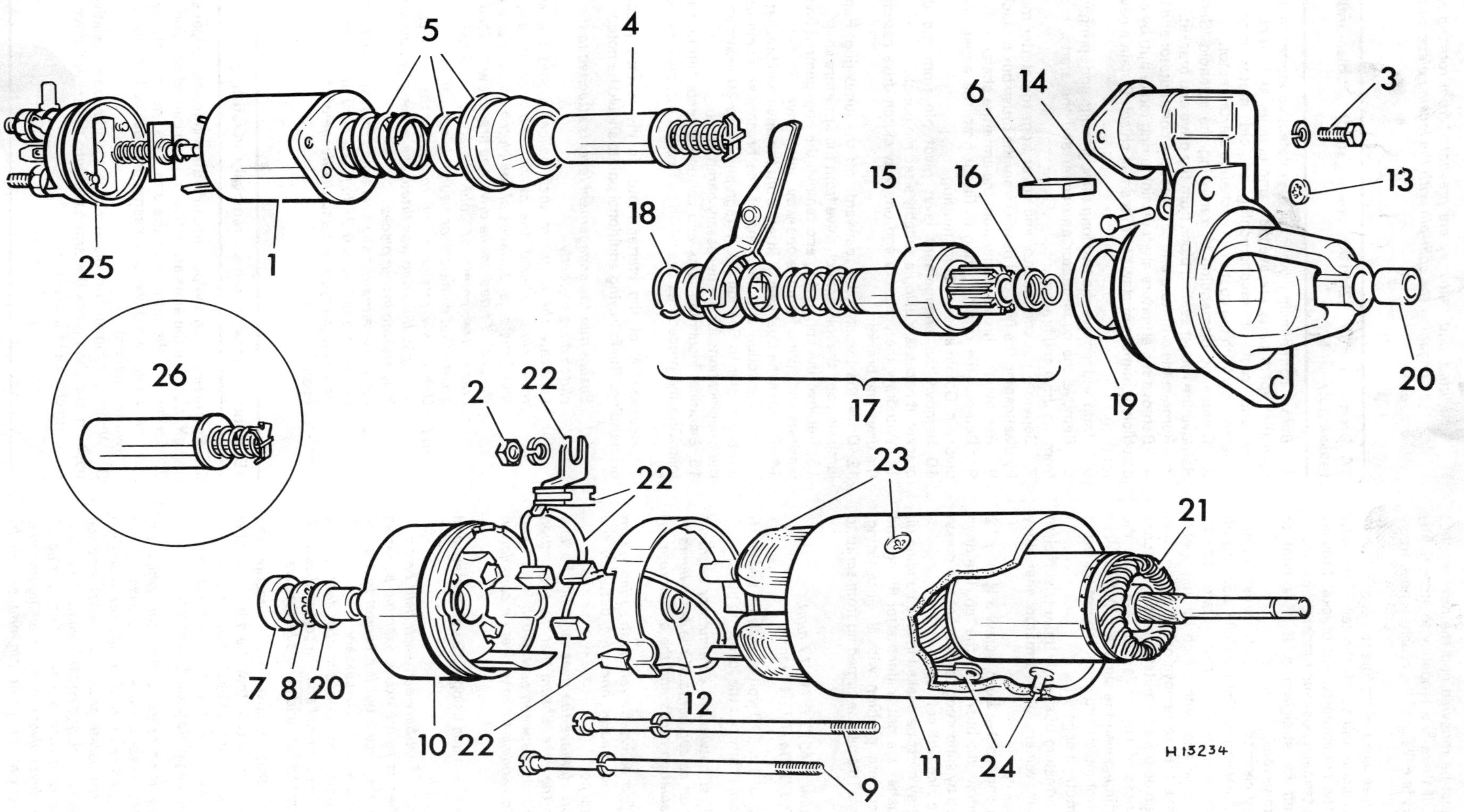

Fig. 10.4 Exploded view of Type 2M100 starter motor (Sec 14)

1 *Solenoid coil*
2 *Terminal nut*
3 *Solenoid securing set-screws*
4 *Solenoid plunger*
5 *Return spring, dust excluder and plunger housing*
6 *Rubber block*
7 *Armature end cap*
8 *Spire nut*
9 *Through bolts*
10 *Commutator end cover*
11 *Yoke*
12 *Armature shaft thrust washer*
13 *Pivot pin retaining ring*
14 *Engagement lever pivot pin*
15 *Roller clutch*
16 *Thrust collar*
17 *Drive assembly*
18 *Spring ring retaining engagement lever, thrust washers and spring*
19 *Dust seal*
20 *Armature bushes*
21 *Armature*
22 *Commutator brushes*
23 *Field winding and retaining screw*
24 *Field winding to yoke retaining rivet*
25 *Solenoid terminal base*
26 *Alternative type of solenoid plunger*

minimum permissible length must be renewed and the new brushes must be tested in their holders. If there is a tendency to stick, lightly polish the brush sides with a very fine file until the brushes move freely in their holders.
15 To renew field winding brushes, cut the flexible tail about ¼ in (6 mm) from the field winding connector and solder the tail of the new brush to the stub remaining. The commutator end bracket brushes must be renewed complete with terminal post and rubber grommet.
16 Examine the armature shaft for straightness. If it is bent or distorted the armature must be renewed.
17 If the armature is satisfactory, clean the commutator brush surface with a petrol-moistened cloth, if necessary first carefully cleaning off any small burnt spots with fine glass paper. If the commutator is badly worn, mount the armature in a lathe and, with a high turning speed, lightly skim the commutator. Take care to remove only the minimum amount of metal required to clean up the commutator, and in any case ensure that the finished thickness of copper is not less than the specified minimum. After skimming, polish the surface with fine glass paper and wipe clean with a petrol-moistened cloth. **Do not** undercut the insulation between the commutator segments.
18 Check the condition of the armature bushes. If there is sufficient wear to permit visible sideways movement of the armature shaft, they should be renewed.
19 To remove the commutator end bracket bush, screw a $\frac{15}{32}$ in or 12 mm tap into the bush and withdraw the bush with the tap. The drive end bracket bush can be removed by supporting the bracket and pressing the bush out, if necessary in a vice with soft jaws.
20 Before fitting new bushes they must be soaked in clean engine oil for at least 20 minutes. The bushes are not reamed after fitting and they must therefore be pressed into position with a close-fitting, polished and shouldered mandrel. This can be turned up from scrap to the following diameters:

(a) Commutator end mandrel: 0.4377 in (11.17 mm)
(b) Driving end mandrel: 0.4729 in (12.011 mm)

21 Electrical insulation tests require a 110 volt ac supply and a 15 watt test lamp. If these are available carry out the following tests:

(a) Armature insulation. Test between a commutator segment and the shaft. Repeat on all segments. If the lamp lights at any time, defective insulation exists and the armature must be renewed
(b) Brushgear insulation. Test between each brush spring in turn and the commutator end bracket and then between the bracket and the terminal post
(c) Field winding insulation. Remove the rivet securing the field winding connection to the yoke. Ensure that the connection is clear of the yoke and test between each brush in turn and the yoke. Refit the field winding-to-yoke rivet on completion of the test

22 Electrical continuity tests can be made using a 12 volt battery and a 12 watt test lamp as follows:

(a) Field windings continuity. Test between each field brush in turn and the yoke. Where a failure exists, examine as far as possible for a break in the circuit, but if it cannot be traced, renewal or a repair by a specialist is necessary
(b) Solenoid windings continuity. Connect the test circuit between the solenoid STA terminal and the solenoid body. If the test is unsatisfactory the solenoid should be renewed.

23 Check that the roller clutch takes up drive in one direction and revolves freely in the other direction. Ensure that the drive assembly moves freely along the armature shaft without binding or sticking.
24 Reassembly of the starter motor is the reverse sequence to dismantling. The following additional points should be noted:

(a) Lubricate the armature shaft helical splines (on which the roller clutch slides), the outer surface of the roller clutch housing and the moving parts of the engagement lever with Shell SB2628 grease (temperate or cold climates), or Shell Retinax A grease (hot climates), or a suitable equivalent
(b) When assembling the drive use a new retaining ring to secure the engagement lever pivot pin. If a retaining ring was not fitted originally, renew the pivot pin and secure by staking in four places
(c) Make sure that the internal thrust washer(s) are fitted to the commutator end of the armature shaft before the armature is fitted, and that they are prevented from rotating separately by engaging the split-pin with the locking piece on the thrust plate

14 Starter motor (2M100 pre-engaged) – dismantling and reassembly

1 Before commencing overhaul of a defective starter motor, make sure that spare parts are available and enquire as to their price. The time and cost of reconditioning an old motor may compare unfavourably with the cost of an exchange starter motor.
2 Commence dismantling by removing the solenoid (Section 12). Remove the rubber sealing block from the drive end bracket.
3 Remove the armature end cap from the commutator end bracket.
4 Remove the armature shaft retaining ring (spire nut) by cutting off a sufficient number of claws with a small chisel. Obtain a new ring for refitting.
5 Undo and remove the two long through bolts and spring washers.
6 Remove the commutator end cover and the starter motor yoke from the drive end bracket.
7 Separate the commutator end cover from the starter motor yoke by disengaging the field coil bushes from the brush box moulding.
8 Remove the thrust washer from the armature shaft.
9 Remove the retaining ring (if fitted) from the engagement lever pivot pin. Obtain a new one for refitting.
10 Remove the engagement lever pivot pin from the drive end bracket, if necessary using a suitably sized pin punch.
11 Withdraw the armature and one-way clutch drive assembly from the drive end bracket.
12 Drive the thrust collar rearwards off the jump ring. Remove the jump ring and the roller clutch drive from the armature shaft.
13 Remove the spring ring and remove the engagement lever, thrust washers and spring from the one-way clutch drive.
14 Remove the dust seal (if fitted) from the drive end bracket.
15 For inspection and servicing of the brushgear, commutator and armature refer to Section 13, paragraphs 13 to 20 inclusive. Refer to the Specifications for the necessary parameters.
16 Electrical insulation tests and continuity tests can be carried out following the procedures described in Section 13, paragraphs 21 and 22.
17 Reassembly of the starter motor is the reverse procedure to dismantling. The following additional points should be noted:

(a) Ensure that the engagement lever is attached to the one-way clutch drive correctly
(b) Lubricate the armature shaft helical splines (on which the roller clutch slides), the moving parts of the engagement lever, the outer surfaces of the clutch housing and the lip of the seal (if fitted) in the drive end bracket with Shell SB2628 grease (temperate or cold climates), or Shell Retinax A grease (hot climates), or a suitable equivalent
(c) Use a new retaining ring to secure the engagement lever pivot pin. If no ring was fitted, use a new pivot pin and secure it by staking in four places
(d) Fit a new armature shaft retaining ring (spire nut) so that the armature endfloat is 0.01 in (0.25 mm) maximum measured between the spire nut and the flange of the commutator end bush

15 Windscreen wiper arms – removal and refitting

1 Before removing a wiper arm, turn the windscreen wiper switch on and off to ensure the arms are in their normal parked position.
2 To remove the arm, pivot the arm back and pull the wiper arm head off the splined drive, at the same time easing back the clip with a screwdriver (photo).
3 When refitting an arm, place it so it is in the correct relative parked position and then press the arm head onto the splined drive until the retaining clip clicks into place.

16 Windscreen wiper blades – changing wiping arc

If it is wished to change the area through which the wiper blades

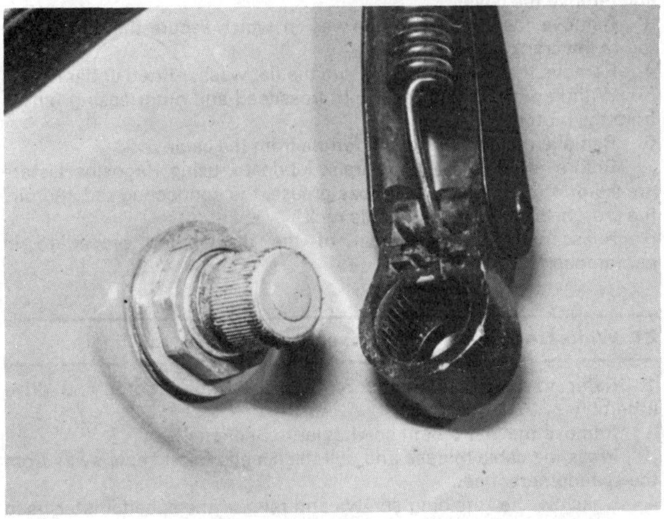

15.2 A windscreen wiper arm removed from its drive spindle

18.8 The windscreen wiper motor

move, this is simply done by removing each arm in turn from its splined drive, and then refitting it on the drive in a slightly different position.

17 Windscreen wiper mechanism – fault diagnosis and rectification

1 Should the windscreen wipers fail, check the number 2 fuse. If the fuse has blown, renew it after having checked the wiring in the circuits serviced by this fuse for short circuits which must be rectified.
2 If the fuse is not blown and the windscreen wiper motor does not run, or works slowly, then check the terminals for loose connections, and make sure that the insulation of the external wiring is not broken or cracked. If this is in order, check the current the motor is taking by disconnecting the wiper rack, see Section 20, connecting an ammeter into the motor supply circuit, and turning on the ignition and wiper switches. Current consumption should be as stated in the Specifications.
3 If no current is passing but the supply to the wiper motor is correct, the motor must be removed for servicing. If there is no supply to the motor, check the wiper switch.
4 Should the wiper take a very high current, check the wiper blades for freedom of movement. If this is satisfactory check the wiper motor and drive cable for signs of damage. Measure the endfloat which should be within the specified limits. The endfloat is set by the thrust screw. Check that excessive friction in the cable connecting tubes, caused by too small a curvature, is not the cause of the high current consumption.
5 If the motor takes a very low current, ensure that the battery is fully charged. Check the brushgear after removing the commutator yoke assembly, and ensure that the brushes are free to move. If necessary, renew the tension springs. If the brushes are very worn they should be renewed. The armature may be checked by substitution.

18 Windscreen wiper motor – removal and refitting

1 Disconnect the battery earth lead and, referring to Section 15, remove the wiper arms and blades.
2 Disconnect the wiring connector from the motor.
3 Unscrew the drive outer casing retaining nut from the motor.
4 Remove the motor strap securing screws.
5 Remove the motor assembly complete with the inner drive cable and mounting pad.
6 Before refitting the windscreen wiper motor and drive, smear the cable rack with Ragosine Listate grease or a good quality general purpose grease.
7 Feed the cable rack into the outer casing, ensuring that the rack engages with the wheelbox gearteeth. Position the mounting pad and the mounting retaining strap.
8 Complete the refitting by reversing the procedure in paragraphs 1 to 4 inclusive (photo).

19 Windscreen wiper motor – dismantling, inspection and reassembly

1 The only repair which can be effectively undertaken by the do-it-yourself mechanic to a wiper motor is brush renewal. Anything more serious than this will mean either exchanging the complete motor or having a repair done by an auto electrician. Spare part availability is really the problem. Brush renewal is described here.
2 Refer to Fig. 10.6 and remove the four gearbox cover retaining screws and lift away the cover. Release the circlip and flat washer securing the connecting rod to the crankpin on the shaft and gear. Lift away the connecting rod followed by the second flat washer.
3 Release the circlip and flat washer securing the shaft and gear to the gearbox body.
4 De-burr the gear shaft and lift away the gear making careful note of the location of the dished washer.
5 Scribe a mark on the yoke assembly and gearbox to ensure correct reassembly and unscrew the two yoke bolts from the motor yoke assembly. Part the yoke assembly including armature from the gearbox body. As the yoke assembly has residual magnetism ensure that the yoke is kept well away from metallic dust.
6 Unscrew the two screws securing the brushgear and the terminal and switch assembly and remove both the assemblies.
7 Inspect the brushes for excessive wear. If the main brushes are worn to the specified limit or the narrow section of the third brush is worn to the full width of the brush fit a new brushgear assembly. Ensure that the three brushes move freely in their boxes.
8 Use Ragosine Listate grease, or a good general purpose grease, to lubricate the gearwheel teeth, armature shaft worm gear, connecting rod and pin, crosshead guide, and cable rack. Use engine oil to lubricate the bearing bushes and sparingly on the armature shaft bearing surfaces and the gearwheel shaft and crankpin. Thoroughly soak the felt washer in the yoke bearing with engine oil.
9 Reassemble, following the dismantling procedure in reverse, and tighten the yoke fixing bolts to the specified torque.

20 Windscreen wiper rack – removal and refitting

1 Refer to Section 18 and remove the windscreen wiper motor and drive assembly.
2 Unscrew the four screws retaining the wiper motor gearbox cover

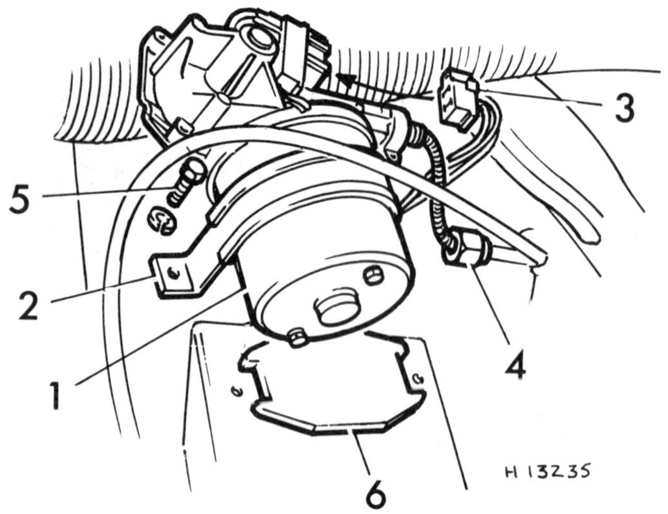

Fig. 10.5 Removing the windscreen wiper motor (Sec 18)

1 Motor assembly
2 Retaining strap
3 Electrical connector
4 Drive outer casing retaining nut
5 Retaining screws
6 Mounting pad

and remove the cover.

3 Remove the circlip and flat washer which secure the connecting rod to the crankpin.

4 Remove the connecting rod and the flat washer fitted under it.

5 Withdraw the cable rack unit crosshead and outer casing ferrule from the motor.

6 Remove the crosshead and ferrule from the cable rack.

7 Before refitting the wiper rack lubricate, using Ragosine Listate grease or a good general purpose grease, the connecting rod and pin, the crosshead slide and the cable rack.

8 Reassembly is a reversal of the dismantling procedure in paragraphs 1 to 6 inclusive.

21 Windscreen wiper wheelboxes – removal and refitting

1 Refer to Section 18 and remove the wiper motor and drive assembly.

2 Remove the instrument cowl, refer to Section 42.

3 Press the cable release and pull the speedometer cable away from the speedometer head.

4 Unscrew the retaining screws and release the speedometer head from the mounting bracket. Note that one screw also retains the earth lead.

5 Remove the instrument bulb holders.

6 Remove the green lead from the B terminal on the voltage stabiliser.

7 Remove the green/blue lead from the temperature gauge.

8 Remove the green/black lead from the fuel gauge.

9 Remove the speedometer head and the ashtray.

10 Remove the passenger side demister hose.

11 Undo the clips which secure the windscreen washer tube.

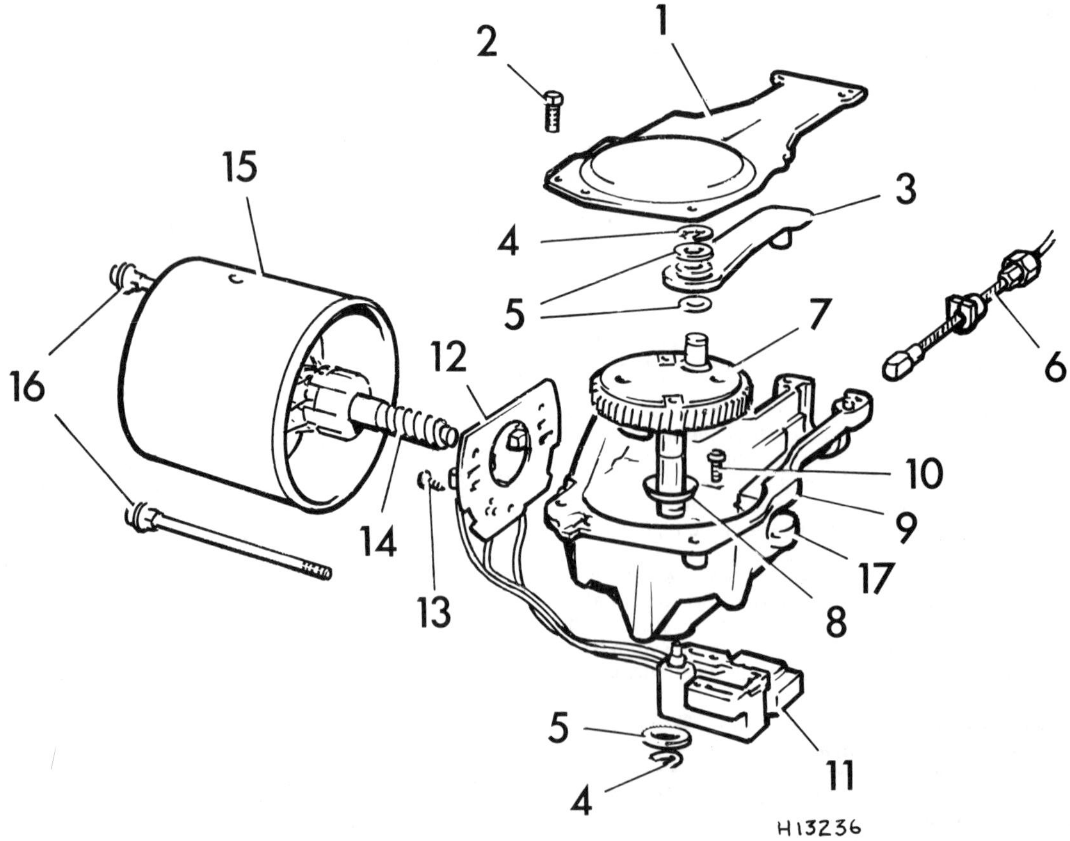

Fig. 10.6 Lucas 14W two-speed wiper motor – exploded view (Sec 19)

1 Gearbox cover
2 Screw for cover
3 Connecting rod
4 Circlip
5 Plain washer
6 Drive cable
7 Gear
8 Dished washer
9 Gearbox
10 Screw for limit switch
11 Limit switch assembly
12 Brush gear
13 Screw for brush gear
14 Armature
15 Yoke assembly
16 Yoke bolts
17 Armature thrust screw

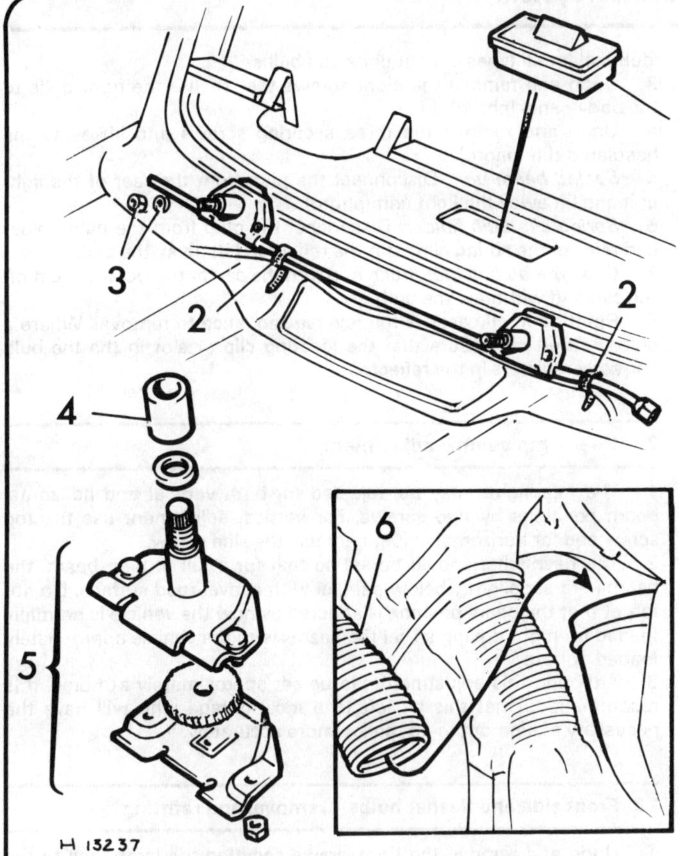

Fig. 10.7 Windscreen wiper wheelboxes removal (Sec 21)

1 *Ashtray*
2 *Washer tube supporting clips*
3 *Wheelbox retaining nut and washer*
4 *Wheelbox spacer*
5 *Wheelbox components*
6 *Passenger side demister hose*

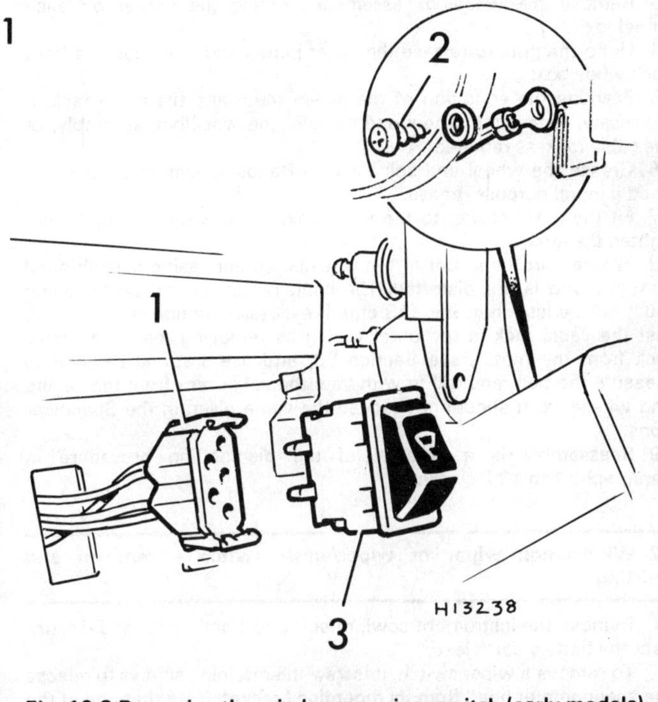

Fig. 10.8 Removing the windscreen wiper switch (early models) (Sec 22)

1 *Electrical harness connector*
2 *Speedometer mounting with earth wire*
3 *Wiper switch*

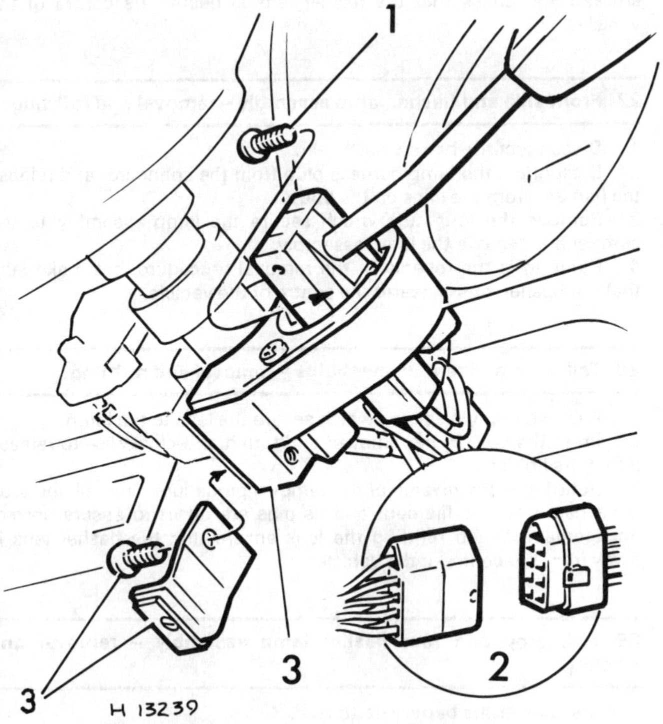

Fig. 10.9 Removing the windscreen wiper/wash switch (later models) (Sec 22)

1 *Retaining screw*
2 *Harness connector*
3 *Retaining screw and cowl bracket*

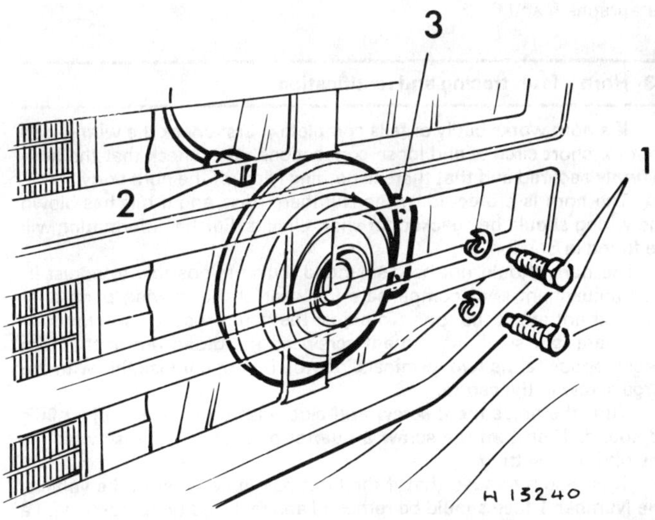

Fig. 10.10 Removing the horn (Sec 23)

1 *Mounting screws*
2 *Electrical connector*
3 *Horn mounting bracket*

12 Undo the retaining nut and remove the washer from each wiper wheelbox.
13 Remove the wheelbox assemblies noting the spacer on each wheelbox.
14 Undo the nuts to release the cover plates and outer casings from each wheelbox.
15 Examine the condition of the wheel teeth and the cable rack. If necessary, renew the wheel and spindle, the wheelbox assembly, or the cable rack, as required.
16 Grease the wheel and spindle with Ragosine Listate grease or a good general purpose grease.
17 Fit the outer casing to the wheelbox, fit the wheelbox plate and tighten the nuts.
18 Ensure that the motor to the wheelbox outer casing is positioned correctly and is not distorted. The bend radius on the outer casing must not be less than 9 in (23 cm). If excessive friction is suspected, test the cable rack in the outer casing by removing the wiper cable rack from the motor, see Section 20, and use a spring balance to measure the pull required to withdraw the cable rack from the casing and wheelbox. It should not exceed the value given in the Specifications.
19 Reassembly is a reversal of the dismantling procedure in paragraphs 1 to 13 inclusive.

22 Windscreen wiper or wiper/wash switch – removal and refitting

1 Remove the instrument cowl, referring to Section 42, and disconnect the battery earth lead.
2 To remove a wiper switch, unscrew the retaining screws to release the speedometer head from its mounting bracket. Note that one of the screws also retains an earth wire.
3 Reach behind the facia and press out the wiper switch. Press the release tabs on the side of the wiring harness connector and remove the connector from the switch.
4 Refitting a wiper switch is the reverse of the removal procedure. Make sure that the earth wire is secured by its retaining screw.
5 To remove a wiper/wash switch, disconnect the harness switch connector from the main harness.
6 Remove the two screws to release the switch and the steering column cowl retaining bracket from the steering column.
7 Refitting of a wiper/wash switch is a reversal of the procedure in paragraphs 5 and 6.

23 Horn – fault tracing and rectification

1 If a horn works badly or fails completely, first check the wiring lead to it for short circuits and loose connections. Also check that the horn is firmly secured and that there is nothing lying on the horn body.
2 The horn is protected by the Number 1 fuse and if this has blown the wiring should be checked for short circuits. Further information will be found in Section 48.
3 The horn should not be dismantled, but it is possible to adjust it. This adjustment is to compensate for wear of the moving parts only and will not affect the tone. To adjust the horn proceed as follows.
4 There is a small adjustment screw on the broad rim of the horn nearly opposite the two terminals (photo). Do not confuse this with the large screw in the centre.
5 Turn the adjustment screw anti-clockwise until the horn just fails to sound. Then turn the screw a quarter of a turn clockwise which is the optimum setting.
6 It is recommended that if the horn has to be reset in the vehicle, the Number 1 fuse should be removed and replaced temporarily with a piece of wire, otherwise the fuse will blow due to the high current required for the horn in continual operation.
7 Should twin horns be fitted, the horn which is not being adjusted should be disconnected while adjustments of the other takes place.
8 Refit the fuse on completion of adjustments.

24 Headlight units – removal and refitting

1 Sealed beam (or renewable bulb) light units are fitted.
2 The method of gaining access to the light unit for replacement is identical for all types of light units and bulbs.
3 Undo and remove the eight screws that secure the front grille to the body panel (photo)
4 Undo and remove the three securing screws and lift away the headlamp rim (photo).
5 *Sealed beam unit:* Disconnect the plug from the rear of the light unit and lift away the light unit (photo).
6 *Spring clip bulb holder:* Disconnect the plug from the bulb holder and release the spring clip from the reflector. Lift away the bulb.
7 *Cap type bulb holder:* Push and turn the cap anti-clockwise. Lift off the cap and withdraw the bulb.
8 Refitting in all cases is the reverse sequence to removal. Where a bulb is fitted make sure that the locating clip or slot in the the bulb correctly registers in the reflector.

25 Headlamp beam – adjustment

1 The headlights may be adjusted for both vertical and horizontal beam positions by two screws. For vertical adjustment use the top screw and for horizontal adjustment use the side screw.
2 The headlights should be set so that for a full or high beam, the beams are set slightly below parallel with a level road surface. Do not forget that the beam position is affected by how the vehicle is normally loaded for night driving so set the beams with the vehicle appropriately loaded (photo).
3 Although this adjustment can be set approximately at home, it is recommended that this be left to a local garage who will have the necessary equipment to do the job more accurately.

26 Front side and flasher bulbs – removal and refitting

1 Undo and remove the two screws securing the lamp lens to the lamp body and lift away the lenses (photo).
2 Either bulb is retained by a bayonet fixing, so to remove a bulb push in slightly and rotate in an anti-clockwise direction (photo).
3 Refitting is the reverse sequence to removal. Take care not to over-tighten the two lens retaining screws as the lenses can be easily cracked and check that the flasher lens is nearer the centre of the vehicle.

27 Front side and flasher lamp assembly – removal and refitting

1 Disconnect the battery earth lead.
2 Disconnect the lamp harness plug from the connector and release the harness from the clips on the body.
3 Remove the four nuts which secure the lamp assembly to the bumper and remove the lamp assembly.
4 Refitting is the reverse of the removal procedure, but make sure that the flasher lens is nearer the centre of the vehicle.

28 Tail, stop and rear flasher bulbs – removal and refitting

1 Remove the two screws which secure the lens to the lamp.
2 Press the bulb to be removed and turn it anti-clockwise to release it from its holder.
3 Refitting is the reverse of the removal procedure. The tail and stop lamp has a double filament and its pins are offset to ensure correct engagement. When refitting the lens ensure that the flasher lens is away from the centre of the vehicle.

29 Tail, stop and rear flasher lamp assembly – removal and refitting

1 Disconnect the battery earth lead.
2 Remove the four nuts which secure the lamp assembly to the bumper.
3 Open the rear door and release the wiring harness grommet from the body on the side being worked on.
4 Disconnect the wiring harness plug from the connector. Thread the plug through the grommet, thread the harness through the body

23.3 The horn adjusting screw (arrowed)

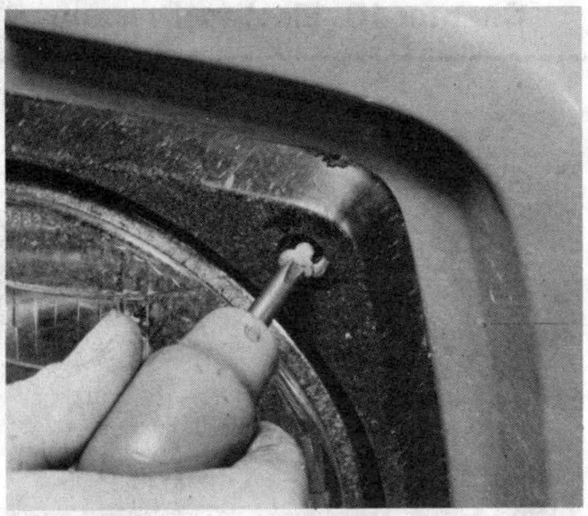

24.3 Undo the grille screws and remove the grille

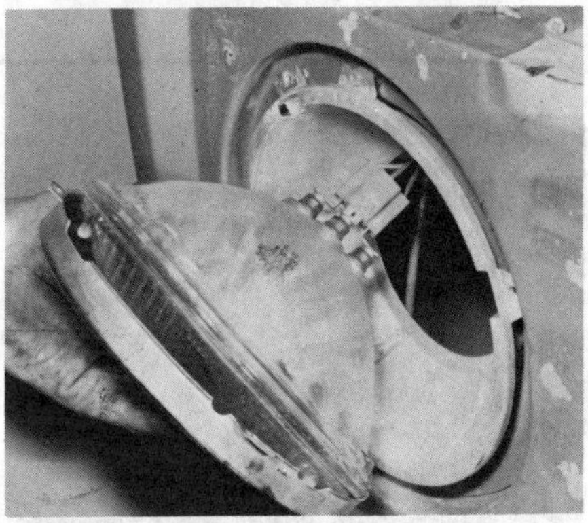

24.4 Lift away the headlamp and ...

24.5 ... disconnect the electrical connector

25.2 Adjusting the headlamp beam up or down

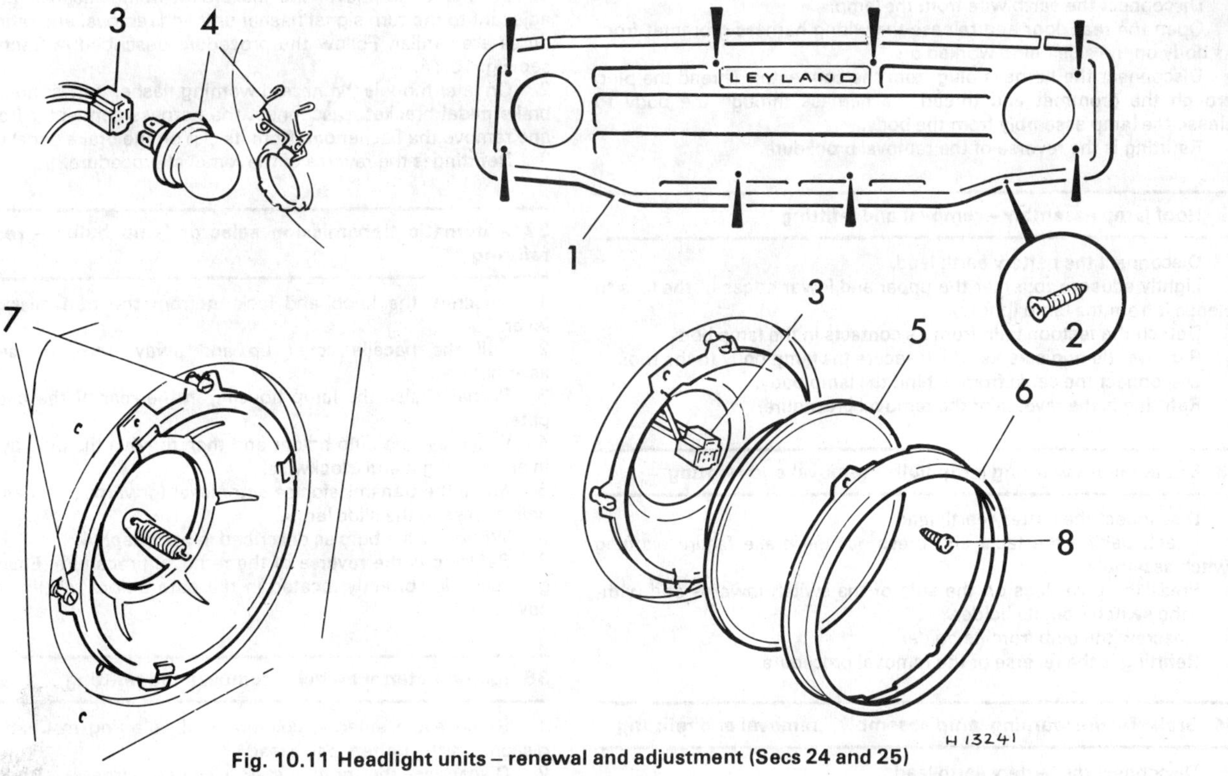

Fig. 10.11 Headlight units – renewal and adjustment (Secs 24 and 25)

1 Grille, attachment screws arrowed
2 Attachment screw
3 Electrical connector
4 Bulb and retaining clip
5 Sealed beam unit (alternative to 4)
6 Headlamp rim
7 Adjusting screws
8 Rim retaining screw

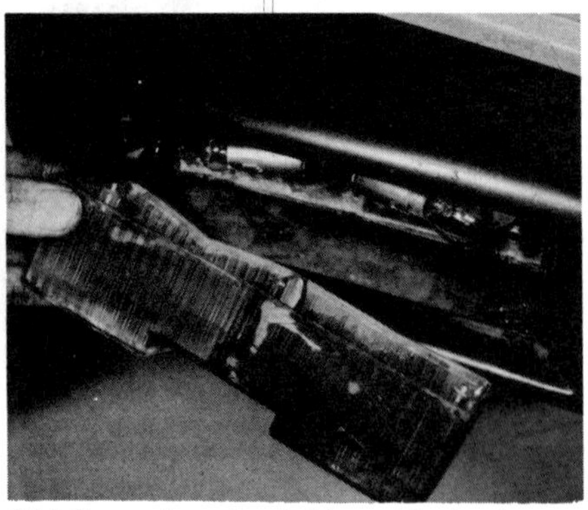
26.1 Removing a front side and flasher lamp lens ...

26.2 ... to reveal the bulbs

30.1 The number plate bulb

and remove the lamp assembly.
5 Refitting is the reverse of the removal procedure but make sure that the flasher lens is fitted furthest from the centre of the vehicle.

30 Number plate lamp bulb – removal and refitting

1 Remove the two screws which secure the lens hood to the body and pull the lens from the lamp (photo).
2 Depress and turn the bulb anti-clockwise to release it from its holder.
3 Refitting is the reverse of the removal procedure.

31 Number plate lamp assembly – removal and refitting

1 Disconnect the battery earth lead.
2 Remove the two screws which secure the lens hood to the body and pull the lamp assembly free from the body.
3 Disconnect the earth wire from the lamp.
4 Open the rear door and release the wiring harness grommet from the body on the side being worked on.
5 Disconnect the harness plug from the connector. Thread the plug through the grommet and thread the harness through the body to release the lamp assembly from the body.
6 Refitting is the reverse of the removal procedure.

32 Roof lamp assembly – removal and refitting

1 Disconnect the battery earth lead.
2 Lightly squeeze together the upper and lower edges of the lens to release it from the lamp (photo).
3 Detach the festoon bulb from its contacts in the lamp body.
4 Remove the two screws which secure the lamp body to the roof.
5 Disconnect the cable from behind the lamp body.
6 Refitting is the reverse of the removal procedure.

33 Brake failure warning lamp bulb – removal and refitting

1 Disconnect the battery earth lead.
2 Reach behind the facia and press out the brake failure warning switch assembly.
3 Press the pivot legs on the side of the switch inwards and withdraw the switch from its holder.
4 Unscrew the bulb from its holder.
5 Refitting is the reverse of the removal procedure.

34 Brake failure warning lamp assembly – removal and refitting

1 Disconnect the battery earth lead.
2 Reach behind the facia and press out the brake failure warning lamp assembly.
3 Disconnect the wires from the terminals on the rear of the lamp assembly and remove the lamp.
4 Refitting is the reverse of the removal procedure.

35 Direction indicator flasher unit – removal and refitting

1 Remove the two screws which secure the steering column cowl (later models have a third screw in the cowl extension), and remove the upper cowl.
2 *Early models:* Withdraw the turn signal flasher unit from its retaining clip in the lower cowl, disconnect the harness connector and remove the flasher unit.
3 *Later models:* Remove the key from the steering lock. Remove the three screws retaining the lower cowl. Disconnect the harness from the flasher unit and remove the flasher unit from its retaining clip in the lower cowl (photo).
4 Refitting is the reverse of the removal procedure.

36 Hazard warning flasher unit – removal and refitting

1 Early models have the hazard warning flasher unit located adjacent to the turn signal flasher unit and removal and refitting procedures are similar. Follow the procedure described in Section 35 and see Fig. 10.14.
2 On later models the hazard warning flasher unit is mounted on the brake pedal bracket. Disconnect the harness connector from the unit and remove the flasher unit from its clip on the brake pedal bracket.
3 Refitting is the reverse of the removal procedure.

37 Automatic transmission selector lamp bulbs – removal and refitting

1 Unscrew the knob and locknut from the transmission selector lever.
2 Pull the nacelle cover up and away from the selector box assembly.
3 Partially raise the lamp housing in the rear of the change speed gate.
4 Withdraw the bulb holder and then remove the bulb by pressing it in and turning it anti-clockwise.
5 Move the transmission selector lever forwards as far as possible to gain access to the slide lamp.
6 Withdraw the bulb as described in paragraph 4.
7 Refitting is the reverse of the removal procedure. Ensure that the gate slide is correctly located in the gate before refitting the nacelle cover.

38 Ignition/starter switch – removal and refitting

1 Remove the steering column cowl, referring to Chapter 11, and disconnect the battery earth lead.
2 Disconnect the ignition/starter switch harness at the three snap connectors and remove the two screws to release the switch from the steering lock.
3 Remove the switch, taking care to retrieve the switch driver that locates on the steering lock spindle.
4 Refitting is the reverse of the removal procedure.

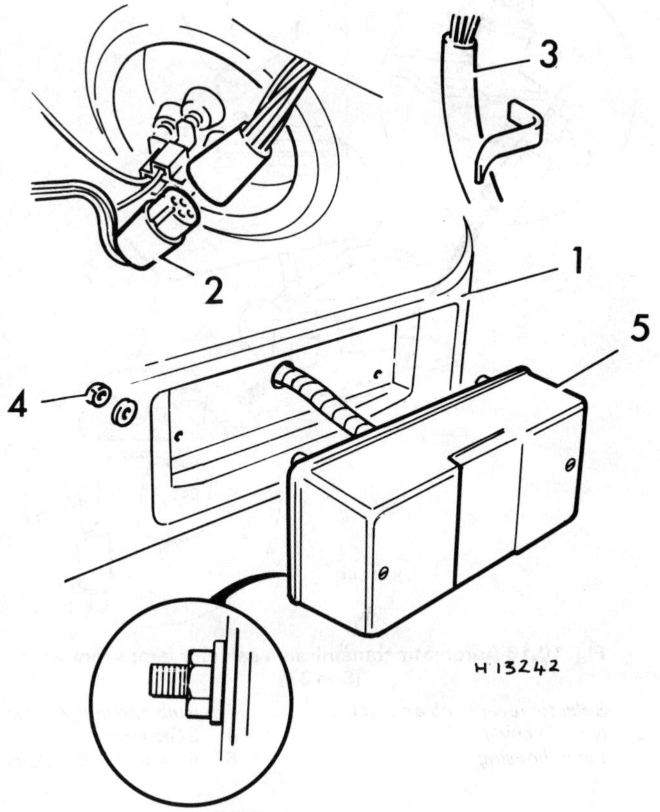

Fig. 10.12 Front flasher and sidelight assembly removal (rear is similar) (Sec 27)

1 *Front bumper*
2 *Electrical connector*
3 *Removing cable from clip*
4 *Retaining nut and washer*
5 *Lamp assembly*

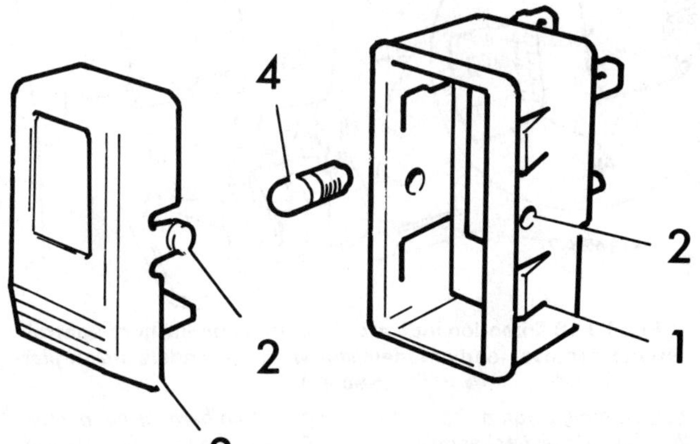

Fig. 10.13 Brake failure warning lamp bulb renewal (Sec 33)

1 *Switch/lamp body*
2 *Pivot legs and location holes*
3 *Switch rocker*
4 *Bulb*

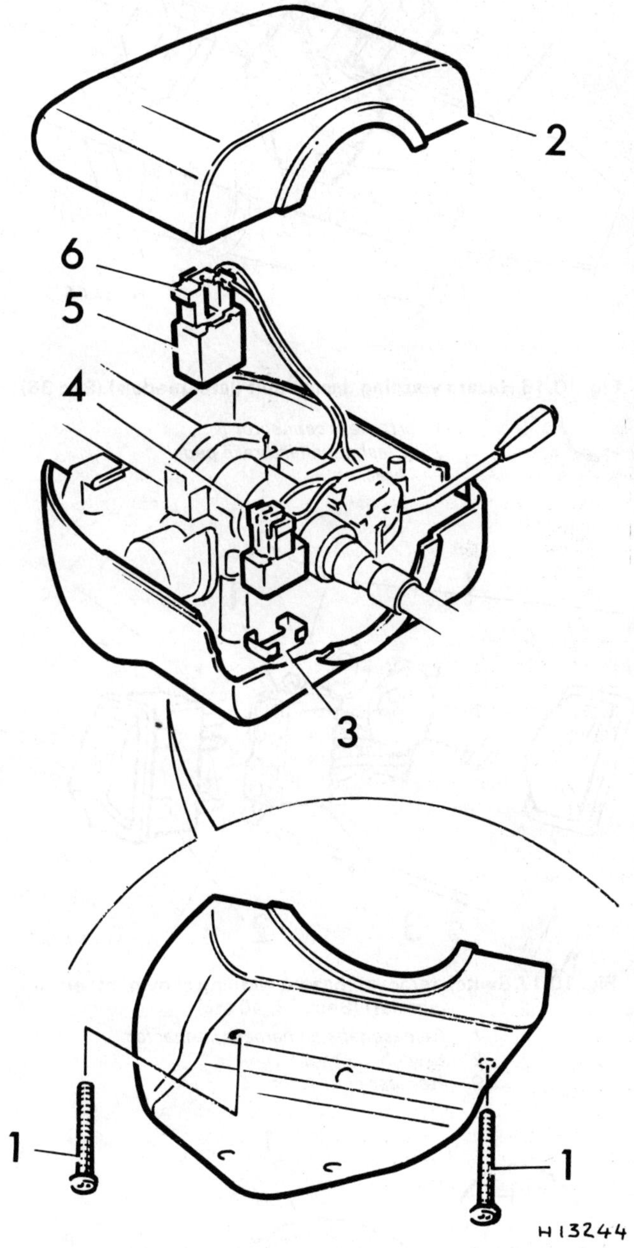

Fig. 10.14 Flasher units (early models) removal (Secs 35 and 36)

1 *Lower cowl retaining screws*
2 *Upper cowl*
3 *Turn signal flasher unit*
4 *Harness connector*
5 *Hazard warning flasher unit*
6 *Harness connector*

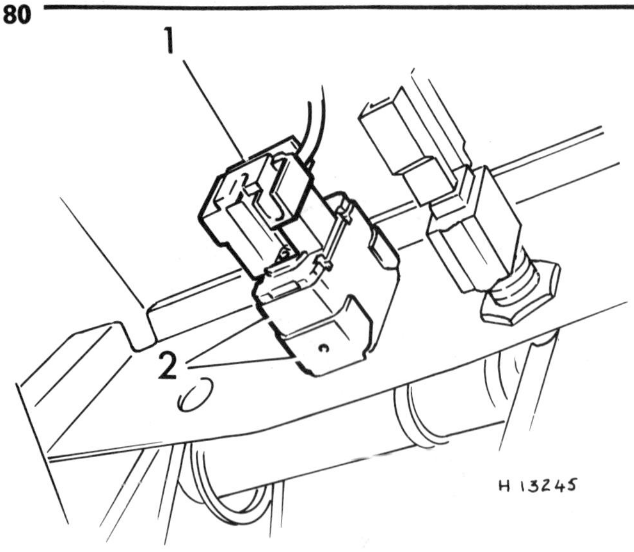

Fig. 10.15 Hazard warning flasher unit (later models) (Sec 36)

1. *Harness connector*
2. *Flasher unit in brake pedal bracket*

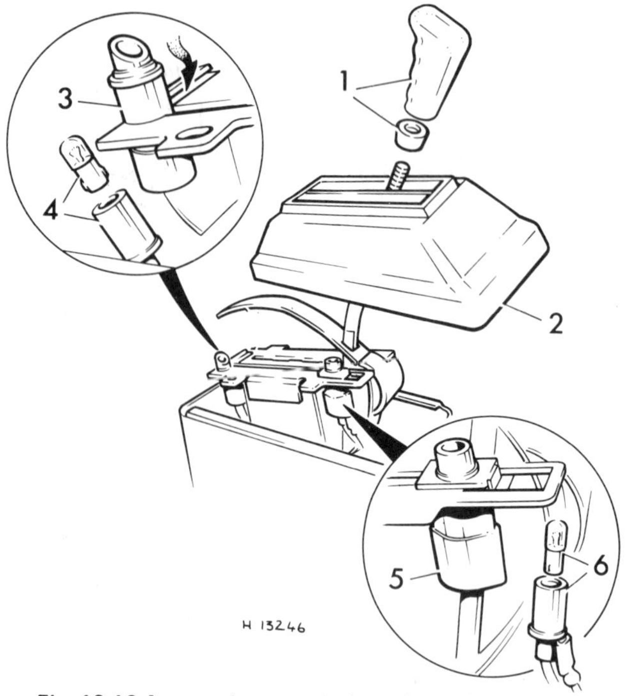

Fig. 10.16 Automatic transmission selector lamps renewal (Sec 37)

1. *Selector lever knob and locknut*
2. *Nacelle cover*
3. *Lamp housing*
4. *Bulb and bulb holder*
5. *Slide lamp*
6. *Bulb and bulb holder*

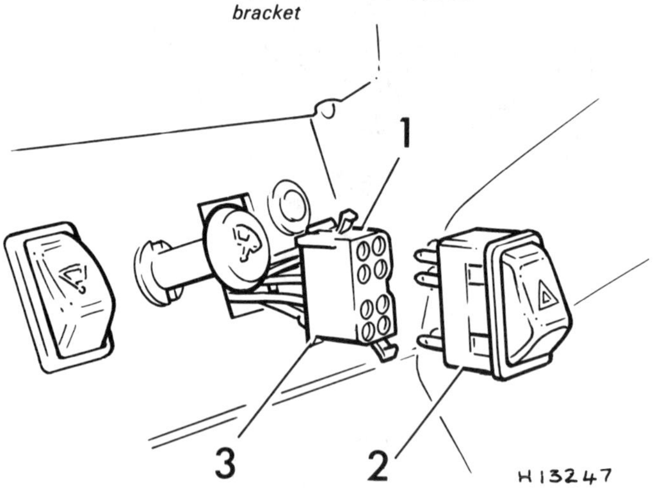

Fig. 10.17 Switch removal (hazard warning shown, others are similar) (Secs 39, 40 etc)

1. *Release tabs on harness connector*
2. *Switch*
3. *Harness connector*

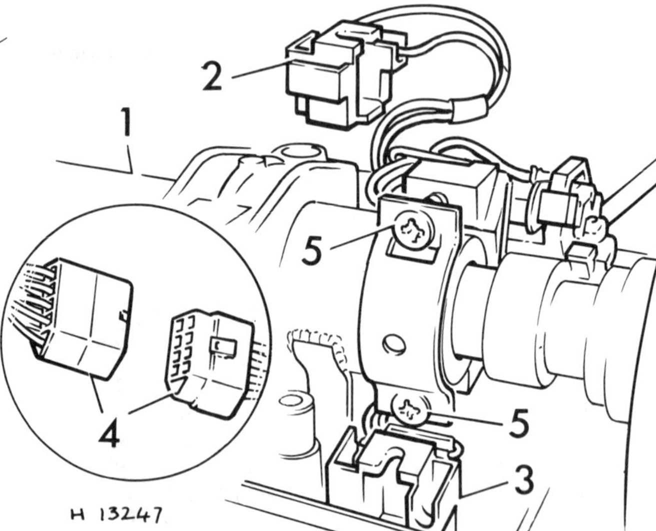

Fig. 10.18 Direction indicator/headlamp beam/horn control switch removal (early models shown, later models are similar) (Sec 41)

1. *Steering column*
2. *Turn signal flasher unit harness connector*
3. *Hazard warning flasher unit harness connector*
4. *Main harness connector*
5. *Switch bracket retaining screws*

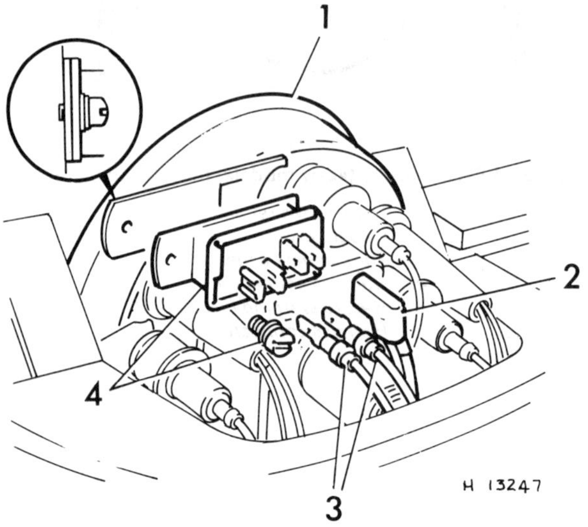

Fig. 10.19 Instrument voltage stabilizer removal (Sec 43)

1. *Speedometer*
2. *Green lead on B terminal*
3. *Light green spade connectors on I terminal*
4. *Retaining screw and voltage stabilizer*

32.2 The roof lamp with the lens removed

35.3 Removing the direction indicator flasher unit

42.1 Removing an instrument cowl screw

39 Lighting switch – removal and refitting

1 Disconnect the battery earth lead.
2 From behind the facia, press out the lighting switch.
3 Press the release tabs on the side of the wiring harness connector and pull the connector from the switch.
4 Refitting is the reverse of the removal procedure.

40 Hazard warning switch – removal and refitting

1 The hazard warning switch can be removed and refitted by following a similar procedure to that described for the lighting switch in Section 39.

41 Direction indicator/headlamp beam/horn control switch – removal and refitting

1 Disconnect the battery earth lead.

Early models

2 Remove the steering column cowl, referring to Chapter 11.
3 Disconnect the wiring harness connectors from the turn signal flasher unit and the hazard warning flasher unit, and also from the main harness.
4 Remove the two screws which secure the switch and bracket to the steering column.
5 Refitting is the reverse of the removal procedure. Ensure that the protrusion on the switch is located in the slot in the steering column before tightening the retaining screws.

Later models

6 Remove the instrument cowl, referring to Section 42, and remove the windscreen wiper/wash switch, referring to Section 22.
7 Disconnect the harness connector from the main wiring harness, and the harness connector from the flasher unit.
8 Remove the two screws which secure the switch and bracket to the steering column.
9 Refitting is the reverse of the removal procedure. Ensure that the protrusion on the switch bracket is located in the recess on the steering column before tightening the retaining screws.

42 Instrument cowl – removal and refitting

1 Remove the screws retaining each side of the cowl and remove the cowl.
2 On later models, remove the screw retaining the cowl extension to the instrument cowl.
3 Refitting is the reverse of the removal procedure.

43 Voltage stabilizer (instruments) – removal and refitting

1 Disconnect the battery earth lead, and remove the instrument cowl as described in Section 42.
2 Disconnect the green lead from the B terminal of the voltage stabilizer.
3 Disconnect the light green spade terminals from the I terminal of the voltage stabilizer.
4 Remove the screw and withdraw the voltage stabilizer from its mounting bracket.
5 Refitting is the reverse of the removal procedure.

44 Speedometer – removal and refitting

1 Disconnect the battery earth lead, and remove the instrument cowl as described in Section 42.
2 Unscrew the retaining screws which secure the speedometer head to its mounting bracket. Note that one screw also retains the earth lead (photo).
3 Press the cable release and withdraw the speedometer cable from the speedometer head (photo).

Early models

4 Remove the warning lamp and panel illumination bulb holders from the speedometer head.
5 Disconnect the following items:

(a) the green and blue lead from the temperature gauge
(b) the green and black lead from the fuel gauge
(c) the green and shrouded lead from the B terminal on the voltage stabilizer

6 Remove the speedometer head.
7 Remove the screw and withdraw the voltage stabilizer from its mounting bracket on the speedometer.
8 Remove the two nuts which secure the fuel gauge to the speedometer, remove the two nuts which secure the temperature gauge to the speedometer, and remove the gauges.

Later models

9 Following paragraph 3, disconnect the wiring multi-connector from the speedometer, and remove the speedometer from the facia (photos).
10 Remove the two screws which secure the retaining brackets to the speedometer.
11 Turn and release the lens from the speedometer. Remove the two screws which secure the speedometer to the outer case and withdraw the speedometer.
12 Remove the nuts and washers which secure the fuel gauge and the temperature gauge to the outer case and remove the gauges.

All models

13 Refitting is the reverse of the removal procedure.

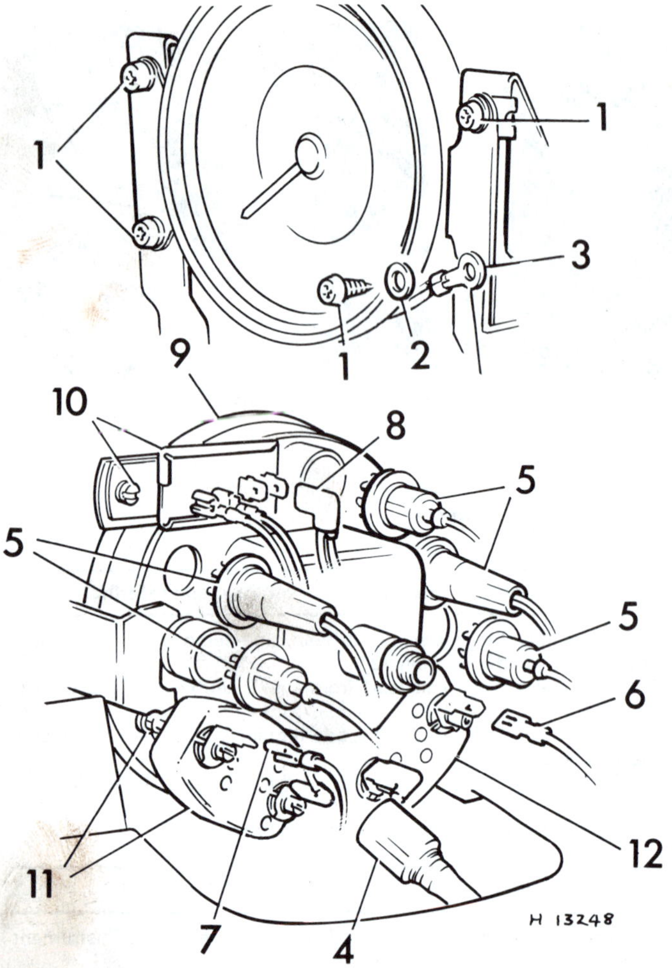

Fig. 10.20 Speedometer (not printed circuit) removal (Sec 44)

1 *Speedometer retaining screws*
2 *Washer*
3 *Earth lead*
4 *Speedometer cable head*
5 *Bulb holders*
6 *Temperature gauge lead (green and blue)*
7 *Fuel gauge lead (green and black)*
8 *Voltage stabilizer B terminal lead (green shroud)*
9 *Speedometer assembly*
10 *Voltage stabilizer and mounting bracket*
11 *Fuel gauge retaining nuts*
12 *Temperature gauge retaining nuts*

45 Instrument printed circuit – removal and refitting

1 Later models are fitted with a speedometer incorporating a printed circuit which can be changed when defective.
2 Follow the procedure for removing the speedometer, as described in Section 44, paragraphs 1 to 3 and 9.
3 After removing the speedometer from the facia, remove all the bulb holders from the speedometer by pulling them free.
4 Remove the gauges as described in Section 44, paragraph 12.
5 Prise out the printed circuit locating peg and remove the printed circuit.
6 Refitting consists of first positioning the printed circuit on the speedometer, fitting the locating peg and bending the tag connections of the circuit into the bulb holder locations in the speedometer. Then refit the gauge retaining nuts and washers and continue by reversing the removal sequence.

46 Gauge units – removal and refitting

1 Either the fuel or temperature gauge can be changed by referring

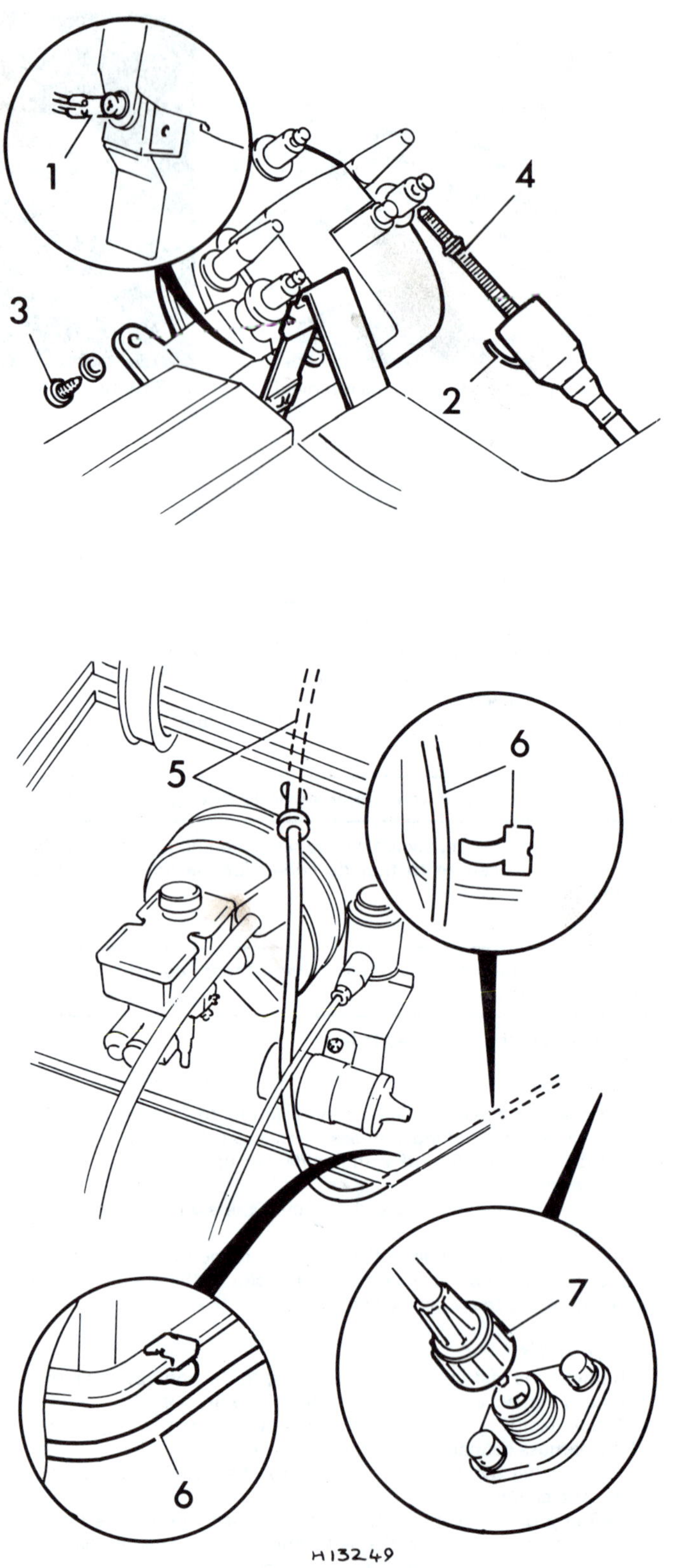

Fig. 10.21 Speedometer cable renewal (printed circuit speedometer cable is similar) (Sec 47)

1 *Earth lead*
2 *Speedometer cable release clip*
3 *Retaining screw*
4 *Inner drive cable*
5 *Cable and grommet route through bulkhead*
6 *Attachment clips*
7 *Cable connection to gearbox pinion*

44.2 Unscrew the retaining screws to release the speedometer head and then ...

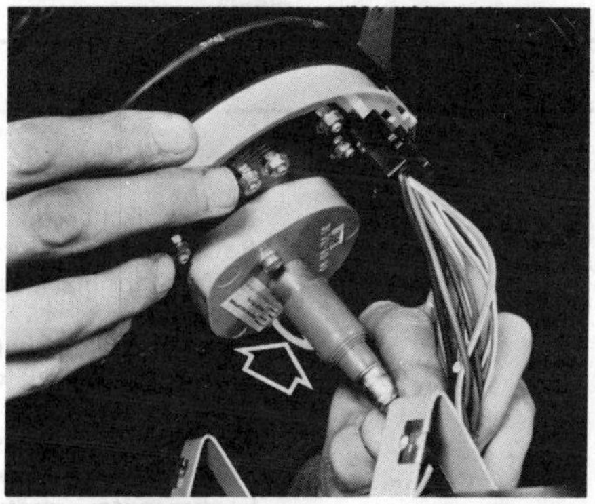
44.3 ... press the speedometer drive cable release clip (arrowed) (printed circuit model shown)

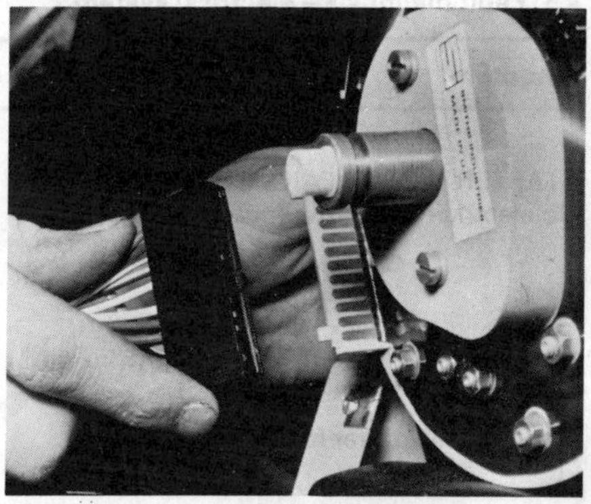
44.9a Disconnect the multi-connector ...

44.9b ... to release the speedometer head

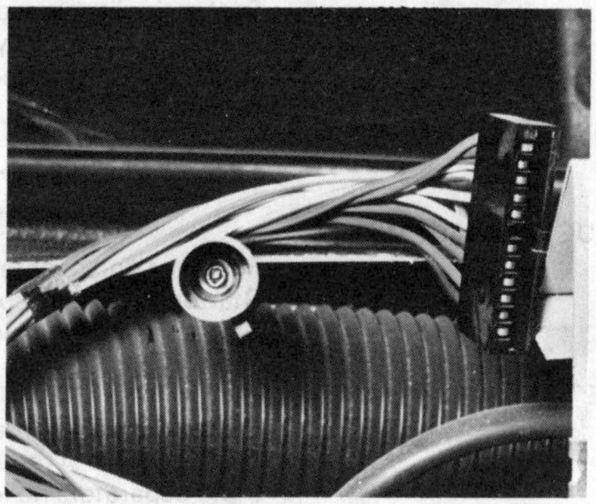
44.9c The drive cable and connector in the facia

48.1 The fuse box (later model) with cover removed

to Section 44 and selecting those operations required.

47 Speedometer cable – removal and refitting

1 Remove the instrument cowl, referring to Section 42.
2 Press the cable release and withdraw the speedometer cable from the speedometer head.
3 Remove the retaining screws which secure the speedometer head to its mounting bracket. Note that one screw retains the earth lead.
4 Pull the inner drive cable from the outer cable.
5 Open the bonnet and withdraw the speedometer cable complete with grommet from the bulkhead. Note that the cable is routed over the top of the demister hose.
6 Release the speedometer cable from the clip in the engine compartment.
7 Unscrew the speedometer cable connector nut from the pinion housing on the right-hand side of the gearbox.
8 Remove the grommet from the speedometer cable.
9 Before refitting the speedometer cable, check the condition of the drives in the gearbox pinion and in the speedometer head.
10 Lightly grease the inner drive cable with general purpose grease and insert it into the outer cable. Withdraw it about 8 inches (200 mm) at the top and wipe off the surplus grease from the exposed drive cable. **Do not** use oil for lubrication.
11 Continue with the refitting by reversing the removal procedure.

48 Fuses

1 On early models the fuse box is located in the cab below the facia panel on the right-hand side. Later models have the fuse box located in the engine compartment on the right-hand side next to the brake servo unit. There are three fuses in circuit and the fuse box cap carries two spare fuses.
2 Some models have in-line fuses to protect individual circuits and this type of fuse is also used with optional accessories.
3 As the circuits which are protected by each fuse vary with different models, reference should be made to the wiring diagrams to identify a circuit with its fuse, if any. Always renew a blown fuse with one of similar capacity and, if a new fuse blows on installation, the cause must be found and rectified. **Do not** bypass a fuse with wire or tin foil without first rectifying the fault which caused the original fuse to blow, or serious damage or fire could result. If a fuse is bypassed in an emergency, always substitute a proper fuse at the earliest opportunity.

49 Fault diagnosis – electrical system

Symptom	Reason/s
STARTER MOTOR FAILS TO TURN ENGINE	
No electricity at starter motor	Battery discharged Battery defective internally Battery terminal leads loose or earth lead not securely attached to body Loose or broken connections in starter motor circuit Starter motor switch or solenoid faulty
Electricity at starter motor: faulty motor	Starter brushes badly worn or sticking Commutator dirty, worn, or burnt Starter motor armature faulty Field coils earthed
STARTER MOTOR TURNS ENGINE VERY SLOWLY	
Electrical defects	Battery in discharged condition Starter brushes badly worn or sticking, Loose wires in starter motor circuit
STARTER MOTOR OPERATES WITHOUT TURNING ENGINE	
Mechanical damage	Pinion or ring gear teeth broken or worn
STARTER MOTOR NOISY OR EXCESSIVELY ROUGH ENGAGEMENT	
Lack of attention or mechanical damage	Pinion or ring gear teeth broken or worn Starter drive main spring broken Starter motor retaining bolts loose
BATTERY WILL NOT HOLD CHARGE FOR MORE THAN A FEW DAYS	
Wear or damage	Battery defective internally Electrolyte level too low or electrolyte too weak due to leakage Plate separators no longer fully effective Battery plates severely sulphated Drivebelt slipping Battery terminal connections loose or corroded Short circuit causing continual battery drain
IGNITION LIGHT FAILS TO GO OUT, BATTERY RUNS FLAT IN A FEW DAYS	
Alternator not charging	Drivebelt slipping or broken Alternator defect
WIPERS	
Wiper motor fails to work	Blown fuse Wire connections loose, disconnected, or broken Brushes badly worn Armature worn or faulty Field coils faulty
Wiper motor works very slowly and takes excessive current	Commutator dirty, greasy, or burnt Drive to wheelboxes too bent or unlubricated Wheelbox spindle binding or damaged Armature bearings dry or unaligned Armature badly worn or faulty
Wiper motor works slowly and takes little current	Brushes badly worn Commutator dirty, greasy, or burnt Armature badly worn or faulty
Wiper motor works but wiper blades remain static	Driving cable rack disengaged or faulty Wheelbox gear and spindle damaged or worn Wiper motor gearbox parts badly worn

Key to Figs. 10.22 and 10.23

1 Alternator
3 Battery
4 Starter solenoid
5 Starter motor
6 Lighting switch
7 Headlamp dipswitch
8 Headlamp dip beam
9 Headlamp main beam
10 Main beam warning lamp
11 RH sidelamp
12 LH sidelamp
14 Panel lamps
15 Number plate illumination lamps
16 Stoplamps
17 RH tail lamp
18 Stoplamp switch
19 Fuse unit
20 Interior light(s)
21 Door switches (if fitted)
22 LH tail lamp
23 Horn
24 Horn push
25 Flasher unit
26 Direction indicator switch
27 Direction indicator warning lamp
28 RH front flasher lamp
29 LH front flasher lamp
30 RH rear flasher lamp
31 LH rear flasher lamp
32 Heater switch
33 Heater motor
34 Fuel gauge
35 Fuel gauge tank unit
36 Windscreen wiper switch
37 Windscreen wiper motor
38 Ignition/starter switch
39 Ignition coil
40 Distributor
42 Oil pressure switch
43 Oil pressure warning lamp
44 No-charge warning lamp
45 Headlamp flasher switch
46 Water temperature gauge
47 Water temperature transmitter
59 Interior light switch (if fitted)
60 Radio (if fitted)
64 Instrument voltage stabiliser
67 Line fuse (if fitted)
71 Overdrive solenoid (if fitted)
72 Overdrive isolation switch (if fitted)
73 Overdrive selector switch (if fitted)
75 Automatic transmission starter inhibitor switch (if fitted)
105 Rear interior light (if fitted)
152 Hazard warning lamp
153 Hazard warning switch
154 Hazard warning flasher unit
159 Brake failure warning lamp and test switch
160 Brake pressure differential switch

Cable colour code (all models)

N	*Brown*	*LG*	*Light Green*
U	*Blue*	*S*	*Slate*
R	*Red*	*W*	*White*
K	*Pink*	*Y*	*Yellow*
P	*Purple*	*B*	*Black*
G	*Green*		

When a cable has two colour code letters the first denotes the main colour and the second denotes the tracer colour

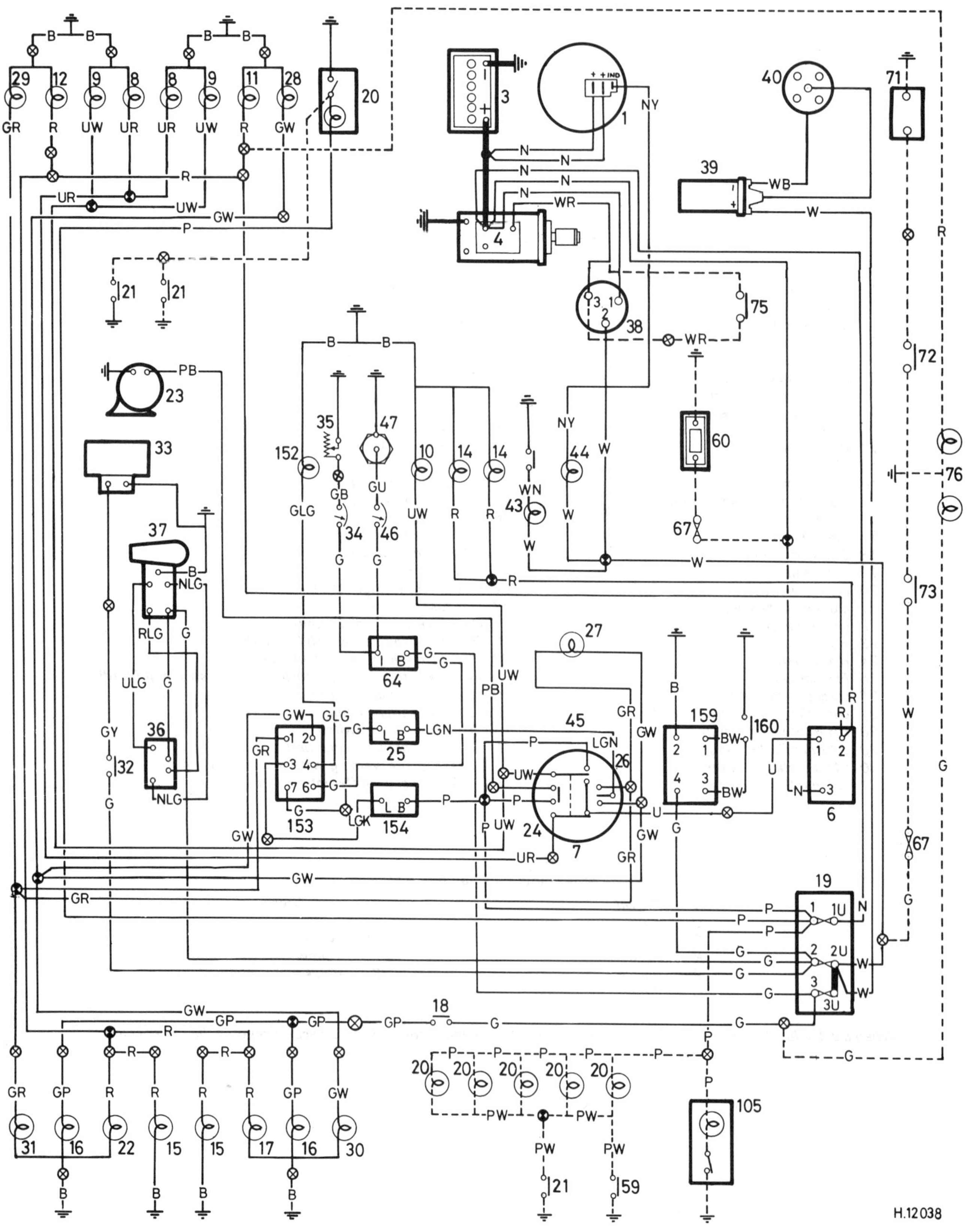

Fig. 10.22 Wiring diagram – 185, 215, 220, 240 and early 250 models, with right-hand steering

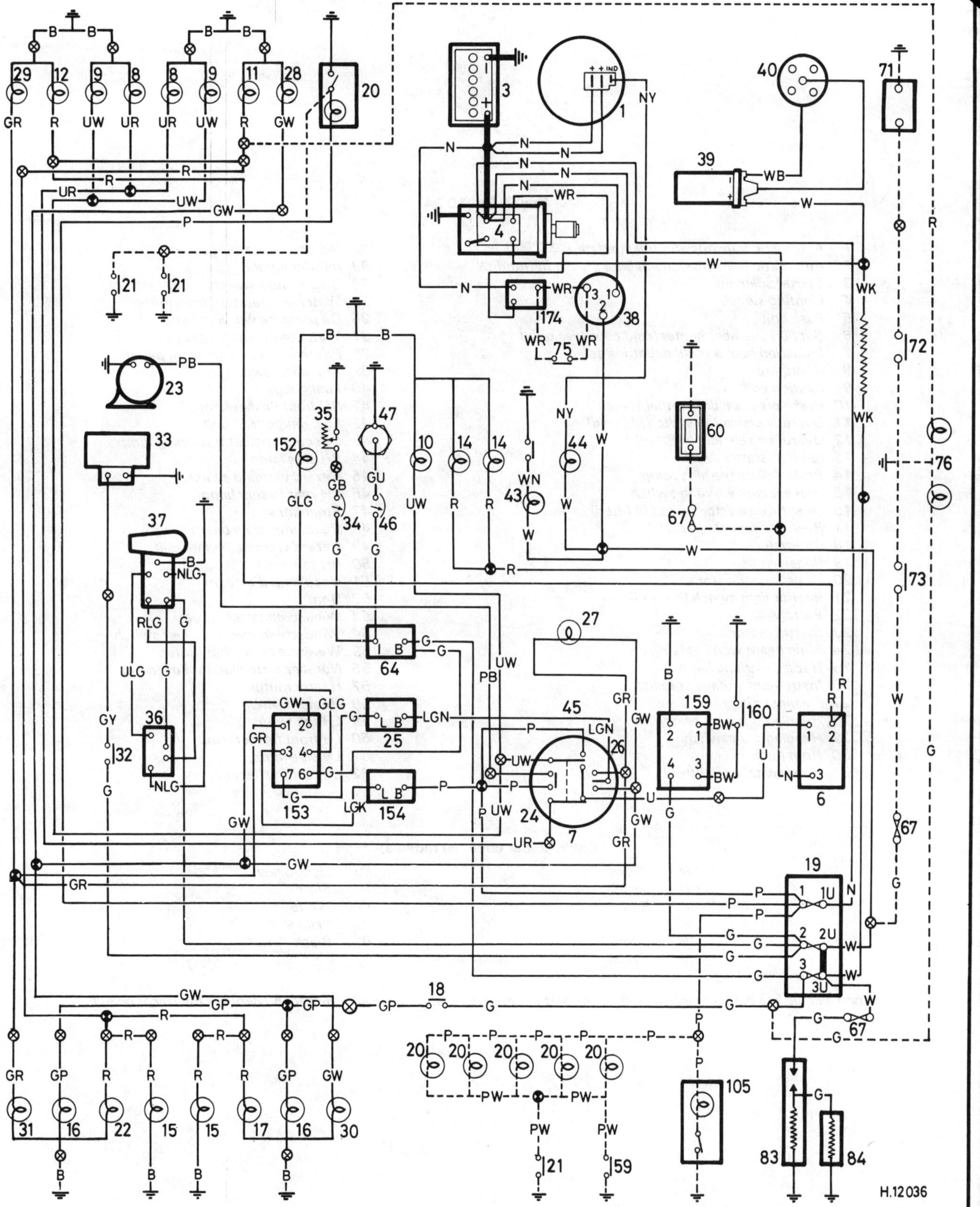

Fig. 10.23 Wiring diagram – 185, 215, 220, 240 and early 250 models, with left-hand steering

Key to Fig. 10.24

1 *Automatic transmission selector lamp (if fitted)*
2 *Automatic transmission inhibitor switch (if fitted)*
3 *Starter solenoid*
4 *Lighting switch*
5 *Fuse unit*
6 *Suction chamber heater (cold countries only)*
7 *Induction heater (cold countries only)*
8 *Distributor*
9 *Ignition coil*
10 *Brake pressure differential switch*
11 *Overdrive isolation switch (if fitted)*
12 *Overdrive solenoid (if fitted)*
13 *Ignition/starter switch*
14 *Brake failure warning lamp*
15 *Brake failure warning switch*
16 *Overdrive selector switch (if fitted)*
17 *Rear interior light (if fitted)*
18 *Alternator*
19 *Panel lamps*
20 *Direction indicator switch*
21 *Interior light switch (if fitted)*
22 *Battery*
23 *Starter motor*
24 *Main beam warning lamp*
25 *Hazard warning lamp*
26 *Instrument voltage stabilizer*
27 *Flasher unit*
28 *Headlamp flasher switch*
29 *Headlamp dipswitch*
30 *Horn push*
31 *Door switches (if fitted)*
32 *No-charge warning lamp*
33 *Interior light(s)*
34 *Oil pressure switch*
35 *Water temperature transmitter*
36 *Oil pressure warning lamp*
37 *Water temperature gauge*
38 *Printed circuit instrument panel*
39 *Stoplamp switch*
40 *Fuel gauge*
41 *RH front flasher lamp*
42 *Fuel gauge tank unit*
43 *Direction indicator warning lamp*
44 *RH side lamp*
45 *Hazard warning switch*
46 *RH rear flasher lamp*
47 *Stoplamps*
48 *Headlamp main beam*
49 *Hazard warning flasher unit*
50 *RH tail lamp*
51 *Headlamp dip beam*
52 *Horn*
53 *Windscreen wiper motor*
54 *Windscreen washer/wiper switch*
55 *Windscreen washer motor*
56 *Number plate illumination lamps*
57 *Heater motor*
58 *Heater switch*
59 *LH tail lamp*
60 *LH front flasher lamp*
61 *LH side lamp*
62 *LH rear flasher lamp*

Cable colour code (all models)

N	*Brown*
U	*Blue*
R	*Red*
K	*Pink*
P	*Purple*
G	*Green*
LG	*Light Green*
S	*Slate*
W	*White*
Y	*Yellow*
B	*Black*

When a cable has two colour code letters the first denotes the main colour and the second denotes the tracer colour

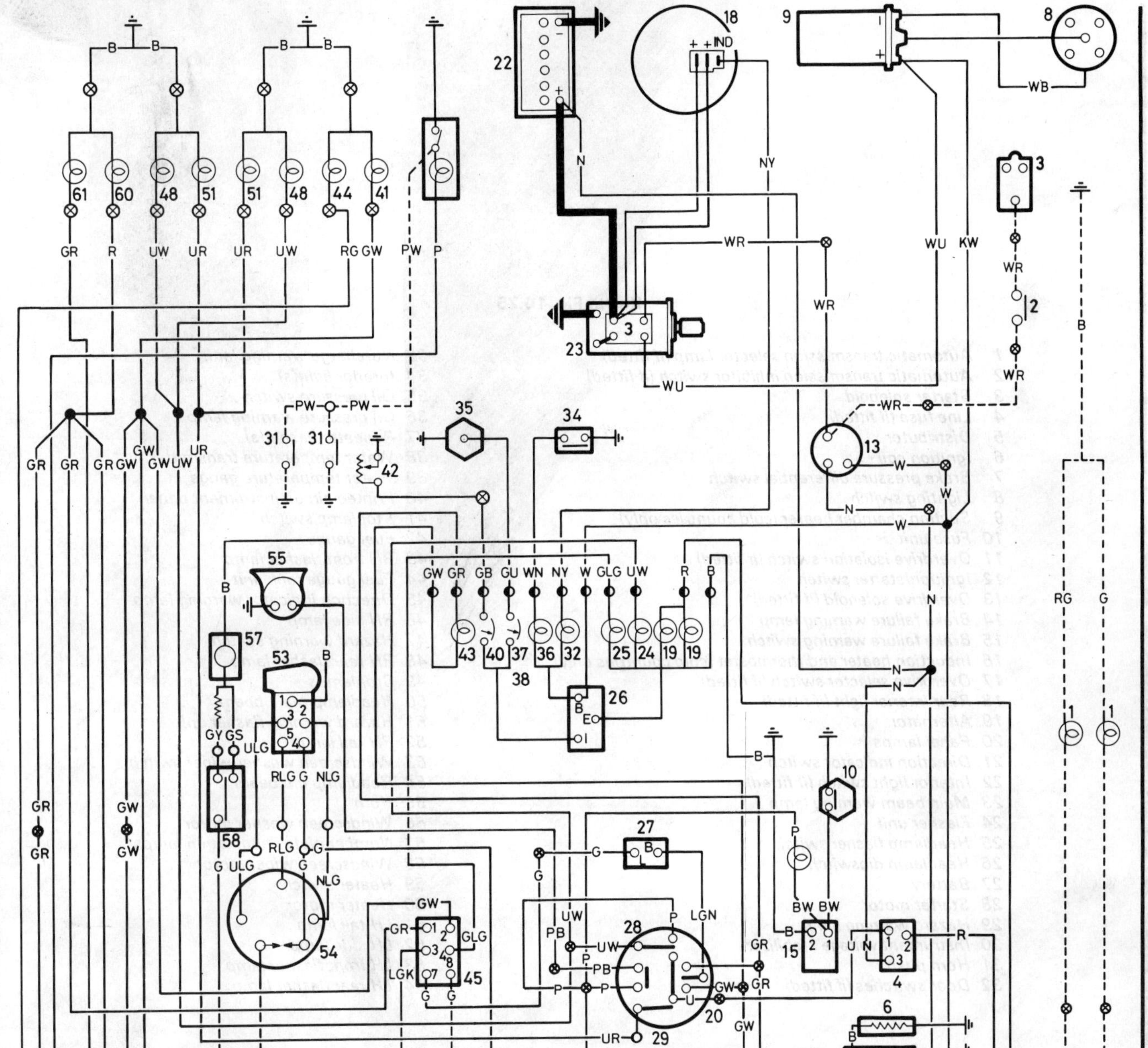

Fig. 10.24 Wiring diagram – 200, 230, later 250 and 255 models, with right-hand steering

Key to Fig. 10.25

1 *Automatic transmission selector lamp (if fitted)*
2 *Automatic transmission inhibitor switch (if fitted)*
3 *Starter solenoid*
4 *Line fuse (if fitted)*
5 *Distributor*
6 *Ignition coil*
7 *Brake pressure differential switch*
8 *Lighting switch*
9 *Suction chamber heater (cold countries only)*
10 *Fuse unit*
11 *Overdrive isolation switch (if fitted)*
12 *Ignition/starter switch*
13 *Overdrive solenoid (if fitted)*
14 *Brake failure warning lamp*
15 *Brake failure warning switch*
16 *Induction heater and thermostat (cold countries only)*
17 *Overdrive selector switch (if fitted)*
18 *Rear interior light (if fitted)*
19 *Alternator*
20 *Panel lamps*
21 *Direction indicator switch*
22 *Interior light switch (if fitted)*
23 *Main beam warning lamp*
24 *Flasher unit*
25 *Headlamp flasher switch*
26 *Headlamp dipswitch*
27 *Battery*
28 *Starter motor*
29 *Hazard warning lamp*
30 *Instrument voltage stabilizer*
31 *Horn push*
32 *Door switches (if fitted)*
33 *No-charge warning lamp*
34 *Interior light(s)*
35 *Oil pressure switch*
36 *Oil pressure warning lamp*
37 *Repeater flasher(s)*
38 *Water temperature transmitter*
39 *Water temperature gauge*
40 *Printed circuit instrument panel*
41 *Stoplamp switch*
42 *Fuel gauge*
43 *RH front flasher lamp*
44 *Fuel gauge tank unit*
45 *Direction indicator warning lamp*
46 *RH side lamp*
47 *Hazard warning switch*
48 *RH rear flasher lamp*
49 *Stoplamps*
50 *Headlamp main beam*
51 *Hazard warning flasher unit*
52 *RH tail lamp*
53 *Windscreen washer/wiper switch*
54 *Headlamp dip beam*
55 *Horn*
56 *Windscreen washer motor*
57 *Number plate illumination lamps*
58 *Windscreen wiper motor*
59 *Heater switch*
60 *Heater motor*
61 *LH tail lamp*
62 *LH side lamp*
63 *LH front flasher lamp*
64 *LH rear flasher lamp*

Cable colour code (all models)

N	Brown
U	Blue
R	Red
K	Pink
P	Purple
G	Green
LG	Light Green
S	Slate
W	White
Y	Yellow
B	Black

When a cable has two colour code letters the first denotes the main colour and the second denotes the tracer colour

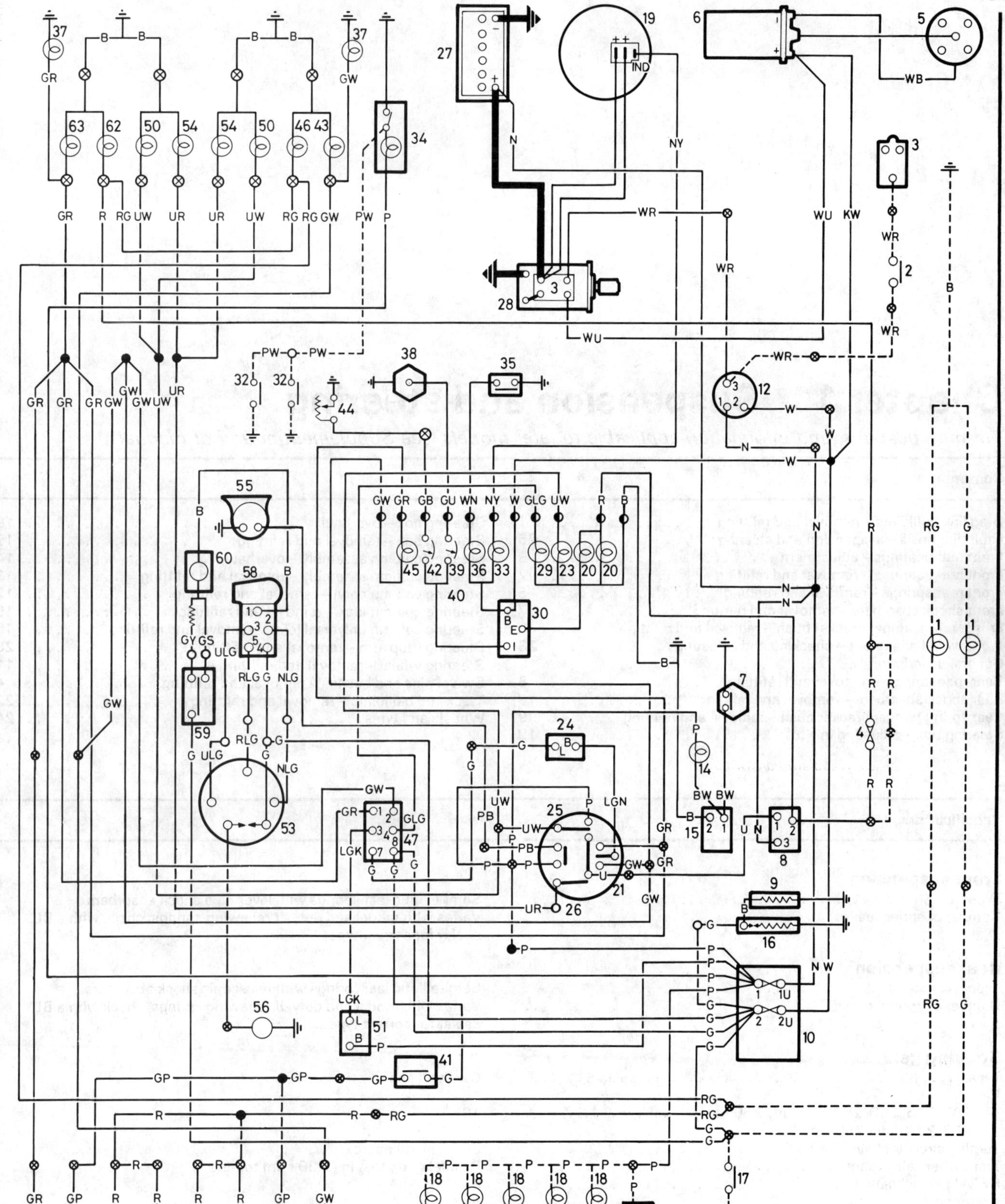

Fig. 10.25 Wiring diagram – 200, 230, later 250 and 255 models, with left-hand steering

Chapter 11 Suspension and steering

For modifications, and information applicable to later models, see Supplement at end of manual

Contents

Specifications

Front suspension

Type	Semi-elliptic leaf springs with lever type shock absorbers
Number of spring leaves	Varies with model and duty. If renewing springs, check with a BL dealer for correct type

Rear suspension

Type	Semi-elliptic leaf springs with telescopic shock absorbers
Number of spring leaves	Varies with model and duty. If renewing springs, check with a BL dealer for correct type

Steering gear

Type	Cam and peg
Ratio:	
185, 215, 220, 240 and early 250	19 : 1
200, 230, later 250, and 255	21 : 1
Turning circle (between kerbs)	36 ft (11 m) approx
Front wheel alignment	Parallel or up to $\frac{1}{16}$ in (1·58 mm) toe-in
Swivel pin inclination	$8\frac{1}{2}°$
Castor angle	5°
Camber angle	$2\frac{1}{2}°$
Steering box worm gear:	
Preload (without oil seal or rocker shaft fitted)	0·005 in (0·127 mm) max.
Torque at steering wheel with preload set	1 to 3 lbf in (0·012 to 0·035 kgf m)
Adjustment	Shims between end covers and casing
Steering box rocker shaft:	
Steering wheel torque at straight ahead position	12 to 22 lbf in (0·138 to 0·254 kgf m)
Adjustment	Screw and locknut on side cover
Rocker shaft bush reamed diameter	1·125 to 1·126 in (28·58 to 28·61 mm)
Steering box oil capacity	$\frac{7}{8}$ pint (0·5 litre)

Swivel pin diameter:	
Early vehicles	0·8127 to 0·8132 in (20·643 to 20·655 mm)
Later vehicles	0·9987 to 0·9992 in (25·367 to 25·380 mm)
Reamed diameter of swivel pin bush:	
Early vehicles	0·8137 to 0·8147 in (20·668 to 20·693 mm)
Later vehicles	0·9993 to 1·0003 in (25·382 to 25·408 mm)
Clearance between top of axle beam and stub axle	0·008 in (0·203 mm) maximum
Stub axle shim sizes:	
Early vehicles	0·052 to 0·057 in (1·321 to 1·448 mm)
	0·058 to 0·063 in (1·473 to 1·600 mm)
	0·064 to 0·069 in (1·676 to 1·753 mm)
Later vehicles	0·046 to 0·050 in (1·168 to 1·270 mm)
	0·051 to 0·055 in (1·295 to 1·397 mm)
	0·056 to 0·060 in (1·422 to 1·524 mm)
	0·061 to 0·065 in (1·599 to 1·651 mm)

Roadwheels

185 model	$4\frac{1}{2}$J x 14, pressed steel disc, 4-stud fixing
All other models	5J x 14, pressed steel disc, 5-stud fixing

Tyres ... See tyre pressure table

Tyre pressures

Plated axle loading weight – tyres cold

Model	Tyre size	Front lbf/in²	Front kgf/cm²	Front bars	Rear lbf/in²	Rear kgf/cm²	Rear bars
185	6.50–14 (6 ply)	32	2·25	2·21	36	2·53	2·48
185	165SR–14 (reinforced)	36	2·53	2·48	40	2·81	2·75
215	185SR–14 (reinforced)	40	2·81	2·75	43	3·02	2·96
215	195–14C (6 ply)	43	3·02	2·96	48	3·37	3·31
220	185–14C (6 ply)	43	3·02	2·96	50	3·51	3·45
220	195–14C (6 ply)	41	2·88	2·82	48	3·37	3·31
240	185–14C (6 ply)	45	3·16	3·10	50	3·51	3·45
240	195–14C (6 ply)	43	3·02	2·96	48	3·37	3·31
250	195–14C (6 ply)	41	2·88	2·82	48	3·37	3·31
200	165SR–14 (reinforced)	37	2·6	2·5	40	2·8	2·8
230	185–14C (6-ply)	38	2·7	2·6	43	3·02	2·96
230	195–14C (6 ply)	43	3·02	2·96	48	3·37	3·31
250	185–14C (6 ply)	42	3·0	2·9	52	3·7	3·6
250	195–14C (6 ply)	43	3·02	2·96	48	3·37	3·31
255	185–14C (6 ply)	45	3·16	3·1	54	3·8	3·7
255	195–14C (6 ply)	45	3·16	3·1	50	3·51	3·45

Approved caravan conversions – tyres cold

Model	Tyre size	Front lbf/in²	Front kgf/cm²	Front bars	Rear lbf/in²	Rear kgf/cm²	Rear bars
215	185–14 reinforced	35	2·46	2·41	45	3·16	3·1
240 *	185–14C (6 ply)	43	3·02	2·96	48	3·37	3·31
230 *	185–14C (6 ply)	38	2·7	2·6	43	3·02	2·96
250 *	185–14C (6 ply)	42	3·0	2·9	52	3·7	3·6

*** If vehicle is used at plated axle loading weight then tyre pressures must be increased to those specified for that condition.***

Torque wrench settings

	lbf ft	Nm
Front axle		
Cotter pin nut	20	27
Steering lever nuts	75	101
Track-rod lever nuts	75	101
Backplate bolts	54	73
Track-rod balljoint pin nuts	41	55
Track-rod locknuts:		
$\frac{3}{8}$ in	14	18
$\frac{3}{4}$ in	90	122
Front suspension		
Spring U-bolt nuts	60	81
Spring front eye bolt:		
$\frac{7}{16}$ in	45	61
$\frac{1}{2}$ in	70	95
Spring shackle pins	45	61
Shock absorber to frame	45	61
Shock absorber link nuts	45	61
Rear axle		
Backplate bolts	35	47
Hub nut	155	210
Rear suspension		
Spring U-bolt nuts:		
185 models	50	68
All other models	60	81

	lbf ft	Nm
Spring front eye bolt:		
$\frac{3}{8}$	45	61
$\frac{1}{2}$ in	70	95
Spring shackle pins	45	61
Shock absorber to axle and frame	40	54
Hanger bracket retaining bolts	30	42
Steering		
Drop arm nut	125	170
Rocker shaft locknut	15	20
Drag link balljoint pin nut	40	54
Drag link locknut	75	101
Steering wheel nut	40	54
Steering box to frame	45	61
Steering box adjustment screw locknut	15	20
Wheels		
Roadwheel nuts	65	88

1 General description

1 The Sherpa front suspension consists of a forged I-section steel beam front axle bolted to semi-elliptic leaf springs which are dampened by lever type shock absorbers.
2 The leaf spring assembly differs between models and has one, five or six leaves according to the vehicle duty requirement. The springs are bolted to the axle with U-bolts and they are connected to the chassis by bushed shackle pins and with shackle plates at the rear.
3 Stub axles are mounted on swivel pins pinned into each end of the axle. The right-hand stub axle on right-hand drive models and the left-hand stub axle on left-hand drive models carries a steering arm which is connected by a drag link to the steering drop arm on the steering box. The cam and peg type steering box is operated by the steering column through a universal joint. An adjustable track-rod links with a track-rod arm in each stub axle.
4 The rear suspension consists of semi-elliptic leaf springs U-bolted to the rear axle and damped by telescopic hydraulic shock absorbers. Again the number of leaves in the springs varies with the type of vehicle duty, each having three, four (two categories) or six leaves. The springs are attached to the chassis spring brackets by bushed shackle pins and shackle plates at the rear.
5 Details of the rear axle and hubs will be found in Chapter 8.

2 Front hub bearings – removal and refitting

1 Jack-up the front of the car and support on firmly based axle stands.
2 Remove the wheel trim and the roadwheels.
3 Refer to Chapter 9 and remove the brake drum.
4 Using a wide blade screwdriver carefully ease off the grease cap.
5 Straighten the split-pin ears and extract the split-pin. Lift away the nut retainer and then undo and remove the hub nut. Withdraw the splined washer. The hub may now be drawn from the axle stub.
6 Remove the outer bearing cone.
7 Using a screwdriver ease out the oil seal noting that the lip is innermost. Lift away the inner bearing cone.
8 If the bearings are to be renewed carefully drift out the bearing cups working from the inside of the hub.
9 Thoroughly wash all parts in paraffin and wipe dry using a non-fluffy rag.
10 Inspect the bearings for signs of rusting, pitting or overheating. If evident, a new set of bearings must be fitted.
11 Inspect the oil seal journal face of the stub axle shaft for signs of damage. If evident, polish with fine emery tape; if very bad a new stub axle will have to be fitted.
12 To reassemble, if new bearings are to be fitted, carefully drift in the new bearing cups using a piece of tube of suitable diameter. Make sure they are fitted the correct way round with the tapers facing outwards.
13 Work some high melting point grease into the inner bearing cone and fit it into the hub.
14 Smear a new oil seal with a little grease and fit it with the lip facing the bearings, using a tube of suitable diameter. The final fitted position should be flush with the flange of the hub.
15 Fit the hub to the axle stub. Work some high melting point grease into the outer bearing cone and fit it into the hub.
16 Refit the splined washer and nut.
17 It is now necessary to adjust the hub bearing endfloat as described in Section 3.
18 Refit the grease cap, brake drum, roadwheel and wheel trim. Lower the front of the vehicle to the ground.

3 Front hub bearings – adjustment

1 Jack-up the front of the vehicle and support on firmly based axle stands.
2 Remove the wheel trim, roadwheel and grease cap.
3 Straighten the split-pin ears and extract the split-pin. Lift away the hub nut retainer.
4 Back off the hub nut and spin the hub. Whilst it is spinning tighten the nut using a torque wrench set to 5 lbf ft (0.69 kgf m).
5 Stop the hub spinning and slacken the nut. Tighten the nut again but this time finger-tight only.
6 Position the nut retainer so that half the split-pin hole is covered by one of the arms of the retainer.
7 Slacken the nut and retainer until the split-pin hole is fully uncovered.
8 Fit a new split-pin, and lock by opening the ears of the split-pin and bending circumferentially around the nut retainer.
9 Fit the grease cap and replace the road wheel and wheel trim.
10 It will be observed that the endfloat setting achieved can cause a

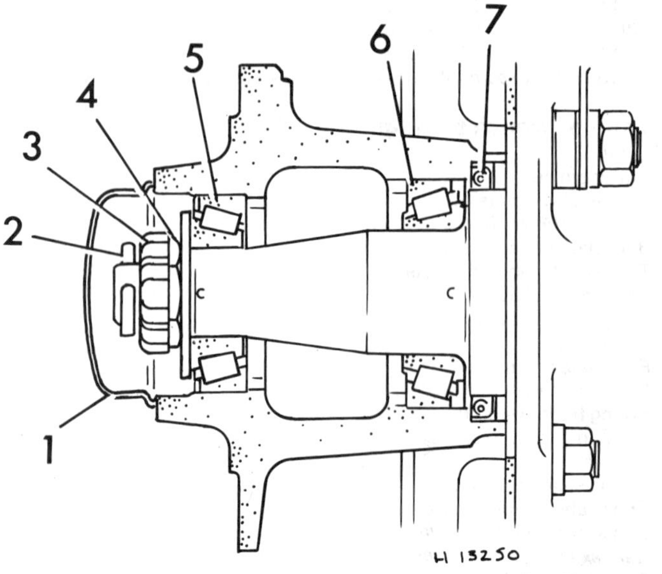

Fig. 11.1 Part sectional view of the front wheel hub (Secs 2 and 3)

1 Grease retainer
2 Split-pin (note legs wrapped round nut retainer)
3 Nut retainer
4 Hub nut
5 Outer bearing
6 Inner bearing
7 Oil seal

considerable amount of movement when the tyre is rocked. Do not reduce the endfloat any further provided it has been set correctly as described. The bearings must not on any account be preloaded.

4 Swivel pins and bushes – removal and refitting

1 Refer to Section 2 and remove the front wheel hubs.
2 Remove the bolt, spacer, spring washer, and nut which secure the brake backplate and the brake hose to the stub axle. Remove the three other bolts, nuts, and spring washers securing the backplate and move the backplate to one side, supporting it to avoid straining the brake hose.
3 On the steering box side, remove the nut which secures the drag link balljoint to the steering arm and detach the balljoint from the steering arm. This will probably require the use of a balljoint separator or wedges.
4 Remove the nut which secures the track-rod balljoint to the track-rod arm and detach the balljoint from the track-rod arm. Again, the use of a balljoint separator or wedges will probably be required.
5 Remove the nut and spring washer from the cotter pin which retains the swivel pin in the axle beam and drive out the cotter pin. Take care not to damage the cotter pin threads: if necessary refit the nut loosely before hitting the pin.
6 Remove the grease nipple and seal from each end of the swivel pin and drift the pin out of the stub axle and the axle beam. Remove the stub axle, the thrust bearing, shims and O-rings from the axle beam.
7 Thoroughly wash all parts in paraffin and wipe dry using non-fluffy rag. Examine the parts for excessive wear and damage. A kit of parts can be obtained from a dealer to renew the bushes, swivel pin, bearing and seals. The bushes will require reaming after fitting and a BL garage will do this work if an adjustable reamer of the right range and a micrometer are not available. If these tools are to hand, proceed as follows.
8 Press out the old bushes using suitably sized socket spanners, pieces of tube or similar makeshift mandrels. If no press is accessible and your vice is not big enough, the bushes can be drifted out taking care not to damage the hub assembly.
9 Fit the new bushes, making sure that there is a clearance of 0.090 to 0.095 in (2.286 to 2.413 mm) between the end of each bush and the stub axle face adjacent to the axle beam (A in Fig. 11.2). The grease grooves in the bushes must lie with the open end towards the axle beam faces.
10 Ream the bushes to the size quoted in the Specifications, lubricating the cutting edges of the tool liberally with clean paraffin.
11 To determine the shim thicknesses required to achieve the correct working clearance between the stub axle and the axle beam, a trial assembly must be made. Locate the stub axle and thrust race in position on the axle beam and insert the swivel pin. The pin must be fitted with the flats on one end at the top. Some pins have one end marked TOP instead of having flats.
12 Measure the clearance between the stub axle and the top of the axle beam (14 in Fig. 11.2) using feeler gauges and note the measurement.
13 Remove the swivel pin, stub axle, and thrust race from the axle beam and fit new 'O' rings to the stub axle at the axle beam ends of the bushes, ie in the clearances left when fitting the bushes (paragraph 9).
14 Coat the thrust race with a specified swivel pin grease or a suitable equivalent, and fit the stub axle and thrust race to the axle beam. Referring to the clearance measured in paragraph 12, select a shim to achieve the clearance listed in the Specifications and fit it between the top face of the axle beam and the stub axle.
15 The rest of the refitting procedure is a reversal of the procedure described in paragraphs 1 to 6 inclusive. When refitting the swivel pin lubricate it first with one of the greases quoted in paragraph 14 and tighten the cotter pin nut to the specified torque. Use a grease gun to inject a general purpose grease into the grease nipples until the grease appears round the seals. Finally check the front wheel alignment, refer to Section 23.

5 Front roadsprings – removal and refitting

1 Slacken the front wheel nuts, jack-up the front of the vehicle and support it firmly on stands under chassis strong points. Support the front axle and adjust the support to relieve the load on the springs.
2 Remove the roadwheels.
3 Remove the four locknuts on the spring U-bolts. Remove the U-bolts, saddles, axle buffer and the packing piece (if fitted).
4 Remove the four nuts and spring washers which secure the shackle plates at the rear end of the spring and remove the shackle plates. Remove the shackle pin from the spring.

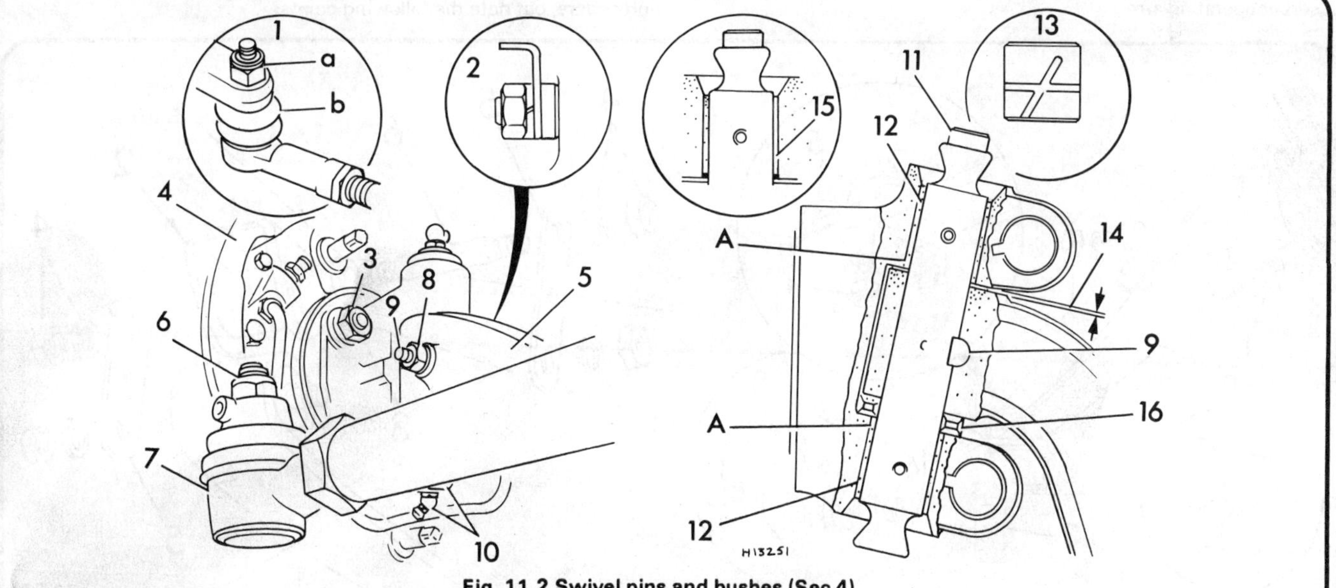

Fig. 11.2 Swivel pins and bushes (Sec 4)

1 Drag link to steering arm connection (on steering box side only)
a Retaining nut
b Balljoint
2 Brake hose bracket and retaining bolt
3 Backplate retaining nut
4 Backplate assembly
5 Axle beam
6 Track-rod balljoint retaining nut
7 Track-rod balljoint
8 Cotter pin retaining nut and spring washer
9 Cotter pin
10 Grease nipples and seals
11 Swivel pin
12 Bushes
13 Upper bush showing sloping grease groove with open end towards axle beam
14 Clearance and shim location
15 O-ring (upper) in stub axle
16 Thrust race
A Clearances for O-rings

5 Remove the nut, spring washer and the spring front eye bolt. Remove the spring from the vehicle.
6 Remove the bush from each spring eye.
7 Refitting a front roadspring is the reverse of the removal procedure, but the following points should be noted:

(a) Refit the axle buffer with the arrow marked CENTRE pointing towards the middle of the vehicle
(b) The vehicle must be in a half-laden state, with all wheels on the ground, before the U-bolts, the front eye bolt, and the rear shackle pin nuts are fully tightened
(c) Tighten all nuts to the specified torque

6 Front spring hanger bracket bush – removal and refitting

1 Slacken the front wheel nuts, jack-up the front of the vehicle and support it firmly on stands under chassis strong points. Support the front axle and adjust the support to relieve the load on the spring.
2 Remove the roadwheel.
3 Remove the four nuts and spring washers which secure the shackle plates. Remove the shackle plates and the shackle pins (photo).
4 Withdraw the bush from the hanger bracket. This can be done by making up an extractor with a nut and a bolt about 6 in x ½ in diameter (150 x 12 mm), with the head of the bolt ground down to ¾ in diameter (19 mm), and a short length of tubing.
5 Refitting of the hanger bracket bush is a reversal of the removal procedure, but note the following points:

(a) To facilitate refitting the bush, lubricate it with Esso Process Oil 'L' or a suitable equivalent
(b) When fitting the bush ensure that there is an equal amount of protrusion at each end
(c) Half load the vehicle with all wheels on the ground, and tighten the shackle pin nuts to the specified torque

7 Front shock absorbers – removal and refitting

1 Slacken the front wheel nuts, jack-up the front of the vehicle and support it firmly on stands under chassis strong points. Support the front axle and adjust the support to relieve the load on the shock absorber operating arm.
2 Remove the roadwheel.
3 Remove the nut and spring washer which secure the link to the shock absorber arm. Detach the link from the arm. This will require the use of a separator or wedges to part the conical joint faces.
4 Remove the two bolts, nuts and spring washers which secure the shock absorber to the chassis and remove the shock absorber (photo).
5 Refitting a shock absorber is the reverse of the removal procedure. Before tightening the fasteners half load the vehicle, with all wheels on the ground, and tighten the link retaining nut to the specified torque.

8 Rear roadsprings – removal and refitting

1 Slacken the rear wheel nuts, chock the front wheels, jack-up the rear of the vehicle and support it firmly on stands under chassis strong points. Support the rear axle and adjust the support to relieve the load on the roadspring to be removed. Remove the rear roadwheel.
2 Remove the four locknuts which secure the U-bolts and remove the U-bolts and clamp plates. Two clamp plates may be fitted and, when removed, these must both be refitted on reassembly (photo).
3 Remove the four nuts and spring washers which secure the shackle plates. Remove the shackle plates and the shackle pin from the spring.
4 Remove the nut, spring washer and front eye bolt from the spring and remove the spring from the vehicle.
5 Refitting is the reverse of the removal procedure. Before tightening the fasteners half load the vehicle with all wheels on the ground, and tighten the nuts to the specified torque.

9 Rear spring hanger bracket bush – removal and refitting

1 Slacken the rear wheel nuts, chock the front wheels, jack-up the rear of the vehicle and support it firmly on stands under chassis strong points. Support the rear axle and adjust the support to relieve the load on the spring.
2 Remove the roadwheel.
3 Remove the four nuts and spring washers which secure the spring shackle plates and remove the plates. Remove the shackle pins.
4 Remove the four bolts and spring washers which secure the hanger bracket to the chassis.
5 Remove the bush from the hanger bracket.
6 Refitting of the hanger bracket bush is a reversal of the removal procedure, but note the following points:

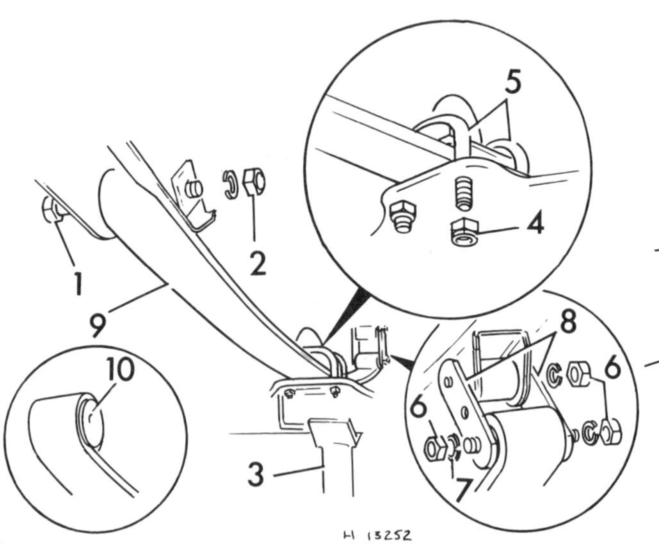

Fig. 11.3 Removing a front roadspring (Sec 5)

1 Front eye bolt
2 Front eye bolt retaining nut
3 Front axle support
4 U-bolt locknut
5 U-bolt
6 Shackle pin retaining nuts
7 Shackle pin
8 Shackle plates
9 Roadspring
10 Bush

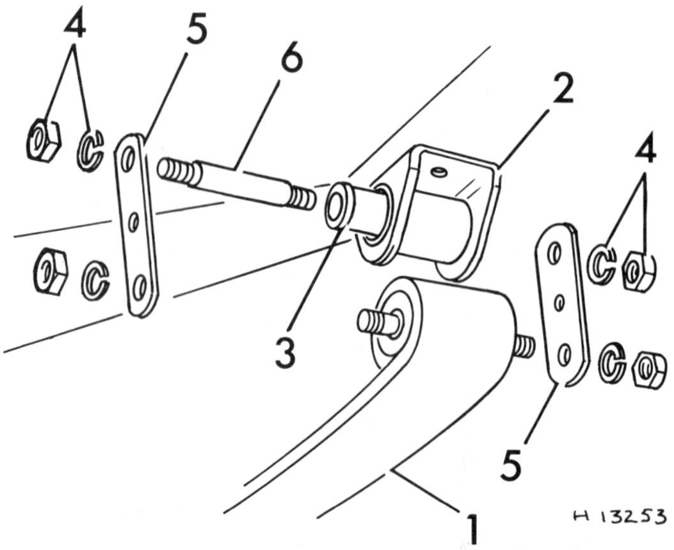

Fig. 11.4 Removing the front spring hanger bracket bush (Sec 6)

1 Roadspring
2 Hanger bracket
3 Bush
4 Retaining nuts and washers
5 Shackle plates
6 Shackle pin

6.3 Front spring shackle plates and hanger bracket

7.4 A front shock absorber

8.2 Removing a rear spring U-bolt locknut

(a) *To facilitate refitting the bush, lubricate it with Esso Process Oil 'L' or a suitable equivalent*
(b) *Fit the bush to the hanger bracket so that there is an equal amount of protrusion at each end*
(c) *Tighten the hanger bracket retaining bolts to the specified torque. Half load the vehicle with all wheels on the ground, before tightening the shackle pins to the specified torque*

10 Rear shock absorbers – removal and refitting

1 Chock the front wheels, jack-up the rear of the vehicle and support it firmly on stands under chassis strong points. Support the rear axle and adjust the support to relieve the load on the shock absorber.
2 Remove the nuts and washers from the shock absorber top and bottom mounting bolts. Remove the bolts and remove the shock absorber from the vehicle.
3 To test the shock absorber alternately compress and extend it throughout its full movement. If the action is jerky or weak, it is an indication that either it is worn or there is air in the hydraulic cylinder. Continue to compress and extend it and if the action does not become more positive a new shock absorber should be obtained. If the shock absorber is showing signs of leaking it should be discarded as it is not possible to overhaul it.
4 Refitting is the reverse of the removal procedure. Tighten the mounting bolts and nuts to the specified torque when the vehicle has been lowered to the ground.

11 Steering wheel – removal and refitting

1 Using a blunt knife carefully prise the safety pad from the centre of the steering wheel.
2 Undo and remove the nut and lockwasher which secure the steering wheel to the inner column (photo).
3 Mark the relative positions of the steering wheel and the inner column so that they may be refitted in the same position.
4 With the palms of the hands beneath the wheel spokes and near the centre hub, thump the steering wheel upwards to remove it from the inner column splines. If it is very tight it will be necessary to use a universal puller fitted with long legs, and a suitable thrust block.
5 Refitting the steering wheel is the reverse sequence to removal, but the following additional points should be noted:

(a) *Realign the marks made before removal. Alternatively, check that the front wheels are in the straight ahead position, fit the wheel with its spokes horizontal and the outer bevels in the pad recess in the spokes pointing down and in*
(b) *Tighten the steering wheel retaining nut to the specified torque*
(c) *Press the pad into position, ensuring that the six legs on the pad engage the clips in the wheel*

12 Steering column cowl – removal and refitting

1 Remove the two screws which secure the upper cowl and remove the upper cowl.
2 Remove the key from the steering lock.
3 Remove the three screws which secure the lower cowl to the steering column, and remove the direction indicator flasher unit from its clip in the lower cowl.

185, 215, 220, 240 and early 250 models

4 Remove the hazard warning flasher unit from its clip in the lower cowl and then remove the cowl.

200, 230, later 250, and 255 models

5 Remove the cowl and then remove the flasher unit retaining clip from the lower cowl.

All models

6 Refitting is a reversal of the removal sequence.

13 Steering column assembly – removal and refitting

1 Refer to Section 11 and remove the steering wheel.
2 Refer to Section 12 and remove the steering column cowl.
3 Remove the two screws which secure the ignition/starter switch to the steering lock, and retrieve the switch driver from the steering lock spindle.
4 Remove the two screws which secure the direction indicator switch and bracket to the steering column.
5 From inside the engine compartment, remove the nut, plain washer and bolt which retain the lower end of the steering column in the universal joint.
6 Drill out or use a proprietary stud extractor to remove the two shear bolts which retain the steering column bracket to the facia and pedal bracket.
7 Withdraw the steering column assembly from the vehicle.
8 Refitting of the steering column is basically the reverse of the removal procedure, but the following additional points require attention:

(a) *Engage the steering lock with the steering column in the straight ahead position. With the wheels in the straight ahead position, engage the steering column with the universal joint so that the line on the end of the column is in the nine o'clock position (viewed from the driver's position). Refit the bolt, plain washer and nut which secure the steering column to the universal joint*
(b) *Use new shear bolts to secure the steering column assembly to the facia and pedal bracket. Tighten the bolts until the heads shear off*

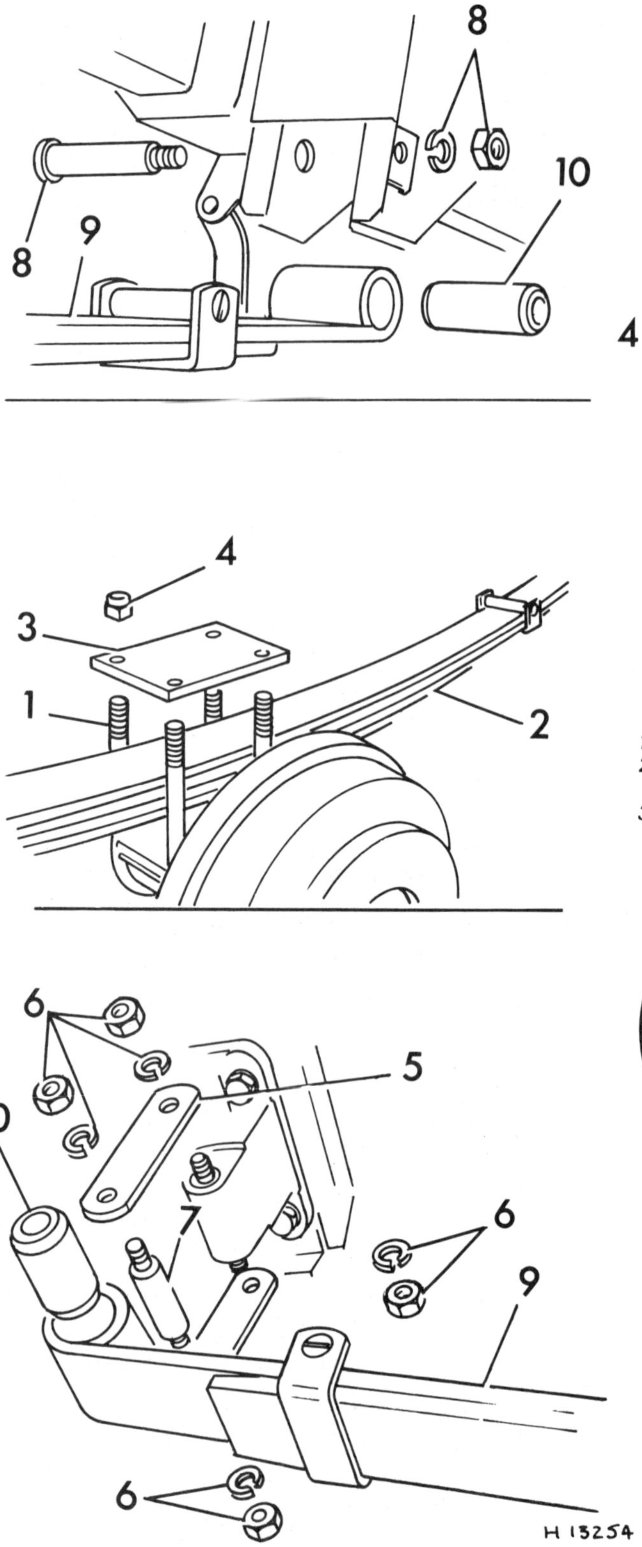

Fig. 11.5 Removing a rear roadspring (Sec 8)

1 *U-bolt*
2 *Roadspring*
3 *Clamp plate*
4 *Locknut*
5 *Shackle plate*
6 *Retaining nuts and washers*
7 *Shackle pin*
8 *Front eye bolt, nut and washer*
9 *Roadspring*
10 *Bush*

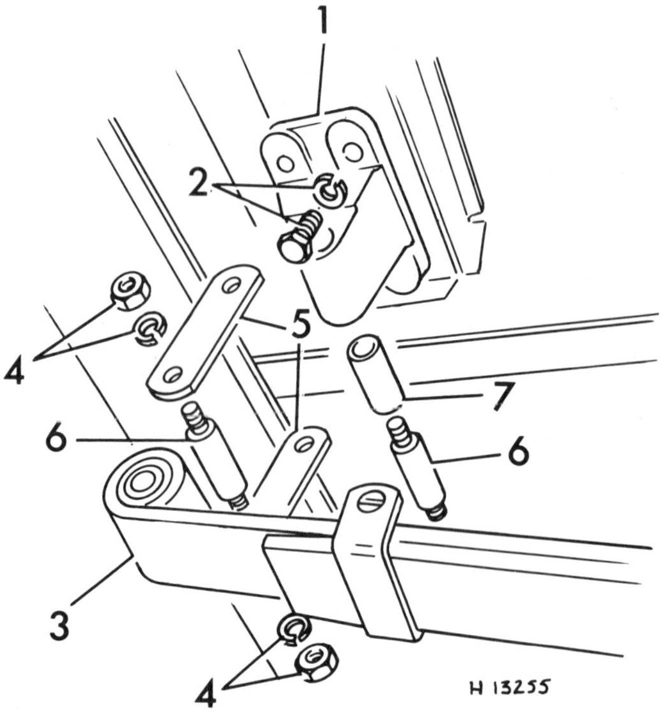

Fig. 11.6 Removing a rear spring hanger bracket bush (Sec 9)

1 *Hanger bracket*
2 *Retaining bolt and spring washer*
3 *Roadspring*
4 *Shackle pin nuts and spring washers*
5 *Shackle plates*
6 *Shackle pins*
7 *Hanger bracket bush*

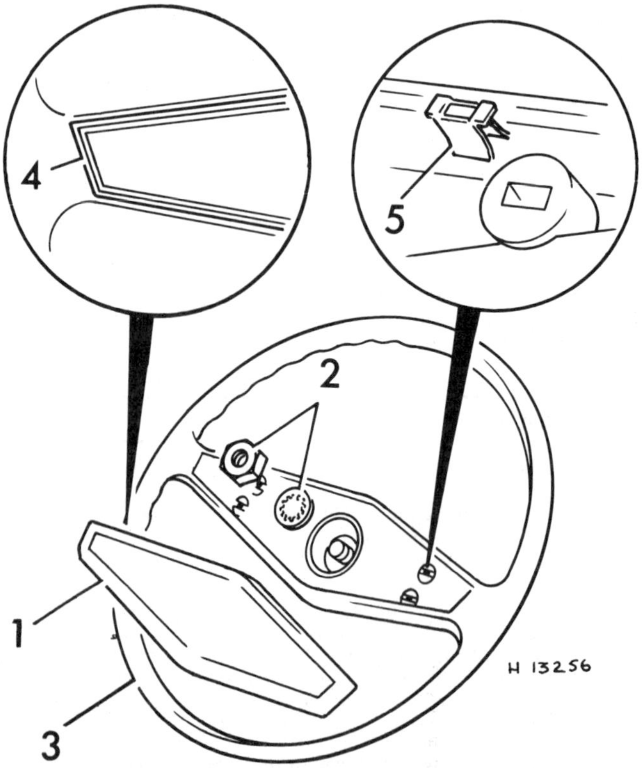

Fig. 11.7 Steering wheel fitting details (Sec 11)

1 *Steering wheel pad*
2 *Retaining nut and lockwasher*
3 *Steering wheel*
4 *Pad fitted, with bevel facing in and down*
5 *Retaining clip*

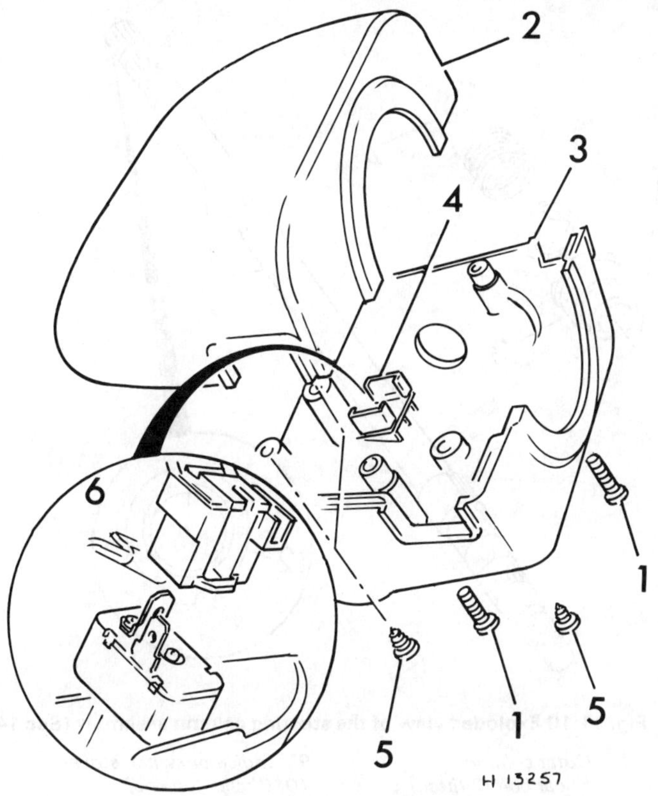

Fig. 11.8 Steering column cowl removal (later models shown, others similar) (Sec 12)

1	*Upper cowl securing screws*	*4*	*Flasher unit retaining clip*
2	*Upper cowl*	*5*	*Lower cowl securing screws*
3	*Lower cowl*	*6*	*Flasher unit*

(c) When refitting the direction indicator switch, make sure that the protrusion on the switch is located in the slot in the column before tightening the retaining screws

(d) Remember to fit the switch driver when refitting the ignition starter switch

14 Steering column assembly – overhaul

1 Refer to Section 13 and remove the steering column assembly from the vehicle.

2 Drill out or use a proprietary stud extractor to remove the two shear bolts which secure the steering lock, and remove the steering lock and clamp plate.

3 Remove the circlip retaining the spring at the bottom end of the outer column, and remove the spring and bearing cup.

4 Withdraw the top bearing from the outer column and remove the outer column from the inner column.

5 Extract the bearing from the bottom end of the outer column.

6 Remove the indicator switch striker from the inner column. Remove the circlip on the top end of the inner column and remove the bearing cup and the bearing from the column.

7 Thoroughly clean all mechanical parts in paraffin and dry with a lint-free rag. Examine for excessive wear, corrosion and damage. Worn bearings and damaged parts should be renewed.

8 On reassembling, fit the top bearing and bearing cup to the top end of the inner column and secure with the circlip.

9 Refit the indicator switch striker to the inner column so that the striker lug is in line with, but opposite to, the line stamped on the bottom end of the column, see Fig. 11.10, item 12.

10 Fit the bottom bearing to the outer column and refit the outer column to the inner column assembly.

11 Refit the bottom bearing cup and spring to the bottom end of the inner column and secure with the circlip.

12 Refit the steering lock using new shear bolts and tighten the bolts

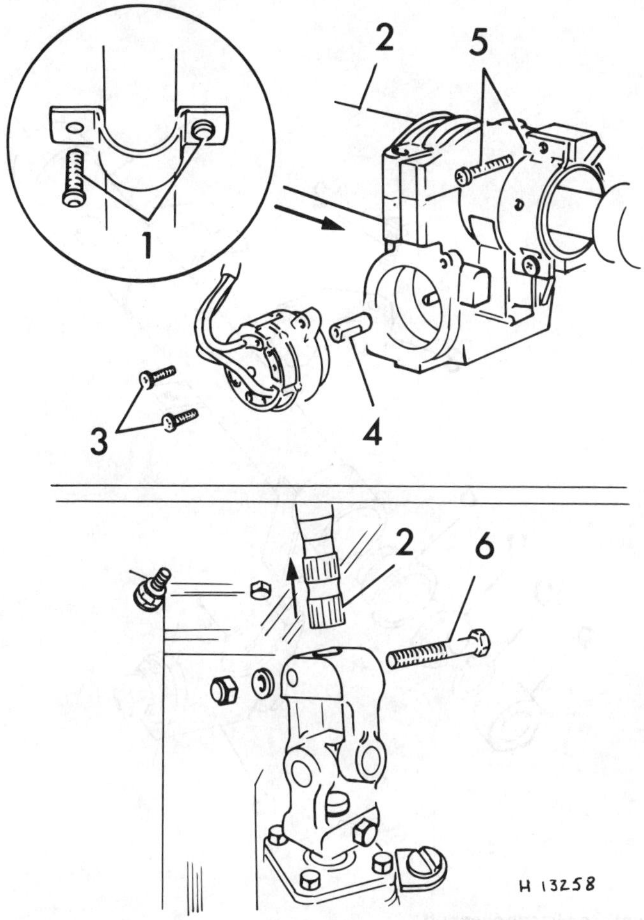

Fig. 11.9 Removing the steering column assembly (Sec 13)

1	*Steering column bracket shear bolts*	*4*	*Switch driver*
2	*Steering column*	*5*	*Direction indicator switch bracket retaining screws*
3	*Ignition/starter switch retaining screws*	*6*	*Steering column/universal joint retaining bolt*

11.2 The steering wheel retaining nut

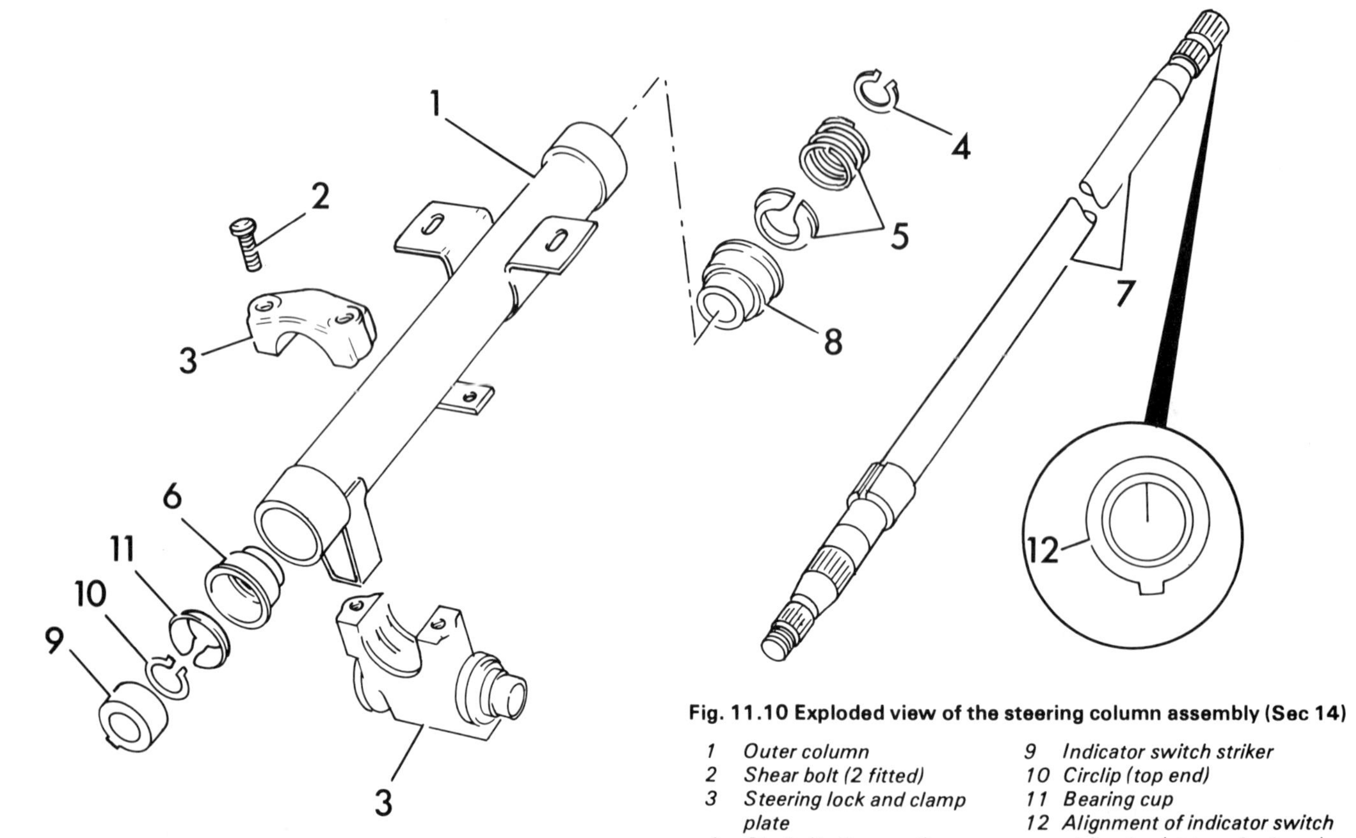

Fig. 11.10 Exploded view of the steering column assembly (Sec 14)

1 Outer column
2 Shear bolt (2 fitted)
3 Steering lock and clamp plate
4 Circlip (bottom end)
5 Spring and bearing cup
6 Top bearing
7 Inner column
8 Bottom bearing
9 Indicator switch striker
10 Circlip (top end)
11 Bearing cup
12 Alignment of indicator switch striker lug (at top of column) and line stamped on bottom of inner column when assembled

until the heads shear off.

13 Refit the steering column to the vehicle, referring to Section 13 for details of procedure.

15 Steering column lock – removal and refitting

1 Refer to Section 12 and remove the steering column cowl.

2 Undo and remove the two screws securing the ignition/starter switch to the steering lock, and move the switch to one side. Retrieve the switch driver from the steering lock spindle.

3 Using either a drill or a stud extractor, remove the two shear bolts which secure the steering column to the facia and pedal bracket.

4 Using the same procedure, remove the two shear bolts which secure the steering lock to the steering column. Remove the lock and the clamp plate (photo).

5 Refitting is basically the reverse of the removal procedure, but note the following special points:

(a) Locate the steering lock with the column and tighten the new shear bolts finger-tight

(b) Use new shear bolts to secure the steering column to the facia and pedal bracket, but only tighten them sufficiently to hold the assembly in place

(c) Check for correct operation of the steering lock, then tighten all four shear bolts until the heads break off

16 Steering column universal joint – removal and refitting

1 Remove the nut, plain washer and bolt which secure the universal joint to the steering box shaft (photo).

2 Remove the nut, plain washer and bolt which secure the universal joint to the steering column.

3 Using a drill or a stud extractor, remove the two shear bolts retaining the steering column to the facia and pedal bracket.

4 Pull the steering column assembly free from the universal joint, and remove the universal joint from the steering box shaft.

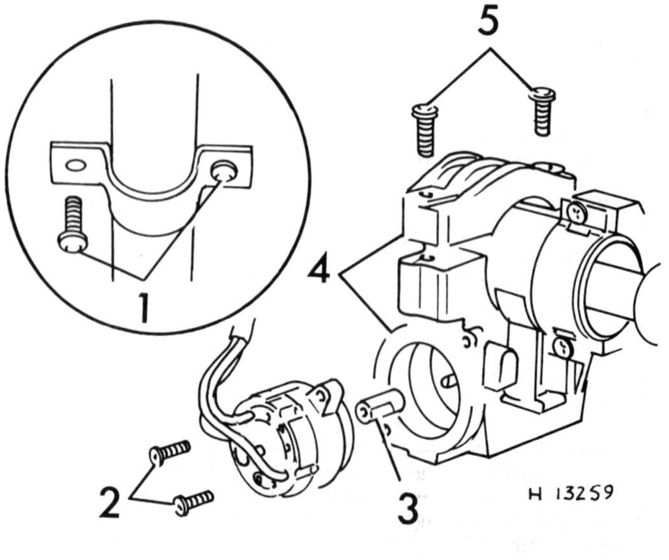

Fig. 11.11 Removing the steering column lock (Sec 15)

1 Steering column bracket shear bolts
2 Ignition/starter switch retaining screws
3 Switch driver
4 Steering lock and clamp plate
5 Shear bolts

15.4 The steering column lock

16.1 The steering column universal joint

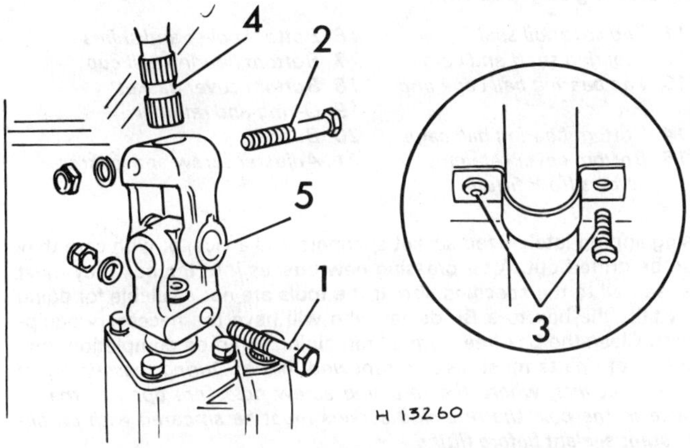

Fig. 11.12 Removing the steering column universal joint (Sec 16)

1 Lower securing bolt
2 Upper securing bolt
3 Steering column bracket shear bolts
4 Steering column
5 Universal joint

5 When refitting the universal joint first fit it to the steering box shaft, fit the bolt, plain washer and nut and tighten them.

6 Engage the steering lock with the steering wheel in the straight ahead position.

7 Check that the roadwheels are in the straight ahead position and assemble the steering column shaft to the universal joint with the line that is stamped on the bottom of the steering column shaft at the nine o'clock position (as viewed from the driver's position).

8 Fit the bolt, plain washer and nut to secure the steering column to the universal joint.

9 Secure the steering column assembly to the facia and pedal bracket, using two new shear bolts, and tighten them until their heads break off.

17 Steering box – removal and refitting

1 Engage the steering lock, with the roadwheels in the straight ahead position.

2 Refer to Section 20 and remove the steering drop arm from the steering box.

3 Remove the nut, plain washer and bolt which secure the universal joint to the steering box shaft.

4 Using a drill or a stud extractor, remove the two shear bolts retaining the steering column assembly to the facia and pedal bracket.

5 Pull the steering column assembly upwards to release the universal joint from the steering box shaft.

6 Remove the three bolts and locknuts securing the steering box to the chassis. Note that there is a distance tube fitted to the bottom bolt. Remove the steering box and the rocker shaft sealing ring from the vehicle.

7 On reassembly check that the sealing ring is fitted to the rocker shaft and bolt the steering box to the body with the three retaining bolts, plain washers and nuts. Remember to fit the distance tube on the bottom bolt between the flanges on the chassis.

8 Refit the steering drop arm to the rocker shaft, referring if necessary to Section 20, and tighten the retaining nut to the specified torque.

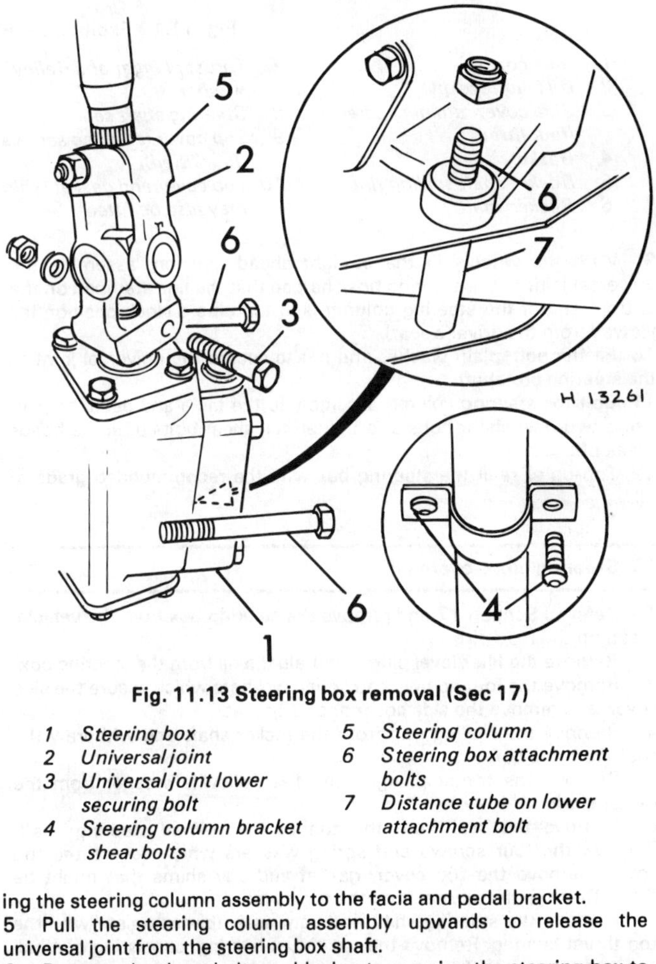

Fig. 11.13 Steering box removal (Sec 17)

1 Steering box
2 Universal joint
3 Universal joint lower securing bolt
4 Steering column bracket shear bolts
5 Steering column
6 Steering box attachment bolts
7 Distance tube on lower attachment bolt

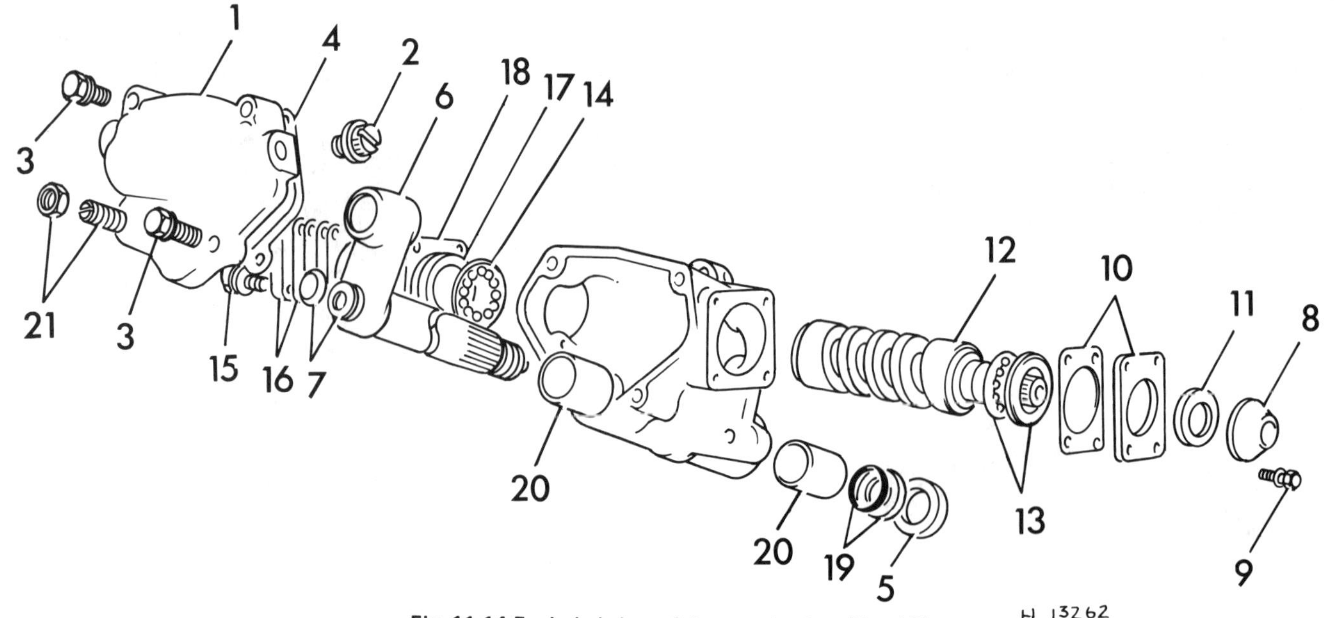

Fig. 11.14 Exploded view of the steering box (Sec 18)

1 *Side cover*
2 *Oil filler/level plug*
3 *Side cover retaining screws (four fitted)*
4 *Gasket*
5 *Rocker shaft sealing ring*
6 *Rocker shaft*
7 *Thrust plunger and Belleville washer*
8 *Steering shaft seal*
9 *Top cover retaining screws (four fitted)*
10 *Top cover and gasket (shims may also be fitted)*
11 *Top cover oil seal*
12 *Steering shaft and cam*
13 *Top bearing ball cage and cup*
14 *Bottom bearing ball cage*
15 *Bottom cover retaining screws (four fitted)*
16 *Bottom cover and shims*
17 *Bottom bearing ball cup*
18 *Bottom cover gasket*
19 *O-ring and retainer*
20 *Bushes*
21 *Adjuster screw and locknut*

9 With the wheels in the straight ahead position, assemble the universal joint to the steering box shaft so that the line stamped on the bottom end of the steering column is in the nine o'clock position (as viewed from the driver's seat).
10 Fit the bolt, plain washer and nut to secure the universal joint to the steering box shaft.
11 Bolt the steering column assembly to the facia and pedal bracket using two new shear bolts and tighten the shear bolts until the heads break off.
12 Top-up or refill the steering box with the recommended grade of oil.

18 Steering box – overhaul

1 Refer to Section 17 and remove the steering box from the vehicle. Clean off the road dirt.
2 Remove the filler/level plug and drain the oil from the steering box.
3 Remove the four screws and spring washers which secure the side cover, and remove the side cover and its gasket.
4 Remove the sealing ring from the rocker shaft and withdraw the shaft from the box.
5 Remove the thrust plunger and flat Belleville washer from the rocker shaft.
6 Remove the seal from the top of the splined steering shaft. Remove the four screws and spring washers which secure the top cover. Remove the top cover, gasket and any shims that might be fitted. Remove the oil seal from the top cover.
7 Remove the steering shaft and cam assembly complete with the top thrust bearing. Remove the top thrust bearing, consisting of a ball cup and a ball cage, from the shaft.
8 Remove the bottom bearing ball cage from the steering box.
9 Remove the four screws and spring washers which secure the bottom cover. Remove the cover, gasket and shims and remove the bottom bearing ball cup from the box.
10 Using a small, sharp chisel, or similar tool, cut away the staking securing the rocker shaft O-ring retainer and remove the retainer and the O-ring.
11 Thoroughly clean all parts in paraffin and dry with lint-free rag. Examine for excessive wear and any apparent abnormality. If wear is evident in the rocker shaft bushes, press the bushes out of the body using appropriately sized socket spanners and a vice, or with care they can be drifted out. After pressing new bushes into the box they must be reamed to the specified size. If the tools are not available for doing this take the box to a BL dealer who will have the necessary equipment. Clean the box free from all remaining swarf on completion. Any other worn parts must also be renewed. **Note:** *During reassembly of the end covers, where the retaining screw holes are open to the oil space in the box, the retaining screws must be smeared with an oil-resistant sealant before fitting.*
12 Smear a new O-ring for the rocker shaft with steering box oil and fit the O-ring with its retainer in the steering box body. Stake the retainer four places, different from those originally used.
13 Insert the bottom bearing ball cups in the steering box, cup side facing inwards. Use a new gasket and fit the bottom cover and the original shims that were fitted. Secure with the four spring washers and retaining screws. Insert the bottom bearing ball cage into the steering box to rest on the ball cup.
14 Assemble the top bearing ball cage to the steering shaft splined end followed by the bearing ball cup, cup face to the ball cage, and insert the shaft and bearing into the steering box.
15 Fit the end cover without its oil seal but with the original shims, if any, and a new gasket. Hold the cover firmly in place and, using feeler gauges, measure the gap, if any, between the cover and the body. This clearance controls the bearing preload and should agree with the figure quoted in the Specifications. If the clearance is incorrect it must be adjusted by inserting or removing shims. There are available in sizes as follows:

0.0024 in (0.061 mm)
0.005 in (0.127 mm)
0.010 in (0.254 mm)
0.020 in (0.508 mm)

16 After setting the cam and shaft bearing preload, remove the top end cover and fit a new oil seal to the cover so that it is flush with the top face of the cover and the seal lip faces inwards.
17 Pack a new dust seal with graphite grease, then fit it to the steering shaft.
18 Fit the thrust plunger and Belleville washer to the rocker shaft and fit the rocker shaft in the steering box.
19 Fit a new sealing ring to the rocker shaft.
20 Position the side cover and a new gasket on the steering box and

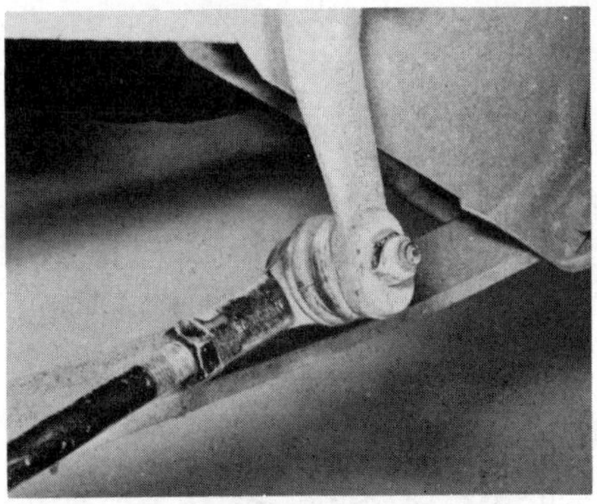

19.3 Drag link balljoint attachment to the steering drop arm

19.4 The adjusting screw and locknut (arrowed) on the steering box

22.2 A track-rod balljoint

secure with the four screws and spring washers. Coat the side cover thrust screw with an oil resistant sealant and insert it, with its locknut loose temporarily, in the side cover.

21 Refit the oil filler plug temporarily. Refit the steering box to the vehicle, refer to Section 17, and refill the box with the approved grade of oil. Adjust the steering box as described in Section 19.

19 Steering box – adjusting in situ

1 Apparent stiffness or slackness in the steering may be due to an incorrectly adjusted steering box. When a new or overhauled steering box is fitted in the vehicle the box must be adjusted. The adjustment procedure is as follows.

2 Apply the handbrake, jack-up the front of the vehicle and support it firmly on stands. Clean off any road dirt from around the steering box oil filler, remove the filler plug and check that the oil level is up to the plug hole.If necessary, top-up with the approved grade of oil. Refit the filler plug.

3 Remove the nut retaining the drag link to the steering drop arm and detach the balljoint from the drop arm. This will probably require the use of a balljoint separator or wedges (photo).

4 Slacken the locknut on the rocker shaft adjustment screw (photo).

5 Measure the torque required to turn the steering wheel through the straight ahead position. This can be done by using a torque spanner on the steering wheel retaining nut but, as the torque is relatively low, a spring balance attached to the steering wheel hub might be more successful. Tie the spring balance securely to the steering wheel hub, it must not slip or the readings will be inaccurate. Measure the distance from the centre of the steering wheel nut to the spring balance attachment point accurately. Pull on the spring balance, keeping it at right angles to the line joining the point of attachment and the centre of the steering column, and note the reading as the wheel moves over the straight ahead position. The spring balance must be pulled round in the same arc of travel as the steering wheel. This procedure, although it might sound complicated, is in fact simple to do. Take several readings and work out an average torque by multiplying the spring balance reading by the distance of attachment from the centre of the wheel.

6 The measured torque must agree with that shown in the Specifications. It can be adjusted by turning the adjusting screw in the steering box side cover; screw the adjuster in to increase torque. After making an adjustment, recheck the torque on the spring balance and repeat the process as necessary.

7 When the torque is correct, tighten the locknut on the adjuster screw to the specified torque. Check the torque at the steering wheel again to make sure that it has not been altered.

8 Reassemble the steering mechanism by reversing the sequence in paragraphs 1 to 3.

20 Steering drop arm – removal and refitting

1 Apply the handbrake, jack-up the front of the vehicle and support it firmly on stands. Remove the steering box.

2 Remove the locknut retaining the drag link balljoint to the steering drop arm and detach the balljoint from the drop arm. This will probably require the use of a balljoint separator or wedges.

3 Remove the nut and spring washer retaining the drop arm to the splined rocker shaft in the steering box.

4 Mark the relative positions of the drop arm and the shaft for refitting. This can be done by carefully centre-punching the two components.

5 Withdraw the drop arm from the shaft using a lever, or a universal puller if it is tight.

6 On refitting, align the marks on the drop arm and shaft. If the original components are not being refitted, the approximate relationship can be obtained by setting the steering box so that the line stamped on the steering box shaft is in the six o'clock position (when viewed from the driver's position) and fitting the drop arm so that it hangs just slightly forward from the vertical at the bottom. If this procedure is adopted make sure that the steering is checked on completion of reassembly for full left and right lock and centralising. If these are incorrect it is probable that the drop arm is wrongly assembled.

7 Secure the drop arm to the steering box shaft with the spring washer and retaining nut and tighten the nut to the specified torque. The remainder of the refitting procedure consists of reversing the removal procedure in paragraphs 1 and 2.

21 Drag link balljoints – removal and refitting

1 Apply the handbrake, jack-up the front of the vehicle and support it firmly on axle stands.

2 Undo and remove the nut retaining the drag link balljoint to the steering drop arm and detach the balljoint from the drop arm. This will probably require the use of a balljoint separator or wedges.

3 Repeat paragraph 2 for the balljoint at the other end of the drag link and remove the drag link assembly from the vehicle.

4 Note the distance between the centres of the balljoint pins. Slacken the locknuts on the drag link and unscrew the balljoints.

5 When reassembling, screw each balljoint onto the drag link for an equal number of turns until the pin centres are at the same distance as noted in the previous paragraph. Secure the balljoints by lightly tightening the locknuts.

6 Engage the steering lock with the steering wheel in the straight ahead position. Fit the drag link to the steering box drop arm and secure it with the retaining nut.

7 With the roadwheels in the straight ahead position, fit the drag link to the steering arm and secure it with the retaining nut. If necessary, adjust the drag link until the balljoint pin fits into the steering arm.

8 Release the steering lock and check that the axle stops limit the steering movement in each direction before reaching the stops in the steering box. If necessary adjust the drag link to achieve this.

9 When satisfactory, hold each drag link balljoint and tighten the locknuts to the specified torque. Then tighten the balljoint pin nuts to the specified torque. Finally lower the vehicle to the ground.

22 Track-rod balljoints – removal and refitting

1 Apply the handbrake, jack-up the front of the vehicle and support it firmly on stands.

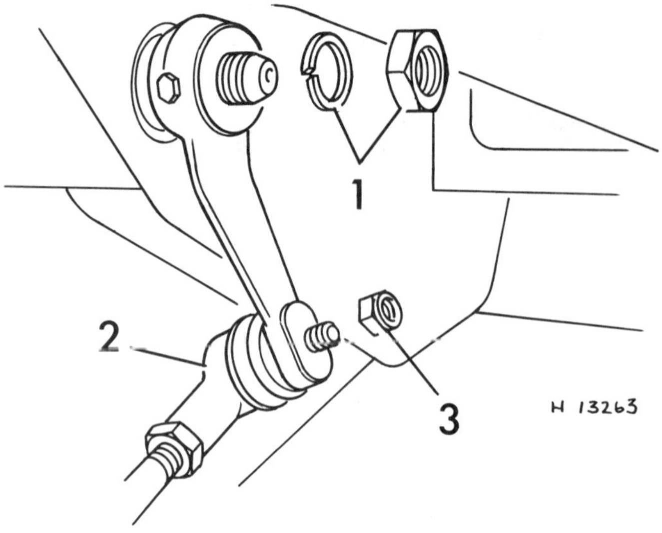

Fig. 11.15 Removing the steering drop arm (Sec 20)

1 Rocker shaft nut and spring washer
2 Drag link balljoint
3 Balljoint locknut

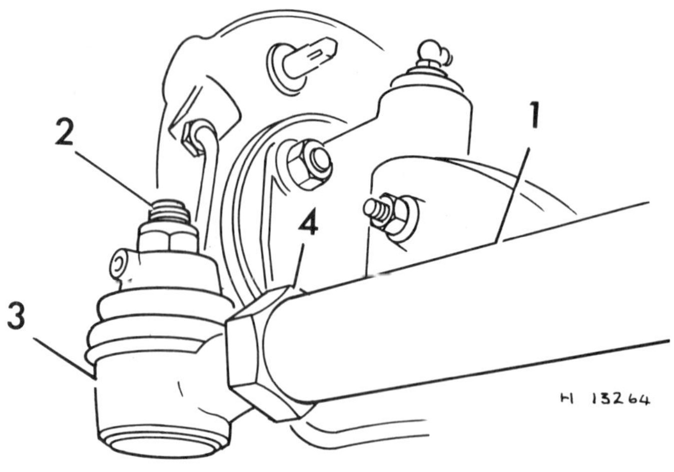

Fig. 11.16 Track-rod balljoint removal (Sec 22)

1 Track-rod
2 Balljoint pin locknut
3 Balljoint
4 Track-rod locknut

2 Remove the nut retaining the track-rod balljoint to the track-rod arm and detach the balljoint from the track-rod arm. This will probably require the use of a balljoint separator or wedges (photo).
3 Repeat the operation in the preceding paragraph on the balljoint at the other end of the track-rod, and remove the track-rod from the vehicle.
4 Note the distance between the centres of the balljoint pins. Slacken the locknuts and unscrew the balljoints from the track-rod. Note that one end of the track-rod has a left-hand thread.
5 On reassembly, screw each balljoint onto the track-rod for an equal number of turns until the pin centres are at the same distance apart as noted in the previous paragraph. Secure the balljoints by lightly tightening the locknuts.
6 Fit the track-rod assembly to the track-rod arms and tighten the balljoint pin nuts to the specified torque.
7 Lower the vehicle to the ground and adjust the front wheel alignment as described in Section 23.

23 Front wheel alignment – checking and adjusting

1 Accurate front wheel alignment is essential for good steering and minimum tyre wear. Adjusting the front wheel alignment on the Sherpa consists of altering the amount of toe-in, ie the amount by which the wheels point inwards at the front. Correct results are more likely to be obtained from using the special equipment held by BL garages, but this is not normally available to the average DIY car mechanic. There are several types of tracking gauges available in car accessory shops and if one of these can be obtained to do the checking, the adjusting of the alignment is relatively simple. In the absence of a checking gauge, a competent enthusiast could devise a means of measuring the amount of toe-in, but it is emphasised that accuracy is essential and it is recommended that, in all cases of non-professional checking and adjusting, the results should be regarded merely as adequate to get to a dealer where an accurate check can be made.
2 Where appropriate equipment is available and can be used with confidence, proceed as follows. Check that the tyres are correctly inflated, that the front wheels are not damaged, the hub bearings are not worn or incorrectly adjusted and that the steering linkage is in good order, without slackness or wear at the joints. Check this last point by observing the steering linkage while as assistant moves the steering wheel.
3 The vehicle must be unladen and on level ground. Using the alignment checking equipment in accordance with the maker's instructions, take a series of measurements and calculate the existing toe-in, if any. **Note:** *If the vehicle is moved at all during the measuring process it must only be moved forwards, as moving it rearwards will alter the toe-in.*
4 To adjust toe-in, slacken the locknuts at each end of the track-rod. Rotate the track-rod which has opposite handed threads at each end and so acts as a turnbuckle, until the correct alignment is obtained, see Specifications.
5 Hold each track-rod balljoint with a spanner at the centre of its arc of travel and tighten the locknuts to the specified torque. Note that different sized locknuts have different tightening torques.
6 After tightening the locknuts, recheck the wheel alignment and readjust if necessary.

24 Wheels and tyres

1 Wheels are normally not subject to servicing problems. If a wheel is suspected of being damaged, change it and have it checked for balance and true running by a specialist.
2 When fitting wheels, do not overtighten the nuts. The correct torque is given in the Specifications at the beginning of this Chapter.
3 On vehicles where cross-ply tyres are fitted as original equipment, replacements may be of the radial type which, although more expensive initially, last longer and give slightly better fuel consumption and handling. However, **do not** mix cross-ply and radial tyres as the design differences could cause dangerous handling and braking characteristics. In any case it is illegal in the UK to mix cross-ply and radial tyres except when radials are fitted to the back wheels and cross-plies to the front.
4 If radial tyres have been fitted as original equipment, cross-ply tyres should not be fitted.

25 Fault diagnosis – suspension and steering

Symptom	Reason/s
STEERING FEELS VAGUE, VEHICLE WANDERS AND FLOATS AT SPEED	
General wear or damage	Tyre pressures uneven Shock absorbers worn Steering gear balljoints badly worn Suspension geometry incorrect Steering mechanism free play excessive Front suspension and rear suspension pick-up points out of alignment or badly worn Front suspension lacking grease
STIFF AND HEAVY STEERING	
Lack of maintenance or accident damage	Tyre pressures too low No oil in steering box No grease in steering balljoints Front wheel toe-in incorrect Suspension geometry incorrect Steering gear adjusted too tightly Steering column badly misaligned
WHEEL WOBBLE AND VIBRATION	
General wear or damage	Wheel nuts loose Front wheels and tyres out of balance Steering balljoints badly worn Hub bearings badly worn Steering gear free play excessive Front springs weak or broken

Chapter 12 Bodywork and fittings

For modifications, and information applicable to later models, see Supplement at end of manual

Contents

1 General description

1 The Sherpa is available in five standard forms: van, crewbus, minibus, pick-up, and chassis/cab. While the van, crewbus and minibus are more or less the same in appearance, it is the interior layout which distinguishes them from one another, together with windows (or lack of them). Both the pick-up and chassis/cab have enclosed cabs, and the chassis/cab is mounted on an all-purpose, parallel ladder-type chassis, making it ideal for specialised body-building purposes.

2 Apart from the chassis/cab configuration, the Sherpa is of integral construction, and in the case of the van and pick-up, incorporates wax-filled box-sections to impede corrosion. Since the engine is forward-mounted, the van, crewbus and minibus are of walk-through design, giving convenient access from the cab to the rear, and across the cab itself.

3 Based on a 114 in (2.9 m) wheelbase (114.9 in in the chassis/cab), the Sherpa is available in many arrangements regarding the doors, which can be either sliding or hinged, and various trims are offered.

4 Because of the differing specialised applications of the Sherpa range, it is possible to cover only the standard, factory-built models in this Chapter. Where non-standard bodies and fittings are used, reference should be made to the relevant bodywork firm for information where necessary.

2 Maintenance – bodywork and underframe

1 The general condition of a vehicle's bodywork is the one thing that significantly affects it value. Maintenance is easy but needs to be regular and particular. Neglect, particularly after minor damage, can quickly lead to further deterioration and costly repair bills. It is important to keep watch on those parts of the car not immediately visible, for instance the underside, inside all the wheel arches and the lower part of the engine compartment.

2 The basic maintenance routine for the bodywork is washing – preferably with a lot of water from a hose. This will remove all the loose solids which may have stuck to the vehicle. It is important to flush these off in such a way as to prevent grit from scratching the finish.

3 The wheel arches and underbody need washing in the same way to remove any accumulated mud which will retain moisture and tend to encourage rust. Paradoxically enough, the best time to clean the underbody and wheel arches is in wet weather when the mud is thoroughly wet and soft. In very wet weather the underbody is usually cleaned of large accumulations automatically, and this is a good time for inspection.

4 Periodically, it is a good idea to have the whole of the underside of the vehicle steam cleaned, engine compartment included, so that a thorough inspection can be carried to see what minor repairs and renovations are necessary. Steam cleaning is available at many

garages and is necessary for the removal of accumulation of oily grime which sometimes is allowed to cake thick in certain areas near the engine, gearbox and back axle. If steam facilities are not available, there are one or two excellent grease solvents available which can be brush applied. The dirt can then be simply hosed off.

5 After washing paintwork, wipe off with a chamois leather to give an unspotted clear finish. A coat of clear protective wax polish will give added protection against chemical pollutants in the air. If the paintwork sheen has dulled or oxidised, use a cleaner/polisher combination to restore the brilliance of the shine. This requires a little effort, but is usually caused because regular washing has been neglected. Always check that the door and ventilator opening drain holes are completely clear so that water can drain out. Bright work should be treated in the same way as paintwork. Windscreen and windows can be kept clear of the smeary film which often appears, if a little ammonia is added to the water. If they are scratched, a good rub with a proprietary metal polish will often clear them. Never use any form of wax or other body or chromium polish on glass.

3 Maintenance – upholstery and matting

Mats and carpets should be brushed or vacuum cleaned regularly to keep them free of grit. If they are badly stained remove them from the vehicle for scrubbing or sponging and make quite sure they are dry before refitting. Seat and interior trim panels can be kept clean by a wipe over with a damp cloth. If they do become stained (which can be more apparent on light coloured upholstery) use a little liquid detergent and a soft nail brush to scour the grime out of the grain of the material. Do not forget to keep the headlining clean in the same way as the upholstery. When using liquid cleaners inside the car do not over-wet the surfaces being cleaned. Excessive damp could get into the seams and padded interior causing stains, offensive odours or even rot. If the inside of the vehicle gets wet accidentally, it is worthwhile taking some trouble to dry it out properly particularly where carpets are involved. **Do not** leave oil or electric heaters inside the car for this purpose.

4 Minor body damage – repair

See photo sequences between pages 32 and 33

Repair of minor scratches in the vehicle's bodywork

If the scratch is very superficial, and does not penetrate to the metal of the bodywork, repair is very simple. Lightly rub the area of the scratch with a paintwork renovator or a very fine cutting paste, to remove the loose paint from the scratch and to clear the surrounding bodywork of wax polish. Rinse the area with clean water.

Apply touch-up paint to the scratch using a thin paint brush; continue to apply thin layers of paint until the surface of the paint in the scratch is level with the surrounding paintwork. Allow the new paint at least two weeks to harden, then, blend it into the surrounding paintwork by rubbing the paintwork in the scratch area with a paintwork renovator, or a very fine cutting paste. Finally apply wax polish.

Where a scratch has penetrated right through to the metal of the bodywork, causing the metal to rust, a different repair technique is required. Remove any loose rust from the bottom of the scratch with a penknife; then apply rust inhibiting paint to prevent the formation of rust in the future. Using a rubber or nylon applicator, fill the scratch with bodystopper paste. If required, this paste can be mixed with cellulose thinners to provide a very thin paste which is ideal for filling narrow scratches. Before the stopper paste on the scratch hardens, wrap a piece of smooth cotton rag around the top of a finger. Dip the finger in cellulose thinners and then quickly sweep it across the surface of the stopper-paste in the scratch; this will ensure that the surface of the stopper-paste is slightly hollowed. The scratch can now be painted over as described earlier in this Section.

Repair of dents in the vehicle's bodywork

When deep denting of the vehicle's bodywork has taken place the first task is to pull the dent out, until the affected bodywork almost attains its original shape. There is little point in trying to restore the original shape completely, as the metal in the damaged area will have stretched on impact and cannot be reshaped fully to its original contour. It is better to bring the level of the dent up to a point which is about $\frac{1}{8}$ in (3 mm) below the level of the surr cases where the dent is very shallow anyway, it is pull it out at all.

If the underside of the dent is accessible, it can be gently from behind, using a mallet with a wooden or Whilst doing this, hold a suitable block of wood firmly impact from the hammer blows to prevent a large area of from being belled-out.

Should the dent be in a section of the bodywork which h double skin or some other factor making it inaccessible from behind different technique is called for. Drill several small holes through th metal inside the dent area – particularly in the deeper sections. Then screw long self-tapping screws into the holes just sufficiently for them to gain a good purchase in the metal. Now the dent can be pulled out by pulling on the protruding heads of the screws with a pair of pliers.

The next stage of the repair is the removal of the paint from the damaged area, and from an inch or so of the surrounding sound bodywork. This is accomplished more easily by using a wire brush or abrasive pad on a power drill, although it can be done just as effectively by hand using sheets of abrasive paper. To complete the preparation for filling, score the surface of the bare metal with a screwdriver or the tang of a file, or alternatively, drill small holes in the affected area. This will provide a really good key for the filler paste. To complete the repair see the Section on filling and re-spraying.

Repair of rust holes or gashes in the vehicle's bodywork

Remove all paint from the affected area and from an inch or so of the surrounding sound bodywork, using an abrasive pad or a wire brush on a power drill. If these are not available a few sheets of abrasive paper will do the job just as effectively. With the paint removed you will be able to gauge the severity of the corrosion and therefore decide whether to renew the whole panel (if this is possible) or to repair the affected area. New body panels are not as expensive as most people think and it is often quicker and more satisfactory to fit a new panel than to attempt to repair large areas of corrosion.

Remove all fittings from the affected area except those which will act as a guide to the original shape of the damaged bodywork (eg headlamp shells etc). Then, using tin snips or a hacksaw blade, remove all loose metal and any other metal badly affected by corrosion. Hammer the edges of the hole inwards in order to create a slight depression for the filler paste.

Wire brush the affected area to remove the powdery rust from the surface of the remaining metal. Paint the affected area with rust inhibiting paint; if the back of the rusted area is accessible treat this also.

Before filling can take place it will be necessary to block the hole in some way. This can be achieved by the use of aluminium or plastic mesh, or aluminium tape.

Aluminium or plastic mesh is probably the best material to use for a large hole. Cut a piece to the approximate size and shape of the hole to be filled, then position it in the hole so that its edges are below the level of the surrounding bodywork. It can be retained in position by several blobs of filler paste around its periphery.

Aluminium tape should be used for small or very narrow holes. Pull a piece off the roll and trim it to the approximate size and shape required, then pull off the backing paper (if used) and stick the tape over the hole; it can be overlapped if the thickness of one piece is insufficient. Burnish down the edges of the tape with the handle of a screwdriver or similar, to ensure that the tape is securely attached to the metal underneath.

Bodywork repairs – filling and re-spraying

Before using this Section, see Sections on dent, deep scratch, rust hole and gash repairs.

Many types of bodyfiller are available, but generally speaking those proprietary kits which contain a tin of filler paste and a tube of resin hardener are best for this type of repair. A wide, flexible plastic or nylon applicator will be found invaluable for imparting a smooth and well contoured finish to the surface of the filler.

Mix up a little filler on a clean piece of card or board – use the hardener sparingly (follow the maker's instructions on the packet), otherwise the filler will set very rapidly.

Using the applicator apply the filler paste to the prepared area; draw the applicator across the surface of the filler to achieve the correct contour and to level the filler surface. As soon as a contour that approximates the correct one is achieved, stop working the paste. If

become sticky and begin to pick up

of filler paste at twenty-minute is just proud of the surrounding

ss can be removed using a metal ely finer grades of abrasive paper rade wet or dry paper. Always rubber cork, or wooden block - not be completely flat. During the wet or dry paper should be ... water - this will ensure that a very smooth finish ...parted to the filler at the final stage.

At this stage the dent should be surrounded by a ring of bare metal, which in turn should be encircled by the finely feathered edge of the good paintwork. Rinse the repair area with clean water, until all of the dust produced by the rubbing-down operation is gone.

Spray the whole repair area with a light coat of grey primer; this will show up any imperfections in the surface of the filler. Repair these imperfections with fresh filler paste or bodystopper, and once more smooth the surface with abrasive paper. If bodystopper is used, it can be mixed with cellulose thinners to form a really thin paste which is ideal for filling small holes. Repeat this spray and repair procedure until you are satisfied that the surface of the filler, and the feathered edge of the paintwork are perfect. Clean the repair area with clean water and allow to dry fully.

The repair area is now ready for spraying. Paint spraying must be carried out in a warm, dry, windless and dust free atmosphere. These conditions can be created artificially if you have access to a large indoor working area, but if you are forced to work in the open, you will have to pick your day very carefully. If you are working indoors, dousing the floor in the work area with water will settle the dust which would otherwise be in the atmosphere. If the repair area is confined to one body panel, mask off the surrounding panels; this will help to minimise the effects of a slight mis-matching in paint colours. Bodywork fittings (eg chrome strips, door handles etc) will also need to be masked off. Use genuine masking tape and several thicknesses of newspaper for the masking operation.

Before commencing to spray, agitate the aerosol can thoroughly, then spray a test area (an old tin, or similar) until the technique is mastered. Cover the repair area with a thick coat of primer; the thickness should be built up using several thin layers of paint rather than one thick one. Using 400 grade wet or dry paper, rub down the surface of the primer until it is really smooth. While doing this, the work area should be thoroughly doused with water. Allow to dry before spraying on more paint.

Spray on the top coat, again building up the thickness by using several thin layers of paint. Start spraying in the centre of the repair area and then, using a circular motion, work outwards until the whole repair area and about 2 inches of the surrounding original paintwork is covered. Remove all masking material 10 to 15 minutes after spraying on the final coat of paint. Allow the new paint at least 2 weeks to harden fully; then, using a paintwork renovator or a very fine cutting paste, blend the edges of the new paint into the existing paintwork. Finally, apply wax polish.

5 Major body damage – repair

1 With the exception of chassis cab versions, the chassis members are spot welded to the underbody, and in this respect can be termed of monocoque or unit construction. Major damage repairs to this type of body combination must of necessity be carried out by body shops equipped with welding and hydraulic straightening facilities.

2 Extensive damage to the body may distort the chassis and result in unstable and dangerous handling, as well as excessive wear to tyres and suspension or steering components.

6 Maintenance – hinges and locks

1 Every six months or 6000 miles (9600 km), the bonnet catch, door locks, door hinges, door check straps, sliding door, and sliding step (where fitted), should be oiled with a few drops of engine oil from an oil can.

2 At the same interval the door striker plates should be given a thin smear of grease.

7 Bumpers – removal and refitting

1 The Sherpa bumpers comprise four separate assemblies, of which one is bolted to each front or rear quarter of the bodywork.

Front quarter bumper

2 Disconnect the battery earth lead.

3 Disconnect the lamp wiring harness at the multi-connector plug and free the harness from the clips on the body.

4 Remove the two screws securing the bumper inner support to the body, and the two screws securing the outer support to the body.

5 Remove the bumper assembly from the vehicle.

6 The supports and lamp assembly, if fit for further service, can be transferred to a new bumper. Refitting is the reverse of the removal sequence, but make sure that the lamp is assembled to the bumper with the drain holes downwards.

Rear quarter bumper

7 Disconnect the battery earth lead.

8 Disconnect the rear lamp wiring harness at the multi-connector plug, free the grommet from the wheel arch and pull the wiring harness through the body.

9 Remove the screw which secures the bumper to each support and remove the bumper assembly from the vehicle. Remove the lamp assembly.

10 Apart from the reference to the supports, paragraph 6 also applies to rear quarter bumpers.

8 Windscreen glass – removal and refitting

1 If you are unfortunate enough to have a windscreen shatter, fitting a new windscreen is one of the few jobs which the average owner is advised to leave to the professional. For the owner who wishes to do the job himself the following instructions are given.

2 Remove the wiper arms from their spindles using a screwdriver to lift the retaining clip from the spindle end and pull away.

3 Using a screwdriver very carefully prise up the end of the filler strip and withdraw it from its slot in the rubber moulding (Fig. 12.2).

4 The assistance of a second person should now be enlisted, ready to catch the glass when it is released from its aperture.

5 Working inside the car, commencing at one top corner, press the glass and ease it from its rubber moulding.

6 Remove the rubber moulding from the windscreen aperture.

7 Now is the time to remove all pieces of glass if the screen has shattered. Use a vacuum cleaner to extract as much as possible. Switch on the ventilating blower fan to fast speed and adjust the air distribution to supply maximum flow to the windscreen. Watch out for flying pieces of glass which might be blown out of the ducting.

8 Carefully inspect the rubber moulding for signs of splitting or deterioration. Clean all traces of sealing compound from the rubber moulding and windscreen aperture flange.

9 To refit the glass, first locate the rubber moulding on the glass so that the joint is at the bottom centre of the aperture and apply sealer between the rubber and the glass. Apply some sealer to the body flange.

10 With the rubber moulding correctly positioned on the glass it is now necessary to insert a piece of strong cord all round the outer channel in the rubber surround which fits over the windscreen aperture flange. The two free ends of the cord should finish at either top or bottom centre and overlap each other by a minimum of 1 ft.

11 Offer up the screen to the aperture and get an assistant to press the rubber surround hard against the body flange. Slowly pull one end of the cord moving round the windscreen so drawing the lip over the windscreen flange on the body. If necessary use a piece of plastic or tapered wood to assist in locating the lip on the windscreen flange.

12 Apply a further layer of sealer to the underside of the rubber to glass groove.

13 The filler strip must next be fitted to the moulding and for this a special tool is required. An illustration of this tool is shown in Fig. 12.3

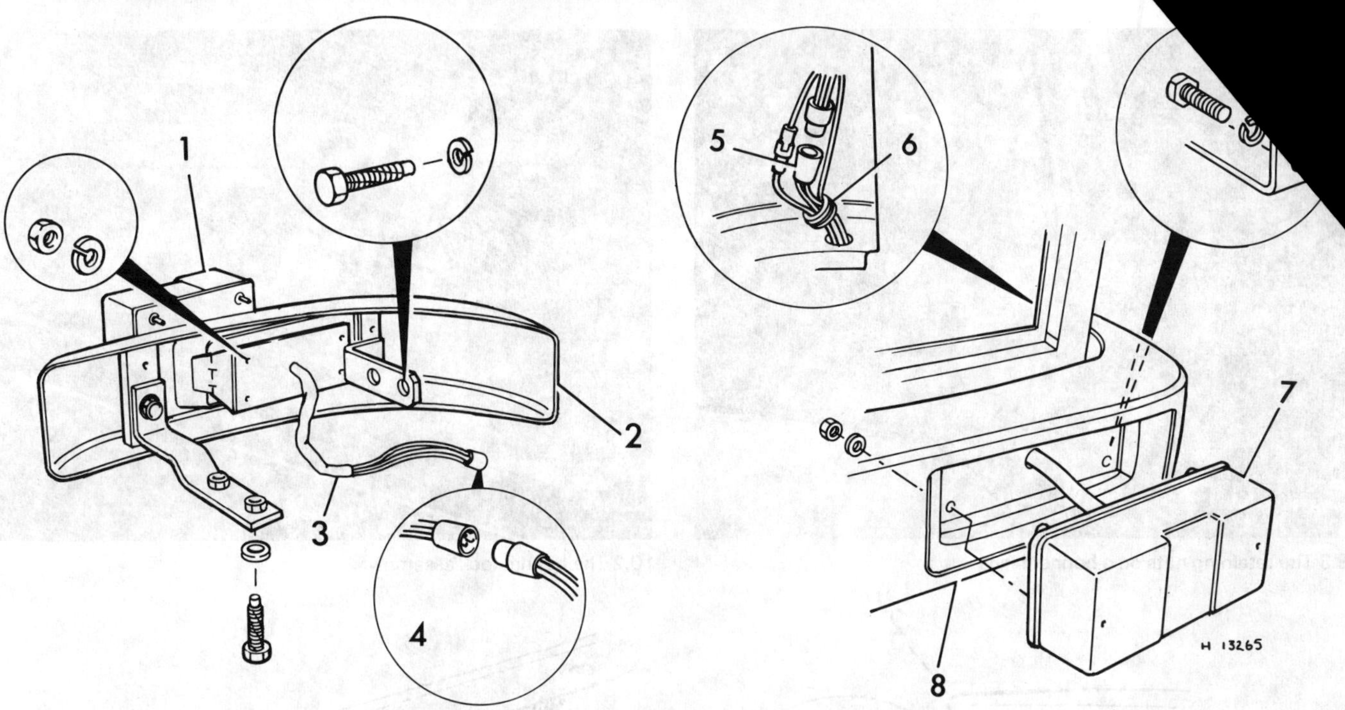

Fig. 12.1 Removing the front and rear bumpers (Sec 7)

1 Front sidelamp assembly
2 Front quarter bumper
3 Electrical wiring harness
4 Multi-connector plug, front
5 Multi-connector plug, rear
6 Grommet
7 Tail lamp assembly
8 Rear quarter bumper

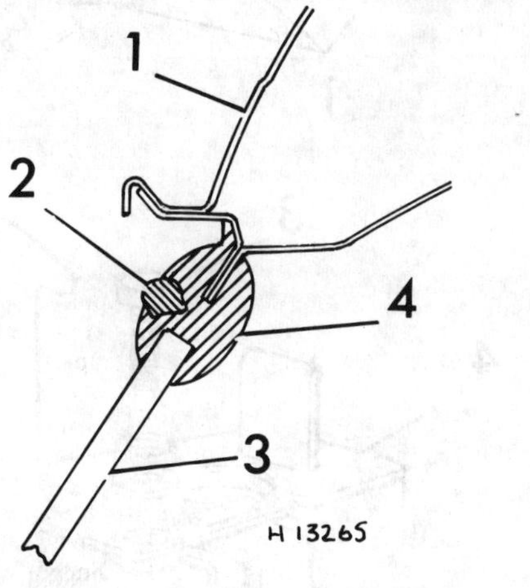

Fig. 12.2 Windscreen surround fitting details (Sec 8)

1 Cab structure
2 Filler strip
3 Windscreen glass
4 Rubber moulding

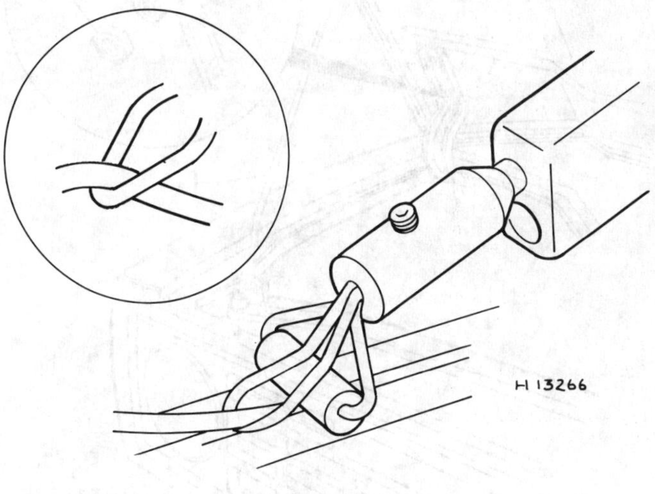

Fig. 12.3 Type of tool required to refit filler strip (Sec 8)

and a handyman should be able to make up an equivalent using wire and a wooden file handle.

14 Fit the eye of the tool into the groove and feed in the filler strip.

15 Push the tool around the complete length of the moulding, feeding the filler into the channel as the eyelet opens it. The back half beds the filler into the moulding.

16 Carefully clean off any surplus sealer before it has a chance to harden and then refit the windscreen wiper arms and blades.

9 Bonnet – removal and refitting

1 Open the bonnet and support it on its stay.

2 With a pencil mark the outline of the hinge on the bonnet to assist correct refitting.

3 An assistant should now take the weight of the bonnet. Undo and remove the bonnet to hinge retaining nuts, spring and plain washers at both hinges. Carefully lift away the bonnet over the front of the vehicle (photo).

4 Refitting is the reverse sequence to removal. Alignment in the body may be made by leaving the securing bolts slightly loose and repositioning by trial and error.

10 Bonnet lock – removal and refitting

1 Open the bonnet and support it on its stay.

2 Slacken the nut and detach the inner, release cable from the trun-

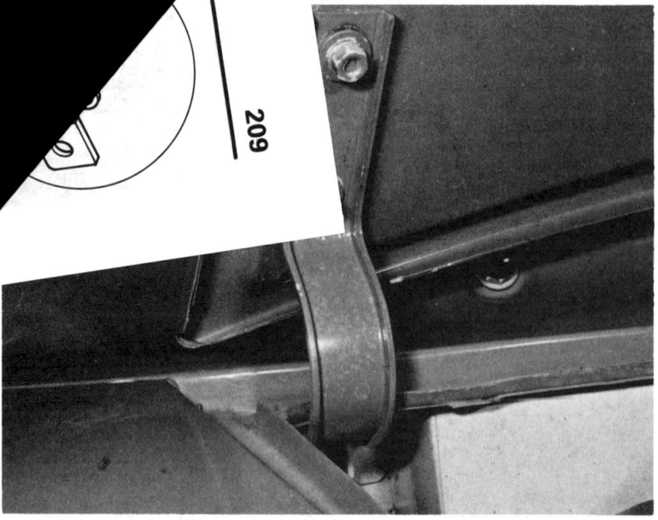

9.3 The retaining nuts on a bonnet hinge

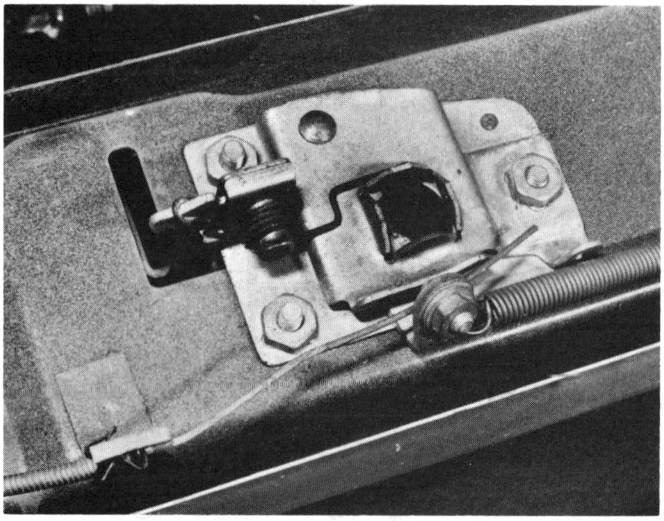

10.2 The bonnet lock assembly

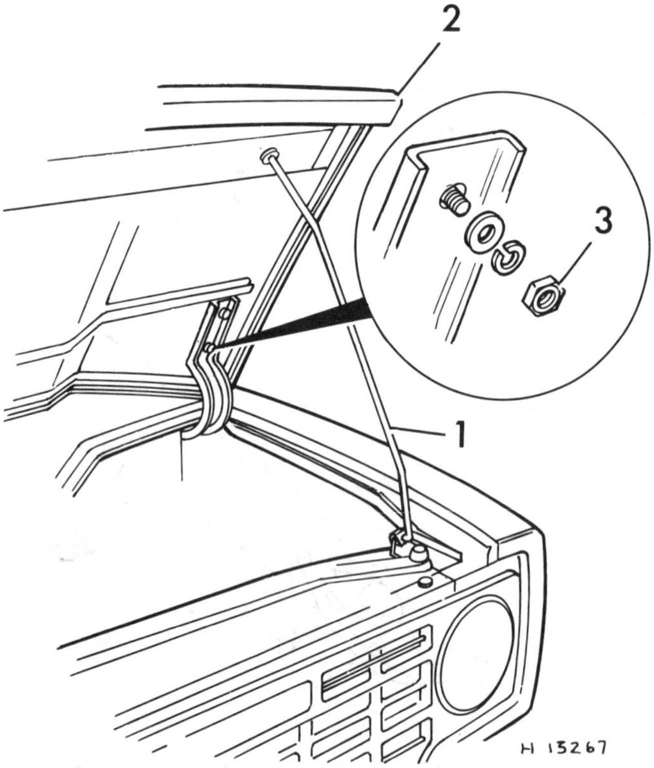

Fig. 12.4 Removing the bonnet (Sec 9)

1 Bonnet stay
2 Bonnet
3 Retaining nut

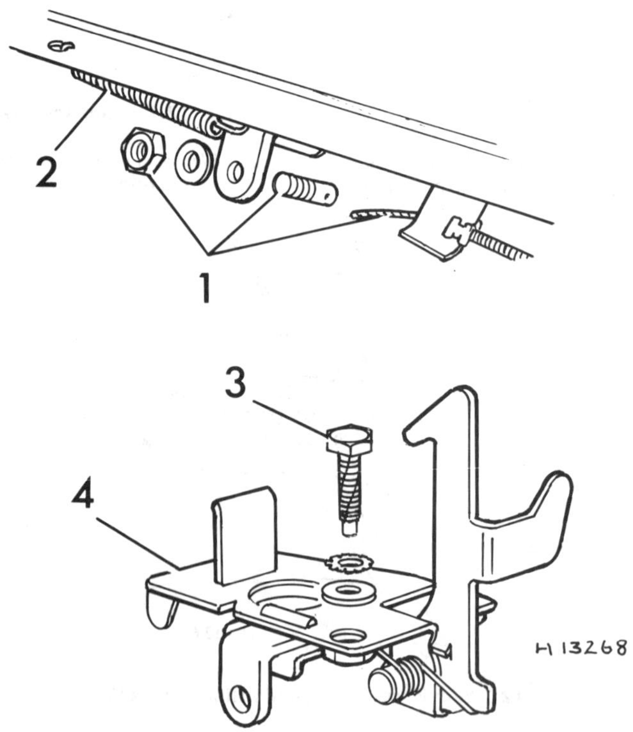

Fig. 12.5 Removing the bonnet lock (Sec 10)

1 Control inner cable and attachment trunnion
2 Return spring
3 Retaining screw
4 Bonnet lock

nion located at the lock lever (Fig. 12.5) (photo).

3 Unhook the return spring from the bonnet lock platform and the bonnet lock lever

4 Undo and remove the three screws which secure the bonnet lock to the bonnet lock platform and remove the bonnet lock from the underside of the platform.

5 Refitting is the reverse sequence to removal, but only fit the lock securing screws finger-tight initially. Centralise the lock in the aperture in the lock platform and close the bonnet to position the lock. Open the bonnet and fully tighten the securing screws.

6 Lubricate the moving parts and finally check the bonnet release operation.

11 Bonnet lock pin – removal, refitting and adjustment

1 Open the bonnet and support it on its stay.

2 Remove the two screws, shakeproof and plain washers securing the bonnet lock pin assembly to the bonnet and remove the lock pin assembly from the bonnet.

3 Slacken the pin retaining nut and unscrew the pin from the reinforcement plate. Unscrew the retaining nut from the pin to release

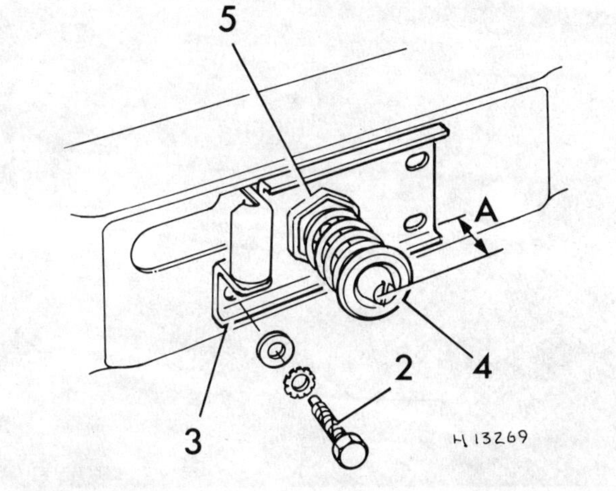

Fig. 12.6 Adjusting or removing the bonnet lock pin (Sec 11)

2	*Retaining screw*	5	*Retaining nut*
3	*Bonnet lock pin assembly*	A	*2.26 to 2.36 in (57 to 59 mm)*
4	*Lock pin*		

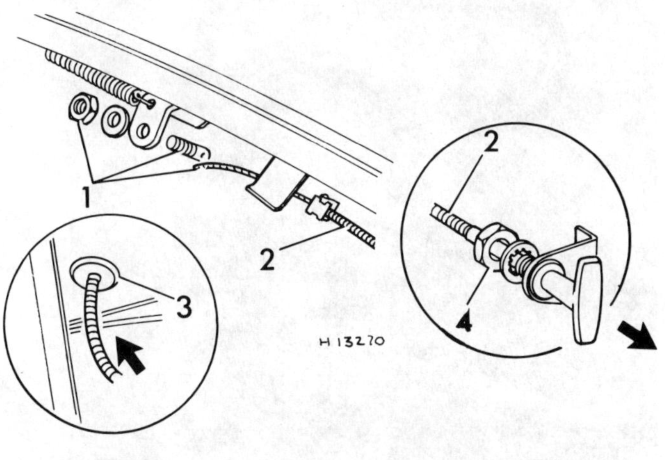

Fig. 12.7 Removing the bonnet lock control cable (Sec 12)

1	*Control inner cable and attachment trunnion*	3	*Grommet*
2	*Outer cable*	4	*Securing nut*

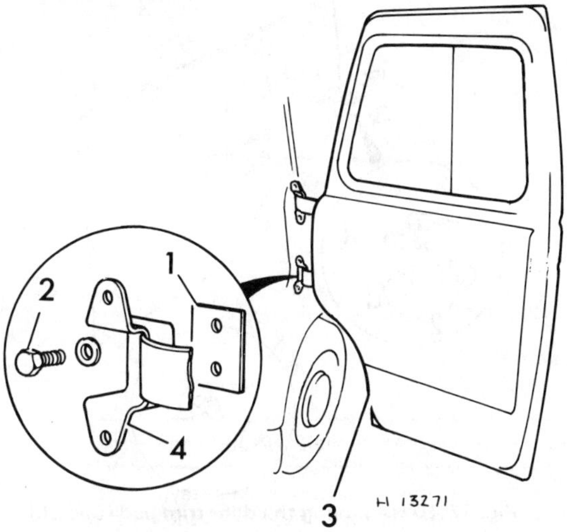

Fig. 12.8 Removing a front hinged door (Sec 13)

1	*Reinforcement plate*	3	*Door*
2	*Setscrew*	4	*Hinge*

the spring and cup.

4 Clean all parts thoroughly in paraffin and dry with lint-free rag. Inspect parts for wear and damage and renew defective parts.

5 Refitting is the reverse of the removal procedure, but do not tighten the retaining nut until the pin has been adjusted.

6 Adjust the pin by screwing it in or out until it extends 2.26 to 2.36 in (57 to 59 mm) from the reinforcing plate, dimension A in Fig. 12.6. Then tighten the retaining nut.

7 Lightly lubricate the moving parts and the head of the pin with a general purpose grease and check the operation by opening and closing the bonnet.

12 Bonnet lock control cable – removal and refitting

1 Open the bonnet and support it on its stay.

2 Slacken the nut and detach the inner, release cable from the trunnion on the bonnet lock lever.

3 Disconnect the release outer cable from its clip on the bonnet lock platform, and release the cable from the clips on the underside of the wing.

4 Undo and remove the nut and shakeproof washer that secure the outer cable to the body side bracket below the facia panel.

5 Carefully withdraw the control cable assembly through the grommet in the bulkhead and through the bracket.

6 Refitting is the reverse of the removal procedure. Before fitting the cable withdraw the inner cable, lightly grease it with a general purpose grease and refit it in the outer cable. When connecting the inner cable to the lock lever trunnion make sure that the lock lever is not preloaded by the release cable.

13 Front hinged side door assembly – removal and refitting

1 Run a sharp pencil around each hinge to mark its position on the door pillar so that door can be refitted in the same position.

2 Remove the eight setscrews which secure the hinges to the door pillar. Note the reinforcement plate for the top hinge.

3 Partially close the door and withdraw it complete with hinges from the door pillar.

4 Refitting is the reverse of the removal procedure. Check the operation of the door lock and striker plate and if necessary adjust to obtain correct functioning.

14 Front side door hinges – removal and refitting

1 Refer to Section 15 and remove the door trim pad.

2 Inside the door mark each hinge position for reassembly.

3 Support the door and remove the six setscrews securing the hinges to the door. Withdraw the door from the hinges.

4 Mark each hinge position on the door pillar. Remove the eight securing setscrews and remove the two hinges. Note the reinforcing plate on the top hinge.

5 Refitting is the reverse of the removal procedure. Check, and if necessary adjust, the door lock and striker plate.

15 Front side door trim pad – removal and refitting

1 Using a flat bladed tool, carefully prise off the two end covers from the door pull handle, and remove the screws from the handle (photos).

2 Slide the upper and lower bezels from the remote control door handle (photo)

3 Using a wide bladed screwdriver, carefully prise the trim pad away from the door frame at the retaining clip points, being careful not to tear the trim base.

4 Refitting the door trim pad is a reversal of the removal procedure.

16 Front side hinged door lock – removal and refitting

1 Refer to Section 15 and remove the door trim pad.

2 Unclip the retainers and free the remote control rods from the lock (photo).

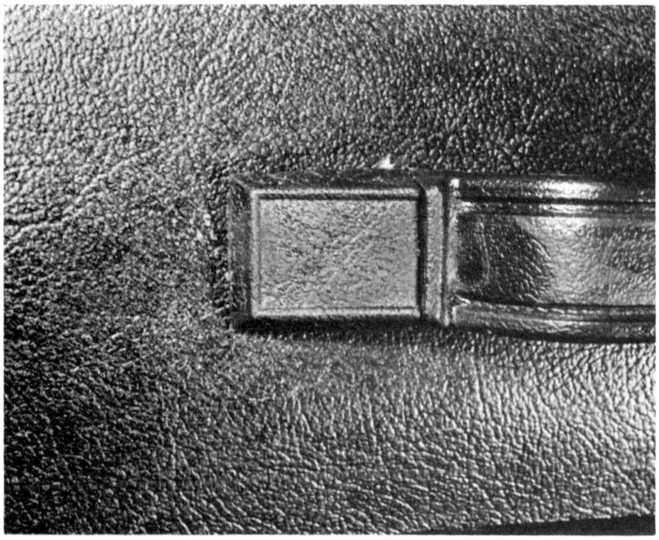

15.1a Prise the end covers ...

15.1b ... off the door pull handle and ...

15.2 ... slide the bezels off the remote control door handle

16.2 A side hinged door lock seen with the door trim pad removed

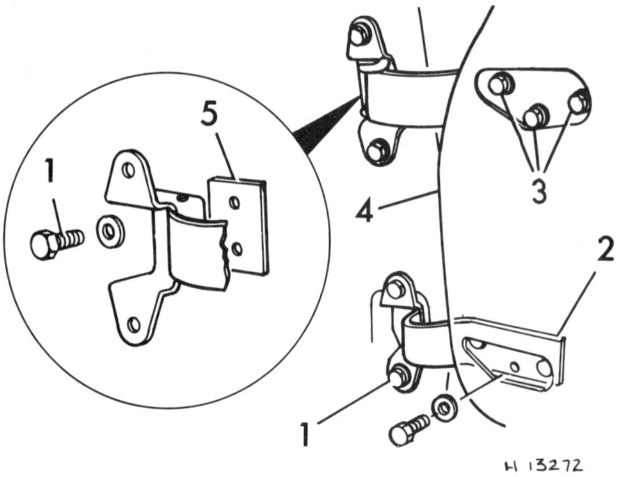

Fig. 12.9 Removing front door hinges (Sec 14)

1 *Setscrews in door pillar*
2 *Hinge in door*
3 *Setscrews in door*
4 *Door*
5 *Reinforcement plate*

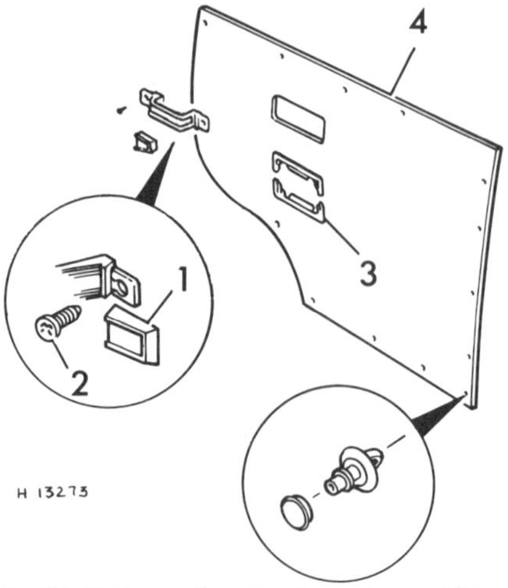

Fig. 12.10 Removing the door trim pad (Sec 15)

1 *Pull handle end cover*
2 *Attachment screw*
3 *Remote control handle bezels*
4 *Trim pad*

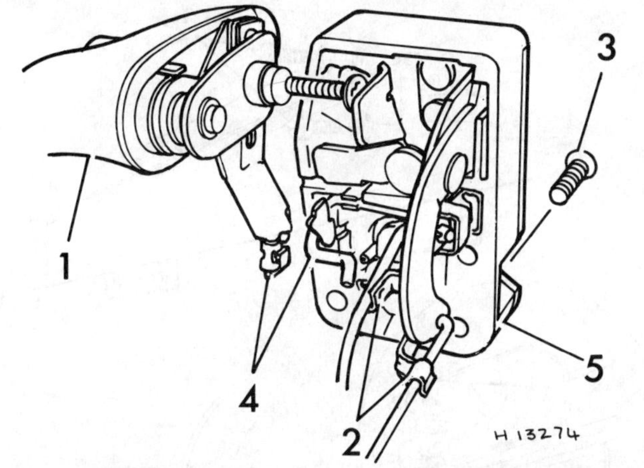

Fig. 12.11 Hinged front door lock removal (Sec 16)

1 Outside handle
2 Remote control rod retainers
3 Securing screws
4 Outside handle locking lever and locking bar cross-shaft
5 Door lock assembly

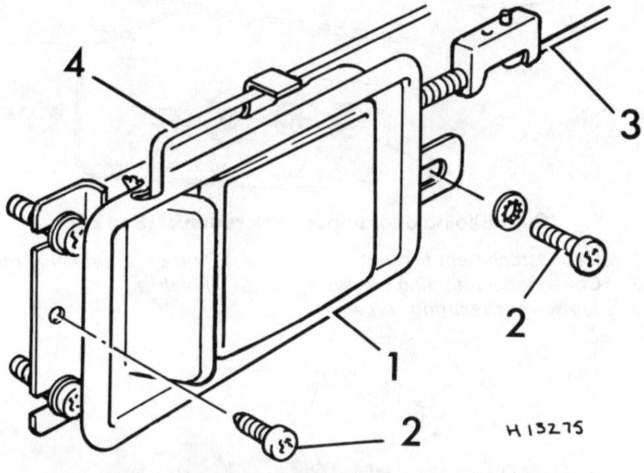

Fig. 12.12 Door lock remote control removal (Sec 17)

1 Remote control assembly
2 Securing screws
3 Latch rod
4 Lock control rod

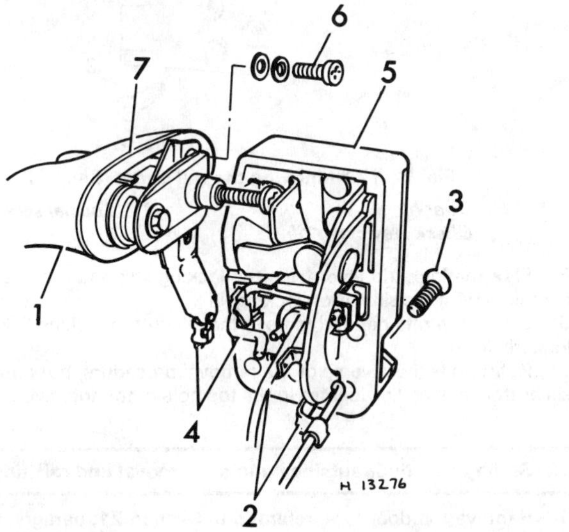

Fig. 12.13 Hinged door outside handle removal (Sec 18)

1 Outside handle
2 Remote control rod retainers
3 Securing screw
4 Outside handle locking lever and locking bar cross-shaft
5 Lock assembly
6 Handle securing screw
7 Gasket

3 Unscrew and remove the four screws which secure the lock to the door.
4 Remove the outside handle locking lever from the locking bar cross-shaft on the door lock and remove the lock.
5 Refitting is the reverse of the removal procedure

17 Front side hinged door lock remote control – removal and refitting

1 Refer to Section 15 and remove the door trim pad.
2 Remove the four screws securing the remote control to the door (photo).
3 Detach the latch rod from the plastic clip and free the lock control rod from the safety locking catch. Remove the remote control from the door.
4 Refitting is the reverse of the removal procedure. Check and adjust the fitted position of the remote control unit and connecting rods as required.

18 Front side hinged door outside handle – removal and refitting

1 Refer to Section 15 and remove the door trim pad.
2 Unclip the retainers and release the remote control rods from the door lock.
3 Remove the four screws which secure the door lock to the door.
4 Release the outside handle locking lever from the locking bar cross-shaft on the door lock and remove the door lock.
5 From inside the door remove the two screws securing the outside door handle to the door. The outside handle can now be removed from the door. Remove the gaskets.
6 Refitting is the reverse of the removal procedure.

19 Sliding side door assembly – removal and refitting

1 Sliding doors are each suspended on two roller assemblies attached to the top flange of the door. They operate on a rail which is attached to the bodywork.
2 The lower edge of each door is retained by a pair of spring roller assemblies which locate with a guide rail.
3 To remove a sliding door, support it and remove the small locating screw in each upper attachment bracket.
4 Undo and remove the four bolts, spring and plain washers which retain the door to the upper attachment brackets.
5 Lower the door to free the roller support brackets at the bottom edge of the door from the lower guide and remove the door from the vehicle.
6 Refitting is the reverse of the removal procedure, but if a different door from the original is being fitted, adjustments will probably be required.
7 After fitting and tightening the four attachment bracket screws in the top edge of the door, close the door and check that there is an even distance between the body and the leading edge of the door. If necessary, re-position the door on the upper attachment brackets.
8 Open and close the door and check that the door locks in the fully open and closed positions. If necessary, adjust the position of the strikers. Also, while opening and closing the door, check that the rollers remain in contact with the lower guide. If necessary, adjust the position of the roller support brackets.
9 After all adjustments have been made, where necessary, recheck the operation of the door and its lock. Drill a new location hole, if required, through each upper attachment bracket, the same size as the original hole, and fit the locating screws.

20 Sliding side door upper track – removal and refitting

1 To remove a sliding door upper track that is excessively worn, damaged or in need of attention, first remove the door, referring to Section 19 for details of procedure.
2 Remove the five screws and washers retaining the cover panel over the door upper track.
3 Remove the four screws, spring and plain washers securing the

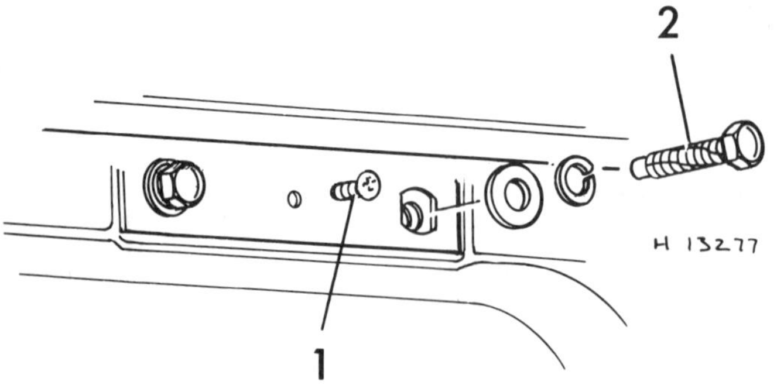

Fig. 12.14 Removing a sliding door (Sec 19)

1 Locating screw 2 Door retaining screw

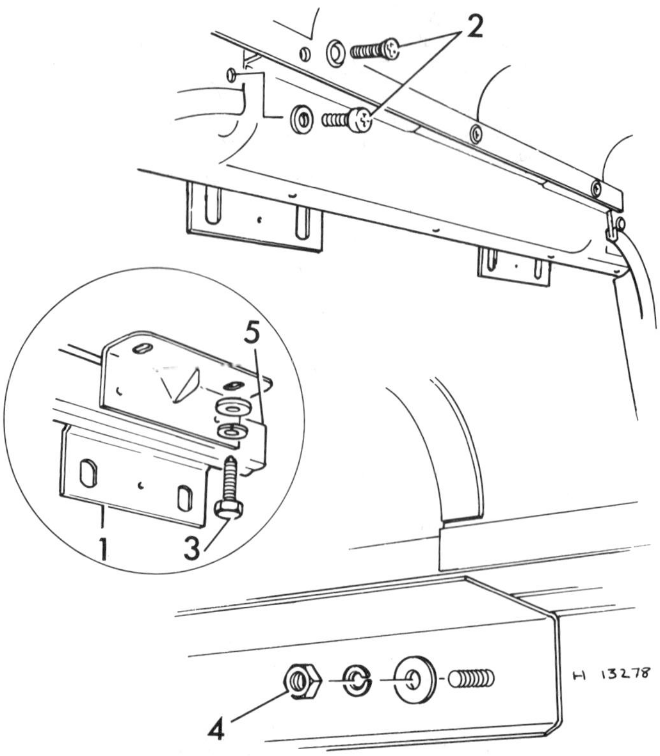

Fig. 12.15 Sliding door upper track removal (Sec 20)

1 Door attachment bracket
2 Cover panel retaining screws
3 Track front securing screw
4 Track rear retaining nut
5 Track

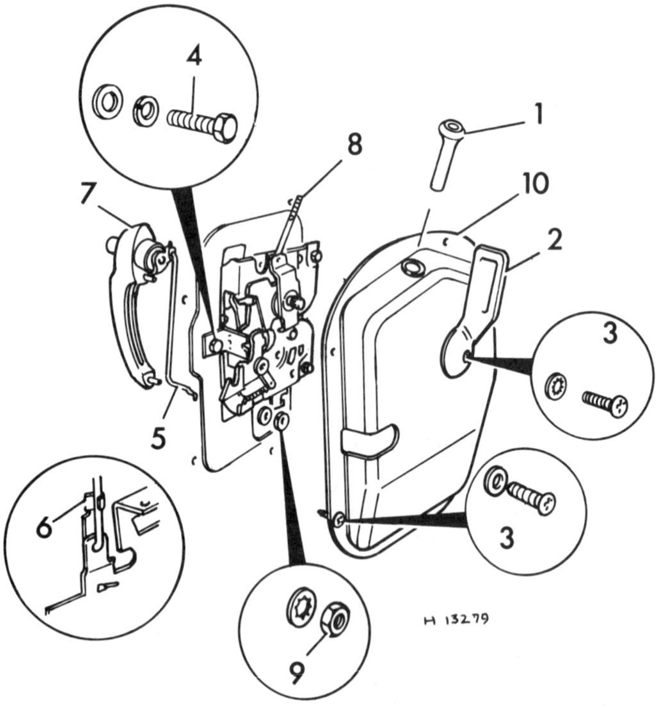

Fig. 12.16 Sliding door lock and outside handle removal (Secs 21 and 22)

1 Locking button
2 Door inner handle
3 Lock cover securing screw
4 Lock securing bolts
5 Locking link
6 Sill button locking link retainer
7 Outside handle
8 Sill button locking link
9 Outside handle retaining nut
10 Lock cover

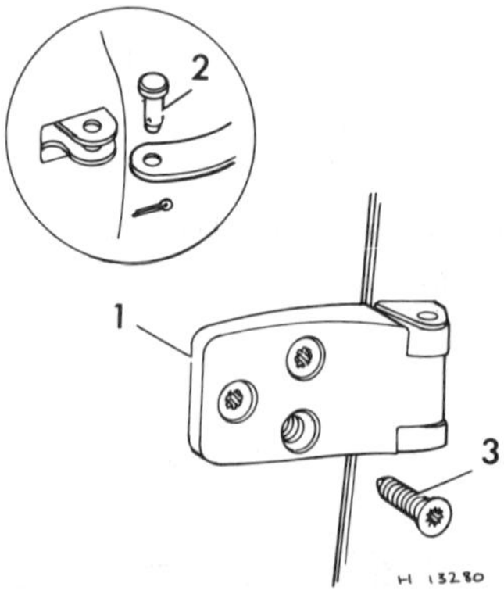

Fig. 12.17 Removing the tail doors (Sec 23)

1 Door hinge
2 Check stay clevis pin
3 Special screw

front half of the track to the roof.

4 Remove the two nuts, spring and plain washers retaining the rear half of the track to the body side and remove the track from the vehicle.

5 Before refitting the upper track, lubricate all moving parts with Duckhams Admax C32 grease or a suitable equivalent. Refitting is the reverse of the removal procedure.

21 Sliding side door lock – removal and refitting

1 Unscrew the sill locking button.

2 Remove the door lock inner handle securing screw and lockwasher and remove the handle.

3 Unscrew and remove the four screws and washers which secure the lock cover to the door panel and remove the cover.

4 Remove the three bolts, spring and plain washers which secure the lock to the door.

5 Free the lock trip lever from the locking link and remove the lock, complete with the sill button locking link.

6 Unclip the retainer and remove the sill button locking link from the lock trip lever.

7 Refitting is the reverse of the removal procedure, but note that the sill button locking link locates in the top hole of the trip lever.

22 Sliding side door outside handle – removal and refitting

1 Remove the door lock, referring to Section 21, paragraphs 1 to 5 inclusive.

2 Remove the two nuts and shakeproof washers which secure the outside handle to the door.

3 Remove the outside handle complete with locking link and seating washers.

4 Refitting is the reverse of the removal procedure, but also refer to paragraph 7 of Section 21.

17.2 The four screws securing the remote control door handle

25.2 The tail right-hand door bottom slam

25.4 ... the centre lock release mechanism

23 Tail doors – removal and refitting

1 Support the door and remove the split-pin and clevis pin retaining the door check stay to the body.
2 Borrow, hire or otherwise obtain a TORX screw wrench which is required to undo the door hinge screws. Remove the three screws which secure each hinge to the door and remove the door assembly.
3 Refitting is the reverse of the removal prodedure, but do not fully tighten the hinge screws initially.
4 Close the door and ensure than an even distance exists between the outer edge of the door and the body. Adjust by moving the door on its hinges.
5 *Left-hand door:* Make sure that the locking bars engage correctly in the striker plates when the door handle is operated.
6 *Right-hand door:* Make sure that the slam locks engage correctly in the striker plates when the door is shut.
7 When all is satisfactory, tighten the door hinge screws using the TORX screw wrench.

24 Tail door hinges – removal and refitting

1 Refer to Section 23, paragraphs 1 and 2, and remove the door.
2 Remove the two nuts and washers which secure each hinge to the body and remove the hinges.
3 Refitting is the reverse of the removal procedure, but do not fully tighten the retaining nuts and screws until the hinge positions on the door and body have been adjusted to let the door shut correctly.

25 Tail door lock, right-hand – removal and refitting

1 Remove the screw and washer which secure the door outside handle to the lock mechanism and remove the outside handle from the door.
2 Remove the four screws and shakeproof washers which secure one of the door slam locks to the door and rotate the lock through 90° to release it from the connecting link (photo).
3 Remove the other slam lock in a similar fashion.
4 Remove the four screws and shakeproof washers to release the centre lock release mechanism from the door (photo).
5 Refitting is the reverse of the removal procedure. Lightly lubricate the moving parts after assembly.

26 Tail door lock, left-hand – removal and refitting

1 Using a small parallel pin punch or a similar tool, tap out the pin which secures the door inside handle to the lock mechanism.
2 Remove the inside handle and the insert from the lock mechanism.
3 Remove the four screws and lockwashers which secure the lock mechanism to the door.

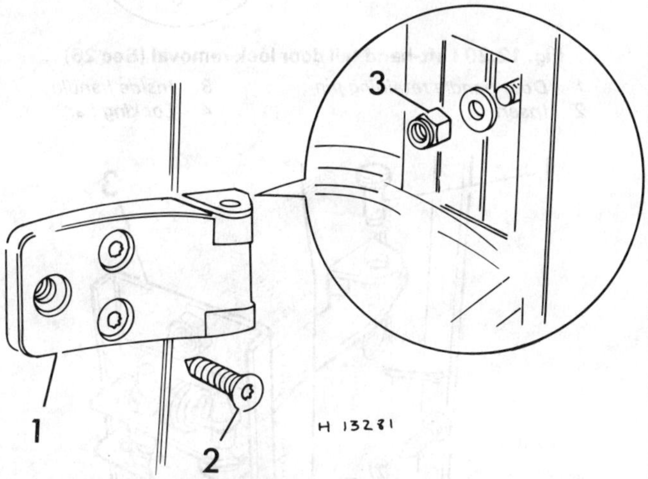

Fig. 12.18 Tail door hinge removal (Sec 24)

1 Door hinge
2 Special screw
3 Hinge retaining nut

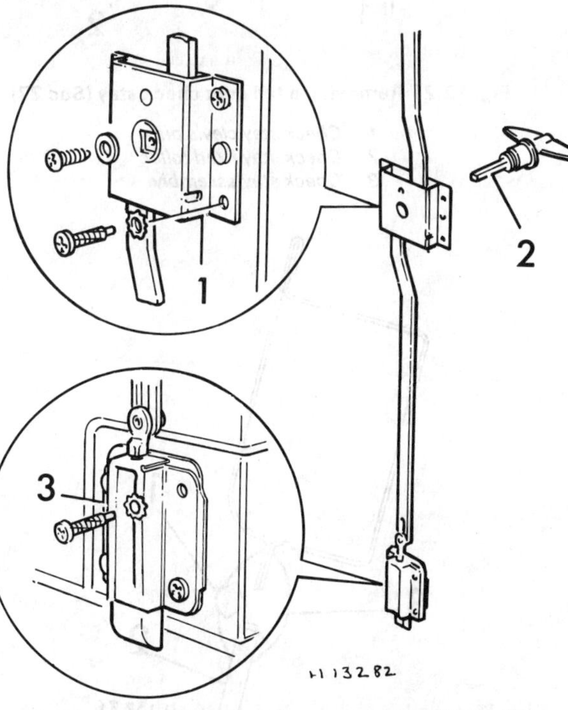

Fig. 12.19 Right-hand tail door lock removal (Sec 25)

1 Door lock centre mechanism
2 Outside handle
3 Bottom slam lock

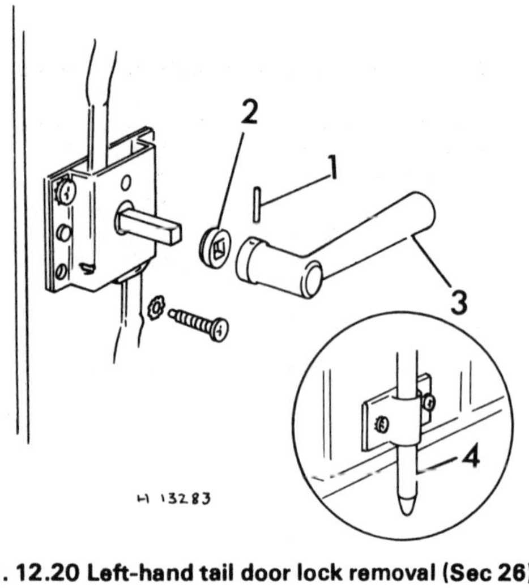

Fig. 12.20 Left-hand tail door lock removal (Sec 26)

1 Door handle retaining pin
2 Insert
3 Inside handle
4 Locking bar

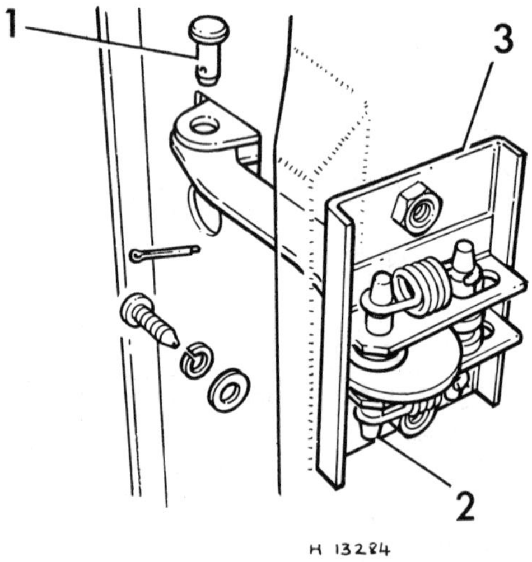

Fig. 12.21 Removing a tail door check stay (Sec 27)

1 Check stay clevis pin
2 Check stay fixed roller
3 Check stay assembly

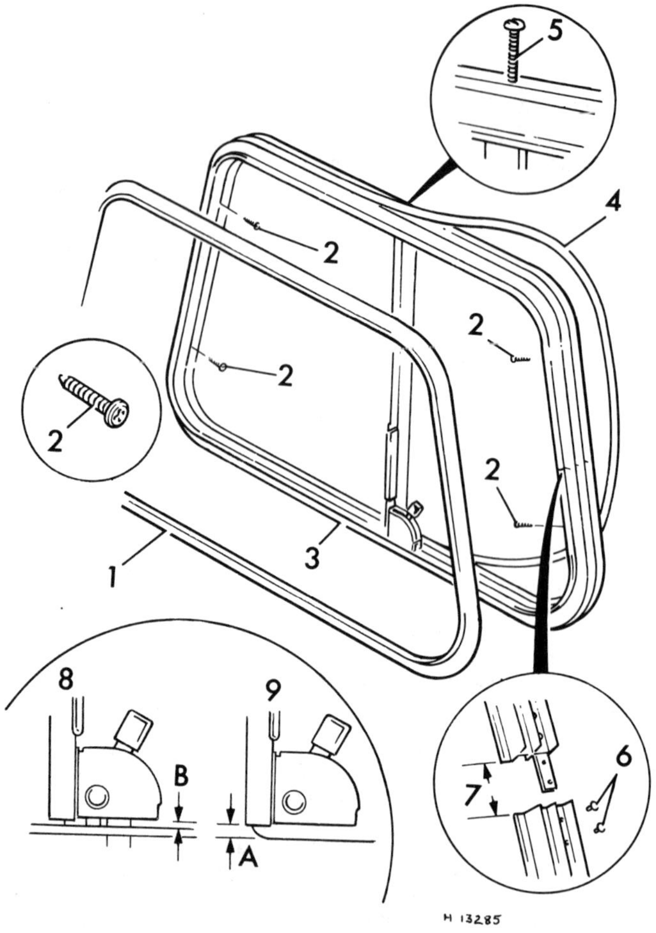

Fig. 12.22 Side door glass removal (Sec 28)

1 Finisher
2 Frame retaining screws
3 Window assembly
4 Outer seal
5 Vertical channel retaining screw
6 Frame joint rivets
7 Spread frame to remove glass
8 Finger-pull catch engaged, B = 0.15 in (3.81 mm) minimum
9 Assemble finger-pull catch to glass so that A = 0.35 in (9 mm)

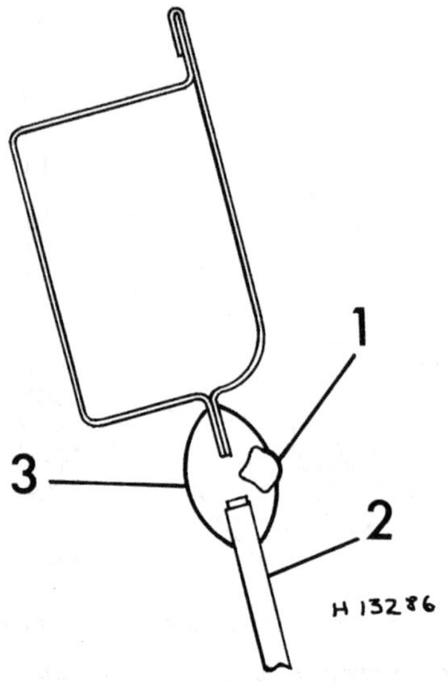

Fig. 12.23 Tail door glass surround detail (Sec 29)

1 Filler strip
2 Glass
3 Rubber moulding

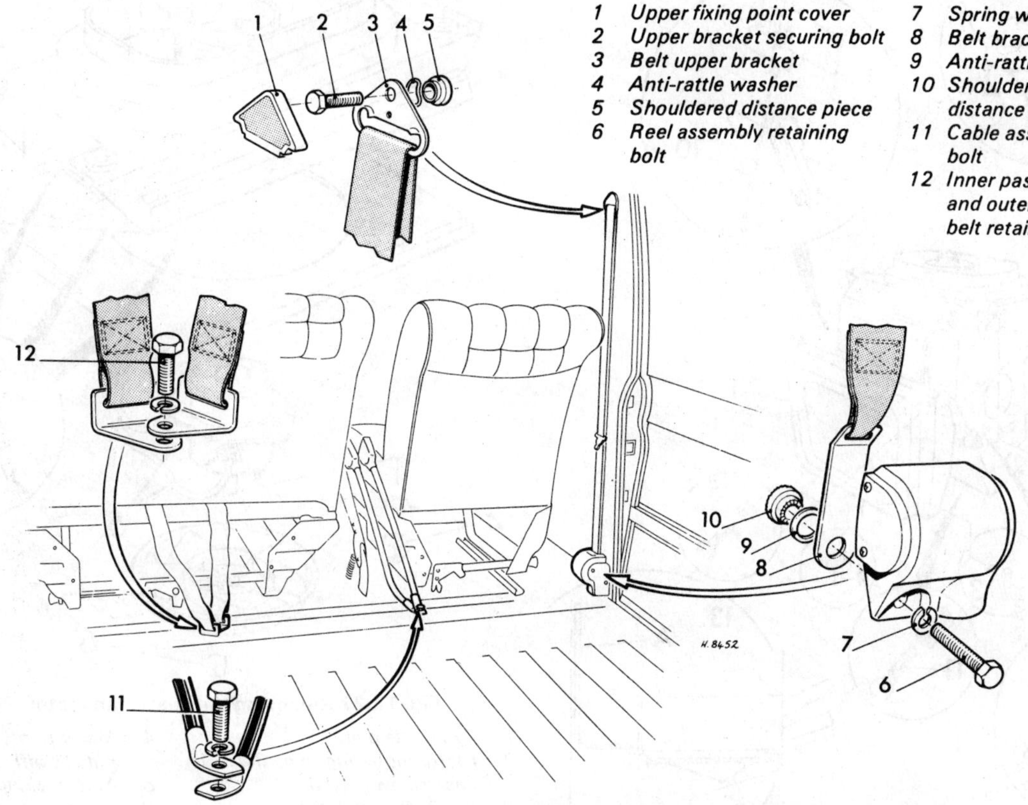

Fig. 12.24 Kangol Automatic front seat belt installation (Sec 30)

1 Upper fixing point cover
2 Upper bracket securing bolt
3 Belt upper bracket
4 Anti-rattle washer
5 Shouldered distance piece
6 Reel assembly retaining bolt
7 Spring washer
8 Belt bracket
9 Anti-rattle washer
10 Shouldered, serrated distance piece
11 Cable assembly retaining bolt
12 Inner passenger long belt and outer passenger short belt retaining bolt

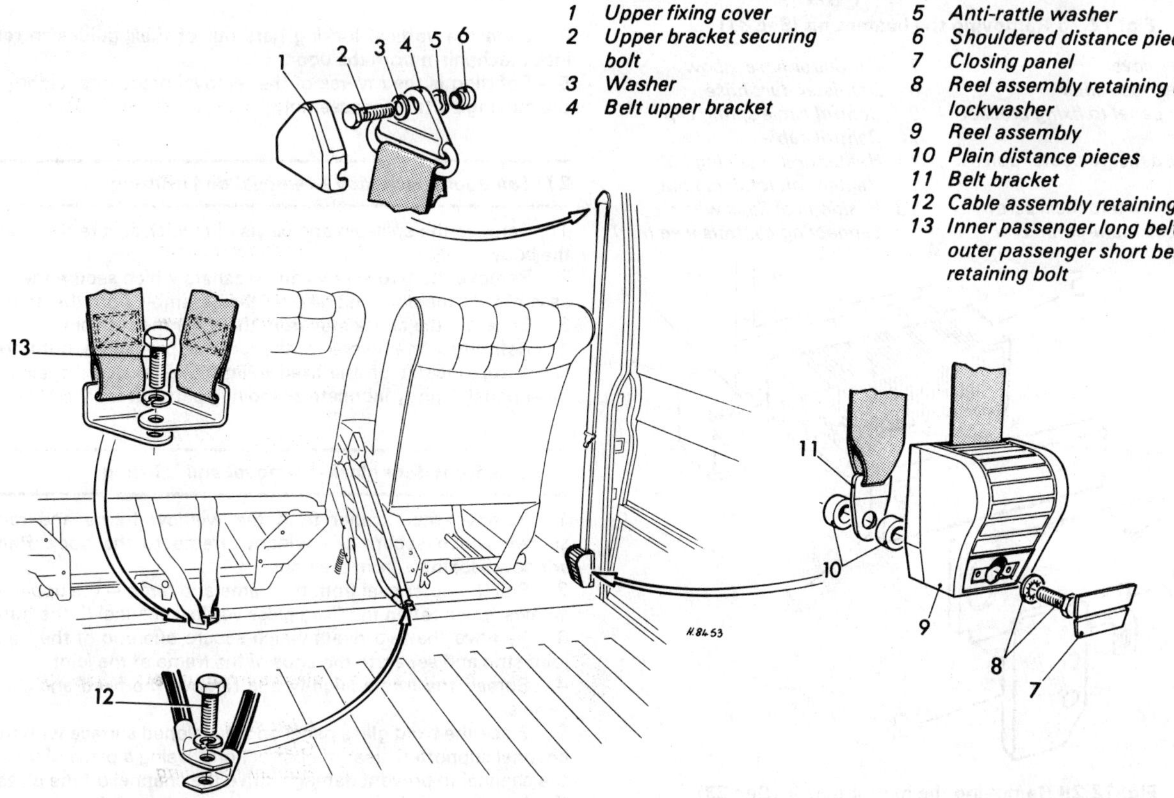

Fig. 12.25 Kangol Euroflex front seat belt installation (Sec 30)

1 Upper fixing cover
2 Upper bracket securing bolt
3 Washer
4 Belt upper bracket
5 Anti-rattle washer
6 Shouldered distance piece
7 Closing panel
8 Reel assembly retaining bolt lockwasher
9 Reel assembly
10 Plain distance pieces
11 Belt bracket
12 Cable assembly retaining bolt
13 Inner passenger long belt and outer passenger short belt retaining bolt

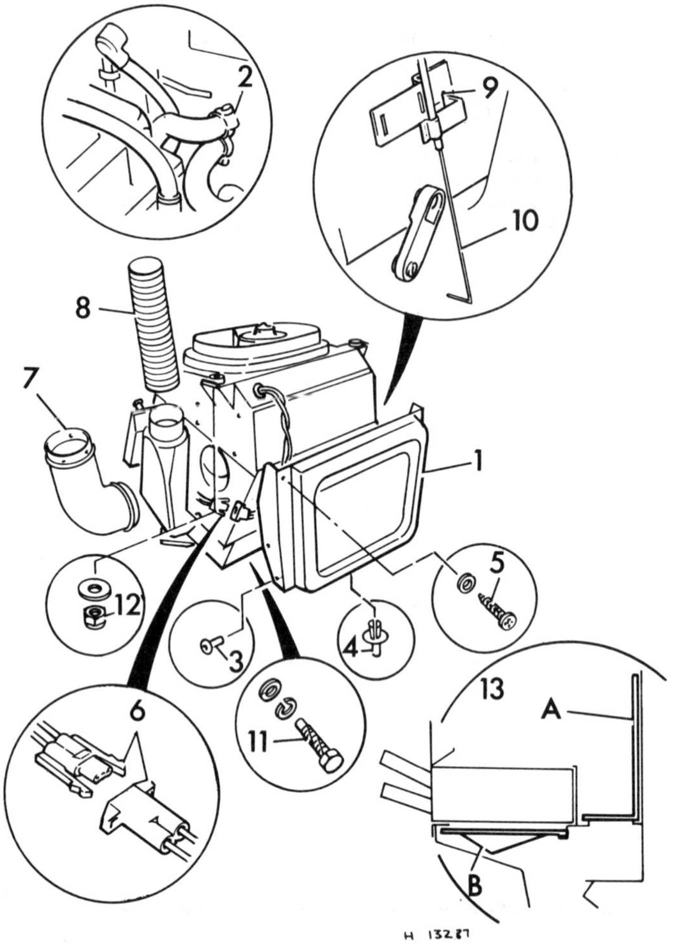

Fig. 12.26 Removing the heater unit (Sec 31)

1 *Heater cover*
2 *Heater hose clips*
3 *Cover panel to fixing bracket rivet*
4 *Cover panel to heater unit rivet*
5 *Cover panel to facia screw*
6 *Harness multi-connector*
7 *Ventilator hose elbow*
8 *Demister duct hose*
9 *Control cable spring clip*
10 *Control cable*
11 *Heater unit retaining bolt*
12 *Heater unit retaining nut*
13 *Positions of flaps when connecting controls (see text)*

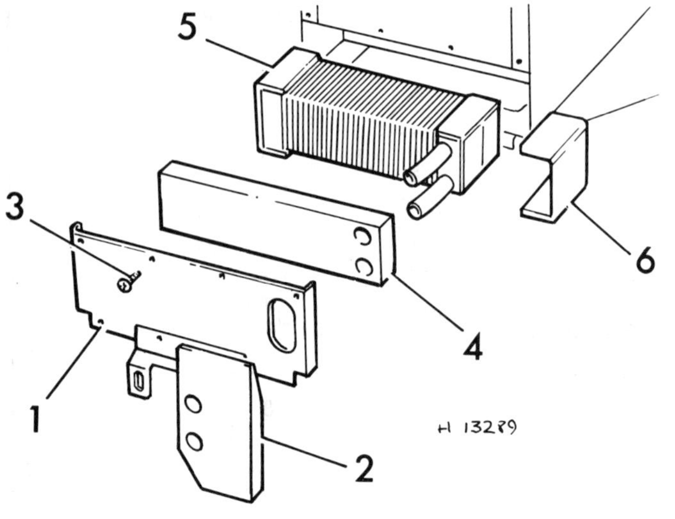

Fig. 12.28 Removing the heater matrix (Sec 33)

1 *Matrix cover-plate*
2 *Sealing block*
3 *Cover-plate securing screws*
4 *Foam packing*
5 *Matrix*
6 *Foam end packings*

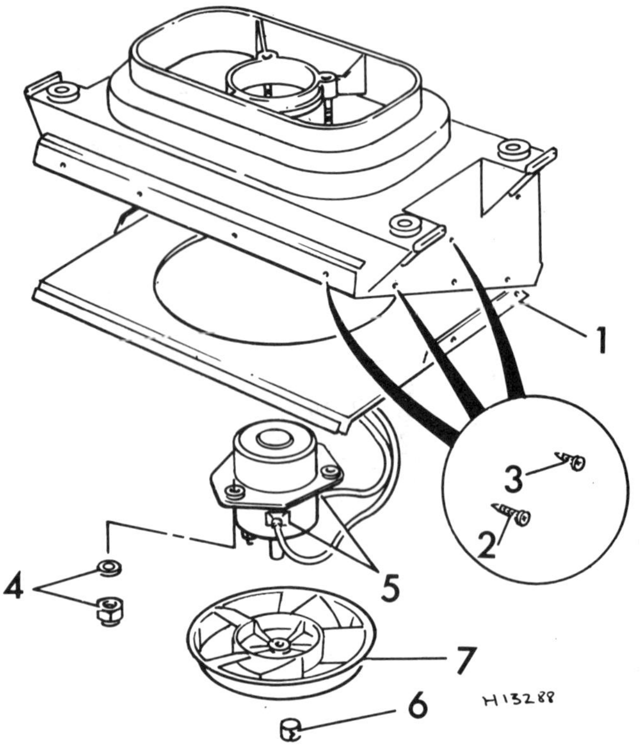

Fig. 12.27 Removing the heater fan motor (Sec 32)

1 *Fan baffle plate*
2 *Motor mounting to heater case screws (10 off)*
3 *Fan baffle plate to motor mounting moulding screws (4 off)*
4 *Motor to moulding securing nuts (3 off)*
5 *Motor wiring harness*
6 *Fan retaining clip*
7 *Fan*

4 Slide the vertical locking bars out of their guides ro release the lock mechanism from the door.
5 Refitting is the reverse of the removal procedure. Lightly lubricate the moving parts after assembly.

27 Tail door check stay – removal and refitting

1 Remove the split-pin and clevis pin which secure the check stay to the body.
2 Remove the two screws and washers which secure the check stay assembly to the door and remove the assembly from the door.
3 Separate the check stay from the roller mechanism.
4 Refitting is the reverse of the removal procedure, but note that the check stay locates on the fixed roller which must be against the door inner panel. Lightly lubricate all moving parts after assembly.

28 Side front door glass – removal and refitting

1 Remove the finisher from the window frame and remove the screws which secure the window frame to the door. Remove the window assembly from the door.
2 Pull the outer seal from the frame and remove the upper and lower screws which retain the fixed glass vertical channel to the frame.
3 Remove the two rivets which secure one end of the frame to the joint strip and separate the ends of the frame at the joint.
4 Spread the frame slightly and remove the fixed and sliding glass panels.
5 Place the fixed glass panel on a cushioned surface with the vertical channel supported clear of obstructions. Using a piece of wood against the channel to prevent damage, drive the channel off the glass panel.
6 In a similar fashion remove the catch and finger-pull assembly from the sliding glass panel.
7 Refitting is the reverse of the removal procedure, but note the following points:

31.5 The Sherpa heater unit

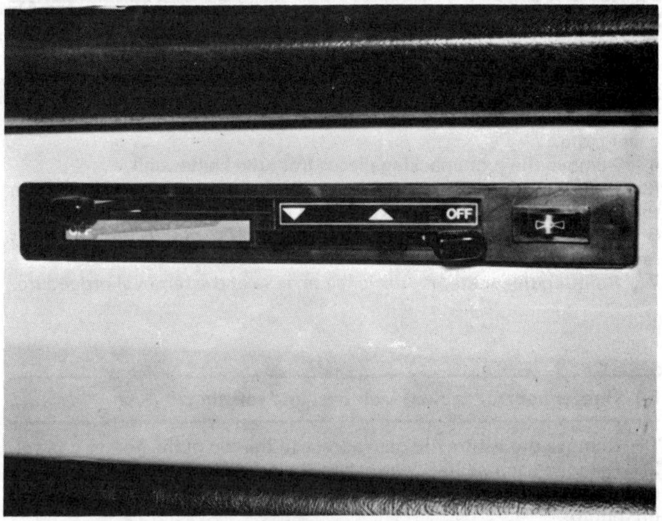

34.2 The heater controls

(a) *Insert the glazing rubber partially into the vertical channel and position the glass centrally on the channel before driving the channel onto the glass in the same way that it was removed*
(b) *When fitting the catch and finger-pull assembly to the sliding glass, ensure that dimension A, Fig. 12.22, is 0.354 in (9 mm) and check that, with the catch engaged, dimension B is not less than 0.15 in (3.81 mm)*

29 Tail door glass – removal and refitting

1 The procedure for removing and refitting a glass panel in the tail doors is very similar to that used when removing or refitting a windscreen. Refer to Section 8 for details.

30 Front seat belts – removal and refitting

Warning: *When seat belts have been in use in a vehicle which has been involved in an accident resulting in a severe impact, the complete belt assemblies must be renewed, including the cable floor anchorage assemblies.*
1 Two types of belt installations are shown in the illustrations; Fig. 12.24 shows the Kangol Automatic type belt, and Fig. 12.25 shows the Kangol Euroflex type of seat belt.
2 Removal of either type of belt is self-explanatory with the aid of the illustrations.
3 When refitting the belts the following points should be noted:

(a) *All distance pieces must make firm contact with the body fixing points*
(b) *The belt brackets, reel assemblies and cable assemblies must be fitted with their components in the exact order shown in the illustrations*
(c) *The top face of each reel housing must be horizontal in both planes*
(d) *Tighten the belt fixing bolts to a torque of 17 lbf ft (23 Nm)*

4 To test the automatic locking device on each belt, with the belts clipped on, drive the vehicle at 5 mph (8 km/h) on a quiet stretch of road, with no other traffic, and brake sharply. The locking device should operate and lock each reel. Make a number of tests and if any doubt exists about the performance of the automatic locking devices refer the matter to a main dealer.

31 Heater unit – removal and refitting

1 Drain the cooling system, if necessary referring to Chapter 2. Remove the battery earth lead.
2 Slacken the two heater hose clips located at the front of the bulkhead and remove the two hoses from the heater matrix pipes.
3 Where fitted, remove the heater cover panel as follows. Remove the two pop rivets which secure the panel to the fixing bracket; press out the centre piece from the rivet which secures the cover to the heater and remove the rivet; unscrew the two screws which secure the cover to the facia and remove the cover.
4 Disconnect the wiring to the heater fan motor at the harness connector.
5 Pull off the face-level ventilator hose elbows from the heater unit. Detach the hoses from the heater unit and the demister ducts (photo).
6 Press the retaining end of each outer cable spring clip away from the cable to release the clips from the bracket on the heater unit.
7 Detach the inner cables from the cleats in the levers on the heater unit.
8 Remove the two bolts and washers which secure the heater unit to the bulkhead and remove the four nuts which secure the heater unit to the heater mounting panel.
9 Remove the heater unit from the vehicle.
10 Replacement is the reverse of the removal procedure, but note the following additional points:

(a) *Fit the temperature control outer cable spring clip to the heater unit after putting the temperature control lever in the OFF position and the temperature control flap in the heater unit in the vertical position as shown at A in Fig. 12.26*
(b) *Fit the air distribution control outer cable spring clip to the heater unit after putting the air distribution control lever in the OFF position and the air distribution flap in the heater unit in the horizontal position as shown at B in Fig. 12.26*

32 Heater fan motor – removal and refitting

1 Refer to Section 31 and remove the heater unit from the vehicle.
2 Unscrew and remove the ten screws retaining the motor mounting assembly to the heater unit case assembly and separate the two assemblies.
3 Unscrew the four screws which secure the fan baffle plate from the motor mounting moulding.
4 Unscrew and remove the three nuts and washers securing the motor to the moulding and release the motor.
5 Take a note of the wiring harness colour coding for reassembly, disconnect the harness from the fan motor and remove the motor complete with fan.
6 Prise the fan retaining clip from the fan hub, hold the fan, and gently drive the motor spindle out of the fan to separate the fan and motor.
7 Refitting is the reverse of the removal procedure.

33 Heater matrix – removal and refitting

1 Refer to Section 31 and remove the heater unit.
2 Remove the sealing block from the matrix pipes.
3 Unscrew and remove the eight screws which secure the matrix cover plate.
4 Remove the foam packing pieces from the heater unit.
5 Withdraw the matrix complete with the foam end packings from the heater unit and detach the packings from the matrix.
6 If the matrix is leaking or blocked refer to Chapter 2, Section 6, for remedial action.
7 Refitting the heater matrix is the reverse of the removal procedure.

34 Heater control cables – removal and refitting

1 Remove the ashtray to gain access to the top of the heater control.
2 Remove each heater control knob by pressing the retaining tongue in the hole in the underside of each knob upwards and pulling the knob off (photo).
3 Gently prise the control bezel, complete with heater fan switch, from the facia panel.
4 Undo and remove the two screws which secure the control assembly to the facia.
5 On the heater unit, press the retaining end of each remote control outer cable spring clip away from the cable to release the clips from the bracket on the heater unit.
6 Detach the control inner cables from the cleats in the levers on the heater unit.
7 Withdraw the control assembly complete with cables through the aperture in the facia.
8 Disconnect the cables from the control lever assembly in the same way as from the heater unit, paragraphs 5 and 6.
9 Refitting is the reverse of the removal sequence, but the following additional points should be noted:

(a) Lubricate the moving parts of controls with Duckhams Keenomax B1 or a suitable equivalent grease
(b) When fitting the control quadrant in the facia, note that the top is marked with a T
(c) Fit the control outer cables to the quadrant clips so that the outer cable protrudes no more than 0.06 in (1.524 mm) measured as shown in Fig. 12.29, item 10

35 Heater fan switch – removal and refitting

1 Disconnect the battery earth lead.
2 Remove the heater control bezel, complete with switch, as described in Section 34, paragraphs 2 and 3.
3 Disconnect the wiring harness from the switch by pulling off the cable tags.
4 Remove the switch from the control bezel by compressing the two side retaining clips inwards and pushing the switch out.
5 Refitting is the reverse of the removal sequence.

36 Demister duct – removal and refitting

1 Detach the hose from the demister duct and the outlet elbow on the heater.
2 The duct is secured to the facia by two screws which are difficult to get at because of the windscreen. The inner screw, being the more difficult, requires the use of a right-angled screwdriver with a cross-point head. If such a tool is available it can be used to remove both screws. The outer screw can be removed using a pair of pliers to turn it from under the facia.
3 After removing the screws, withdraw the demister duct from beneath the facia and recover the duct bezel from the top of the facia.
4 Refitting is the reverse of the removal sequence.

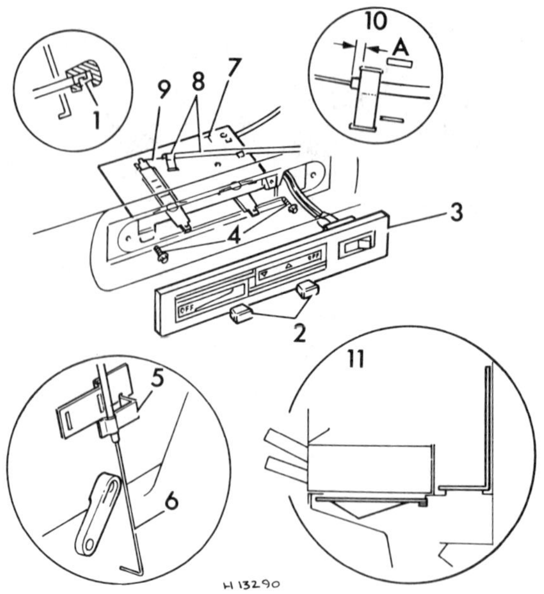

Fig. 12.29 Removing the heater control cables (Sec 34)

1 Control knob retaining tongue
2 Control knobs
3 Control bezel
4 Control retaining screws
5 Cable spring clips
6 Inner cable
7 Control quadrant
8 Cable and spring clip
9 Inner cable
10 Outer cable to control quadrant assembling dimension A = 0.06 in (1.524 mm) maximum
11 Positions of flaps when connecting controls

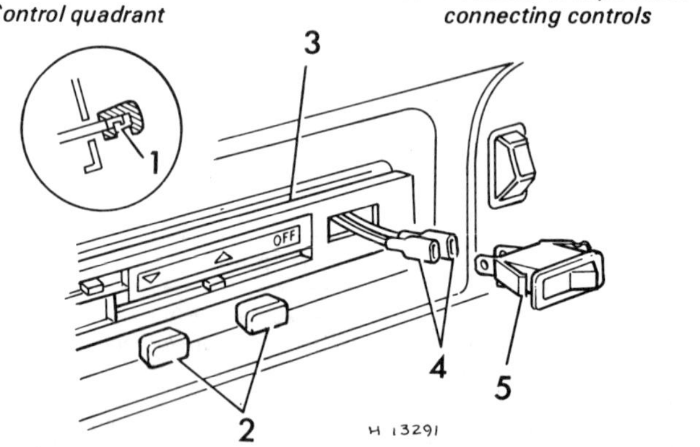

Fig. 12.30 Heater fan switch removal (Sec 35)

1 Control knob retaining tongue
2 Control knobs
3 Control bezel
4 Wiring harness connectors
5 Switch retaining clip

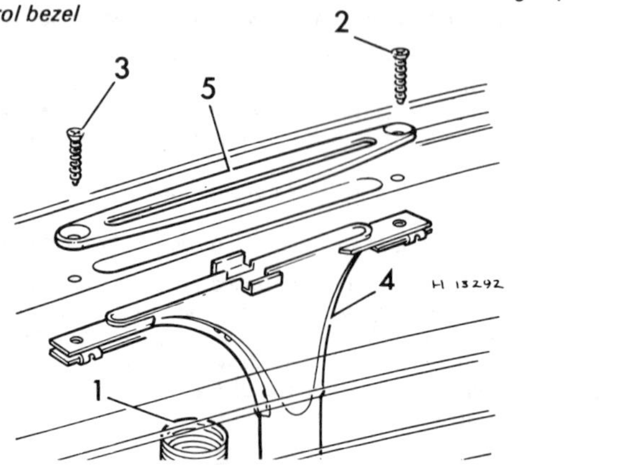

Fig. 12.31 Removing a demister duct (Sec 36)

1 Demister duct hose
2 Inner securing screw
3 Outer securing screw
4 Demister duct
5 Demister duct bezel

Chapter 13 Supplement: Revisions and information on later models

Contents

1 Introduction

This Supplement covers the modifications and additions made to the Sherpa range since 1980. The Sections of the Supplement are arranged in the same order as the Chapters in the main part of the manual. The Specifications are all grouped together for ease of reference, but they too follow the order of the main Chapters.

To use the Supplement to its best advantage, it should be consulted before commencing any particular task on the vehicle. If no information to the contrary is given, the appropriate procedure in the main Chapters will apply.

Project vehicle

The vehicle seen in most of the photographs in this Supplement is a 1987 model 255 chassis cab, loaned to us by Freight Rover.

2 Specifications

The Specifications below are supplementary to, or revisions of, those at the beginning of the preceding Chapters.

Engine – 20V (1980 on)

General

Stroke	3.504 in (89.0 mm)
Capacity	121.68 cu in (1994 cc)

Valve timing

Inlet opens	19° BTDC
Inlet closes	41° ABDC
Exhaust opens	61° BBDC
Exhaust closes	15° ATDC

Tappet clearances

Nominal	0.012 in ± 0.001 in (0.30 ± 0.03 mm)
Adjust only if less than	0.008 in (0.20 mm) inlet or exhaust

Engine – 20V (1980 to 1982)

Pistons

Clearance in bore:	
Below oil control groove	0.0008 to 0.0024 in (0.02 to 0.06 mm)
At bottom of skirt	0.0004 to 0.0012 in (0.01 to 0.03 mm)

Engine – 17V and 20V (1982 to 1986)

Pistons

Clearance in bore:	
Below oil control groove – 17V	0.0051 to 0.0063 in (0.13 to 0.16 mm)
Below oil control groove – 20V	0.0055 to 0.0071 in (0.14 to 0.18 mm)
At bottom of skirt (17V and 20V)	0.0008 to 0.0020 in (0.02 to 0.05 mm)

Piston rings

Ring-to-groove clearance:	
Top	0.0020 to 0.0031 in (0.05 to 0.08 mm)
Second	0.0016 to 0.0028 in (0.04 to 0.07 mm)
Fitted gap (oil control ring)	0.015 to 0.023 in (0.38 to 0.58 mm)

Tappet clearances

Nominal	0.12 in ± 0.001 in (0.30 ± 0.03 mm)
Adjust only if less than	0.008 in (0.20 mm) inlet or exhaust

Engine – 17V and 20V, 02 Series (1987 model year)

General

Compression ratio:	
Standard (low compression)	8.0 : 1
Optional (high compression)	9.0 : 1
Maximum power:	
17V (LC)	70 bhp (71 PS) @ 4800 rpm
20V (LC)	90 bhp (91 PS) @ 4800 rpm
20V (HC)	98 bhp (99 PS) @ 5450 rpm
Maximum torque:	
17V (LC)	91 lbf ft (123 Nm) @ 2850 rpm
20V (LC)	112 lbf ft (152 Nm) @ 2700 rpm
20V (HC)	117 lbf ft (158 Nm) @ 3400 rpm

Main bearings

Running clearance	0.0010 to 0.0027 in (0.025 to 0.069 mm)

Big-end bearings

Running clearance	0.0010 to 0.0027 in (0.025 to 0.069 mm)

Pistons

Clearance in bore:	
Below oil control groove	0.0008 to 0.0024 in (0.02 to 0.06 mm)
At bottom of skirt	0.0004 to 0.0015 in (0.010 to 0.038 mm)

Piston rings

Compression:	
Ring-to-groove clearance	0.0016 to 0.0028 in (0.04 to 007 mm)
Fitted gap	0.012 to 0.020 in (0.3 to 0.5 mm)
Oil control:	
Fitted gap	0.013 to 0.055 in (0.33 to 1.4 mm)

Lubrication system

Oil pressure:	
Idling	10 lbf/in² (0.7 bar) minimum
Above 2500 rpm	55 to 80 lbf/in² (3.8 to 5.5 bar)
Oil pump clearances (see Chapter 1, Section 101):	
Outer rotor to housing (A)	0.004 in (0.10 mm) maximum
Inner to outer rotor (B and C)	0.004 in (0.10 mm) maximum
Outer rotor endfloat (D)	0.003 in (0.08 mm) maximum
Oil capacity	7 pints (4 litres) approx

Torque wrench settings

	lbf ft	Nm
Camshaft cover bolts (oiled threads)	12 to 15	16 to 20
Camshaft sprocket bolt	45 to 50	61 to 68
Cylinder head bolts (see text):		
Stage 1	35	47
Stage 2	60	87
Stage 3	Tighten through a further 90°	Tighten through a further 90°
Main bearing cap bolts (oiled threads)	70 to 85	95 to 115
Big-end cap nuts (oiled threads)	38 to 43	52 to 58
Crankshaft pulley bolt	60 to 65	81 to 88
Flywheel bolts*	59 to 66	80 to 90
Sump drain plug	30 to 33	40 to 45
Knock sensor	7 to 11	10 to 15
Oil filter	6 to 9	8 to 12
Thermistor (temperature sensor)	7 to 11	10 to 15
Oil pressure switch	9	12
Spark plugs	11 to 15	15 to 20
Oil pump backplate screws	3 to 5	5 to 7

**If thread locking compound was found when dismantling, renew the bolts and clean the threads in the crankshaft*

Cooling system – 02 Series

Water pump/alternator drivebelt

Type	Flat, ribbed
Deflection midway between pulleys under 10 lbf (4.5 kgf) load	0.28 to 0.35 in (7 to 9 mm)

Fuel and exhaust systems

Fuel grade

All models except 20V high compression	91 RON (UK 2-star)
20V high compression	97 RON (UK 4-star)

Carburettor (1980 to 1986)

17V engine:	
Make and type	SU HIF 6
Specification:	
Manual gearbox	FZX 1320
Automatic transmission	FZX 1321
Piston spring colour	Red
Jet and needle	0.100 in/BEU
20V engine:	
Make and type	SU HIF 44
Specification:	
Manual gearbox	FZX 1299
Automatic transmission	FZX 1300
Piston spring colour	Yellow
Jet and needle	0.100 in/BEU

Carburettor (1987 model year)

17V engine:	
Make and type	SU HIF 44
Specification	FZX 1429
Piston spring colour	Yellow
Needle	BGJ
20V engine:	
Make and type	SU HIF 44
Specification:	
Low compression	FZX 1485
High compression	FZX 1487
Piston spring colour	Yellow
Needle:	
Low compression	BGN
High compression	BGM

Idle adjustment data (1980 on)

Idle speed (hot):	
17V	650 ± 50 rpm
20V	750 ± 50 rpm
Fast idle speed (hot):	
Models up to 1986	1100 rpm
1987 models	1300 ± 50 rpm
Exhaust CO level at idle:	
17V up to 1986	2.5 ± 1%
17V, 1987 model	2.25 ± 0.75%
20V up to 1986	2.5 ± 1%
20V, 1987 model	2.25 ± 1.5%

Fuel pump (1987 model year)

Make and type	Pierburg, electric

Fuel tank (1980 on)

Capacity	50 to 105 litres (11 to 23 gallons) depending on model and option selected

Ignition system (1980 on)

General

System type:	
Early models (precise date unknown)	Contact breaker and coil (as Chapter 4)
Mid-period models (up to 1986)	Constant energy electronic (breakerless) system
1987 models	Programmed ignition with electronic control and anti-lock sensor

Spark plugs

Make and type:	
Models up to 1986	Champion S9YC or equivalent
1987 models	Champion N9YC or equivalent
Electrode gap:	
Models up to 1986	0.035 in (0.90 mm)
1987 models	0.039 in (1.00 mm)

Ignition timing*

Models up to 1986:	
17V, 91 octane fuel	13° BTDC
17V, 97 octane fuel (optional setting)	20° BTDC
20V, 91 octane fuel	8° BTDC
20V, 97 octane fuel (optional setting)	15° BTDC
1987 models	Not adjustable (see text)

**At 1500 rpm, vacuum advance pipe disconnected*

Torque wrench settings

	lbf ft	Nm
Spark plugs:		
Models up to 1986	7	10
1987 models	11 to 15	15 to 20
Knock sensor (1987 models)	7 to 11	10 to 15
Thermistor (temperature sensor) – 1987 models	7 to 11	10 to 15

Manual gearbox (5-speed)

General

Gearbox type	Constant mesh. Five forward speeds (all synchromesh) and one reverse
Maker's designation	77 mm

Ratios

1st	3.985 : 1
2nd	2.519 : 1
3rd	1.507 : 1
4th	1.000 : 1
5th	0.831 : 1
Reverse	3.701 : 1

Lubrication

Lubricant type	See *'Recommended lubricants and fluids'*
Lubricant capacity	3.15 pints (1.79 litres) approx*

**Quantity specified in operator's handbook may be less – this is incorrect*

Tolerances and adjustments

Synchro baulk ring clearance:	
Minimum value	0.025 in (0.64 mm)
Adjustment method	Renewal
1st gear bush clearance:	
Permitted value	0.0002 to 0.0022 in (0.005 to 0.056 mm)
Bush thicknesses available	1.5821 in (40.185 mm) 1.5841 in (40.236 mm) 1.5860 in (40.285 mm) 1.5880 in (40.335 mm) 1.5900 in (40.386 mm) all ± 0.0010 in (0.025 mm)
5th gear clearance on mainshaft:	
Permitted value	0.0002 to 0.0022 in (0.005 to 0.056 mm)
Washer thicknesses available	0.2008 to 0.2221 in (5.10 to 5.64 mm), in steps of 0.0024 in (0.061 mm)
Mainshaft endfloat:	
Permitted value	0.0004 to 0.0024 in (0.01 to 0.061 mm)
Washer thicknesses available	0.0595 to 0.1091 in (1.51 to 2.77 mm), in steps of 0.0024 in (0.061 mm)
Layshaft endfloat/preload:	
Permitted value	0 ± 0.001 in (0.025 mm)
Washer thicknesses available	0.0665 to 0.0996 in (1.69 to 2.53 mm), in steps of 0.0024 in (0.061 mm)

Torque wrench settings

	lbf ft	Nm
Oil pump to rear extension	5 to 7	7 to 10
Clutch release lever pivot	5 to 7	7 to 10
Rear extension to main casing	16 to 21	22 to 28
Selector housing to rear extension	16 to 21	22 to 28
Clutch slave cylinder	16 to 21	22 to 28
Front cover to main casing	16 to 21	22 to 28
5th gear selector fork support bracket	16 to 21	22 to 28
Yoke to selector shaft	16 to 21	22 to 28
Release lever pin to centre plate	35 to 40	47 to 54
Bellhousing to main casing	48 to 59	65 to 80
Detent spring plug	16 to 21	22 to 28
Oil filler/level and drain plugs	18 to 26	25 to 35
Breather	5 to 8	7 to 11
Output flange nut	120	163
5th gear layshaft nut	160	217

Automatic transmission

Torque wrench setting

	lbf ft	Nm
Brake band locknuts	35	47

Rear axle – 300 Series (including 285)

General

Maker's designation	15 HA or 16 HA
Type:	
15 HA	Hypoid drive, three-quarter floating
16 HA	Hypoid drive, fully floating
Application:	
15 HA	All 285 models, and 310 models with single rear wheels
16HA	310 models with twin rear wheels, and all 350 models

Ratios

15 HA	4.55 : 1
16 HA – colour-coded:	
Blue/black	4.45 : 1
Red/black	4.89 : 1
Pink/black	5.22 : 1

Lubrication

Lubricant type	See *'Recommended lubricants and fluids'*
Lubricant capacity:	
15 HA	3 pints (1.7 litres) approx
16 HA	4.3 pints (2.4 litres) approx

Pinion turning torque

Pinion only	30 to 40 lbf in (3.4 to 4.5 Nm)
Pinion and differential	35 to 55 lbf in (4.0 to 6.2 Nm)
Complete axle	37 to 65 lbf in (4.2 to 7.3 Nm)

Torque wrench settings

	lbf ft	Nm
15 HA:		
Brake backplate nuts and bolts	55	75
Differential bearing cap bolts	50	68
Differential case bolts	46	62
Ring gear bolts	82	111
Cover plate screws	17	23
16 HA:		
Brake backplate bolts	55	75
Differential bearing cap bolts	98	133
Differential case bolts	30	40
Ring gear bolts	117	159
Cover plate screws	12	16
Hub bearing locknuts	115	156
Shaft flange bolts	30	40

Braking system (later models)

General

Front brakes (introduction date unknown)	Disc, with 2-piston or 4-piston calipers
Rear brakes:	
Up to 1984	Single leading shoe drum, manually adjusted (as in Chapter 9)
1985 on	Single leading shoe drum, automatically adjusted
Handbrake (all models)	Mechanical to rear wheels

Disc brakes

Disc diameter	10.87 in (276.1 mm) nominal
Disc thickness (refinishing limit)	0.53 in (13.5 mm)
Pad lining minimum thickness	0.12 in (3.1 mm)

Drum brakes

Drum internal diameter:	
200 Series	9 in (229 mm) nominal
300 Series	10 in (254 mm) nominal
Shoe lining minimum thickness	0.06 in (1.5 mm)

Torque wrench settings

	lbf ft	Nm
Disc-to-hub nuts	30	40
Caliper mounting bolts:		
200 Series	55	75
300 Series	47	64
Caliper union banjo bolts	22	30

Torque wrench settings (continued)

	lbf ft	Nm
Bleed screws:		
7 mm	4	5.5
10 mm	7	9
Master cylinder to servo	11	15
Load sensing valve to bracket	11	15

Electrical system (later models)

Alternator

Make and type	Lucas A115 or A127
Maximum output	45 A @ 14V and 6000 rpm
Rotor resistance @ 68°F (20°C)	3.3 Ω
Brush length:	
New	0.8 in (20 mm)
Wear limit	Not known
Brush spring tension (brushes flush with brush box)	7.4 ozf (2.1 N)

Starter motor

Make and type	Lucas 9M90 or M79
Overhaul data – 9M90:	
Brush length	0.4 in (10 mm) minimum
Brush spring tension	32 to 36 ozf (9 to 10 N)
Commutator thickness	0.08 in (2 mm) minimum
Overhaul data – M79	No information available

Wiper motor

Make and type	Lucas 28W
Operating speed (wet screen):	
Low speed	36 to 46 rpm
High speed	58 to 72 rpm

Fuse applications – *circa* 1983

Fuse No	Rating (A)	Circuit(s) protected
A2	5	LH side, tail and number plate lights; switch and instrument illumination
A3	5	RH side, tail and number plate lights
A4	10	Interior lights
A5	15	Hazard warning flasher, horn, clock, headlight flasher
A6	10	Cigar lighter
B3	10	LH dipped beam
B4	10	RH dipped beam
B5	10	LH main beam
B6	10	RH main beam
C2	10 or 15*	Heater blower
C3	5	Rear foglights
C4	3	Radio
C5	10	Reversing lights, direction indicators, windscreen washer
C6	10	Windscreen wiper, stop-lights, automatic transmission selector light
Line	2	Tachograph (when fitted)
Line	15	Manifold heater (when fitted)

**15A may be fitted if 10A blows for no reason*

Fuse applications – *circa* 1985

Fuse No	Rating	Circuit(s) protected
A1	5	RH side and tail lights
A2	10	LH side and tail lights; switch and instrument illumination
A3	15	Hazard warning flasher
A4	10	Clock and interior lights
A5	15 max	Spare (battery fed)
A6	10	Cigar lighter
A7	15 max	Spare (battery fed)
B1	25	Manifold heater (when fitted)
B2	10	Rear foglights
B3	10	LH dipped beam
B4	10	RH dipped beam
B5	10	LH main beam
B6	10	RH main beam
B7	20 max	Spare (battery fed)
C1	15 max	Spare (ignition fed)
C2	25	Headlight washer pump

C3	15	Heater blower
C4	3	Radio
C5	15	Direction indicator, reversing lights, stop-lights
C6	15	Windscreen wiper and washer
C7	20 max	Spare (ignition fed)

Suspension and steering (later models)

General

Front shock absorber type	Telescopic

Steering geometry (300 Series)

Camber	+ 1°30′
Castor	+ 5°00′
Swivel pin inclination	+ 9° 30′
Front wheel alignment (toe)	Parallel to 0.06 in (1 mm toe-in)

Roadwheels

200, 230, 250, 255, 310 (twin rear wheels) and 350	5J x 14
280, 285 and 310 (single rear wheels)	5½J x 14

Tyre pressures (cold)*

Model	Tyre size	Front lbf/in²	Front bar	Rear lbf/in²	Rear bar
City/200	165 14R	37	2.6	42	2.9
230	185 14R	38	2.6	43	3.0
	195 14C	43	3.0	48	3.3
250	185 14C	42	2.9	52	3.6
	195 14C	43	3.0	48	3.3
255	185 14C	45	3.1	54	3.7
	195 14C	45	3.1	50	3.5
280	185 14C	48	3.3	58	4.0
285	205 14C	48	3.3	53	3.7
310	185 14C	44	3.0	46	3.2
	205 14C	48	3.3	64	4.4
350	185 14C	46	3.2	48	3.3
PSV/Minibus	Not known	51	3.5	54	3.7

**Maker's recommended pressures for plated axle loading weight. For special applications, motor caravan conversion etc, consult a tyre specialist.*

Torque wrench settings – 200 Series

	lbf ft	Nm
Front shock absorber mounting:		
3/8 in UNF	30	40
1/2 in UNF	34	46
Rear shock absorber mountings	34	46
Wheel nuts:		
7/16 in UNF	59	80
1/2 in UNF	92	125

Torque wrench settings – 300 Series

	lbf ft	Nm
Steering column to facia bracket	15	20
Steering drop arm nut	144	195
Steering drag link ball-pin nut	47	64
Steering drag link pinch-bolt	21	29
Steering universal joint clamp nut	20	27
Steering box to frame	52	70
Steering wheel nut	40	54
Front axle cotter pin nut	12	16
Steering lever nuts	87	118
Track rod lever nuts	87	118
Track rod ball-pin nuts	47	64
Track rod pinch-bolt	21	29
Anti-roll bar to front axle	57	77
Anti-roll bar link to bracket	45	61
Anti-roll bar link to bar	50	68
Anti-roll bar bracket to frame	52	70
Shock absorber mountings (front and rear)	33	45
Spring eye bolts (front and rear)	60	81
Spring shackle bolts (front and rear)	60	81
Spring U-bolts (front and rear)	62	84
Wheel nuts *(left-hand thread on left-hand side)*:		
285 and 310	130 to 150	176 to 203
350	170 to 190	231 to 258

3 Buying spare parts and vehicle identification

Buying spare parts

References to 'BL dealer' throughout the manual should now read 'Freight Rover dealer'.

Vehicle identification numbers

Later models have a vehicle identification number (VIN) plate attached to the left-hand stepwell (photo). The plate carries information about axle loadings, paint colour and trim levels, as well as the VIN itself.

Definition of 200/300 Series

The term '200 Series' as used in this Chapter covers models up to and including 280, from 1983 model year unless otherwise specified. '300 Series' covers 285, 310 and 350 models.

4 Jacking, towing and wheel changing

Jacking points (later models)

Instead of the screw jack shown at the beginning of the manual, some later models have a pillar jack with a side lifting arm. The lifting arm must be inserted into the jacking point nearest the wheel to be raised. Each jacking point is protected by a bung.

Towing (later models)

If it becomes necessary to tow a Sherpa fitted with a 5-speed gearbox, the propeller shaft should be removed or disconnected.

Some models are fitted with a front towing eye which is hinged out of the way when not in use. Slacken the clamp nut to release the eye, swivel it forwards and retighten the nut.

Wheel changing

If possible, park the vehicle on firm level ground. On soft ground a board or plank may be used to spread the load under the foot of the jack. Jacking on a slope is dangerous and extra care should be taken.

Chock both sides of the wheel diagonally opposite the one to be removed. Engage 1st or reverse gear (or 'P' on automatics) and set the handbrake. Slacken the securing nuts of the wheel to be removed by half a turn each. **Note:** On 300 Series models, the wheel nuts on the left-hand side of the vehicle have a **left-hand thread**, ie they are undone **clockwise**.

Engage the jack in the correct jacking point. Raise the vehicle until the wheel is clear of the ground. Remove the wheel nuts. Mark the relationship of the wheel to the hub if it is wished to refit the wheel in the same position (for example if the wheels have been balanced on the vehicle). Remove the wheel.

On twin rear wheel models, make sure that all the wheel studs are fitted with split collars, with the flat faces of the collars towards the brake drum. The wheel mating faces must be clean and dry. Transfer the inner wheel valve extension if necessary; position the inner and outer wheel valves on opposite sides of the hub when refitting.

Fit the wheel over the studs. If possible apply a smear of grease or

3.0 Later type VIN plate

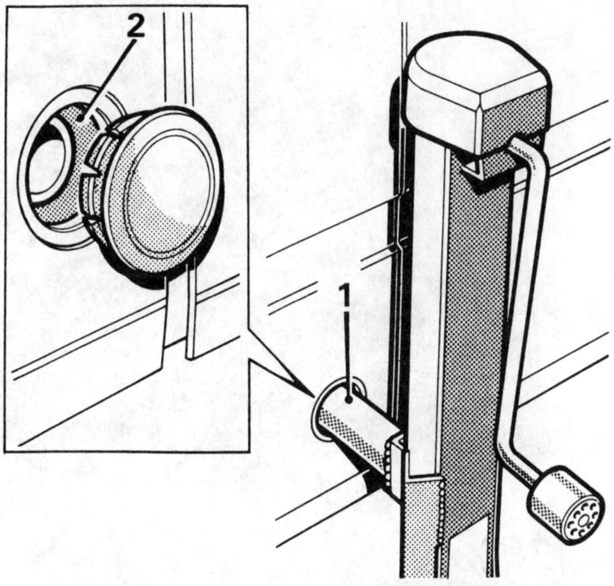

Fig. 13.1 Pillar jack supplied with certain models (Sec 4)

1 Lifting arm *2 Jacking point*

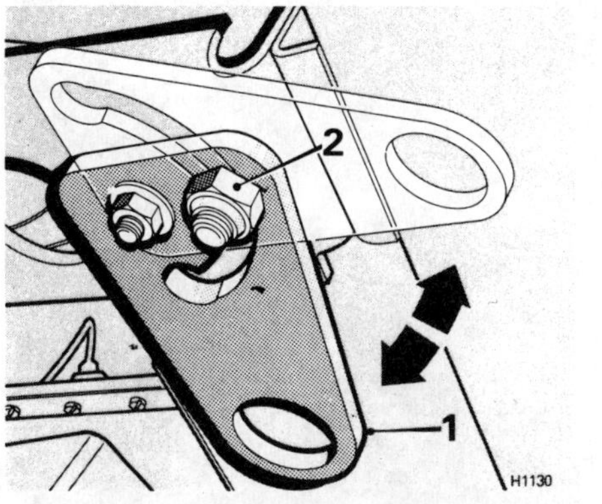

Fig. 13.2 Hinged towing eye (1) and clamp nut (2) (Sec 4)

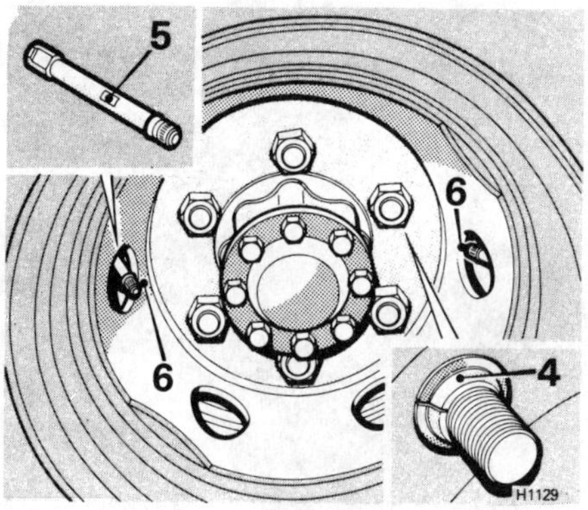

Fig. 13.3 Special features of twin rear wheels (Sec 4)

4 Split collars
5 Inner wheel valve extension
6 Two valve holes

1 Coolant filler/pressure cap
2 Brake fluid filler cap
3 Brake delay valve
4 Brake servo
5 Clutch fluid filler cap
6 Servo non-return valve
7 Brake pressure failure warning switch
8 Air intake temperature sensor
9 Plenum chamber
10 Speedometer cable
11 Fuel pipe
12 Anti-run-on valve harness
13 Breather hoses
14 Engine oil filler cap
15 Heater hoses
16 Washer bottle bracket
17 Starter motor positive lead
18 Ignition control unit
19 Earth strap
20 Ignition vacuum connection
21 Battery positive terminal cover
22 Ignition coil
23 Starter motor
24 Breather/oil separator
25 Engine mounting
26 Alternator
27 Alternator/motor pump drivebelt
28 Bonnet prop
29 Distributor cap
30 Thermostat housing
31 Bonnet catch
32 Radiator top hose
33 Air cleaner
34 Spark plug cap
35 Knock sensor
36 Engine oil dipstick tube

Under-bonnet view of a Sherpa with the 17V 02 Series engine. Windscreen washer bottle has been removed

anti-seize compound to each stud. Fit the wheel nuts, taper inwards, and tighten them moderately. Lower the vehicle to the ground and carry out the final tightening of the wheel nuts in a criss-cross sequence.

Remove the wheel chocks and stow the jack and other equipment. If the spare wheel has just been brought into use, check the tyre pressure and adjust as necessary.

5 Routine maintenance

Maintenance schedules for later models are given below. The home mechanic may consider it prudent to observe shorter intervals for some tasks, particularly engine oil changing and brake pad inspection. Vehicles operating under adverse conditions (eg climatic extremes, full-time towing or constant stop-start work) will also need more frequent attention.

When the prescribed mileage is not covered in the stated time, follow the time interval to determine when service is due. This is necessary because some fluids and systems deteriorate with time as well as with use.

Some maintenance tasks previously specified (eg greasing the propeller shaft joints) no longer apply because the components are now sealed for life.

Weekly maintenance remains as specified at the beginning of the manual, except for the battery electrolyte check – see Section 17A.

Maintenance schedule – circa *1982*

Every 6000 miles (10 000 km) or six months, whichever comes first.

Inspect front brake pads for wear and check condition of brake discs; renew as necessary
Adjust rear brakes (when applicable)
Lubricate swivel pins and propeller shaft (when possible)
Check condition of exhaust system
Check condition and tension of alternator drivebelt
Check clutch hydraulic fluid level
Lubricate door locks, hinges etc
Thoroughly inspect wheels and tyres

Every 12 000 miles (20 000 km) or 12 months, whichever comes first

Check condition and security of seats, mountings and seat belts
Check operation of footbrake and handbrake (road test or rolling road)
Check security of steering column and steering intermediate shaft joints
Inspect rear brake shoes for wear and check condition of brake drums; renew as necessary
Adjust front wheel bearings
Check adjustment of steering box
Check gearbox oil level
Drain engine oil (hot), refill with fresh oil
Renew engine oil filter
Check rear axle oil level
Lubricate brake load sensing valve pivot and spring (when so equipped)
Inspect brake pipes and hoses
Inspect fuel and clutch hydraulic pipes and hoses
Lubricate handbrake linkage and cables
Check security and condition of steering and suspension joints, fittings etc
Inspect steering box and shock absorbers for fluid leaks
Top up steering box oil level
Lubricate pedal pivots, control linkages etc
Top up carburettor piston damper (dashpot)
Renew air cleaner element
Renew spark plugs
Check battery electrolyte level (if possible)
Clean battery terminals, apply petroleum jelly
Check condition and security of brake servo vacuum hose
Clean HT leads, coil tower and distributor cap
Check strength of antifreeze in coolant
Check condition of cooling system hoses and clips
Lubricate distributor (contact breaker type)
Check operation of throttle and choke controls
Check ignition timing
Check idle speed and mixture adjustments
Check automatic transmission fluid level
Clean and lubricate spare wheel carrier (under-floor type)
Check front wheel alignment
Check headlight beam alignment
Road test vehicle, then check for fluid leaks

Every 18 000 miles (30 000 km) or 18 months, whichever comes first

Renew brake fluid

Every 24 000 miles (40 000 km) or two years, whichever comes first

Drain gearbox oil, clean filter and refill with fresh oil – overdrive gearbox only
Check adjustment of brake load sensing valve (when so equipped)
Check condition and tension of timing belt
Renew timing belt cover rubbers
Renew alternator drivebelt
Renew coolant
Renew engine oil filler cap
Renew contact breaker points (when applicable)
Check valve clearances, adjust if necessary

Every 36 000 miles (60 000 km) or three years, whichever comes first

Renew braking system rubber seals and hoses

Every 48 000 miles (80 000 km) or four years, whichever comes first

Renew timing belt

Maintenance schedule – circa *1985*

The schedule is basically as given for 1982 vehicles, except that the 6000-mile service is abandoned and the items in it are carried forwards to the 12 000-mile service. References to adjusting the rear brakes, and to renewing contact breaker points, can be ignored when appropriate.

Maintenance schedule – circa *1987*

No formal schedule was available at the time of writing. Periodic checking of the ignition timing is no longer required.

6 Engine

20V (1994 cc) engine – description

1 The 20V engine, available from 1980, is a larger version of the 17V engine described in Chapter 1, Part B. The increase in capacity is achieved by lengthening the stroke, the bore remaining unchanged.

2 Except as noted in this Chapter, operations on the 20V engine are the same as those described in Chapter 1 for the 17V.

O Series Commercial engine – description

3 'O Series Commercial' is the designation given by Freight Rover to the ohc petrol engine fitted to Sherpa models up to 1986. The engine is described in Chapter 1, Part B.

02 Series Commercial engine – description

4 '02 Commercial' is the designation given by Freight Rover to the ohc petrol engine fitted to Sherpa models from 1987. Externally the 02 Series engine is easily distinguished by having the intake and exhaust manifolds on the right-hand side, where previously they were on the left. There are changes to the cylinder head posting and to the oil pump.

5 Other associated changes include the provision of an electric fuel pump instead of a mechanical one, an intake manifold heater and a programmed ignition system. These ancillaries are considered in the appropriate Sections.

6 Camshaft endfloat is now controlled by a semi-circular steel insert in the camshaft cover, which runs in a groove in the camshaft rear bearing.

7 Valve stem oil seals are now fitted to both inlet and exhaust valves.

8 Except as specified in this Chapter, dismantling and repair operations on the 02 Series engine are as described in Part B of Chapter 1.

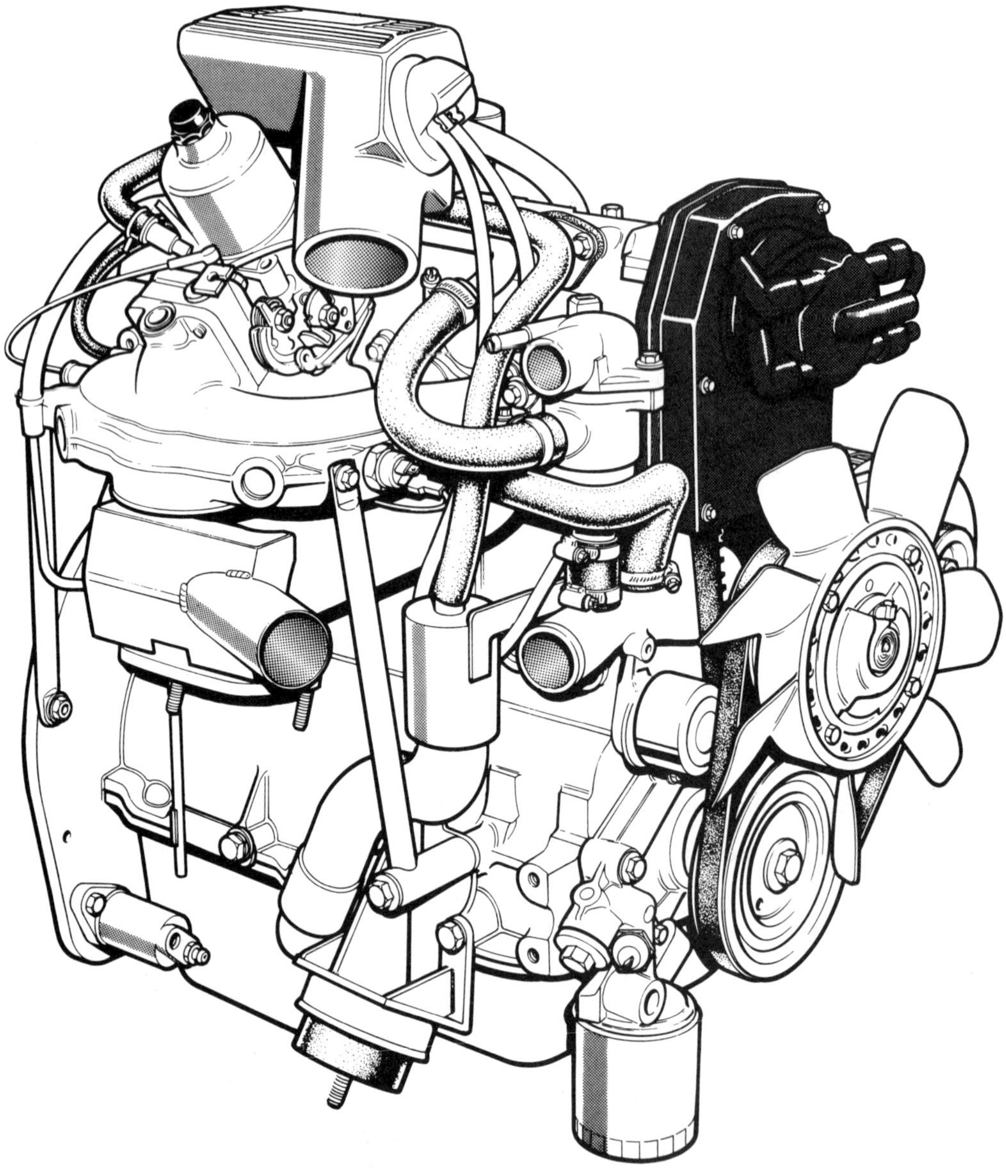

Fig. 13.4 The 02 Series engine (Sec 6)

Oil filter and oil pump (02 Series) – description

9 The oil filter and oil pump are basically the same as those fitted to the original O Series engine, but the oil pump housing has been modified. The oil filter now faces upwards on the right-hand side of the engine (photo).

Cylinder head (02 Series) – removal and refitting

10 The procedure is essentially as described in Chapter 1, Sections 77 and 113, but note the following points:

- (a) *There is no need to disconnect the fuel pump*
- (b) *Follow the sequence in Fig. 13.6 for tightening the cylinder head bolts*
- (c) *Note that there are now three stages of tightening for the bolts; the third stage is defined by angular rotation, not by torque – see Specifications. Retightening after warm-up is not required*

Camshaft cover (02 Series) – removal and refitting

11 Turn the engine to 90° BTDC, No 1 cylinder on compression.

12 Remove the distributor cap, rotor arm and shield. Refer to Section 10 if necessary.

13 Remove the timing belt front cover, and the plate at the rear end of the camshaft cover.

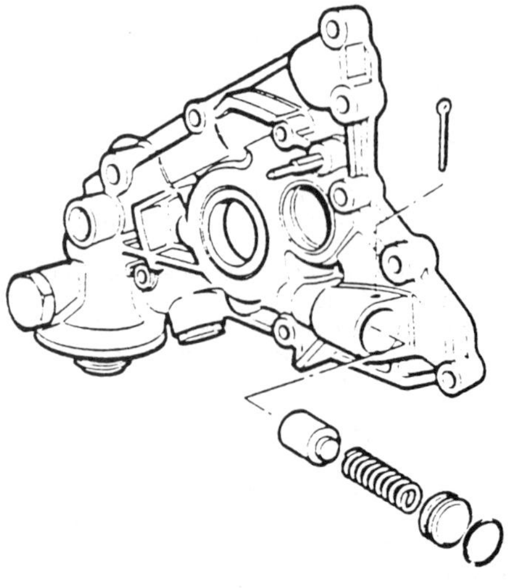

Fig. 13.5 02 Series oil pump and pressure relief valve (Sec 6)

6.9 Oil filter on 02 Series engine

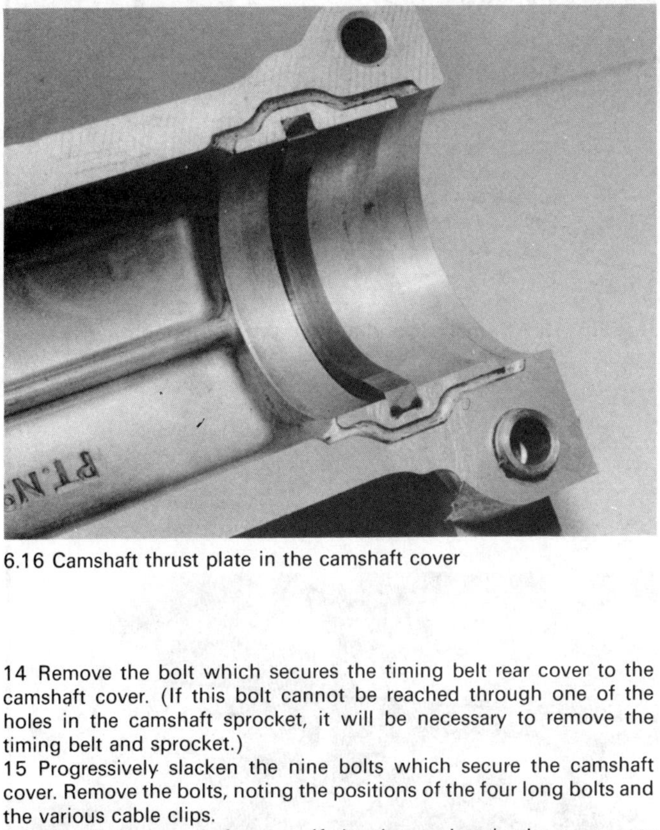
6.16 Camshaft thrust plate in the camshaft cover

14 Remove the bolt which secures the timing belt rear cover to the camshaft cover. (If this bolt cannot be reached through one of the holes in the camshaft sprocket, it will be necessary to remove the timing belt and sprocket.)
15 Progressively slacken the nine bolts which secure the camshaft cover. Remove the bolts, noting the positions of the four long bolts and the various cable clips.
16 Lift off the camshaft cover. If the thrust plate in the cover rear journal is loose, recover it and keep it safe (photo).
17 Clean the old sealant off the mating surfaces. Apply fresh RTV sealant when reassembling, keeping it off the bearing surfaces.
18 Refit by reversing the removal operations. Oil the threads of the camshaft cover bolts and tighten them progressively to the specified torque. Remember to use a new screw to secure the rotor arm.

Camshaft (02 Series) – removal and refitting

19 Remove the camshaft cover as just described. If not done during cover removal, also remove the timing belt and the camshaft sprocket.
20 Remove the timing belt rear cover. The camshaft can now be removed.
21 Refit by reversing the removal operations. Renew the camshaft oil seals unless they are known to be in perfect condition.

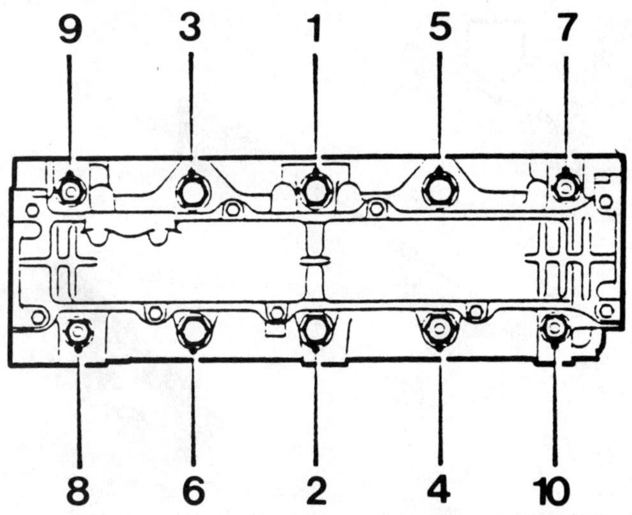

Fig. 13.6 Cylinder head bolt tightening sequence – 02 Series (Sec 6)

Camshaft rear oil seal (02 Series) – renewal (engine in vehicle)

22 Remove the camshaft as previously described.
23 Remove the rear oil seal from the camshaft.
24 Lubricate all surfaces of the new seal and fit it to the camshaft, lips inwards. Make sure that the seal is fitted square and that the lip does not roll back.
25 Refit the camshaft, being careful not to nip either seal as the camshaft cover is tightened down.

Engine/gearbox adaptor plate (02 Series) – sealing

26 Instead of a gasket, a bead of RTV jointing compound is used to seal the adaptor plate to the engine.
27 Make sure that the raised areas on either side of the rear main bearing cap are inside the bead of sealant when refitting the plate.

Valve clearances (all 0 and 02 Series) – checking and adjustment

28 The following points are additional to Chapter 1, Section 114. (For 02 engines, see also paragraphs 11 to 18 in this Section.)
29 There is no need to remove the timing belt when checking the clearances if the special retaining brackets (tool 18G 1301 or 1301A) or equivalent are used. The timing belt need only be removed if the clearances must be adjusted.
30 Equivalents of the special brackets are easily made using short lengths of angle iron and some scraps of wood or plastic (photo). See Fig. 13.9 for the dimensions.
31 In use, the brackets are secured using some of the camshaft cover bolts and positioned so that the wooden or plastic pads rest on the camshaft bearings (photo). Tighten the bolts lightly at first so that the camshaft can still be rotated.
32 Turn the crankshaft using a socket or spanner on the pulley bolt until Nos 1 and 3 cam lobes are vertical, with their heels above the valves. Tighten the bracket retaining bolts so that the camshaft bearings are fully seated in their journals.
33 Using feeler gauges, measure and record the clearance between Nos 1 and 3 cam lobes and their respective tappets.
34 Slacken the bracket retaining bolts, turn the crankshaft until Nos 4 and 7 cam lobes are vertical then tighten the brackets again. Measure and record the clearance, then repeat the procedure for Nos 2 and 5, and Nos 6 and 8 cam lobes.
35 If all the clearances were correct, remove the brackets and refit the camshaft cover and other disturbed components.
36 If adjustment is necessary, bring the engine to 90° BTDC (No 1 on compression), then remove the brackets, the timing belt and the camshaft. Adjust the clearances as described in Chapter 1, Section 114.

6.30 Home made brackets for retaining the camshaft. Note cut-out (arrowed) in rear bracket

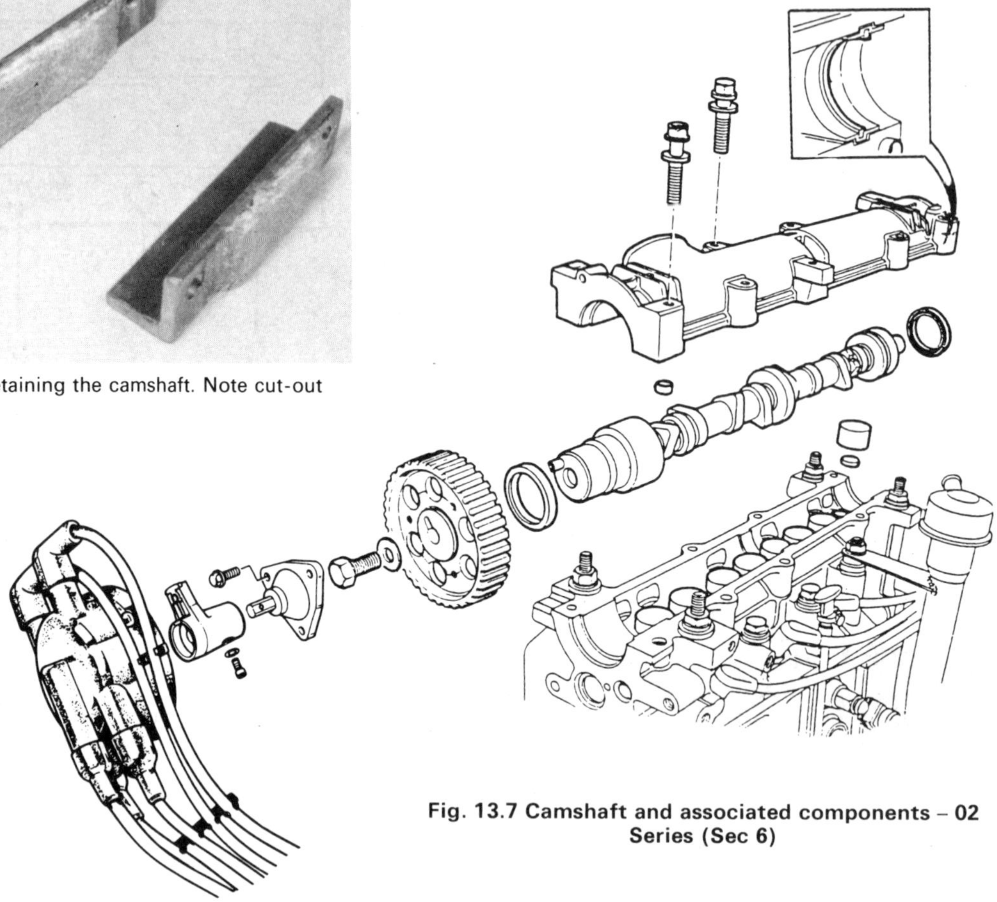

Fig. 13.7 Camshaft and associated components – 02 Series (Sec 6)

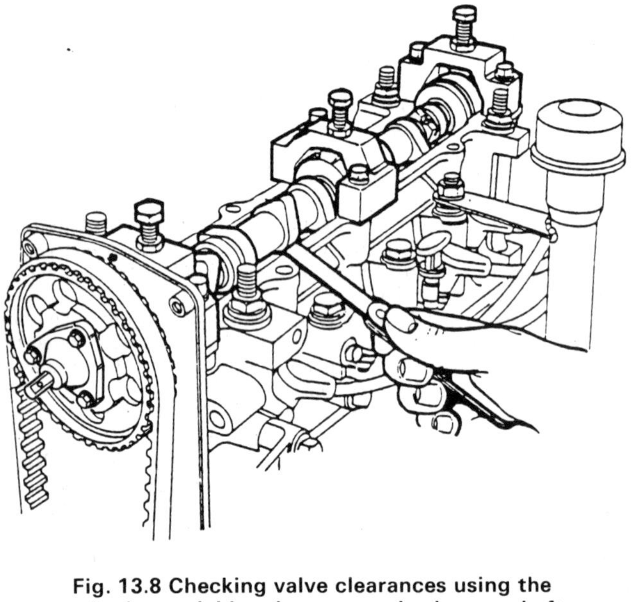

Fig. 13.8 Checking valve clearances using the maker's special brackets to retain the camshaft (Sec 6)

6.31 Camshaft retaining brackets in use

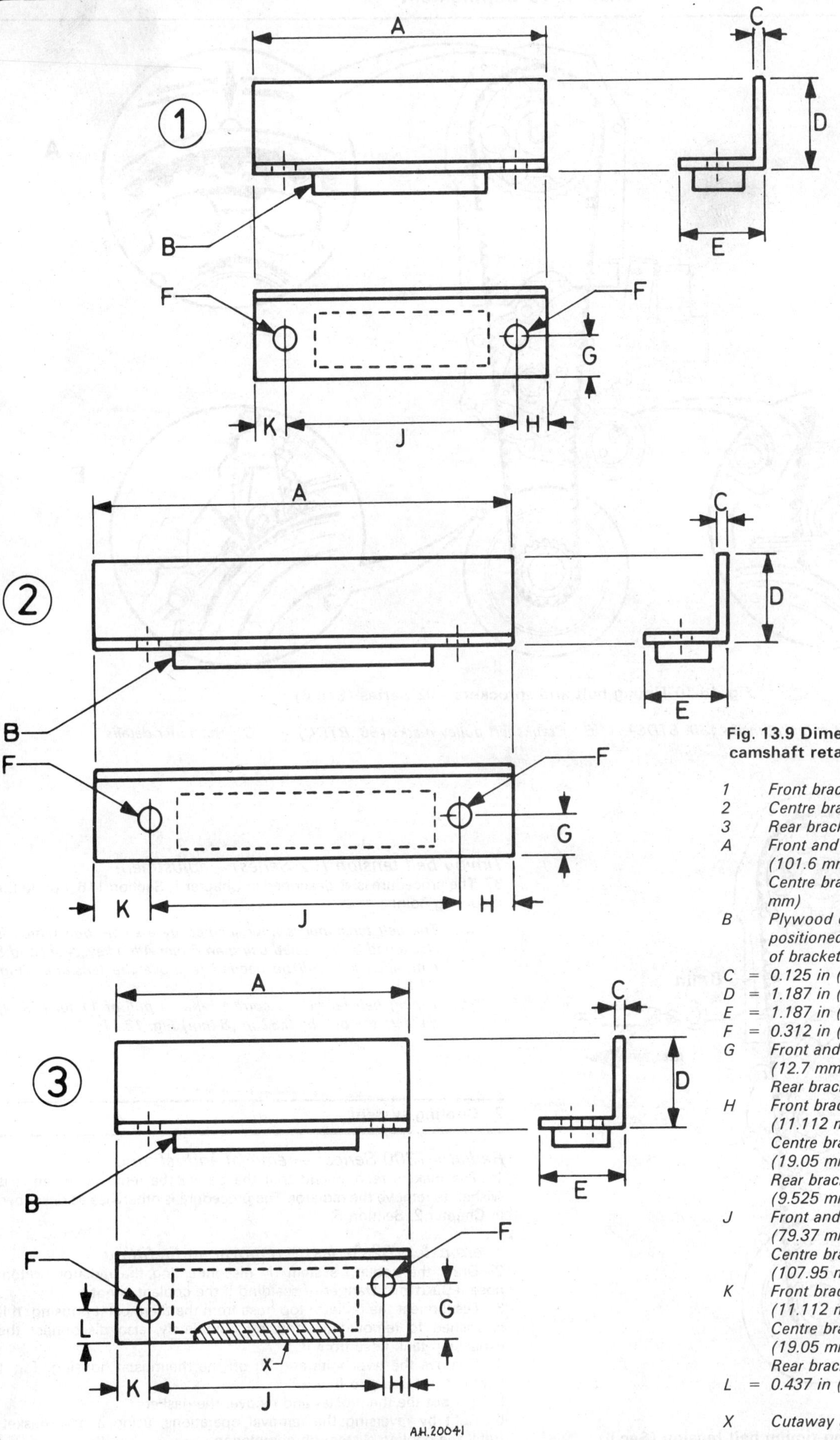

Fig. 13.9 Dimensions for home-made camshaft retaining brackets (Sec 6)

1	*Front bracket*
2	*Centre bracket*
3	*Rear bracket*
A	*Front and rear bracket – 4.00 in (101.6 mm)*
	Centre bracket – 5.75 in (146.05 mm)
B	*Plywood or other soft material positioned centrally on base of bracket*
C	*= 0.125 in (3.175 mm)*
D	*= 1.187 in (30.162 mm)*
E	*= 1.187 in (30.162 mm)*
F	*= 0.312 in (7.937 mm)*
G	*Front and centre brackets – 0.5 in (12.7 mm)*
	Rear bracket – 0.75 in (19.05 mm)
H	*Front bracket – 0.437 in (11.112 mm)*
	Centre bracket – 0.75 in (19.05 mm)
	Rear bracket – 0.375 in (9.525 mm)
J	*Front and rear bracket – 3.125 in (79.37 mm)*
	Centre bracket – 4.25 in (107.95 mm)
K	*Front bracket – 0.437 in (11.112 mm)*
	Centre bracket – 0.75 in (19.05 mm)
	Rear bracket – 0.5 in (12.70 mm)
L	*= 0.437 in (11.112 mm)*
X	*Cutaway area (determine in use)*

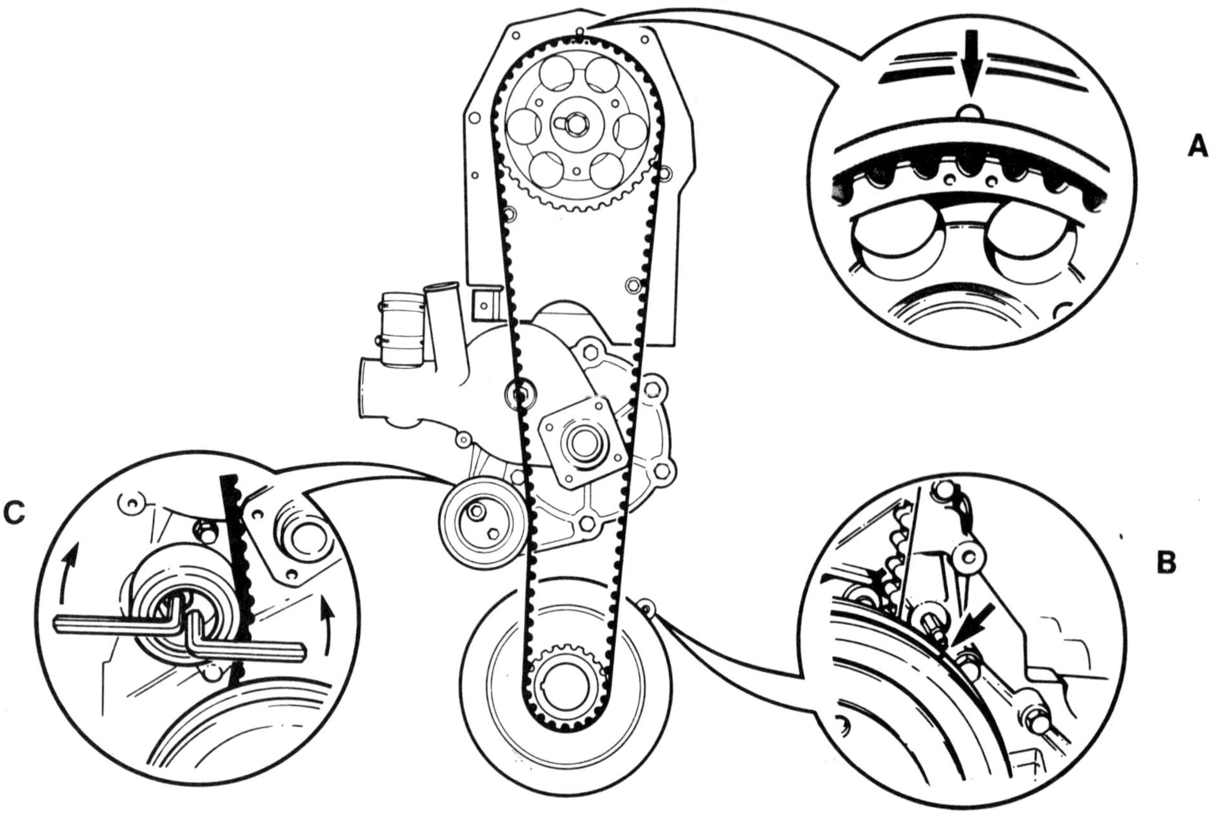

Fig. 13.10 Timing bolt and sprockets – 02 Series (Sec 6)

A Camshaft sprocket marks (90° BTDC) *B Crankshaft pulley marks (90° BTDC)* *C Tensioner details*

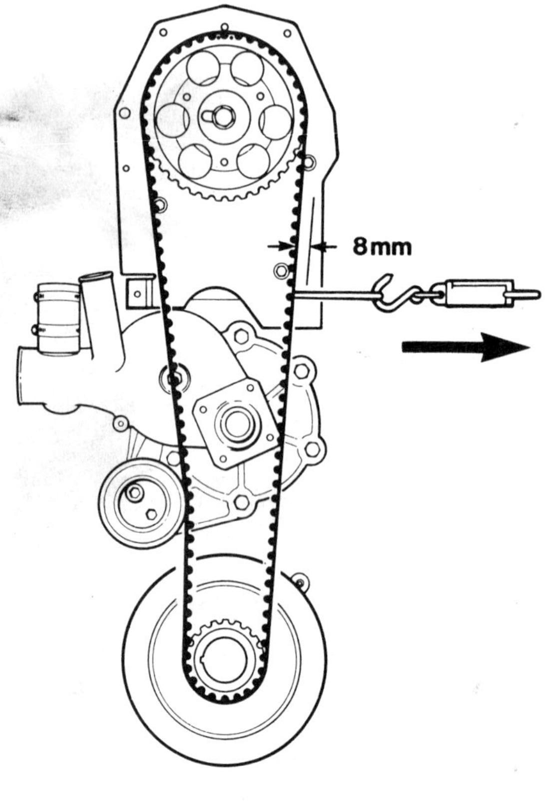

Fig. 13.11 Checking timing belt tension (Sec 6)

Timing belt tension (02 Series) – adjustment

37 The procedure is as described in Chapter 1, Section 116, noting the following points:

(a) *The belt tensioner is now secured by a single bolt which is slackened or tightened using an 8 mm Allen key. A second 8 mm Allen key will be needed to move the tensioner (Fig. 13.10)*

(b) *Timing belt tension is correct when a pull of 11 lbf (49 N) deflects the belt by 0.32 in (8 mm) Fig. 13.11).*

7 Cooling system

Radiator (300 Series) – removal and refitting

1 The makers recommend that the bonnet be removed when it is wished to remove the radiator. The procedure is otherwise as described in Chapter 2, Section 5.

Thermostat (02 Series) – removal and refitting

2 Drain the cooling system by disconnecting the radiator bottom hose. (**Caution:** Danger of scalding if the coolant is hot.)

3 Disconnect the radiator top hose from the thermostat housing. If it is wished to remove the housing completely, also disconnect the expansion tank hose from it.

4 Remove the two bolts and lift off the thermostat housing. Tap it lightly if necessary to free it.

5 Lift out the thermostat and recover the gasket.

6 Refit by reversing the removal operations, using a new gasket. Refill the cooling system on completion.

7.7 Front view of the viscous coupled fan

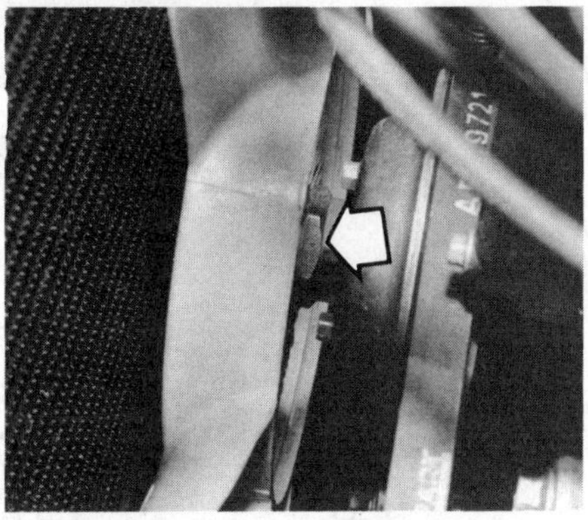

7.13 Viscous coupling securing nut (arrowed). Remember it has a left-hand thread

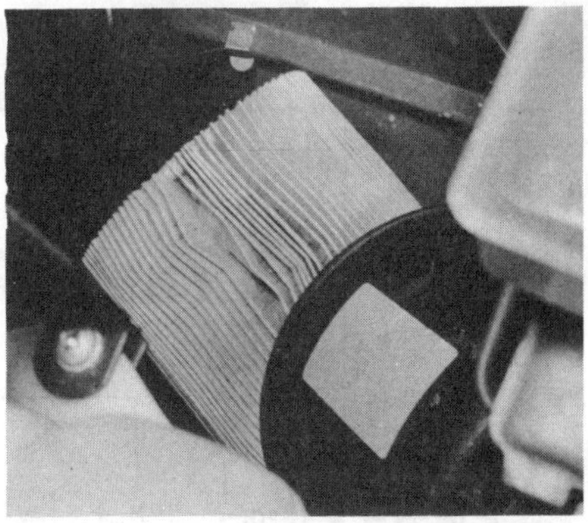

8.3 Air cleaner element renewal (02 Series)

Viscous coupled fan – description

7 The fan fitted to later models is bolted to a viscous coupling, which is in turn attached to the water pump pulley. The coupling carries a bi-metallic spring on its front face (photo).

8 When the coupling is cold, little power is transmitted from the pulley to the fan. The fan speed is low, regardless of engine speed.

9 As the radiator warms up, hot air causes the bi-metallic spring to unwind. This sets on the coupling and allows the fluid within it to transmit drive from the pulley to the fan. The fan speed will now tend to increase with engine speed.

10 The advantages of the viscous coupled fan are that it is less noisy, and consumes less energy unnecessarily, than the simple direct fan.

Viscous coupled fan – removal and refitting

11 Disconnect the battery negative lead.

12 Depending on engine and model, it may be necessary to remove (or at least displace) the radiator. On the vehicle seen in the photographs (a 200 Series chassis cab with a 17V 02 Series engine) the fan could be removed without disturbing the radiator.

13 Undo the nut which secures the viscous coupling to the water pump pulley (photo). This nut has a **left-hand thread**, ie it is undone **clockwise**. A slim cranked spanner, $1^{1}/_{4}$ in/32 mm AF, will be needed to undo the nut; an ordinary spanner (or even a slim adjustable) can be used if a couple of the water pump pulley bolts are removed. Stop the pulley turning as the nut is undone by pressing on the drivebelt.

14 Remove the fan and coupling, being careful not to damage the radiator if it is still in place.

15 The fan blades may now be unbolted from the coupling if mislaid.

16 Refit by reversing the removal procedure. The use of thread locking compound on the fan hub nut is recommended.

8 Carburation and fuel system (02 Series)

Air cleaner element – renewal

1 Disconnect the air cleaner-to-plenum chamber trunking to improve access.

2 Unclip the air cleaner lid, which is down beside the right-hand wing. Move the lid aside without straining the vacuum pipe.

3 Remove the old element, wipe clean inside the lid and housing and fit the new element (photo).

4 Refit the lid and secure it with the clips.

5 Refit and secure the trunking.

Fuel pump – removal and refitting

6 The fuel pump is located under the vehicle, on the left-hand chassis member (photo).

7 Disconnect the battery negative lead.

8 Disconnect the multi-plug from the pump.

9 Unbolt the pump from its bracket. Disconnect and plug the fuel lines; be prepared for fuel spillage. Make sure that the line from the tank is well plugged, or raise it above the level of the tank, otherwise all the fuel will be lost.

10 A defective pump must be renewed. (Note however that the pump electrical supply is via a relay – see Fig. 13.12 and Section 17A – so check first that the relay is not at fault.)

11 Refit by reversing the removal operations.

Carburettor – description

12 The HIF 44 carburettor fitted to the 02 Series engine is a development of earlier HIF types. The main difference is the inclusion of an anti-run-on valve, which functions as follows.

13 In normal running the valve is not energised and its plunger is extended by spring pressure. Air enters the float chamber through an air passage in the valve. The depression in the venturi draws fuel through the jet.

14 When the ignition is switched off, the valve is energised and its plunger is retracted. The float chamber air vent is now open to the venturi instead of to atmosphere. Since the depression in the float chamber and venturi are the same, no fuel will flow from the jet and the engine will stop.

15 The valve is energised via a timer relay, which supplies current for approximately ten seconds after the ignition is switched off.

Carburettor – removal and refitting

16 Disconnect the battery negative lead.

17 Disconnect the air cleaner trunking from the plenum chamber (photo).

18 Disconnect the vacuum hoses from the air temperature control sensor (photo).

19 Remove the plenum chamber, which is secured by two screws, and its bracket (photo).

8.6 The electric fuel pump

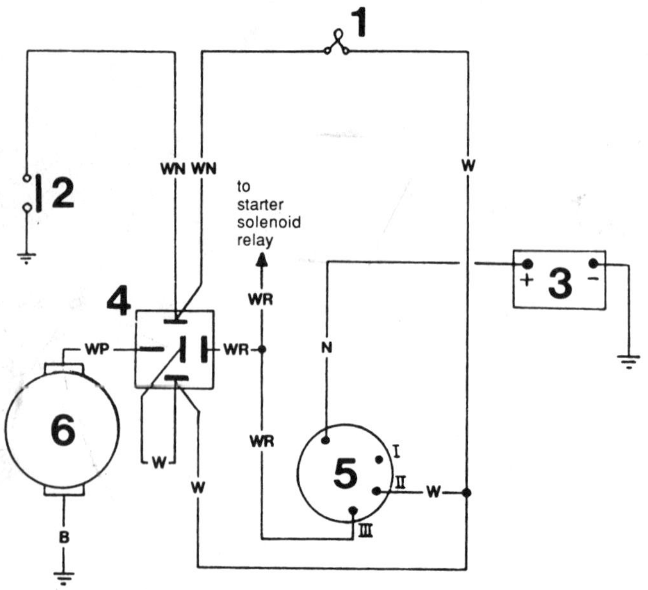

Fig. 13.12 Circuit diagram for electronic fuel pump (Sec 8)

1	*Oil pressure warning light*	*6*	*Fuel pump*
2	*Oil pressure switch*	*N*	*Brown*
3	*Battery*	*P*	*Purple*
4	*Fuel pump relay*	*R*	*Red*
5	*Ignition/starter switch*	*W*	*White*

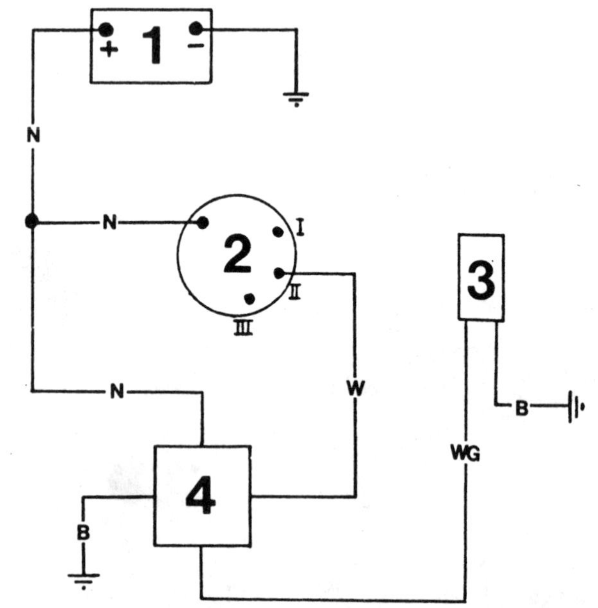

Fig. 13.14 Circuit for anti-run-on valve (Sec 8)

1	*Battery*	*B*	*Black*
2	*Ignition/starter switch*	*G*	*Green*
3	*Anti-run-on valve*	*N*	*Brown*
4	*Timer*	*W*	*White*

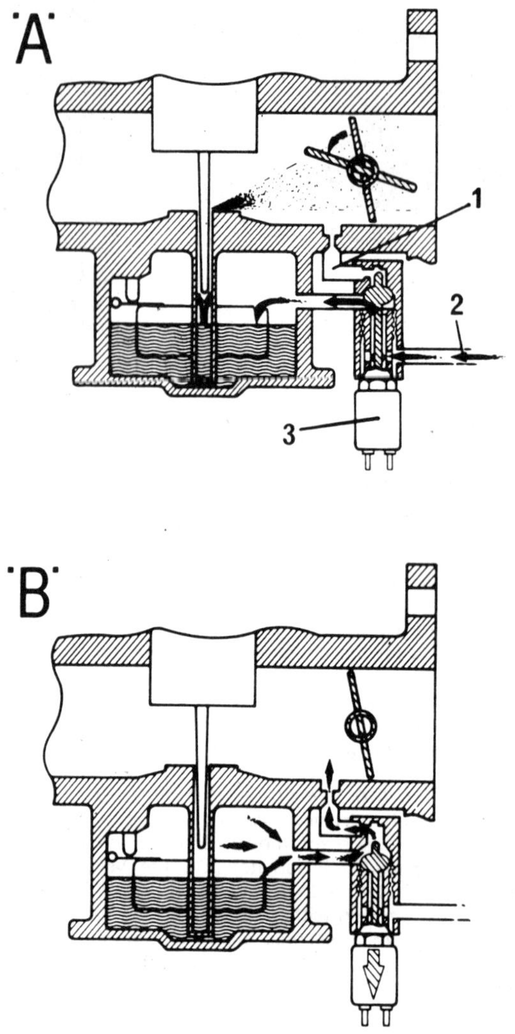

Fig. 13.13 Operation of anti-run-on valve with engine running normally (A) and after switching off (B) (Sec 8)

1	*Passage to venturi*	*3*	*Valve*
2	*Passage to air*		

20 Disconnect the choke and throttle cable inners from the carburettor (photo).
21 Disconnect the vacuum advance pipe and the crankcase ventilation hose.
22 Disconnect and plug the fuel pipe.
23 Disconnect the two wires from the anti-run-on valve.
24 Remove the two flange nuts and lift the carburettor off the manifold studs (photo). Recover the gasket.
25 Refit by reversing the removal operations, using a new gasket.

Carburettor – dismantling and reassembly

26 The procedure is similar to that described in Chapter 3, Section 16. Additionally the anti-run-on valve should be removed and a new O-ring fitted (photos).
27 The anti-run-on valve may be tested by applying 12 volts across its terminals. With voltage applied, the plunger should retract; if not, renew the valve.

Carburettor – adjustments

28 Proceed as described in Chapter 3, Section 17. Refer to Fig. 13.15 for the adjustment points.

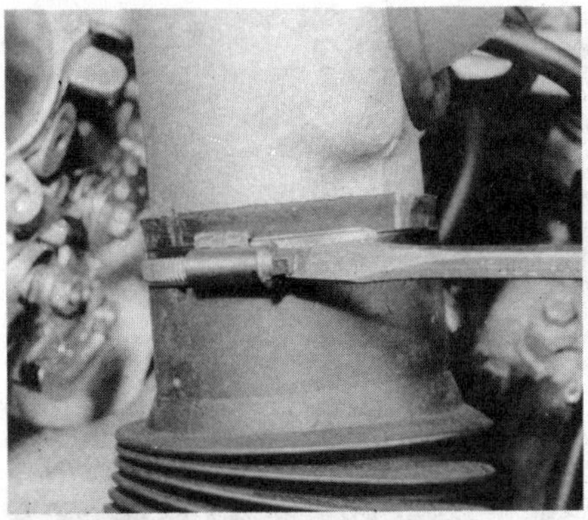

8.17 Releasing the trunking from the plenum chamber

8.18 Air temperature control sensor vacuum hoses

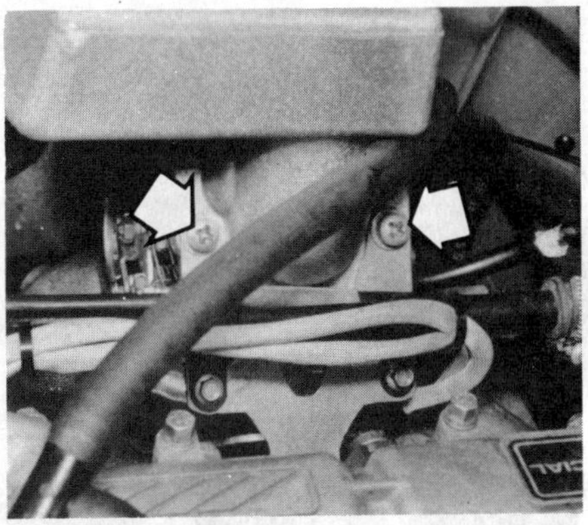

8.19 Plenum chamber securing screws (arrowed)

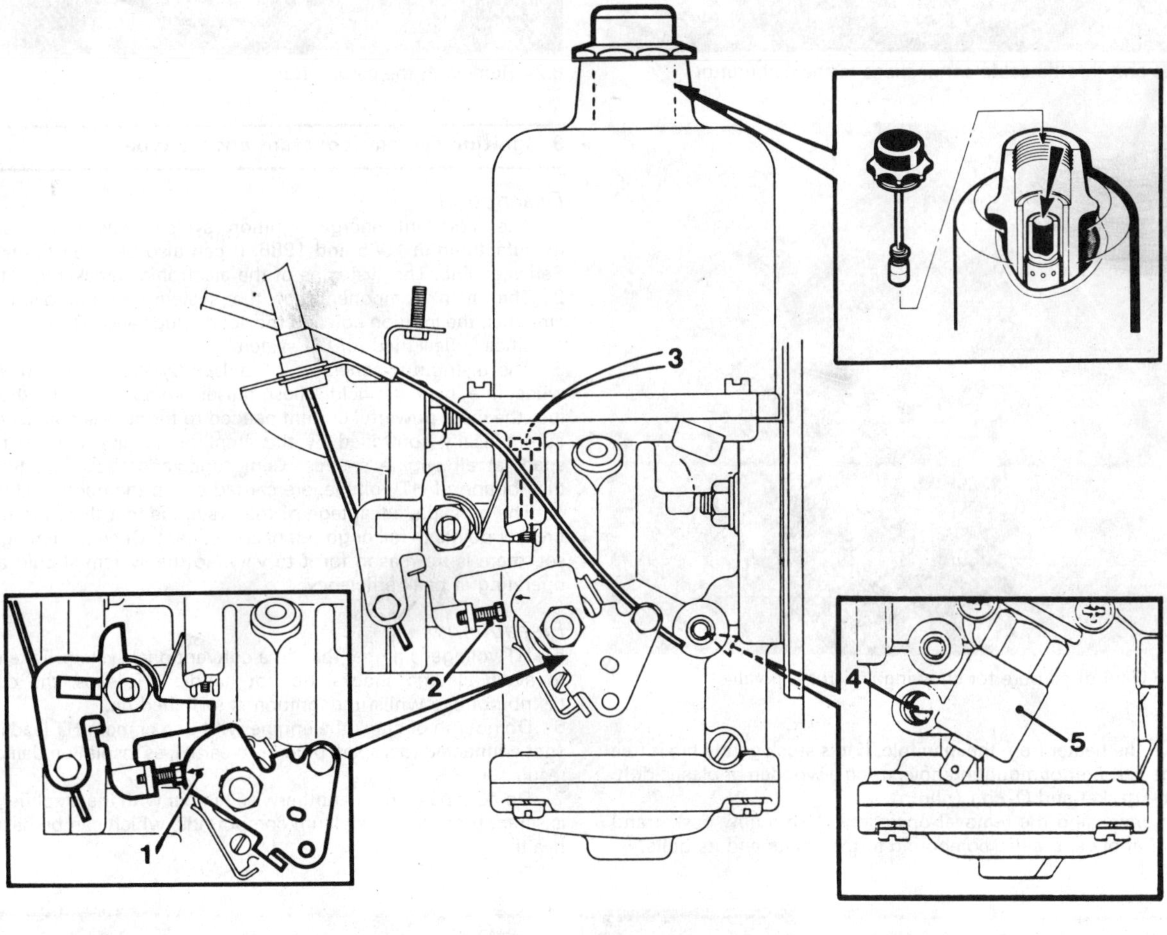

Fig. 13.15 Details of carburettor fitted to 02 Series engine (Sec 8)

1 Cam position for fast idle adjustment
2 Fast idle adjustment screw
3 Idle speed adjusting screw
4 Mixture adjusting screw
5 Anti-run-on valve

Intake manifold heating systems – description

29 The intake manifold is heated both electrically and by engine coolant. This gives good cold driveability and rapid warm-up.

30 The electric heater is thermostatically controlled. When the coolant flowing through the manifold reaches a temperature of approximately 55°C (131°F), a temperature-sensitive switch in the manifold opens and interrupts the supply to the heater relay.

Intake manifold heater – removal and refitting

31 Disconnect the battery negative lead.

32 Remove the carburettor.

33 Remove the heat shield from below the carburettor. This is secured by two screws (photo).

34 Remove the three bolts which secure the heater to the underside of the manifold. Disconnect the wire from the front of the heater.

8.20 Choke and throttle cable connections to the carburettor

8.24 Removing the carburettor

8.26A Side view of carburettor showing anti-run-on valve

35 Remove the heater from the manifold. If it is stuck gently tap it free through the carburettor mounting hole, using a wooden or plastic drift. Recover the gasket and O-ring (photo).
36 Refit by reversing the removal operations. Use a new gasket and O-ring, and apply anti-seize compound to the heater and its bolts.

9 Ignition system (constant energy type)

Description

1 The 'constant energy' ignition system was fitted to vehicles manufactured in 1985 and 1986. It can also be fitted to any earlier O Series engine. The system is of the electronic (breakerless) type.
2 The main components of the system are the distributor, the amplifier, the ignition coil and the spark plugs and HT leads. The coil is specifically designed for the system.
3 The distributor generates LT pulses by the rotation of a toothed wheel in a magnetic field. These pulses are converted by the amplifier into the more powerful current needed to trigger the coil. Dwell time is automatically controlled by the amplifier to give the most effective spark at all engine speeds. Centrifugal and vacuum advance, and distribution of HT voltage, are carried out in the traditional way.
4 The principal advantage of the system is that there are no contact breaker points to fail or go out of adjustment. Once the timing has been set, there is no reason for it to vary, so the system should always be operating at peak efficiency.

Precautions

5 HT voltage is higher than in a conventional system. Take extra care to avoid electric shocks: do not handle HT leads, the coil or the distributor cap whilst the ignition is switched on.
6 Do not run or crank the engine with one or more HT leads 'floating' (not connected to a spark plug or to earth), as insulation damage could result.
7 Do not open, crush or otherwise tamper with the amplifier, even if it is scrap. It contains beryllium compounds, which can be hazardous to health.

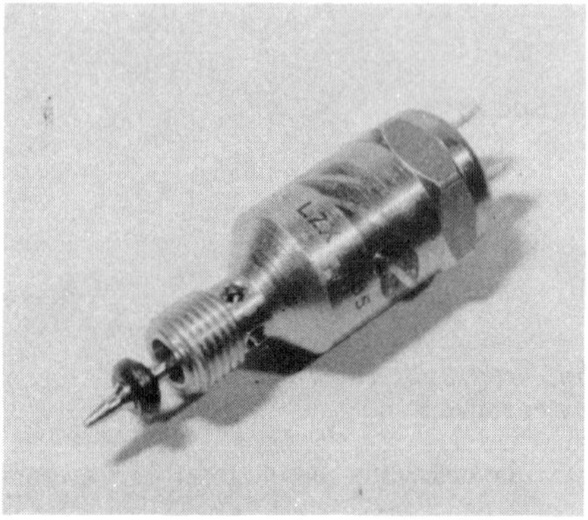

8.26B Anti-run-on valve removed. Note O-ring at tip

8.33 Manifold heat shield screws (arrowed)

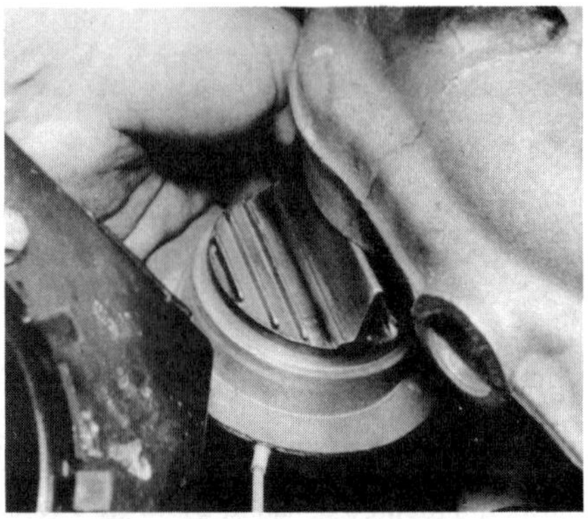

8.35 Removing the intake manifold heater

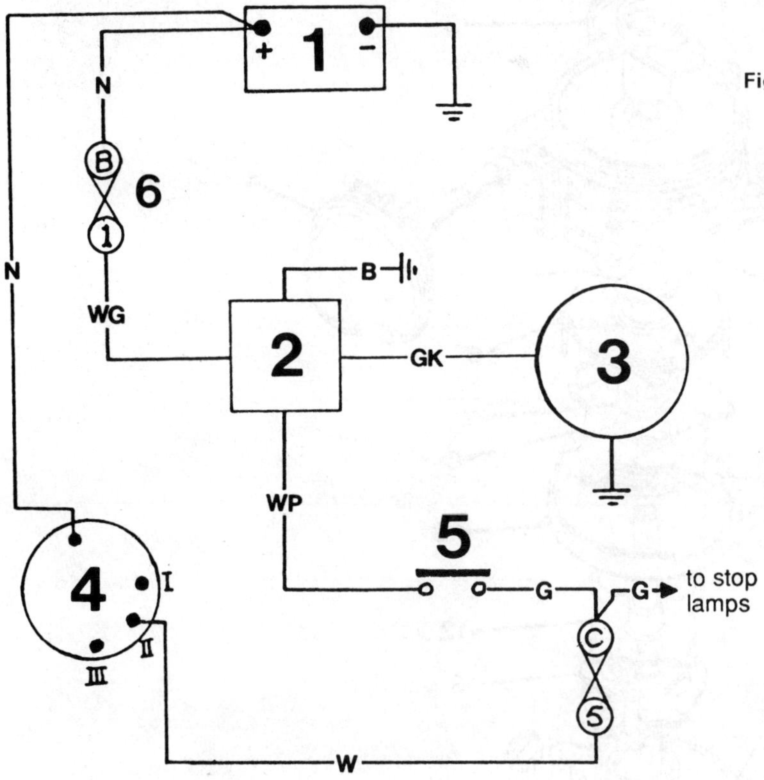

Fig. 13.16 Circuit diagram for manifold heater (Sec 8)

1 *Battery*
2 *Heater relay*
3 *Heater*
4 *Ignition/starter switch*
5 *Temperature-sensitive switch*
6 *Fuse*
B *Black*
G *Green*
K *Pink*
N *Brown*
P *Purple*
W *White*

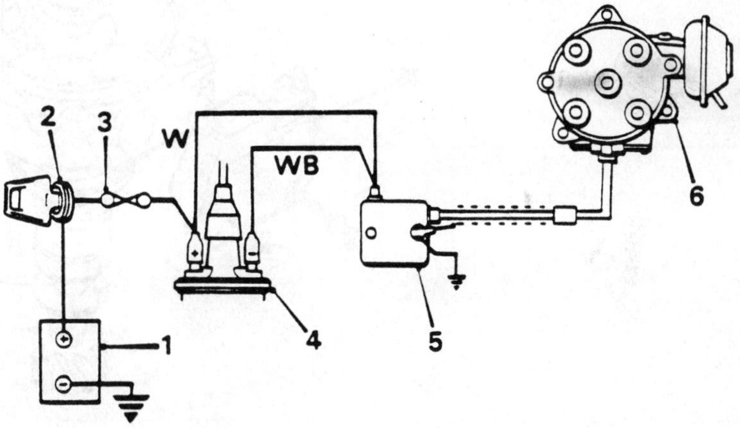

Fig. 13.17 Circuit diagram for constant energy ignition system (Sec 9)

1 *Battery*
2 *Ignition/starter switch*
3 *Fuse*
4 *Ignition coil*
5 *Amplifier (remote, or on distributor)*
6 *Distributor*
B *Black*
W *White*

8 Protect the system against over-voltage which could arise when using electric welding equipment or from the careless use of rapid battery charges.

Maintenance

9 Maintenance is confined to periodic renewal of the spark plugs, and cleaning and inspection of the HT leads, coil tower and distributor cap. The ignition timing should only need checking after removal and refitting of the distributor, or after timing belt renewal.

Distributor – removal and refitting

10 Proceed as in Chapter 4, Section 6, paragraph 10 onwards, with the following differences:

(a) *Set the engine to 90° BTDC (as for timing belt removal) – not to the firing point – before removing the distributor*
(b) *Note that the amplifier module (when mounted on the distributor) is uppermost*
(c) *Check the ignition timing on completion*

Distributor – overhaul

11 Normally the only components requiring attention will be the distributor cap and rotor arm. These should be inspected for cracks, burning, evidence of 'tracking' or other damage, and renewed if necessary. It is prudent to renew them every couple of years on a preventive basis in any case.
12 Before embarking on a complete strip of the distributor, check the cost and availability of spare parts.
13 No specific dismantling instructions are available. Refer to Fig. 13.18 and observe the following points:

(a) *Do not attempt to separate the reluctor from the shaft*
(b) *Apply silicone heat sink compound between the amplifier and the distributor body*
(c) *Shaft endfloat should be 0 to 0.010 in (0 to 0.25 mm)*

Fault diagnosis – constant energy ignition system

14 If a fault develops in the ignition system first check the condition of the spark plugs, HT leads, distributor cap (inside and out), rotor arm and coil tower.
15 Disconnect the coil-to-distributor HT lead from the distributor cap. Connect the lead to a spare spark plug which can rest on the engine, or support it so that its bare end is no more than 0.25 in (6 mm) from bare metal on the engine. Do not hold the lead directly in the hand. Crank the engine and check for a spark; if there are no sparks, this suggests that the coil connections, the HT lead or the coil itself must be faulty.

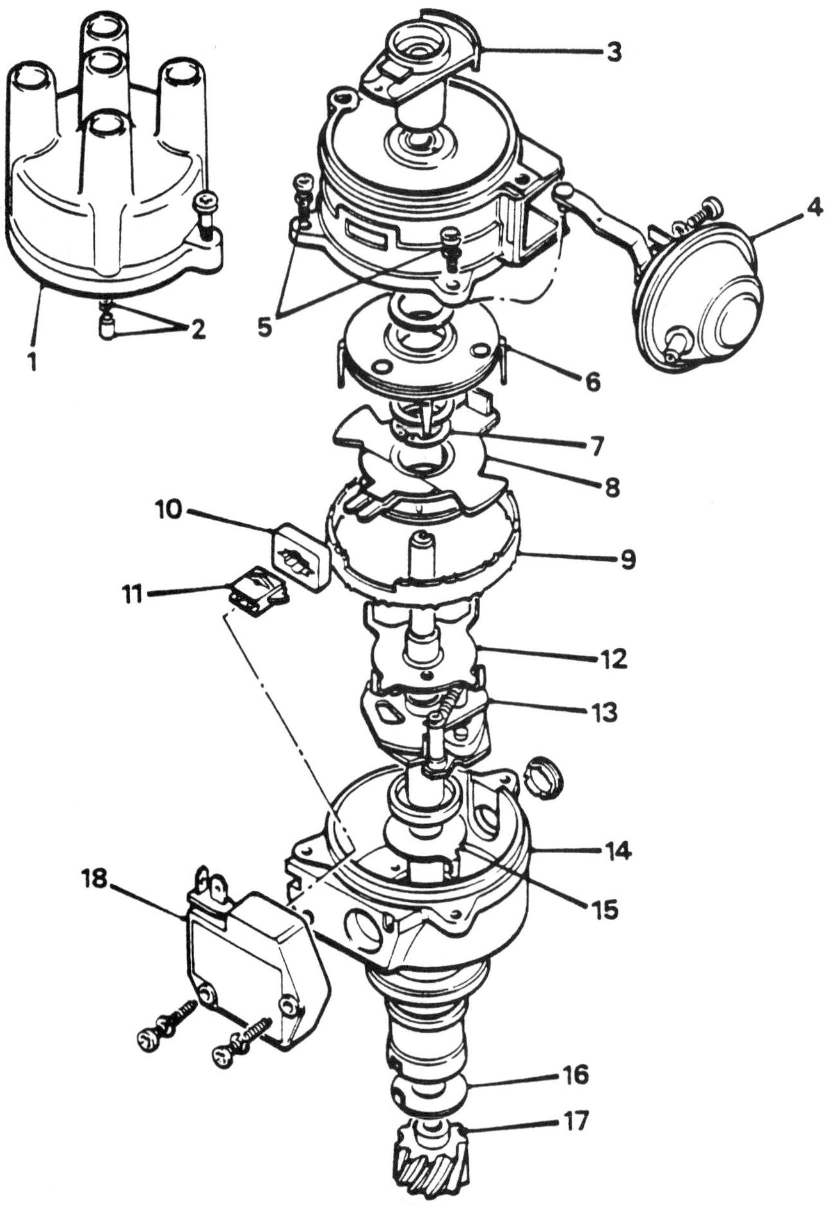

Fig. 13.18 Exploded view of constant energy system distributor (Sec 9)

1 Cap
2 Carbon brush and spring
3 Rotor arm
4 Vacuum unit
5 Screws
6 Stator
7 Circlip
8 Pick-up coil
9 Clamp ring
10 Grommet
11 Connector block
12 Reluctor (toothed wheel)
13 Centrifugal advance mechanism
14 Body
15 Tab washer
16 Thrust washer
17 Gear
18 Amplifier (distributor-mounted type)

16 If the preceding test gave satisfactory sparks, or if the coil has been changed for one known to be good without success, proceed as follows. With the ignition on, connect a voltmeter (0 to 15V) between earth and each of the four points shown in Fig. 13.19. Possible results are as follows:

Point 1 – *More than 11.5V is OK*
Less than 11.5V – battery discharged

Point 2 – *Up to 1V below Point 1 voltage is OK. A drop of more than 1V shows a fault in the ignition switch, fuse or wiring*

Point 3 – *Up to 1V below Point 1 voltage is OK. A drop of more than 1V shows a fault in the ignition coil or the amplifier unit*

Point 4 – *Zero to 0.1V is OK. More than 0.1V shows a bad amplifier earth connection*

17 If the voltage checks are satisfactory, switch off the ignition. Disconnect the amplifier from the distributor pick-up coil, either by unplugging the harness (remote amplifier) or by physically removing it from the distributor body. Measure the pick-up coil resistance, which should be between 950 and 1150– renew the pick-up coil if not. (In the case of the remote amplifier, check first that the incorrect reading is not due to a short-circuit or open-circuit in the distributor-to-amplifier harness).

18 Further checks can only be made by substitution of known good items.

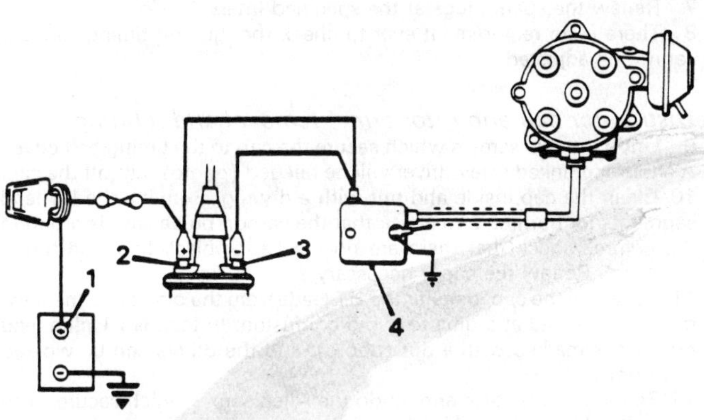

Fig. 13.19 Constant energy ignition system test points. For 1 to 4 see text (Sec 9)

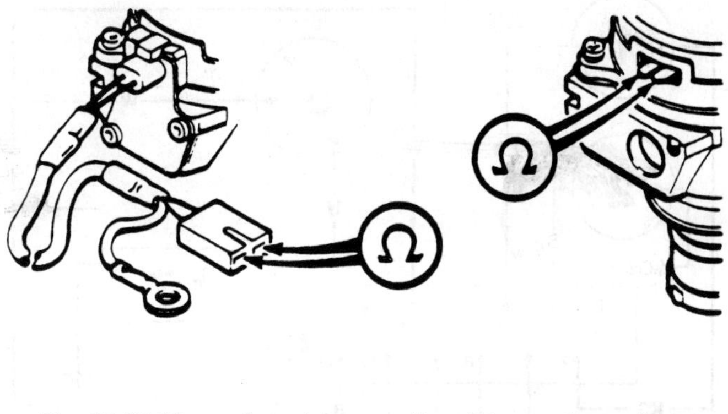

Fig. 13.20 Measuring pick-up coil resistance through the remote amplifier lead (left) and directly (right) (Sec 9)

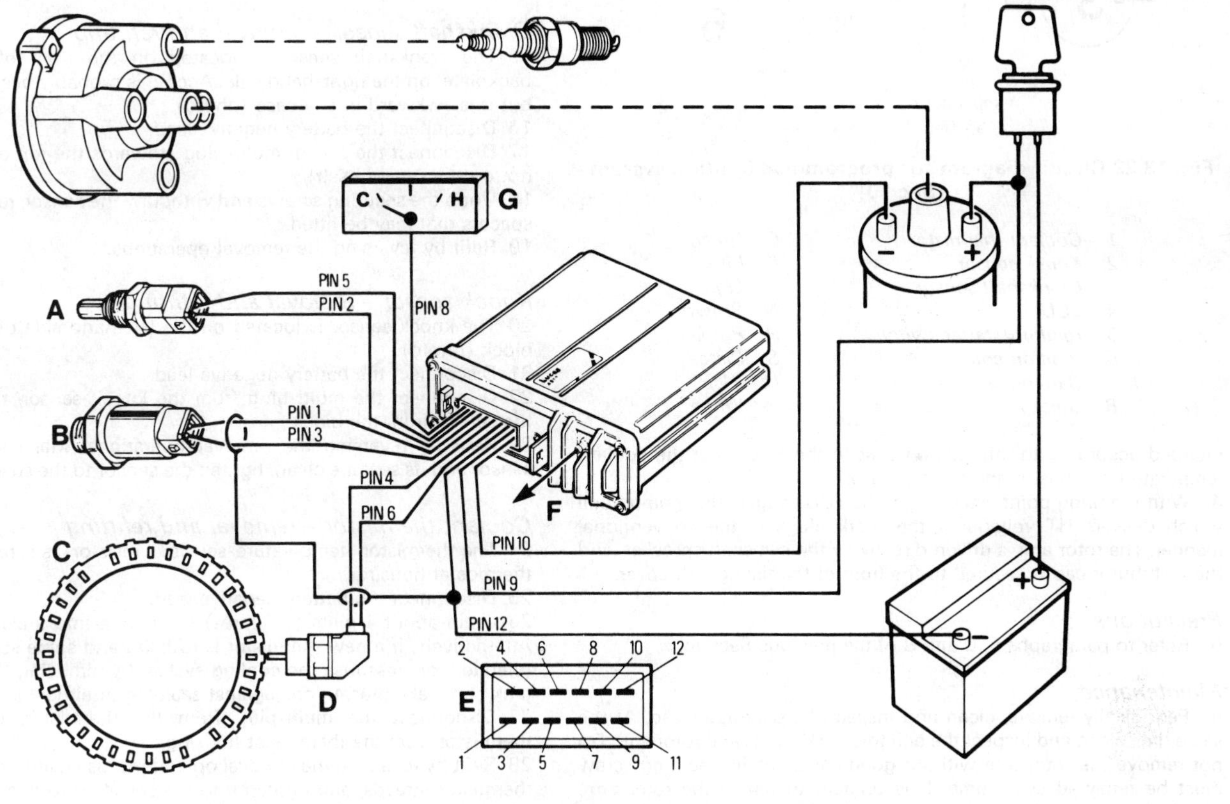

Fig. 13.21 Programmed ignition system components (Sec 10)

A Coolant thermistor
B Knock sensor
C Reluctor ring
D Crankshaft sensor
E Terminal block (on ECU)
F Vacuum signal
G Temperature gauge

10 Ignition system (programmed type)

Description

1 The 'programmed' ignition system is fitted to all models with the 02 Series engine. Like the constant energy system it is breakerless, but it incorporates some much more advanced features. It cannot be fitted to earlier models.

2 A reluctor ring on the periphery of the engine flywheel, and a crankshaft sensor whose inductive head runs between the reluctor ring teeth, replace the operation of the contact breaker points in a conventional system. The reluctor ring utilizes 34 teeth spaced at 10° intervals with two spaces 180° apart corresponding to TDC for Nos 1 and 4 pistons and Nos 2 and 3 pistons respectively. As the crankshaft rotates, the reluctor ring teeth pass over the crankshaft sensor which transmits a pulse to the ignition electronic control unit (ECU) every time a tooth passes over it. The ECU recognises the absence of a pulse every 180° and consequently establishes the TDC position. Each subsequent pulse then represents 10° of crankshaft rotation. This, and the time interval between pulses, allows the ECU to determine crankshaft position and speed.

3 A small bore pipe connecting the inlet manifold to a pressure transducer within the ECU supplies the unit with information on engine load. From these constantly changing data, the ECU selects a particular advance setting from a range of ignition characteristics stored in its memory. This basic setting can be further advanced or

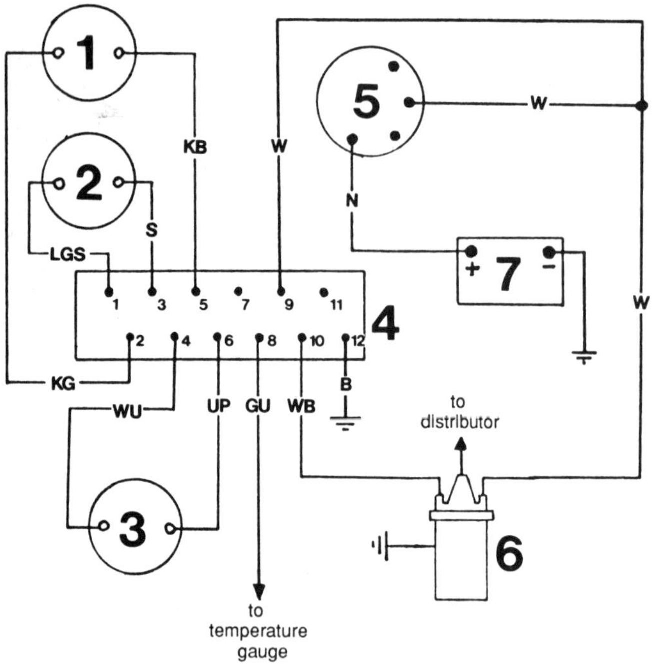

Fig. 13.22 Circuit diagram for programmed ignition system (Sec 10)

1	*Coolant thermistor*	G	*Green*
2	*Knock sensor*	K	*Pink*
3	*Crankshaft sensor*	L	*Light*
4	*ECU*	N	*Brown*
5	*Ignition/starter switch*	P	*Purple*
6	*Ignition coil*	S	*Slate*
7	*Battery*	U	*Blue*
B	*Black*	W	*White*

retarded according to information sent to the ECU from the coolant temperature thermistor and knock sensor.

4 With the firing point established, the ECU triggers the ignition coil which delivers HT voltage to the spark plugs in the conventional manner. The rotor arm is driven directly by the camshaft sprocket, and the distributor cap is secured to the front of the timing belt cover.

Precautions

5 Refer to paragraphs 5, 6 and 8 of the previous Section.

Maintenance

6 Periodically remove, clean and inspect the distributor cap. At the same time clean and inspect the coil tower, HT leads and rotor arm. Do not remove the rotor arm without good cause, as its securing screw must be renewed every time. It is prudent to renew the rotor arm, distributor cap and HT leads every couple of years in any case.

7 Renew the spark plugs at the specified intervals.

8 There is no requirement ever to check the ignition timing, since it cannot be adjusted.

Distributor cap and rotor arm – removal and refitting

9 Undo the two screws which secure the cap to the timing belt cover. A short or cranked screwdriver will be needed (photo). Lift off the cap.

10 Clean the cap inside and out with a dry rag. Examine the HT lead segments for burning and check that the carbon brush stands proud in the centre. Check that there are no cracks, or black lines indicating 'tracking'. Renew the cap if necessary.

11 To renew the cap, transfer the HT leads from the old cap to the new, moving one lead at a time to avoid confusion. In fact, No 1 plug lead position is marked with a dot (photo), and the others can be worked out from that.

12 To remove the rotor arm, undo the Allen screw which secures it to its shaft (photo). A new screw must be used when refitting. Renew the rotor arm if it is cracked or burnt.

13 Fit the rotor arm onto its shaft – it will only go on one way – and secure it with a new screw.

14 Refit the distributor cap and secure it with the two screws.

Crankshaft sensor – removal and refitting

15 The crankshaft sensor is located on the front of the engine backplate, on the right-hand side. Access is probably best from below, but it is awkward in any case (photo).

16 Disconnect the battery negative lead.

17 Disconnect the sensor multi-plug (towards the top of the engine, not on the sensor itself).

18 Undo the securing screws and withdraw the sensor, recovering any spacers that may be fitted.

19 Refit by reversing the removal operations.

Knock sensor – removal and refitting

20 The knock sensor is located on the left-hand side of the cylinder block (photo).

21 Disconnect the battery negative lead.

22 Disconnect the multi-plug from the knock sensor, then unscrew the sensor from the block.

23 Refit by reversing the removal operations. Make sure that the sensor and its seat are clean; tighten the sensor to the specified torque.

Coolant thermistor – removal and refitting

24 The thermistor (temperature-sensitive resistor) is screwed into the thermostat housing.

25 Disconnect the battery negative lead.

26 Drain about 4 pints (2.3 litres) of coolant into a clean container. Alternatively, if a new thermistor is to hand and some spillage can be tolerated, depressurise the cooling system by removing the expansion tank cap. Take precautions against scalding in either case.

27 Disconnect the multi-plug from the thermistor. Unscrew the thermistor from the thermostat housing.

28 Refit by reversing the removal operations; use a little sealant on the thermistor threads, and tighten it to the specified torque.

29 Refill or top up the cooling system on completion.

10.9 Undoing a distributor cap screw

10.11 Dot on distributor cap marks No 1 lead

10.12 Allen screw secures rotor arm

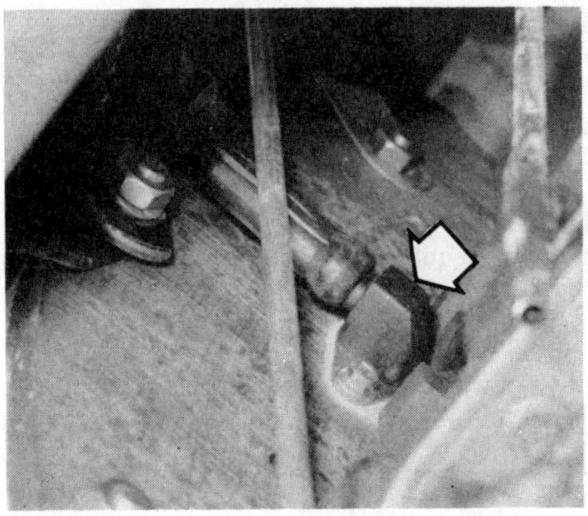
10.15 Crankshaft sensor seen from below (arrowed)

10.20 Ignition system knock sensor

10.31 Disconnecting the ECU multi-plug ...

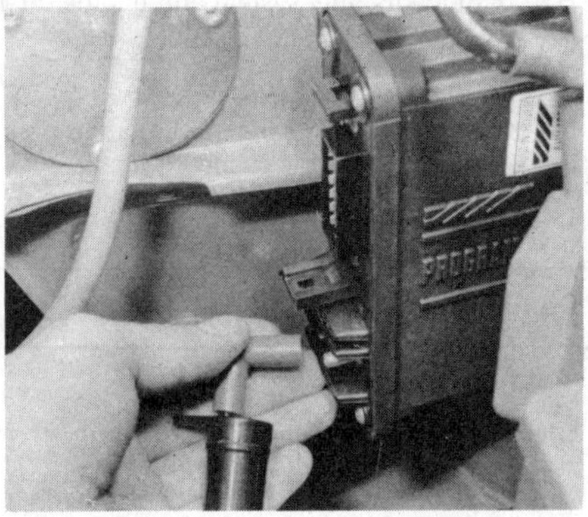
10.32 ... and the vacuum connector

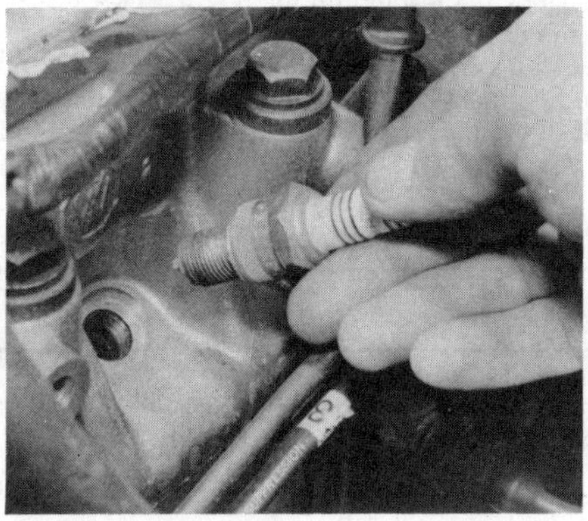
10.43 Fitting a spark plug

Electronic control unit (ECU) – removal and refitting

30 Disconnect the battery negative lead.
31 Release the spring clip and disconnect the multi-plug from the ECU (photo).
32 Carefully pull the vacuum connector off the ECU (photo).
33 Remove the securing screw and lift the ECU off its bracket.
34 Refit the reversing the removal operations.

Ignition coil – general

35 The ignition coil is mounted on the battery tray. Although it looks the same as coils used with contact breaker systems, it is specifically designed for this system and other types should not be substituted.
36 If the coil is suspected of malfunctioning, the simplest test is by substitution of a known good unit. If a multi-meter is available, the resistance of the primary and secondary windings may be checked. Typical values are 1 ohm or less (primary), and 4 to 10 kilohms (secondary).
37 To ensure correct polarity, make sure that the coil LT feed leads are connected to the appropriate terminals (white lead to positive and white/black to negative).

Spark plugs – removal and refitting

38 Identify the plug leads if necessary, then disconnect them from the plugs. Pull on the plug caps, not on the leads.
39 Brush or blow away any dirt from the spark plug seats.
40 Unscrew the plugs, using a proper spark plug socket or box spanner. Be careful not to apply sideways pressure as this may crack the insulator. Access to the plugs is quite good on this engine and most types of the plug spanner may be used.
41 Finish unscrewing the plugs by hand and remove them.
42 Check the electrode gap before refitting and adjust if necessary by carefully bending or tapping the earth (side) electrode. This applies to new plugs as well as old.
43 Apply a smear of anti-seize compound to the threads of one plug and offer it to its hole (photo). Start to screw it in by hand. If it is not possible to screw the plug most of the way home by hand, it is probably cross-threaded. Do not force it in with the spanner, but unscrew it and try again. The aluminium threads in the head are easily damaged.
44 Finish tightening the plug with the spanner or socket and (ideally) a torque wrench. In the absence of a torque wrench, plugs with washers (as fitted to the 02 Series engine) should be tightened no more than one quarter-turn beyond the point where the washer contacts the plug seat. Taper seat plugs (as fitted to the old 0 Series engine) must be tightened no more than one-sixteenth of a turn beyond the point of contact.
45 Fit and tighten the other three plugs. Make sure that the screwed adaptors on the tops of the plugs are secure and the right way round, then refit the plug leads, pushing the plug caps home.
46 Run the engine to verify that the leads have been fitted correctly.

Fault diagnosis – programmed ignition system

47 Faults on the HT distribution side should be traced as described in Chapter 4, Section 13, paragraphs 1 to 6. Remember to take extra precautions against electric shock.
48 Further testing should ideally be carried out by a Freight Rover dealer or other specialist, using dedicated test equipment. If a multi-meter is available, however, the home mechanic may attempt the tests below.

Engine fails to start

Test	Remedy
1 Connect a voltmeter across pins 9(+) and 12(–) of the electronic control unit (ECU) wiring connector. Does the voltmeter indicate battery voltage 10 seconds after switching on the ignition?	Yes: Proceed to test 2 No: Check the wiring between the ignition switch and pin 9, and between pin 12 and earth. Rectify as required
2 Connect a voltmeter across pins 10(+) and 12(–) of the ECU wiring connector. Does the voltmeter indicate battery voltage 10 seconds after switching on the ignition?	Yes: Proceed to test 3 No: Check the wiring between the ignition switch and coil (+) terminal and between pin 10 and the coil (–) terminal. Rectify as required
3 Connect an ohmmeter across the coil terminals. Is the coil primary winding resistance between 0.73 and 0.83 ohms?	Yes: Proceed to test 4 No: Renew the coil
4 Connect a voltmeter between the battery (+) terminal and the coil (–) terminal. Does the reading on the voltmeter increase when the engine is cranking?	Yes: Engine should start. If not check ignition. HT components, fuel system and engine internal components No: Proceed to test 5
5 Switch ignition off and connect an ohmmeter across terminals 4 and 6 of the ECU. Does the ohmmeter register 1.5 k ohms approximately?	Yes: Probable ECU fault No: Check crankshaft sensor wiring and connections. If satisfactory, sensor is suspect

Engine misfires and performance is unsatisfactory

Test	Remedy
1 Highlight ignition timing marks (notch on crankshaft pulley, corresponding notch on water pump bracket) with white chalk, connect a stroboscopic timing light, disconnect ECU vacuum pipe at manifold and start engine. Does the pulley mark advance as engine speed is increased?	Yes: Proceed to test 2 No: Probable ECU fault
2 With the engine operating as in test 1 above, apply suction to the end of the disconnected ECU vacuum pipe. Does the pulley mark advance as suction is applied?	Yes: Ignition system is satisfactory, fault lies elsewhere No: Check for leaks in vacuum pipe and connections. If satisfactory, ECU is faulty

Fig. 13.23 Clutch components (5-speed models) (Sec 11)

1 Flywheel
2 Lockwasher
3 Disc (driven plate)
4 Pressure plate assembly
5 Release bearing
6 Release lever
7 Ballhousing
8 Dowel

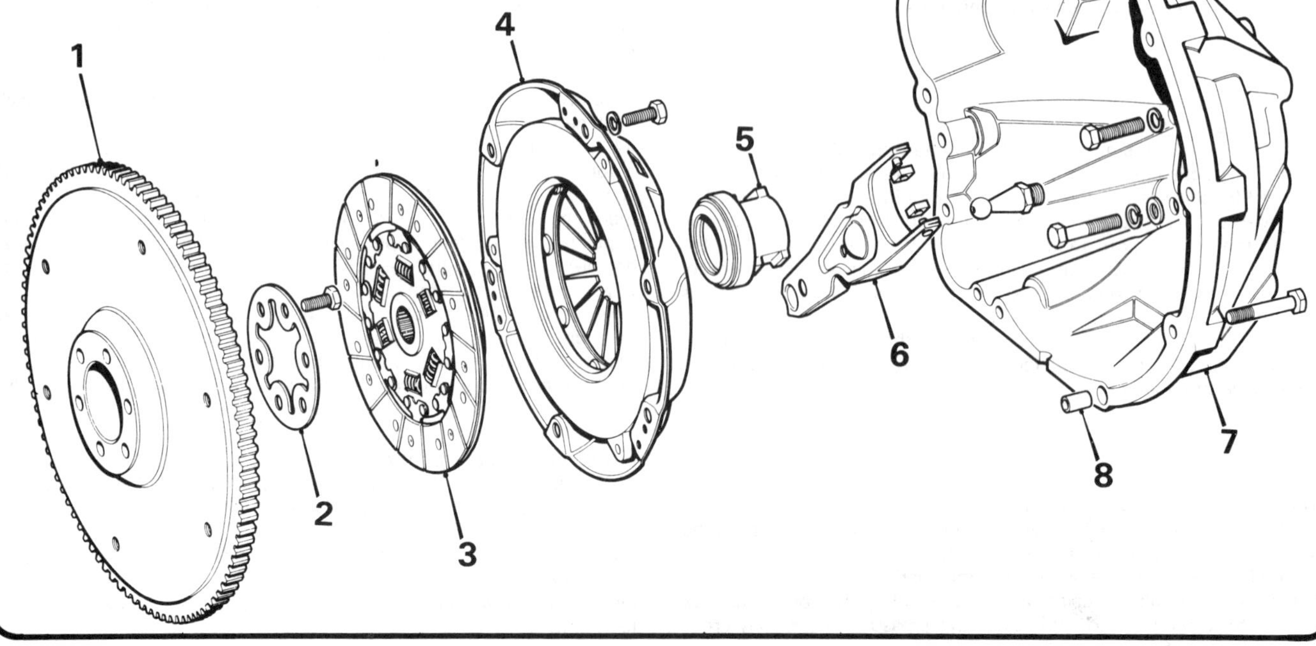

11 Clutch (5-speed models)

Description

1 On models with a 5-speed gearbox, a diaphragm spring clutch is fitted. The diaphragm spring takes the place of the coil springs, release levers (toggles) and ring described in Chapter 5.

Examination

2 When examining the diaphragm spring type clutch, inspect the spring fingers for cracks or distortion. Also inspect the tips of the fingers for wear where the release bearing acts on them. Renew the pressure plate assembly if wear or damage is evident.

Release bearing – removal and refitting

3 Remove the engine or the gearbox, as desired. Unclip the clutch pushrod from the release lever.

4 Using a 'crow's foot' or cranked open-ended spanner (19 mm/$^3/_4$ in AF), unscrew the release lever pivot from the gearbox front cover. (It is possible simply to pull the release lever off the pivot, but this risks straining the spring clip). Remove the pivot, lever and bearing together (photo).

5 Unhook the release lever from the bearing. Recover the bearing pads from the lever fork (photo).

6 Refit by reversing the removal operations, noting the following points:

(a) *Lubricate the bearing pads, pivot and pushrod up with a smear of molybdenum-based grease*

(b) *Tighten the pivot to the specified torque (see 5-speed gearbox Specifications)*

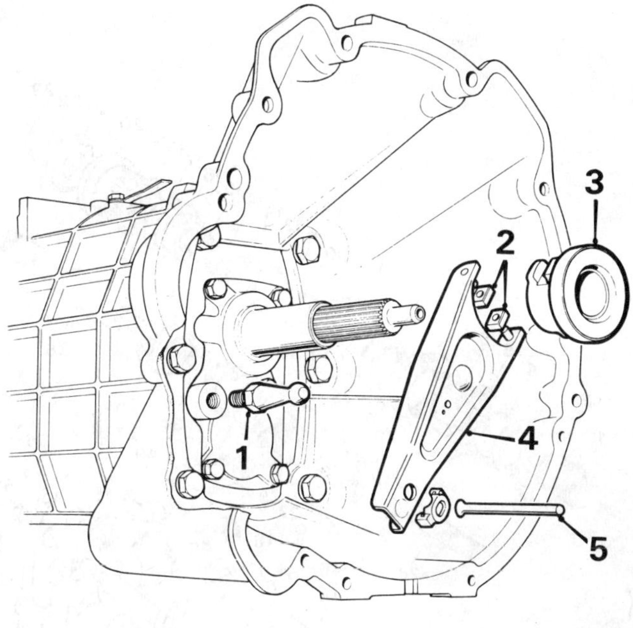

Fig. 13.24 Clutch release components (5-speed models) (Sec 11)

1 Release lever pivot
2 Bearing pads
3 Bearing
4 Release lever
5 Pushrod

12 Manual gearbox (5-speed)

Description

1 The gearbox is of the usual constant mesh type, with synchromesh on all forward gears. the 5th gear components are on the far side of a plate which carries the mainshaft and laygear bearings. The laygear is in one piece with the layshaft. Taper roller bearings are used.

2 Gear selection is by a single rod, along with the associated forks and interlock components.

3 A positive lubrication system is employed, the oil pump being driven from the tail of the laygear. This imposes restrictions on towing the vehicle, as the oil pump will not operate if the engine is not running. Refer to Section 4.

4 This gearbox is essentially the same as that fitted to the Range Rover, but the bellhousing and rear extension housing are different.

Maintenance

5 Check the oil level at the specified intervals, or if leakage or excessive noise become evident. The oil filler/level plug is located on the left-hand side of the rear extension housing (photo).

6 With the vehicle parked on level ground, wipe clean around the filler/level plug and unscrew and remove it. The oil level should be just up to the bottom of the plug hole. If not, top up with clean oil of the specified type. Allow any excess to drip out, then refit and tighten the plug.

7 Periodic oil changing is not specified by the makers, but a drain plug is fitted for the owner who wishes to use it. The drain plug is on the right-hand side of the main casing; it is magnetic and the filings adhering to it should be cleaned off before refitting it (photos).

Gearbox – removal and refitting

8 The procedure is similar to that described in Chapter 6, Section 2. Note the following points:

(a) *Gear lever removal and refitting is described later in this Section.*

(b) *The reversing light switch leads must be disconnected during removal.*

(c) *The gearbox rear mountings are slightly different. Slacken the locknut before undoing the centre mounting bolt.*

(d) *The speedometer cable is secured by a bolt and clamp plate (photo).*

(e) *The clutch pushrod is captive on the release lever (photo) and tends to drop during refitting. Have an assistant guide it into its hole.*

11.4 Removing the clutch release components (5-speed)

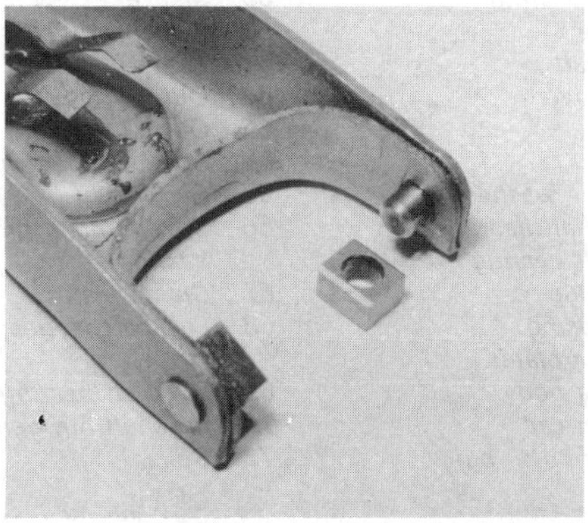

11.5 Clutch release lever with a bearing pad removed

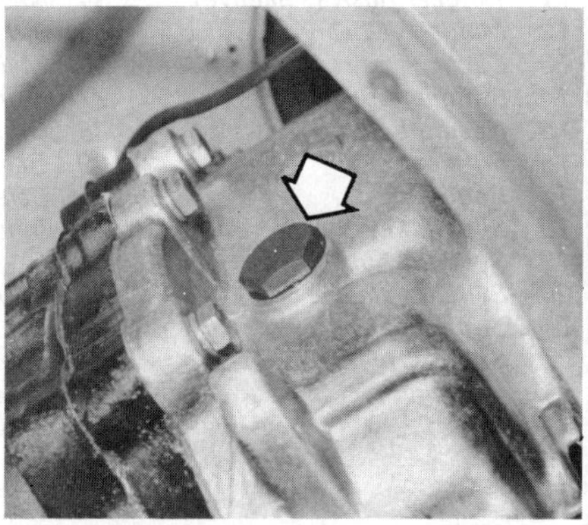

12.5 Gearbox filler/level plug (arrowed)

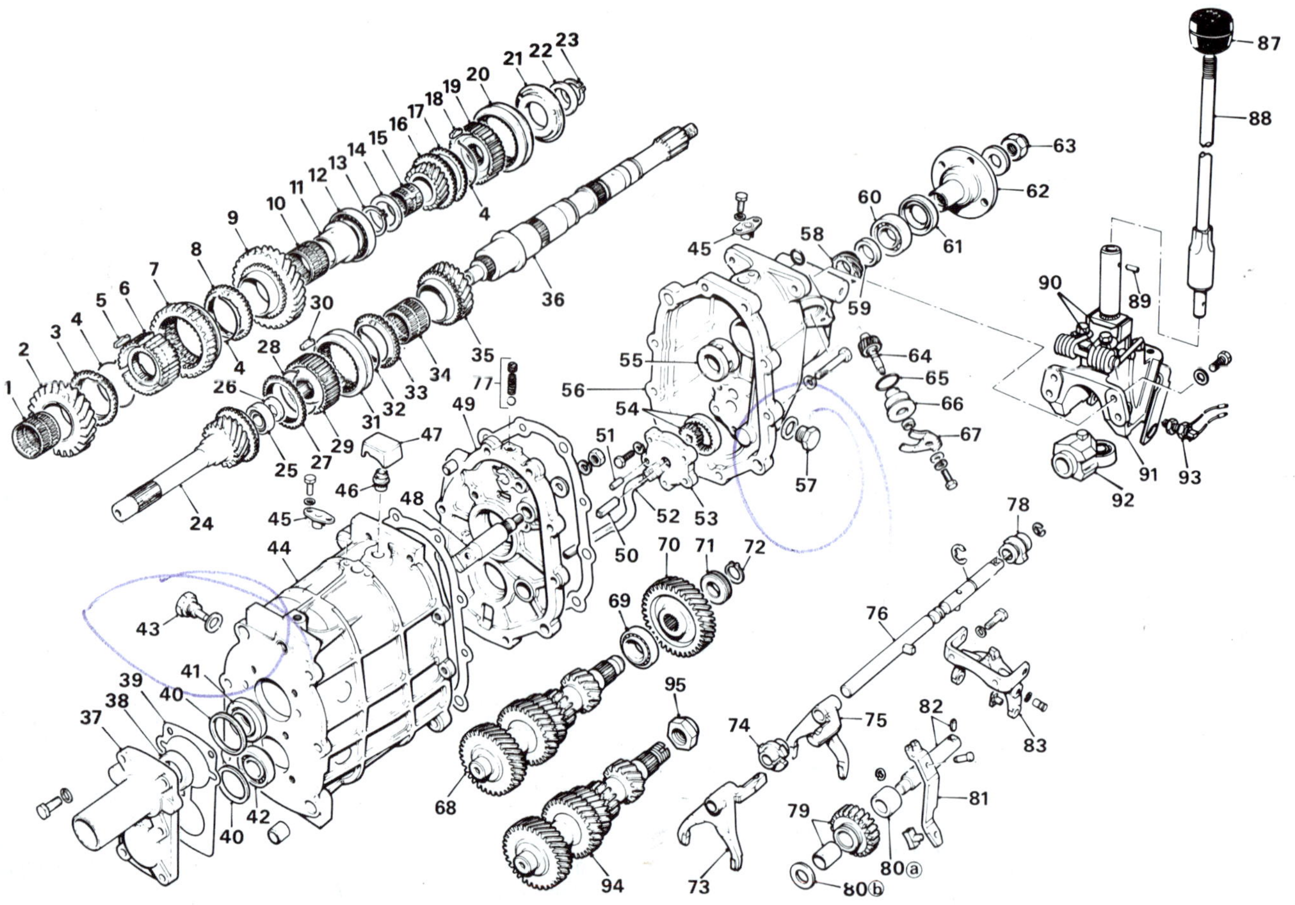

Fig. 13.25 Exploded view of the 5-speed gearbox (Sec 12)

1 *Needle bearing*
2 *2nd gear*
3 *Baulk ring*
4 *Synchro springs*
5 *Sliding key*
6 *Synchro hub*
7 *Synchro sleeve*
8 *Baulk ring*
9 *1st gear*
10 *Needle bearing*
11 *1st gear bush (selective)*
12 *Mainshaft bearing*
13 *Circlip*
14 *5th gear collar*
15 *Needle bearing*
16 *5th gear*
17 *Baulk ring*
18 *Sliding key*
19 *Synchro hub*
20 *Synchro sleeve*
21 *Support plate*
22 *Support plate washer (selective)*
23 *Circlip*
24 *Input shaft*
25 *Pilot bearing*
26 *Spacer*
27 *Baulk ring*
28 *Synchro spring*
29 *Synchro hub*
30 *Sliding key*
31 *Synchro sleeve*
32 *Synchro spring*
33 *Baulk ring*
34 *Needle bearing*
35 *3rd gear*
36 *Mainshaft*
37 *Front cover*
38 *Oil seal*
39 *Gasket*
40 *Selective washers*
41 *Input shaft bearing*
42 *Layshaft bearing*
43 *Drain plug*
44 *Main casing*
45 *Interlock plates*
46 *Breather body*
47 *Breather cap*
48 *Reverse cover pin*
49 *Centre plate*
50 *Oil pump driveshaft*
51 *Dowel*
52 *Oil pump pick-up pipe*
53 *Oil pump cover*
54 *Oil pump gears*
55 *Oil feed ring*
56 *Rear extension*
57 *Oil filler/level plug*
58 *Speedometer drivegear*
59 *Sleeve*
60 *Rear bearing*
61 *Rear oil seal*
62 *Output flange*
63 *Output flange nut*
64 *Speedometer pinion*
65 *O-ring*
66 *Speedometer pinion bearing*
67 *Clamp plate*
68 *Layshaft/laygear (early type)*
69 *Layshaft bearing*
70 *Layshaft 5th gear*
71 *Collar*
72 *Circlip*
73 *3rd/4th selector fork*
74 *Interlock spool*
75 *1st/2nd selector fork*
76 *Selector shaft*
77 *Detent plug, spring and ball*
78 *Interlock spool*
79 *Reverse idler gear and bush*
80a *Reverse idler spacer*
80b *Reverse idler spacer*
81 *Reverse lever*
82 *Reverse lever pivot pin*
83 *5th selector fork and bracket*
87 *Gear knot*
88 *Gear lever*
89 *Pin*
90 *Bias springs*
91 *Selector housing*
92 *Selector yoke*
93 *Reversing light switch*
94 *Layshaft/laygear (later type)*
95 *5th gear layshaft nut*

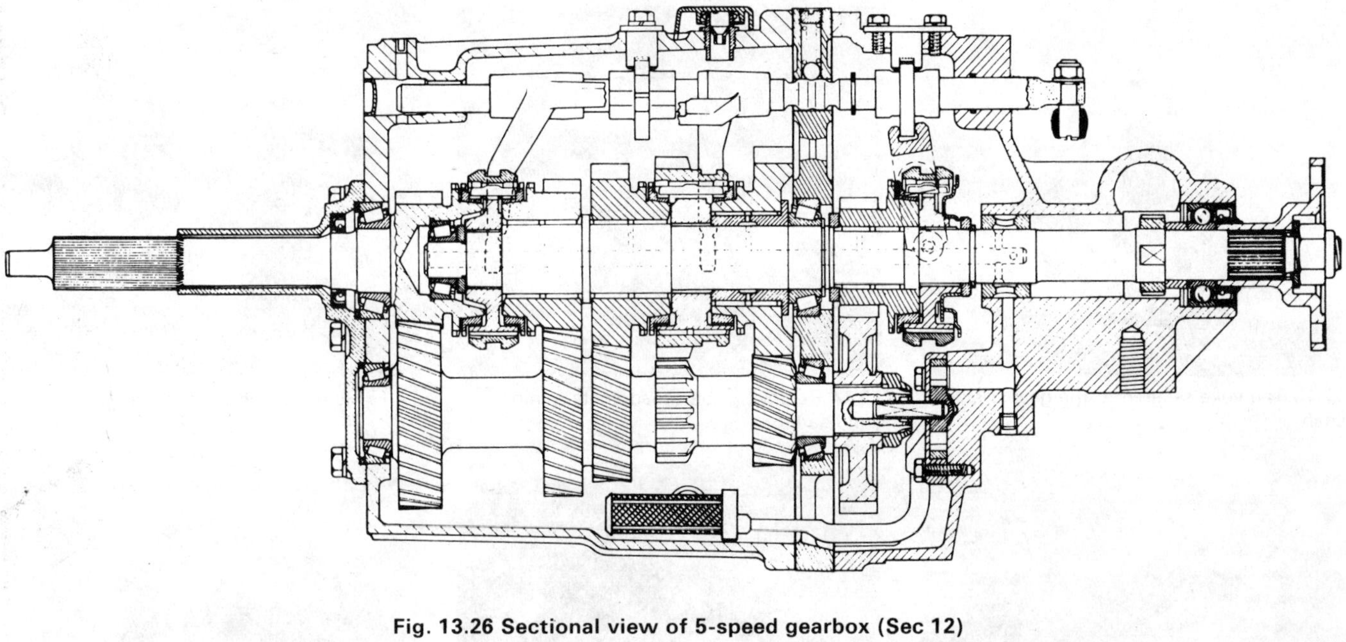

Fig. 13.26 Sectional view of 5-speed gearbox (Sec 12)

12.7A Gearbox drain plug

12.7B Clean the filings off the drain plug before refitting

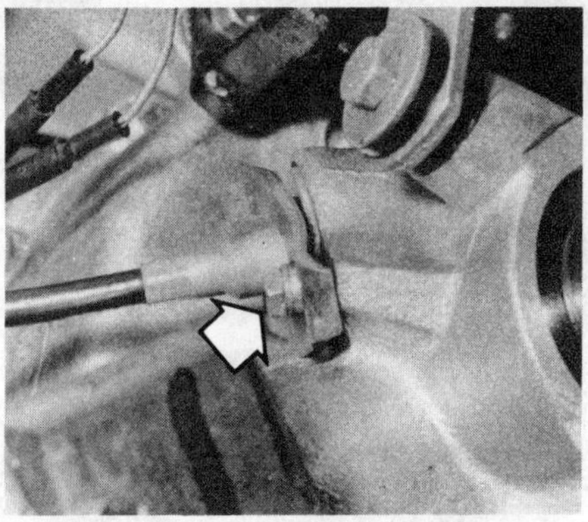

12.8A Bolt (arrowed) securing speedometer cable clamp plate

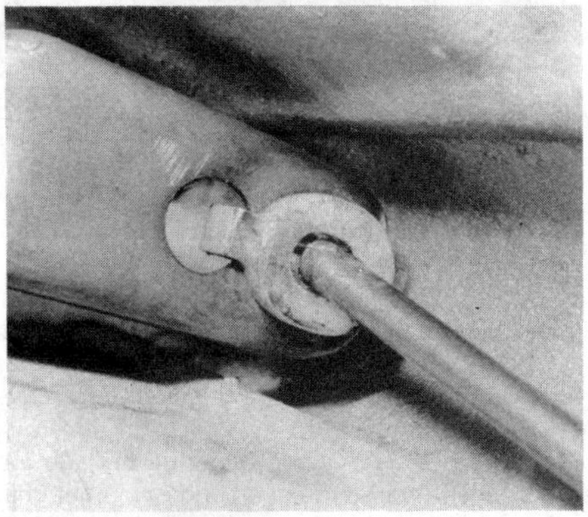

12.8B Clutch pushrod clipped to release lever

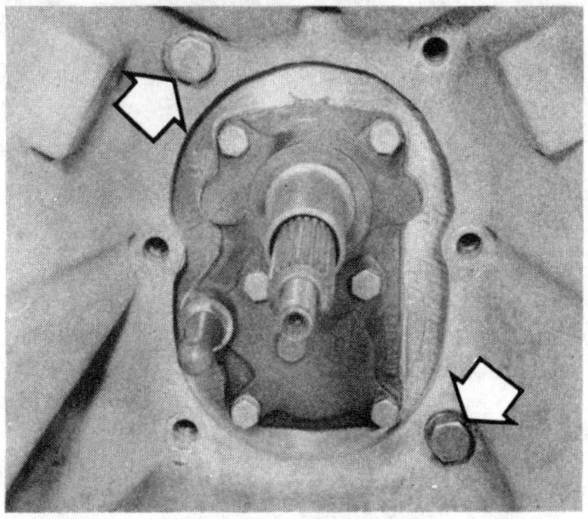

12.11 Two longer bots (arrowed) must be fitted in these positions

12.12 Unbolting the gear selector housing

Gearbox – dismantling into major assemblies

9 Drain the oil from the gearbox (if not already done). Clean the gearbox with paraffin or a grease solvent and dry it.

10 Remove the clutch release bearing and lever – see Section 11.

11 Remove the six bolts which secure the bellhousing to the main casing. Note the position of the two longer bolts, which are fitted with both spring and plain washers (photo). Remove the bellhousing.

12 Unbolt the gear selector housing from the rear extension. Note the arrangement of the spacers. Remove the housing complete with reversing light switch (photo).

13 Remove the gear selector yoke, which is secured to the selector shaft by a single nut and bolt (photo).

14 Remove the speedometer pinion and bearing (photo).

15 Undo the output flange nut. This nut is very tight: restrain the

12.13 The yoke secured to the gear selector shaft

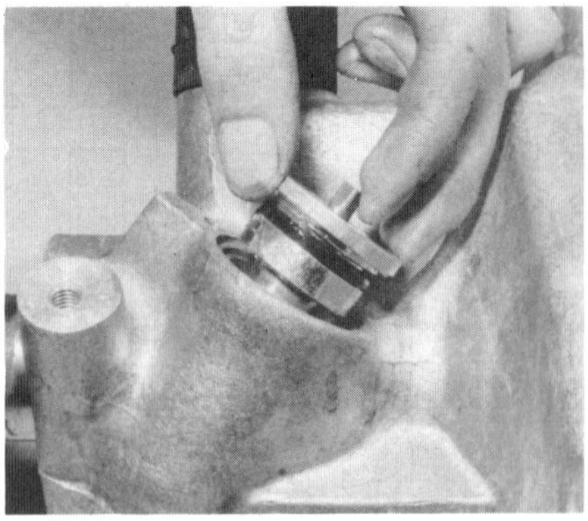
12.14 Removing the speedometer pinion and bearing

12.15 Removing the output flange

12.16 Removing the interlock plate from the rear extension

12.21 5th selector spool correctly fitted

12.24 Selector shaft circlip (arrowed)

12.25 5th synchro secured by its circlip

12.27A Removing 5th gear needle roller bearings ...

12.27B ... and 5th gear collar

flange with a bar and a couple of old nuts and bolts. Remove the nut, washer and flange (photo). If the flange is stuck, use a puller – **do not** hammer it off.

16 Remove the interlock plate from the top of the rear extension. It is secured by two bolts (photo).

17 Remove the ten bolts which secure the rear extension to the main casing. Two of the bolts (at opposite corners) are longer – they are fitted where there are locating dowels in the centre plate.

18 Separate the rear extension from the centre plate by hitting the lugs on the extension with a wooden or hide mallet. Make sure that the centre plate stays with the main casing.

19 Remove the rear extension complete with speedometer drivegear, rear bearing and seal, and oil pump. Recover the oil pump driveshaft from the pump or laygear.

20 Fit two of the rear extension bolts to the centre plate and nip them up to secure the plate to the main casing.

21 Remove the two bolts which secure the 5th selector fork bracket. Remove the bracket, fork and spool: note which way round the spool is fitted (photo).

22 On early models, layshaft 5th gear is retained by a circlip. Remove the circlip.

23 On later models, layshaft 5th gear is retained by a nut. Engage 5th gear by sliding the synchro sleeve towards the gear, and any other gear by twisting and pulling the selector shaft: this will lock the gearbox shafts. Relieve the staking on the nut, have an assistant steady the gearbox and unscrew the nut. Return the gears to the neutral position.

24 On all models, remove the selector shaft circlip (photo).

25 Remove 5th gear synchro circlip (photo).

26 Remove 5th synchro unit complete with support plate washer and 5th gear. A puller will be needed: be careful that it does not foul the selector shaft, which can be bent quite easily.

27 Remove 5th needle roller bearing and 5th gear collar from the mainshaft (photos).

28 On early models with a circlip instead of a nut retaining 5th gear,

12.29 Pulling off layshaft 5th gear

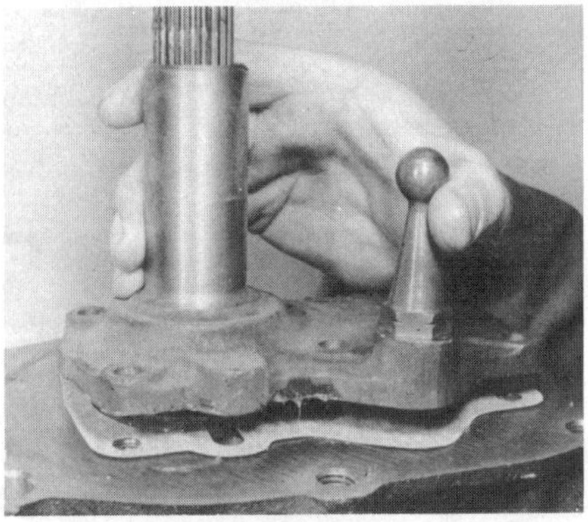

12.30 Removing the front cover. Clutch release pivot is still in place

12.31A Remove the selective washers (arrowed) ...

12.31B ... the layshaft bearing outer track ...

12.31C ... and the input shaft bearing outer track

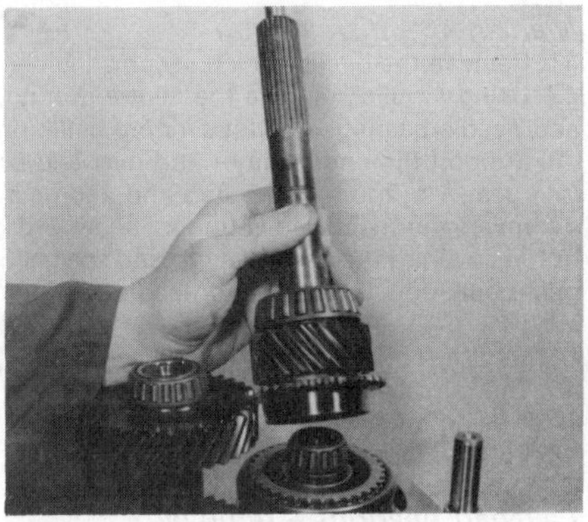

12.35A Remove the input shaft ...

12.35B ... and 4th gear baulk ring

12.36 Removing the reverse lever pivot pin circlip

12.41 Removing the centre bearing circlip

pull off the collar which sits between layshaft 5th gear and the circlip.
29 On all models, pull of layshaft 5th gear. Take the thrust of the puller by screwing the nut back on a few turns, or with a piece of scrap metal (photo).
30 Place the gearbox on a stand or in a vice so that the input shaft is uppermost. Unbolt and remove the front cover (photo).
31 Recover the gasket, the selective washers and the bearing outer tracks (photos).
32 Remove the selector shaft front interlock plate, which is secured by two bolts. One of the bolts also secures the reversing light switch wire bracket.
33 Unscrew the detent plug from the centre plate. Remove the plug, spring and ball; fish out the ball with a pencil magnet if it is stuck.
34 Remove the two bolts which were fitted to retain the centre plate (paragraph 20). Carefully lift the main casing off the centre plate, leaving the geartrains and selector mechanism on the plate.
35 Lift off the input shaft and 4th gear baulk ring (photos).
36 Remove the circlip from the reverse lever pivot pin (photo). Remove the pivot pin, reverse lever and the slipper pad.
37 Remove the reverse idler shaft downwards. Recover the two spacers and reverse idler gear.
38 Have an assistant tip the mainshaft away from the laygear. Lift out the laygear, being careful not to damage the bearings by forcing it away from the mainshaft.
39 Rotate the selector shaft so that the pin below the plate will pass through the slot in the plate. Lift out the mainshaft, selector shaft and forks together, then separate the forks from the gears.
40 The gearbox is now reduced to its major components. Further dismantling will depend on the work to be done. Check the cost and availability of spare parts before proceeding.

12.42 Pulling off 1st gear, bush and bearing

12.45 Extracting the input shaft pilot bearing track

12.54 Driving a bearing track out of the centre plate

Mainshaft – dismantling

41 Remove the circlip which secures the centre bearing (photo).
42 Using a puller, remove the centre bearing, 1st gear bush, 1st gear and needle bearing, and 1st synchro baulk ring (photo).
43 Support the underside of 2nd gear and press the mainshaft out of 2nd gear, 1st/2nd synchro hub and 2nd gear needle bearing.
44 Invert the mainshaft. Support under 3rd gear and press the shaft out, recovering the pilot bearing and spacer, 3rd/4th synchro hub, 3rd gear baulk ring, 3rd gear and needle bearing.

Input shaft – dismantling

45 The pilot bearing track can be pulled out of the rear of the input shaft using an extractor tool (photo).
46 Pull the taper roller bearing from the front of the input shaft.

Layshaft bearings – removal

47 The bearings can be removed from the layshaft/laygear assembly using a puller.

Housings and centre plate – dismantling

Main casing

48 Remove the plastic oil trough from the main casing. The trough must be renewed during reassembly.
49 Remove the breather from the top of the main casing if desired.

Rear extension

50 Remove the three bolts which secure the oil pump. Remove the pump cover, gears and oil pick-up pipe.
57 Drive or lever out the oil feed ring and the rear oil seal.
52 Drive out the rear bearing. Recover the speedometer drivegear and sleeve.
53 Remove the selector shaft O-ring.

Centre plate

54 Support the centre plate and drive out the bearing tracks (photo).
55 If it is intended to remove the reverse lever pin, mark its position first so that it can be refitted in the same position.
56 Remove the oil pump strainer, which is secured by a circlip. A new strainer must be obtained for reassembly.

Synchro units – dismantling

57 Make alignment marks between the synchro sleeves and hubs so that they can be reassembled in the same positions. Dismantle one unit at a time so that the components do not get mixed up.
58 On 5th gear synchro only, lever off the support plate.
59 Unhook the springs, withdraw the sliding keys and separate the hub and sleeve.

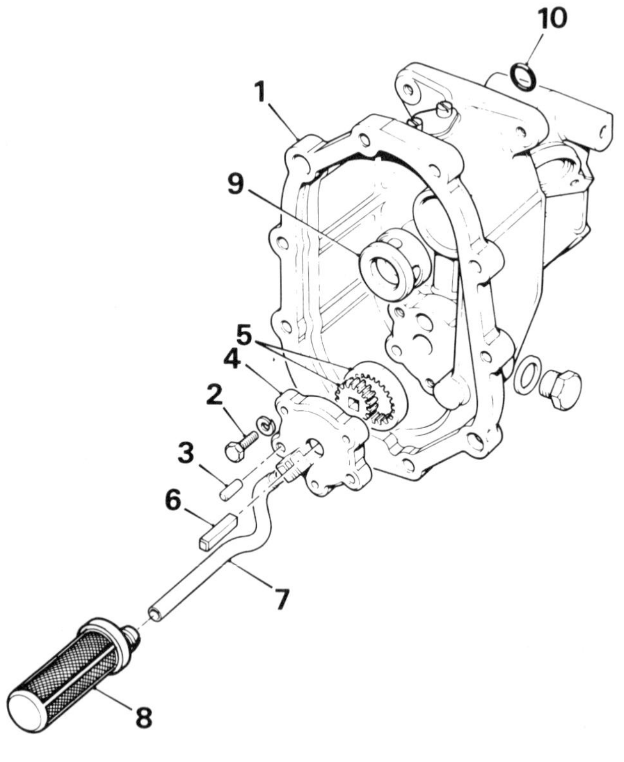

Fig. 13.27 Gearbox oil pump and rear extension (Sec 12)

1 Extension housing
2 Oil pump bolt
3 Dowel
4 Oil pump cover
5 Oil pump gears
6 Oil pump driveshaft
7 Oil pump pick-up pipe
8 Oil pump strainer (on far side of centre plate)
9 Oil feed ring
10 Selector shaft O-ring

Gearbox components – inspection

60 Clean all components in solvent, then lubricate the bearing races with clean ATF. Do not spin the bearings whilst they are unlubricated.
61 Renew gaskets, oil seals, O-rings and circlips as a matter of course. Also renew the plastic trough in the main casing, the oil pump strainer, the rear extension oil feed ring and (when applicable) the layshaft 5th gear nut.
62 Inspect the bearing tracks and races for scoring, cracks or other damage. Spin the lubricated races and check for roughness. Renew as necessary.
63 Inspect the synchro components for wear or damage. Offer the baulk rings to their hubs and check the clearance between them using feeler blades (photo). The minimum acceptable clearance is given in the Specifications; renew the baulk rings if it is not met.
64 Inspect the gears, needle roller bearings and spacers for wear, physical damage and signs of overheating. Renew the needle roller bearings as a matter of course unless they are obviously in perfect condition.
65 Check the selector forks, detent and interlock components for wear and damage and renew as necessary. Check the selector shaft for

12.63 Checking the clearance between a synchro hub and baulk ring

12.66 Oil pump gears and cover

12.69 An assembled synchro unit

straightness. Do not attempt to remove the pins or the captive selector fork from the shaft.
66 Inspect the oil pump gears and cover for obvious wear or damage (photo). No limits are specified.

Gearbox reassembly – general

67 Read through the procedures first to see what is involved. Depending on the work carried out, a dial test indicator, a selection of bearing spacers, a 1st gear endfloat spacer and layshaft support components may be needed. See Fig. 13.28 and 13.29 for the dimensions of locally made items.
68 Lubricate all components with clean ATF as reassembly proceeds. Apply a smear of grease to gasket faces and use a general purpose sealant on bolt threads unless otherwise instructed.

Synchro units – reassembly

69 Refit the synchro sleeves to their hubs, observing the alignment marks made when dismantling. Refit the sliding keys and insert the springs (photo). Note that the springs run in opposite directions (Fig. 13.30).
70 Refit the 5th synchro support plate. The tag on the plate locates in a slot in the hub.

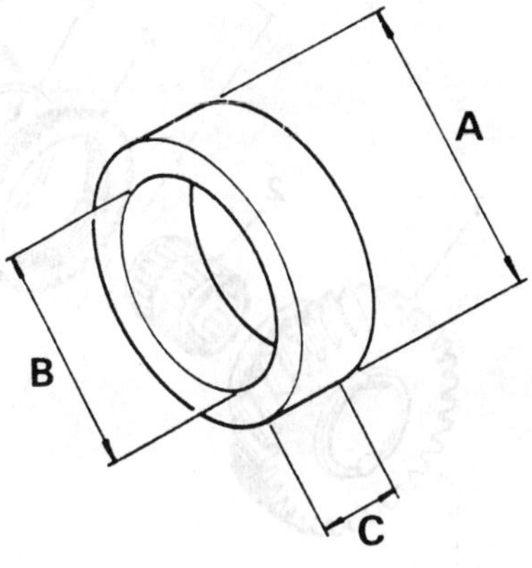

Fig. 13.28 Dimensions of 1st gear endfloat spacer (Sec 12)

A 2.10 in (53.34 mm)
B 1.500 $^{+0.002}_{-0.000}$ in
(38.10 $^{+0.05}_{-0.00}$ mm)
C 0.650 $^{+0.002}_{-0.000}$ in
(16.51 $^{+0.05}_{-0.00}$ mm)

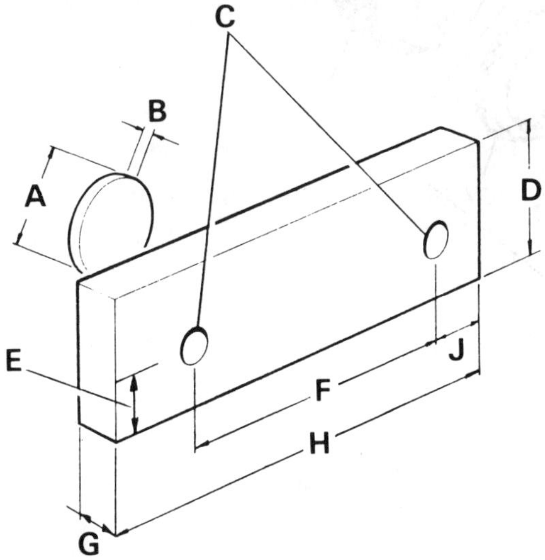

Fig. 13.29 Layshaft support plate and disc. Most dimensions are not critical (Sec 12)

A 1 in (25.4 mm)
B 0.13 in (3.3 mm)
C 0.354 in (9.0 mm)
D 1.5 in (38.1 mm)
E 0.625 in (15.9 mm)
F 3.375 in (85.7 mm)
G 0.5 in (12.7 mm)
H 5 in (127 mm)
J 0.562 in (14.3 mm)

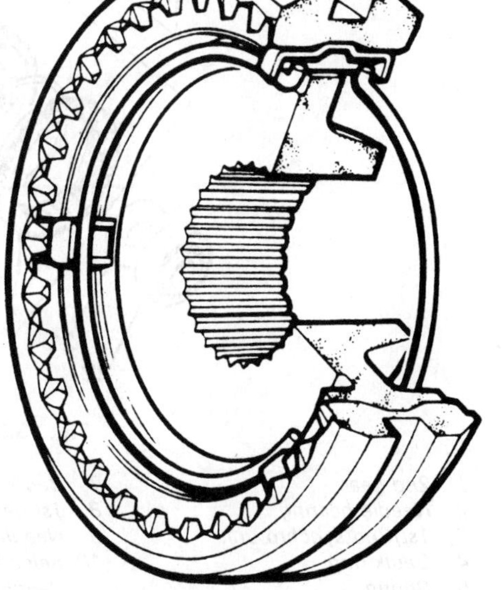

Fig. 13.30 Cutaway view of synchro unit showing the two springs (Sec 12)

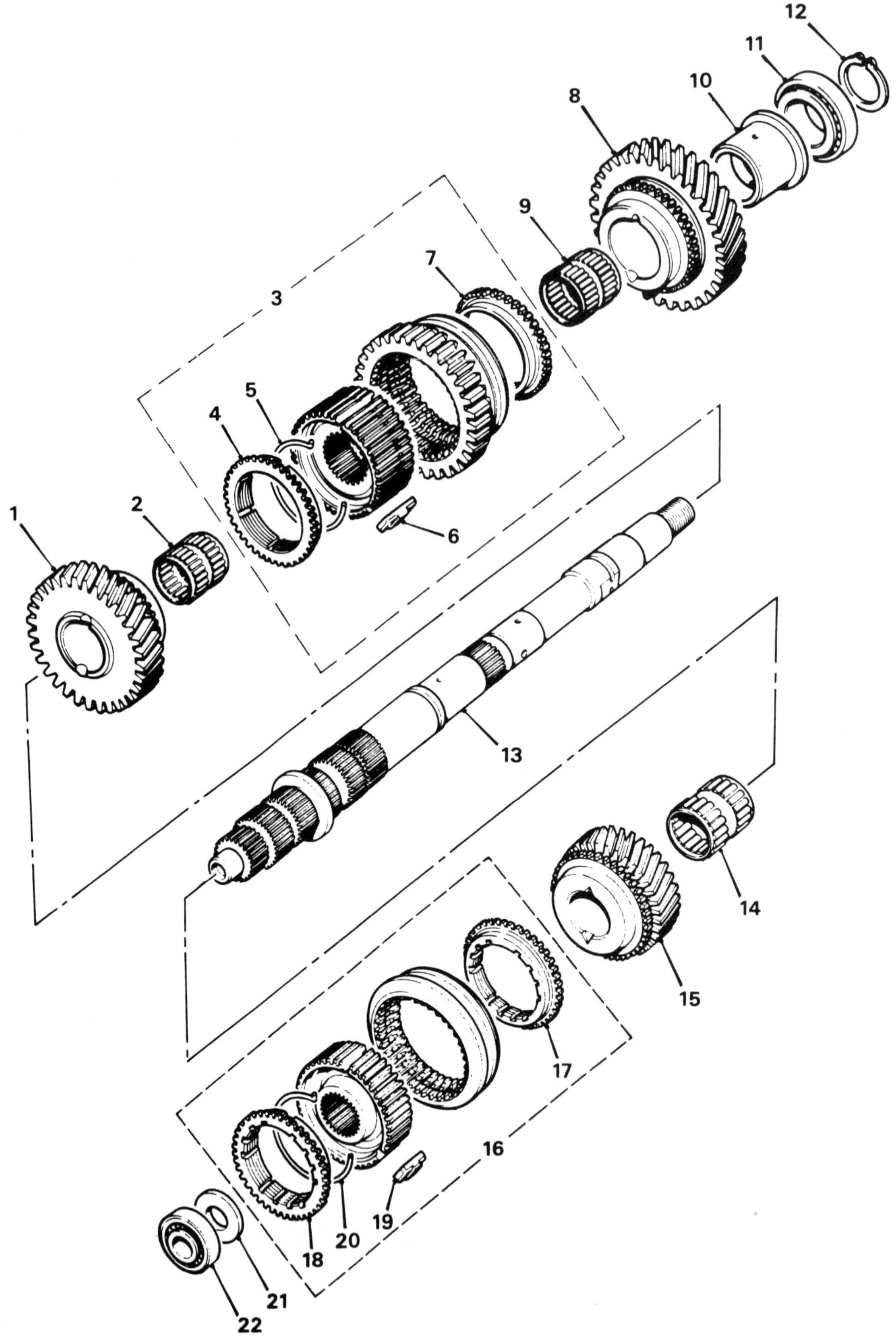

Fig. 13.31 Mainshaft components (Sec 12)

1	*2nd gear*	*7*	*Baulk ring*	*13*	*Mainshaft*	*18*	*Baulk ring*
2	*Needle bearing*	*8*	*1st gear*	*14*	*Needle bearing*	*19*	*Sliding key*
3	*1st/2nd synchro unit*	*9*	*Needle bearing*	*15*	*3rd gear*	*20*	*Spring*
4	*Baulk ring*	*10*	*Selective bush*	*16*	*3rd/4th synchro unit*	*21*	*Spacer*
5	*Spring*	*11*	*Bearing*	*17*	*Baulk ring*	*22*	*Bearing*
6	*Sliding key*	*12*	*Circlip*				

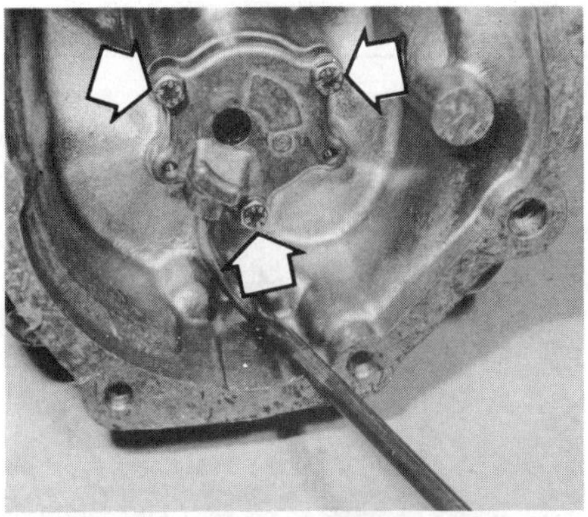
12.71 Oil pump cover bolts (arrowed)

12.75A Fitting 2nd gear needle bearing ...

12.75B ... 2nd gear ...

12.75C ... and 2nd gear baulk ring

12.76 1st/2nd synchro unit

12.79 Fitting 1st gear components to the mainshaft

12.80 Driving the bearing race home

12.81A Fit 3rd gear needle bearing ...

12.81B ... 3rd gear ...

12.81C ... 3rd gear baulk ring ...

12.81D ... and 3rd/4th synchro unit

12.82A Fit the pilot bearing spacer ...

12.82B ... and the pilot bearing race

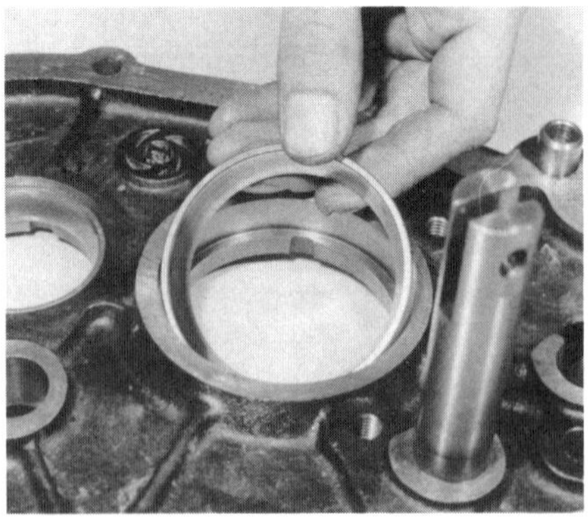
12.87 Fitting a bearing track to the centre plate

12.88A Mainshaft and selector assembly fitted to the centre plate

12.88B Close-up of the front spool correctly fitted

12.88C The front spool with the selector shaft turned to show the locating pin

12.89 Laygear meshed with mainshaft gears

Rear extension – reassembly

71 Refit the oil pump gears, cover and pick-up pipe. Secure the cover with the three bolts (photos).
72 Fit a new oil feed ring, making sure that the oil holes are correctly aligned.
73 Fit a new selector shaft O-ring.
74 Do not fit the rear bearing or oil seal yet.

Mainshaft – reassembly

75 Lubricate and fit 2nd gear needle bearing, 2nd gear and baulk ring. The 2nd gear baulk ring has larger slots than the others (photos).
76 Fit 1st/2nd synchro unit with the selector fork groove away from 2nd gear (photo).
77 Fit 1st gear bush, the endfloat spacer (paragraph 67) and a new circlip. Press the spacer against the circlip and measure the clearance between the spacer and the bush. The desired result is given in the Specifications; different thickness bushes are available if needed.
78 Remove the circlip, spacer and bush. Lubricate 1st gear needle roller bearing and fit it to 1st gear.
79 Fit the selector 1st gear bush and the 1st gear baulk ring to 1st gear. Fit the assembled gear to the mainshaft (photo).
80 Fit the centre bearing race and drive it into position with a tube on the inner track (photo). When the bearing is flush with the bush flange, secure it with the circlip.
81 Invert the mainshaft. Lubricate and fit 3rd gear needle roller bearing, 3rd gear, 3rd gear baulk ring and 3rd/4th synchro unit. The longer boss of the synchro hub faces the front of the gearbox (photos).
82 Fit the spacer and the pilot bearing to the front of the mainshaft, using a piece of tube to seat them if necessary (photos).

Input shaft – reassembly

83 If they were removed, drive the pilot bearing track and input shaft bearing race into position on the shaft.

Layshaft bearings – refitting

84 Fit the bearings to each end of the layshaft/laygear and drive them home with a tube on the inner track.

Centre plate, geartrains and main casing – reassembly

85 If the reverse lever pin was removed, refit it in the same position and tighten its nut to the specified torque.
86 Fit a new oil pump strainer to the plate and secure it with a circlip.
87 Place the centre plate on a stand or stout box so that the shafts can protude below it. Fit the bearing tracks (photo).
88 Lubricate the selector shaft and forks. Make sure that the front spool is correctly located between the forks. Engage the forks with their synchro sleeves on the mainshaft and fit the mainshaft/selector assembly to the centre plate, turning the selector shaft to allow the pin to pass through the slot in the plate (photo).
89 Fit the layshaft/laygear, if necessary having an assistant displace the mainshaft slightly to allow the gears to mesh (photo).
90 Rotate the selector shaft so that the reverse lever forks are aligned with the reverse lever pin. Fit the reverse lever, locating its end with the selector spool, and secure it with the pivot pin and a new circlip.
91 Fit the slipper pad to the reverse lever with the slot in the pad nearest the centre plate.
92 Fit the reverse gear spacer, reverse idler gear and (from below) the idler shaft (photos). The idler shaft roll pin must align with the slot in the plate. If the reverse lever pin has been disturbed, it may be necessary to adjust its position slightly so that the slipper pad engages snugly with the idler gear.
93 Lubricate the detent ball and spring and apply sealant to the plug threads. Fit the ball, spring and plug (photos). Screw in the plug until it is flush with the plate, then stake it to secure it.
94 Fit 4th gear baulk ring onto 3rd/4th synchro hub, followed by the input shaft.
95 Fit the small spacer to the reverse idler shaft.
96 Fit a new gasket to the centre plate, using a smear of grease to retain it (photo).
97 Fit a new plastic oil trough to the main casing, making sure it is the right way up (photo). Also refit the breather if it was removed.
98 Carefully lower the main casing onto the centre plate, making sure that the shafts and dowels enter their respective holes. Be particularly careful not to displace the reverse idler shaft. No force should be necessary.

12.92A Fit the reverse gear spacer ...

12.92B ... reverse idler gear ...

12.92C ... and the idler shaft. The slipper pad (arrowed) must engage snugly

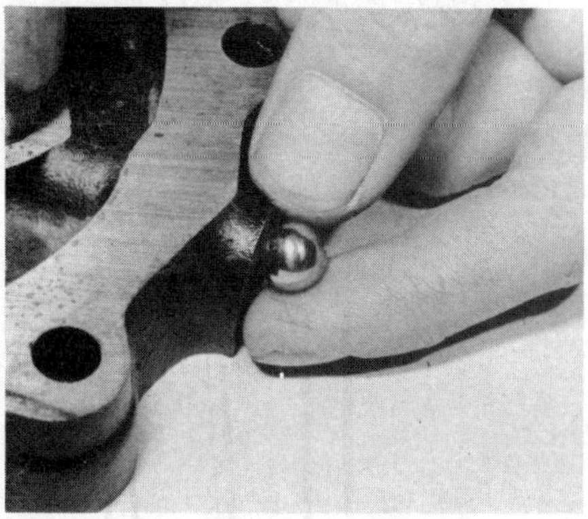
12.93A Fit the detent ball ...

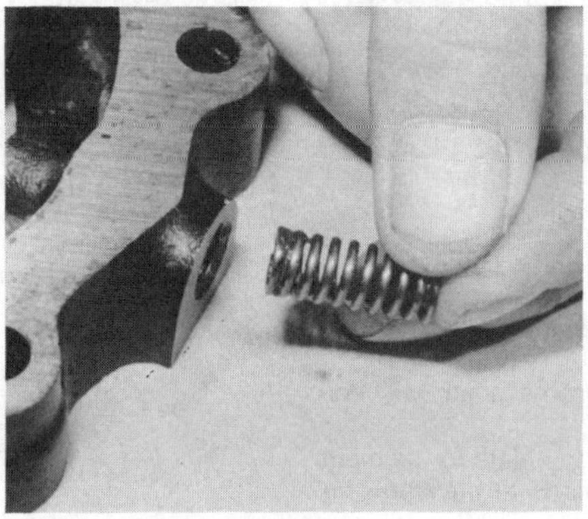
12.93B ... detent spring ...

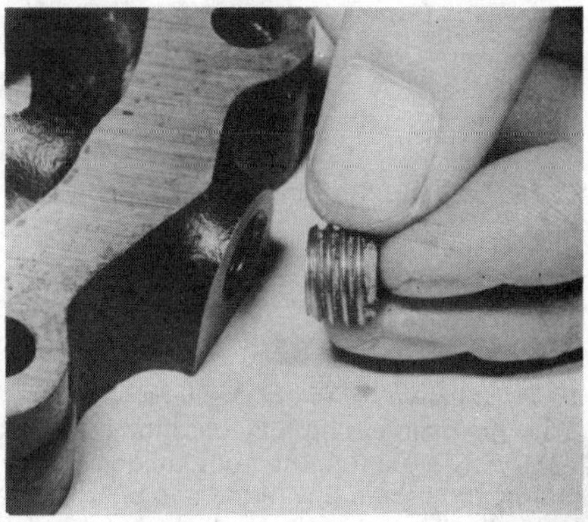
12.93C ... and detent plug

12.96 Centre plate with a new gasket fitted

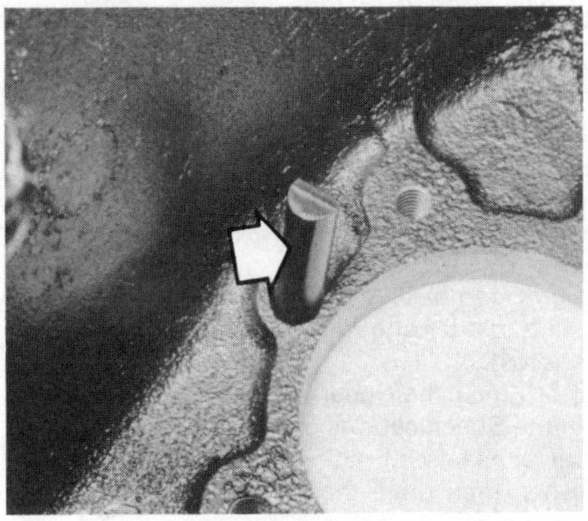
12.97 Oil trough (arrowed) in the main casing

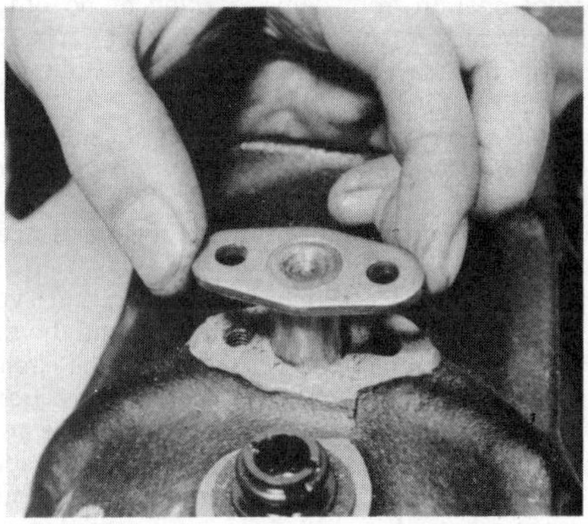
12.101 Fitting the front interlock plate

12.103 Fitting layshaft 5th gear

12.107A Refit mainshaft 5th gear ...

12.107B ... 5th gear baulk ring ...

12.107C ... and 5th synchro unit

12.108 Measuring 5th gear clearance

99 Fit the layshaft and input shaft bearing outer tracks to the main casing.
100 Using two of the rear extension bolts and some plain washers, draw the main casing fully onto the centre plate.
101 Apply sealant to the front interlock plate. Fit the plate to the main casing and secure it with the two bolts. Remember to fit the reversing light wire bracket under the right-hand bolt (photo).
102 The layshaft support plate and disc (Fig. 13.29) will now be needed. Bolt the plate to the front of the main casing so that it presses the disc against the nose of the layshaft. This will protect the layshaft front bearing from damage when fitting 5th gear. The support plate should also retain both bearing outer tracks.
103 Fit the layshaft 5th gear, using a tube or socket to drive it home (photo).
104 On early models, fit 5th gear collar and circlip.
105 On later models, fit a new nut to the layshaft but do not tighten it yet.
106 On all models, fit 5th gear collar and needle roller bearing to the mainshaft.
107 Fit 5th gear, synchro baulk ring and sychro unit complete with support plate washer. Drive them along the mainshaft just far enough to expose the circlip groove, then fit the circlip (photos).
108 Measure the clearance between the front of 5th mainshaft gear and the collar (photo). The desired value is given in the Specifications. Various thicknesses of support plate washer are available. If adjustment is necessary, remove the circlip and washer, then drive the synchro unit forwards or draw it rearwards, according to whether a thicker or thinner washer is to be fitted. Fit the chosen washer and circlip, then recheck the clearance.

Mainshaft and layshaft endfloat assessment

109 Remove the support plate and disc. Wrap about ten turns of masking tape around the input shaft below the splines to prevent side movement giving false readings.
110 Fit selective washers, of thickness 0.0402 in (1.02 mm), to the input shaft and layshaft bearing tracks. Fit the front cover and gasket and secure it with the six bolts and spring washers, tightened to the specified torque. Make sure that the input shaft can still move back and forth, but not from side to side.
111 Check that the two bolts holding the centre plate to the main casing are tight.
112 Position the gearbox with the input shaft downwards. Rotate the mainshaft to settle the bearings, then position a dial test indicator to record mainshaft endfloat. Place the indicator probe on a ball-bearing which will rest on the end of the mainshaft.
113 Move the mainshaft up and down and record the endfloat. To the value obtained add the thickness of the washer presently fitted (paragraph 110) and subtract the desired endfloat (see Specifications). For example:

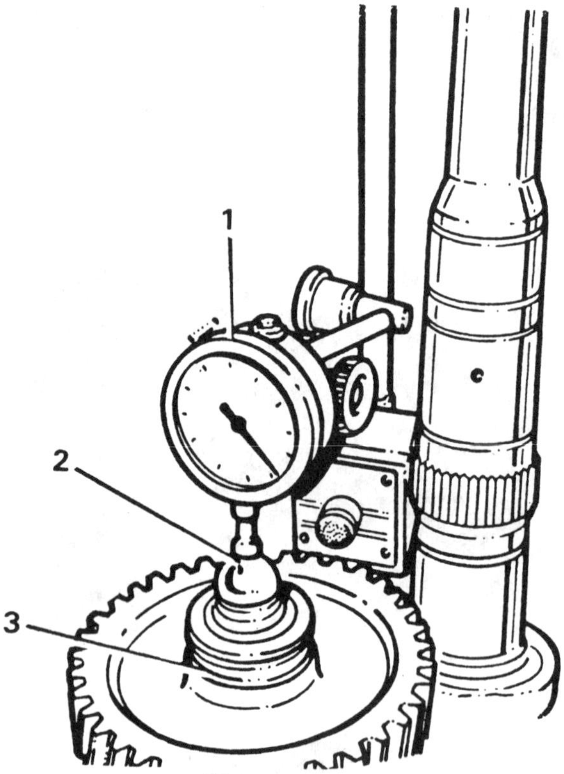

Fig. 13.32 Measuring layshaft endfloat (Sec 12)

1 Dial test indicator
2 Ball-bearing
3 Layshaft

Reading obtained (A) = 1.19 mm
Washer thickness (B) = 1.02 mm
Endfloat required (C) = 0.01 to 0.6 mm
A + B − C = 2.15 to 2.20 mm
Washer to be fitted = 2.17 mm (closest available).

114 Transfer to ball-bearing and indicator probe to the end of the layshaft. Lever the layshaft up and down with a couple of screwdrivers under 5th gear and record the endfloat.

115 The desired layshaft endfloat is zero – neither endfloat nor preload – so the calculation of washer thickness required is simply recorded endfloat plus washer thickness presently fitted. Washer thicknesses available are listed in the Specifications; the astute reader will notice that the steps in thickness of the washers are more than double the specified endfloat/preload tolerance. In practice the closest available thickness should be selected.
116 Remove the ball-bearing and dial test indicator. Invert the gearbox, remove the front cover and the temporary selective washers.
117 Fit the selective washers of the calculated thicknesses. Refit the front cover and tighten the bolts to the specified torque.
118 Recheck the mainshaft and layshaft endfloat.
119 Remove the front cover again. Fit a new input shaft oil seal to the front cover and lubricate it with gear oil. Remove the masking tape from the input shaft. Put a turn of tape round the input shaft splines to protect the oil seal, refit the front cover and remove the tape. Tighten the front cover bolts to the specified torque.

Final assembly

120 On models where layshaft 5th gear is secured by a nut, lock up the geartrains by selecting 5th gear and any other (paragraph 23). Tighten the nut to the specified torque whilst an assistant holds the gearbox. Stake the nut collar into the layshaft groove using a blunt instrument (photo); support the gear with a block of wood when doing this. Return the gears to neutral.
121 On all models, fit the 5th gear selector spool, fork and bracket to the selector rod and centre plate. Tighten the bracket bolts to the specified torque. Make sure that the spool is the right way round (photo).
122 Fit a new circlip to the selector shaft.
123 Position the gearbox horizontally. Remove the two bolts from the centre plate. Fit a new gasket to the centre plate, using a smear of grease to hold it in position.
124 Insert the oil pump driveshaft into the end of the layshaft, using a little petroleum jelly to keep it in position (photo).
126 Rotate the centre of the oil pump so that it will accept the driveshaft. Offer the rear extension to the gearbox, making sure that the oil pick-up pipe enters the strainer and the oil pump driveshaft enters the pump. (It may be helpful to have an assistant turn the input shaft back and forth slightly if the driveshaft does not quite line up).
126 Fit the ten bolts which secure the rear extension, remembering that the long bolts are fitted where the dowels are. Tighten the bolts to the specified torque.
127 Refit the interlock plate, applying sealant to its mating face, and secure it with the two bolts.
128 Fit the speedometer drivegear, with the indents facing rearwards and the flats on the gear aligned with those on the mainshaft.
129 Fit the sleeve and the rear bearing. Seat the bearing with the output flange and a piece of tube (ie do not hammer on the flange).
130 Remove the output flange. Lubricate a new oil seal with gear oil and fit it to the rear extension, lips inwards (photo). Tap the seal home with a piece of tube. The outer face of the seal must not be recessed by more than 0.08 in (2 mm).
131 Apply some sealant to the splines, then refit the output flange, washer and nut. Tighten the nut to the specified torque, using a bar and a couple of nuts and bolts through the flange holes to restrain the flange (photo).
132 Fit the gear selector yoke. Secure it with the nut and bolt, tightened to the specified torque.
133 Refit and secure the gear selector housing.
134 Refit the speedometer pinion and bearing, using a new O-ring.
133 Fit the bellhousing to the main casing and secure it with the six bolts, remembering that the longer bolts (with spring and plain washers) go into the dowelled holes. Tighten the bolts to the specified torque.
136 Refit the clutch release components (Section 11).
137 Refit the drain plug, using a new fibre washer, and tighten it to the specified torque.
138 The gearbox may be refilled with oil now or after refitting to the vehicle. Check the level after refitting if it is refilled now.

12.120 Staking 5th gear nut. Gear should be supported when doing this

12.121 5th gear selector components correctly fitted

12.124 Oil pump driveshaft in position

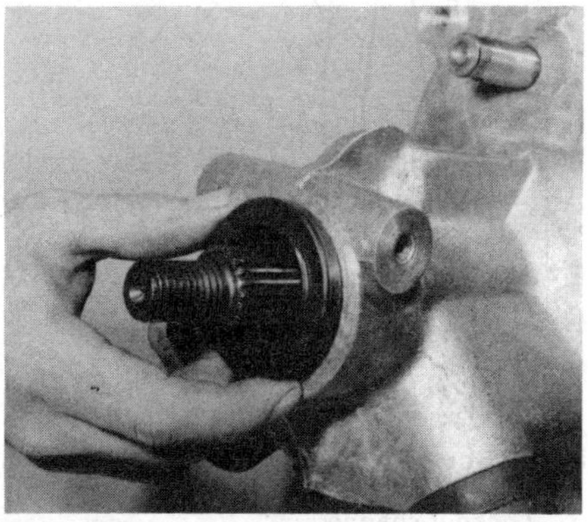
12.130 Fitting a new rear extension oil seal

12.131 Tightening the output flange nut

12.145 Spring leg in contact with cross-pin (arrowed)

12.148 Checking the gear lever clearance in 1st gear

Gear lever – removal and refitting

139 Remove the six screws which secure the gear lever gaiter to the cab floor. Peel back the gaiter.
140 Note the arrangement of the springs and washers, then remove the three bolts which secure the gear lever. Lift out the gear lever, springs and washers.
141 Refit by reversing the removal operations. Check the gear lever adjustment if necessary, as described below, before refitting the gaiter.

Gear lever – adjustment

142 Remove the six screws which secure the gear lever gaiter. Peel back the gaiter.
143 Move the gear lever into 3rd or 4th gear. Slacken the adjuster bolt locknuts and turn the adjuster bolts until the spring legs on both sides are clear of the cross-pin by approximately 0.020 in (0.5 mm).
144 Push the gear lever over to the left. The right-hand spring leg should be just in contact with the cross-pin. Adjust the right-hand bolt if necessary to achieve this.
145 Push the gear lever to the right and similarly adjust the left-hand bolt so that the left-hand spring leg contacts the cross-pin (photo).
146 Screw both adjusting bolts inwards (clockwise) by an equal amount, half a turn at a time, until radial play in the gear lever is just eliminated. Tighten the locknuts.
147 Select neutral and check that the gear lever returns to the 3rd/4th place after being pushed across the gate.
148 Select 1st gear. A feeler blade 0.010 in (0.25 mm) thick should just fit between the gear lever base and the bias spring bracket (photo). Slacken the three securing bolts if necessary and move the gear lever to obtain the correct clearance, then tighten the bolts.
149 Check the operation of the reversing light switch and adjust if necessary (Section 17A).
150 Refit the gear lever gaiter.

13 Automatic transmission

Fluid draining

1 Not all models are provided with a sump drain plug. In the absence of a drain plug, the fluid can be drained from the transmission by undoing the dipstick/filler tube nut where the tube enters the sump.

Brake bands – adjustment

2 This is not a routine operation, but may be attempted if transmission slip is evident (rear brake band) or if the 1–2 shift is erratic (front brake band). Always check the fluid level first.
3 The front brake adjuster is located on the left-hand side of the transmission, and the rear adjuster on the right-hand side. The procedure is the same for both.
4 Slacken the adjuster locknut. Tighten the adjuster screw to 5 lbf ft (7 Nm), then back it off by exactly three-quarters of a turn.
5 Hold the adjuster stationary and tighten the locknut to the specified torque.

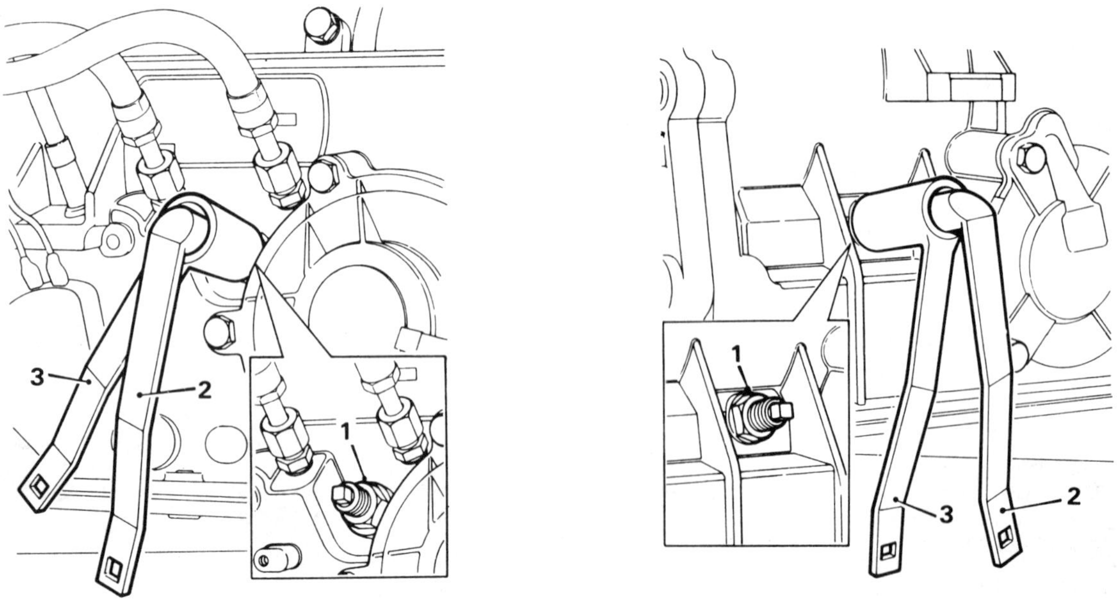

Fig. 13.33 Brake band adjustment – front (left) and rear (right) (Sec 13)

1 Adjuster and locknut *2 Adjuster spanner* *3 Locknut spanner*

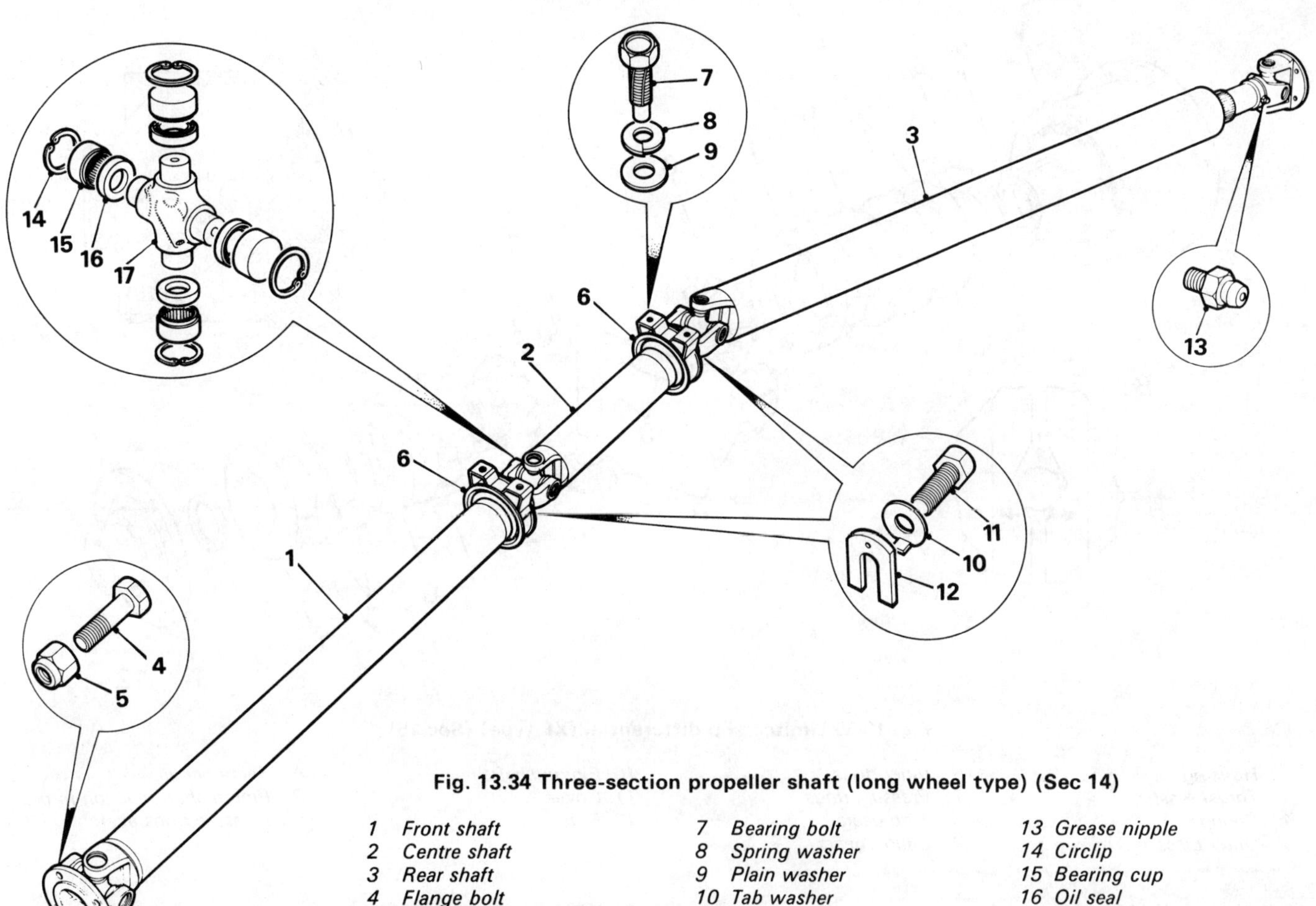

Fig. 13.34 Three-section propeller shaft (long wheel type) (Sec 14)

1 Front shaft
2 Centre shaft
3 Rear shaft
4 Flange bolt
5 Flange nut
6 Centre bearings
7 Bearing bolt
8 Spring washer
9 Plain washer
10 Tab washer
11 Bolt
12 C-washer
13 Grease nipple
14 Circlip
15 Bearing cup
16 Oil seal
17 Spider

14 Propeller shaft

Description (long wheelbase models)

1 On long and extra-long wheelbase models, a three-section shaft with two intermediate bearings is used.

Lubrication (later models)

2 Later models are not fitted with grease nipples on the universal joints, periodic lubrication is therefore not possible.

15 Rear axle

Limited slip differential – description

1 The limit slip or 'self-locking' differential is available as an optional extra on some models. Its purpose is to provide improved traction on loose or slippery surfaces.

2 Drive is transmitted to each bevel gear by a series of friction clutch plates. If one rear wheel begins to spin faster than the other, the loading on the clutch plates alters to reduce traction to the faster-spinning wheel and increase traction to the other wheel. In extreme circumstances, such as when one wheel is in a ditch and the other on the road, the differential will behave like a solid axle. It may be necessary to apply the handbrake momentarily to create an artificial load on the spinning wheel.

Limited slip differential – precautions

3 To obtain maximum life and satisfactory performance from the limited slip, observe the following precautions:

(a) Do not spin one rear wheel whilst the other is on the ground (eg for wheel balancing)
(b) Do not apply power to the differential with one wheel raised and one on the ground: the wheel on the ground will turn, and the vehicle will be pulled off its supports.

4 No special oil is specified for the limited slip differential.

Limited slip differential – overhaul

5 As with the ordinary differential, overhaul should be left to a specialist, or an exchange unit be fitted.

Rear axle (300 Series) – description

6 The rear axle fitted to 300 Series models is designated 15 HA or 16 HA. Externally the two axles are distinguished by the number of wheels: 15 HA carries single rear wheels and 16 HA carries twin wheels.

7 Internally the construction of the axles is similar to the smaller version described in Chapter 8. Access to the differential is via a rear cover plate instead of a nose cover. Pinion bearing preload is set by means of a collapsible spacer; this means that care must be taken when renewing the pinion oil seal.

8 The 16 HA axle is 'fully floating', ie the half shafts do not run directly in bearings, which makes them very easy to renew.

9 Special equipment is required even to remove the differential from these axles. In case of malfunction, consult a specialist for repair, or fit a new or reconditioned axle.

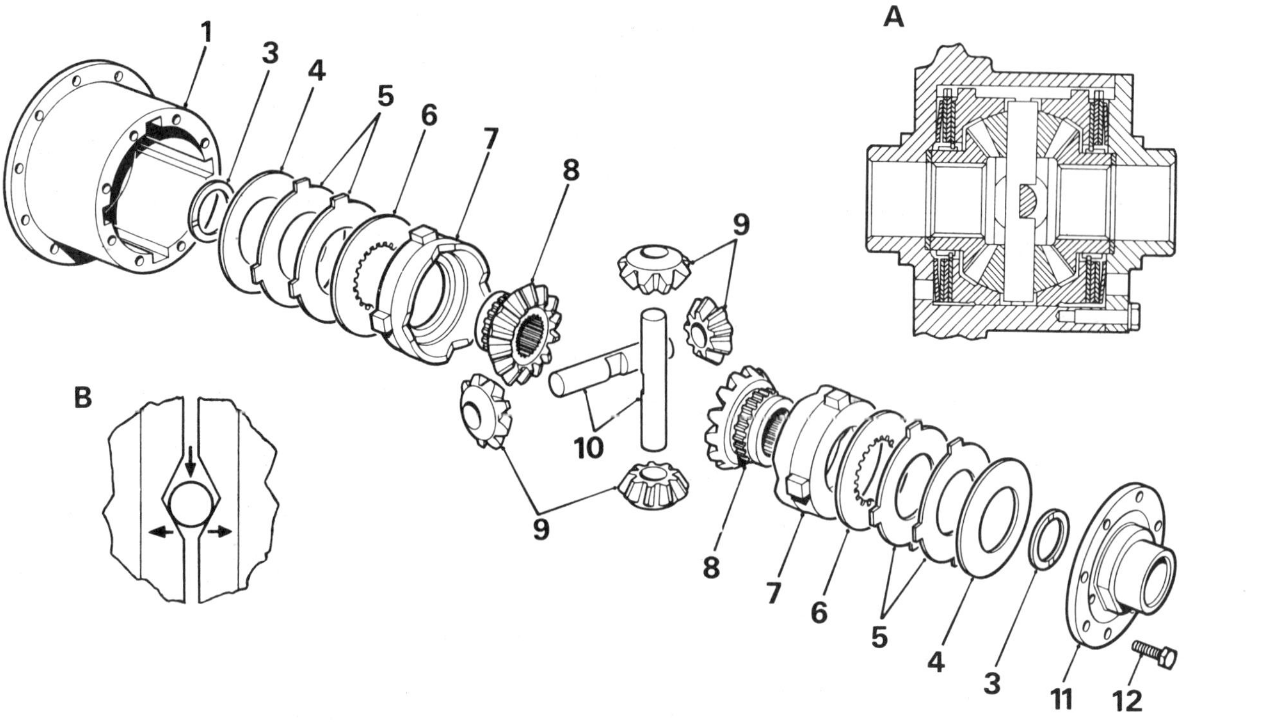

Fig. 13.35 Limited slip differential (XF type) (Sec 15)

1 Housing
3 Thrust washers
4 Springs
5 Outer discs
6 Inner discs
7 Pressure rings
8 Side gears
9 Pinion gears
10 Pinion shafts
11 Cover
12 Bolt
A Sectional view
B Pinion shaft reaction forces pressure rings apart

Half shaft (15 HA) – removal and refitting

10 Chock the front wheels. Slacken the rear wheel nuts (left-hand thread on the left-hand side), raise and support the rear of the vehicle and remove the rear wheel.
11 Release the handbrake. Remove the two screws which secure the brake drum and remove the drum.
12 Disconnect the handbrake cable from the brake backplate and bracket.
13 Disconnect the rear wheel cylinder hydraulic pipe. Plug the cylinder and cap the pipe.
14 Remove the six nuts and bolts which secure the brake backplate and hub bearing retainer to the axle tube. Turn the half shaft to gain access to the nuts and bolts through the hole in the flange.
15 Place tray under the hub to catch any oil spillage. Withdraw the half shaft using a slide hammer if necessary, together with the brake backplate, hub bearing and associated components.
16 Extract the hub inner oil seal from the axle tube and renew it.
17 If it is wished to renew the hub bearing and outer oil seal, a press will be needed for successful removal of the old bearing and fitting of the new one. Consult a Freight Rover dealer or other specialist.
18 Commence refitting by inserting the splined end of the half shaft into the axle tube, making sure that the inner oil seal is not damaged. Slide the shaft home, checking that the bearing, spring ring and outer seal enter the axle tube. Locate the bearing retainer and the brake backplate on the axle tube flange studs.
19 Tighten the brake backplate nuts and bolts to the specified torque.
20 Reconnect the rear wheel cylinder hydraulic pipe and the handbrake cable. Refit and secure the brake drum.
21 Bleed the appropriate hydraulic circuit (Section 16).
22 Refit the roadwheel, lower the vehicle and tighten the wheel nuts.
23 Check the axle oil level and top up if necessary.
24 Make a number of stops in reverse to adjust the brakes, then check the operation of the foot and hand brakes.

Half shaft (16 HA) – removal and refitting

25 Remove the eight bolts which secure the half shaft flange to the rear hub.
26 Withdraw the half shaft and recover the flange gasket.
27 When refitting, place a new gasket on the hub flange. Insert the half shaft and secure it with the flange bolts, tightened to the specified torque.

Pinion oil seal (15 HA/16 HA) – renewal

28 A spring balance or a special low-reading torque wrench will be required for this procedure. Read through it first to see what is involved.
29 Chock the front wheels, release the handbrake and raise and support the rear of the vehicle.
30 Make alignment marks between the propeller shaft and pinion flanges, then disconnect them by removing the four nuts and bolts.
31 Establish the pinion turning torque, apply a piece of string through one of the flange bolt holes and a spring balance, or (if available) a low-reading torque wrench. Turning torque values are given in the Specifications. Note the reading obtained.
32 Restrain the pinion flange and unscrew the flange nut, noting the number of turns needed to remove it. Obtain a new nut for reassembly.
33 Place an oil drain pan below the pinion flange, then pull off the flange.
34 Remove the dust cover (if fitted) and the oil seal.
35 Grease the lips of a new oil seal. Fit the seal, lips inwards, and drive it home with a piece of tube.
36 Refit the pinion flange, being careful not to damage the new oil seal. Screw on the new nut but do no tighten it yet.
37 Tighten the flange nut a little at a time – the number of turns taken for removal may be a guide – and measure the pinion turning torque after each tightening. Ideally the same turning torque should be achieved as existed before dismantling (plus a small amount for the drag of the new seal). **Caution:** If the maximum specified turning torque is exceeded, the axle will have to be removed and a new collapsible spacer fitted by a Freight Rover dealer or other specialist.
38 Refit the propeller shaft, observing the flange alignment marks, then lower the vehicle.
39 Check the axle oil level and top up if necessary.

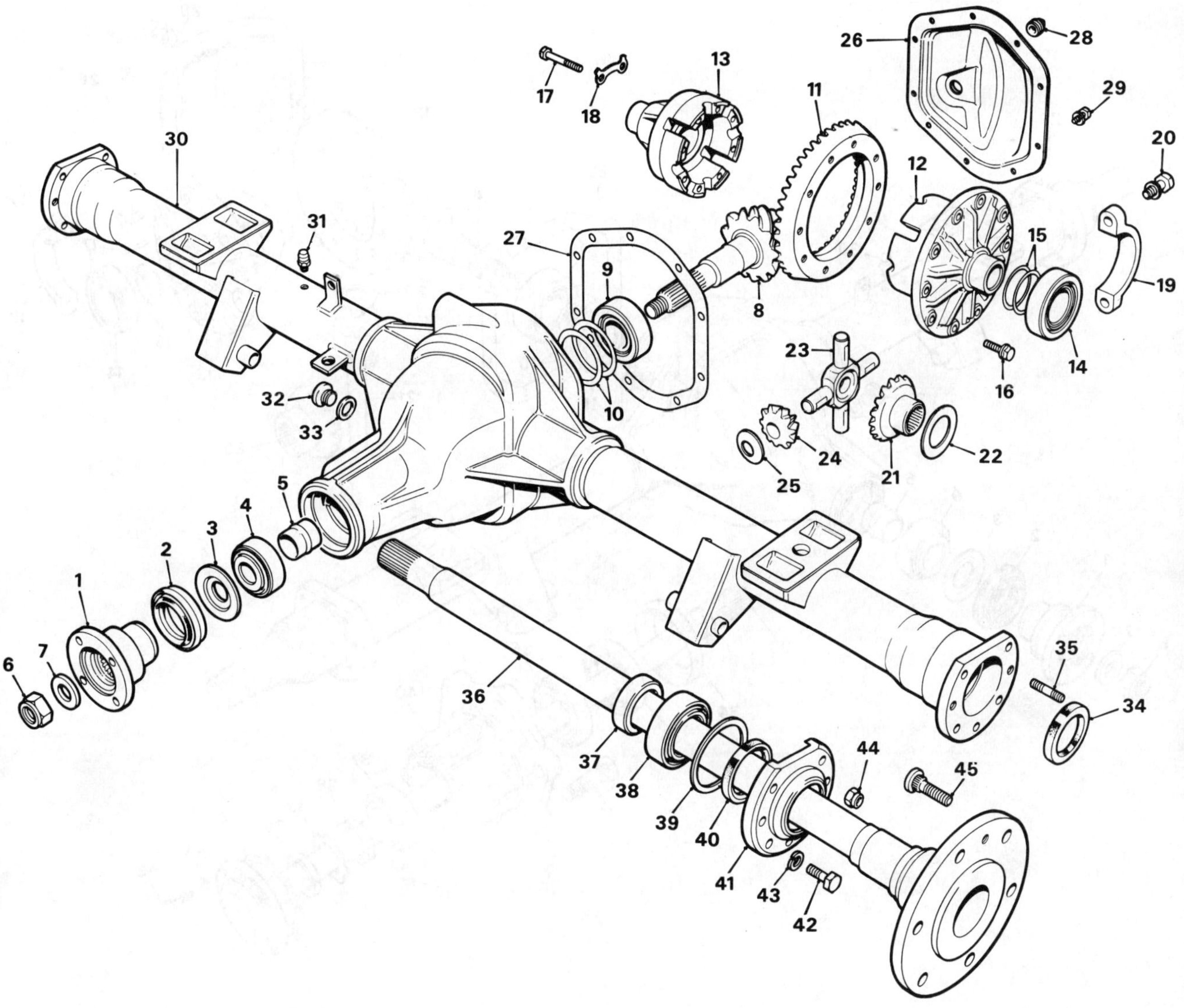

Fig. 13.36 Exploded view of 15 HA (single rear wheel) rear axle (Sec 15)

1 Pinion flange
2 Pinion oil seal
3 Oil thrower
4 Pinion tail bearing
5 Collapsible spacer
6 Flange nut
7 Washer
8 Pinion
9 Pinion head bearing
10 Shims
11 Crownwheel
12 Differential casing
13 Differential casing
14 Differential bearing
15 Shims
16 Crownwheel bolt
17 Casing bolt
18 Locktab
19 Bearing cap
20 Cap bolt
21 Side gear
22 Thrust washer
23 Pinion spider
24 Pinion gear
25 Thrust washer
26 Cover plate
27 Gasket
28 Filler/level plug
29 Cover plate bolt
30 Axle casing
31 Breather
32 Drain plug
33 Fibre washer
34 Hub inner oil seal
35 Bearing retainer plate stud
36 Half shaft
37 Bearing collar
38 Hub bearing
39 Spring
40 Hub outer oil seal
41 Bearing retainer plate
42 Retainer plate bolt
43 Spring washer
44 Retainer plate nut
45 Wheel stud

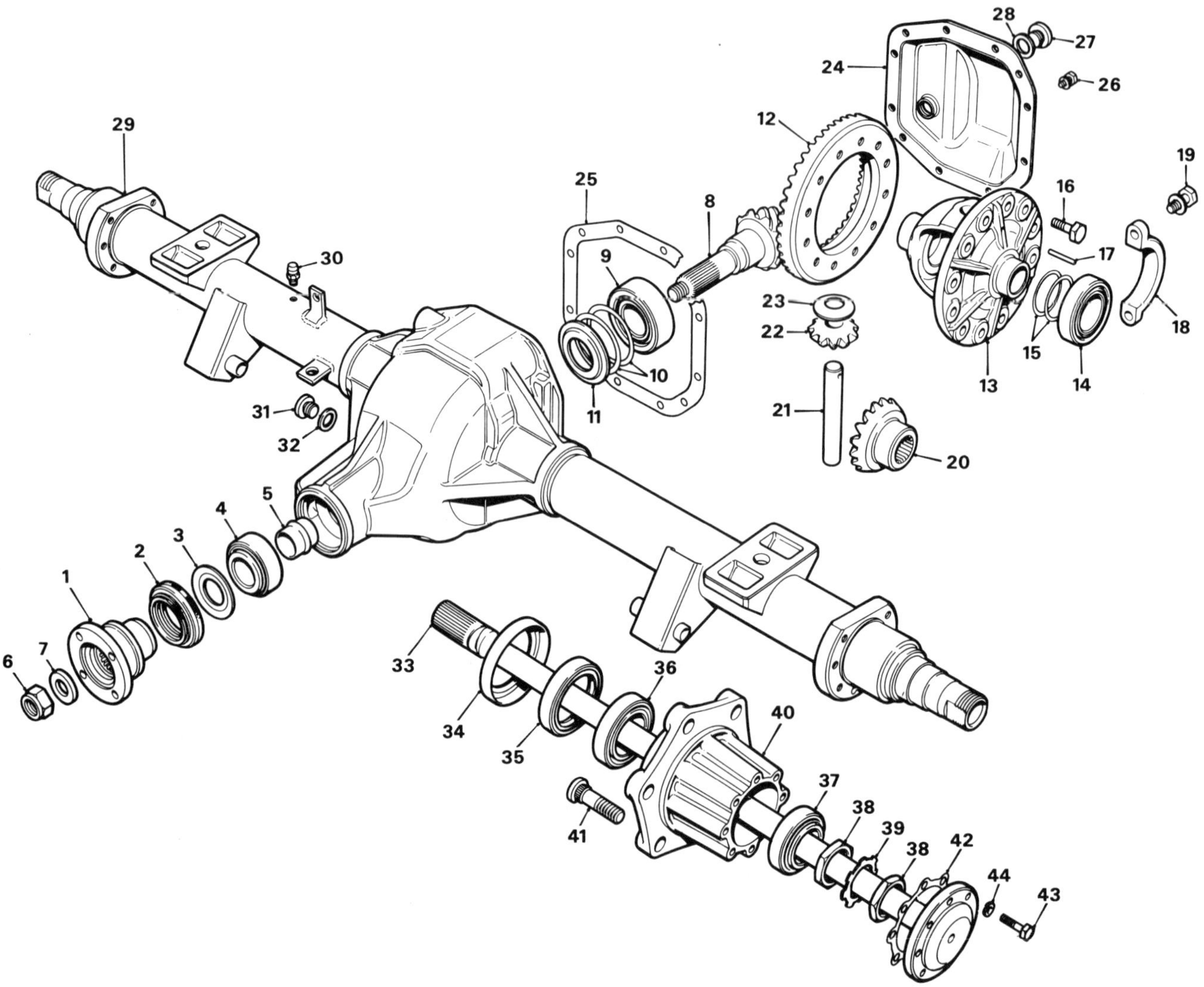

Fig. 13.37 Exploded view of 16 HA (twin rear wheel) rear axle (Sec 15)

1 *Pinion flange*
2 *Pinion oil seal*
3 *Oil thrower*
4 *Pinion tail bearing*
5 *Collapsible spacer*
6 *Flange nut*
7 *Washer*
8 *Pinion*
9 *Pinion head bearing*
10 *Shims*
11 *Oil baffle*
12 *Crownwheel*
13 *Differential casing*
14 *Differential bearing*
15 *Shims*
16 *Crownwheel bolt*
17 *Roll pin*
18 *Bearing cap*
19 *Cap bolt*
20 *Side gear*
21 *Pinion shaft*
22 *Pinion gear*
23 *Thrust washer*
24 *Cover plate*
25 *Gasket*
26 *Cover plate bolt*
27 *Filler/level plug*
28 *Fibre washer*
29 *Axle casing*
30 *Breather*
31 *Drain plug*
32 *Fibre washer*
33 *Half shaft*
34 *Oil thrower*
35 *Hub oil seal*
36 *Hub inner bearing*
37 *Hub outer bearing*
38 *Hub locknuts*
39 *Lockwasher*
40 *Hub*
41 *Wheel stud*
42 *Gasket*
43 *Half shaft bolt*
44 *Spring washer*

16 Braking system

Introduction

1 The main changes to the braking system on later models are the introduction of disc brakes on the front wheels, and self-adjusting brakes on the rear wheels. Two types of self-adjusting rear brake are fitted.

2 Minor changes have also been made to the hydraulic system, including the introduction on some models of delay and load sensing valves.

Disc brakes – description

3 The disc brakes fitted to the front wheels of later models function as follows.

4 Each front hub has a brake disc attached to it. The disc runs between the jaws of a caliper which is attached to the stub axle. The caliper jaws contain two friction-lined pads, one each side of the disc; when the brake pedal is pressed, hydraulic pressure acts on pistons in the caliper and forces the pads against the disc. Friction between the pads and the disc slows the disc down, the energy being dissipated as heat.

5 When the brake pedal is released, the elasticity of the caliper seals pulls the pistons back slightly, so withdrawing the pads from the disc.

6 As the pad linings wear, the pistons come further out of their bores to compensate. The brakes are thus self-adjusting in use. The hydraulic fluid level in the master cylinder reservoir will fall as the linings wear.

7 The main advantages of the disc brake are its greater effectiveness and the ease of access to all components, compared with the drum brake.

Disc brake pads – inspection and renewal

8 The disc pads must be inspected periodically, and renewed when the friction lining on any one pad is worn down to the specified minimum.

9 Chock the rear wheels. Slacken the front wheel nuts, raise and support the vehicle and remove the front wheels.

10 The edges of the pads can now be seen through the opening in the caliper. If the linings are obviously worn to the limit, or for closer inspection, remove the pads as follows.

11 Remove the two pad retaining pins. These may be large split pins, or 'dumb-bell' pins which must be driven out. Recover the anti-rattle springs from under the pins, noting which way round they are fitted (photo). Obtain new pins (of the same type) and anti-rattle springs for reassembly.

12 Push the pads away from the disc – do not lever against the disc when doing this – and withdraw the pads from the jaws of the caliper (photo). Identify the pad position if they are to be refitted.

13 Remove any anti-squeal shims from the backs of the pads, noting which way round they are fitted.

14 Clean the caliper recesses using a soft cloth or brush and clean brake fluid or methylated spirit. **Warning:** Do not blow out the dust, it may contain asbestos. Inspect the caliper pistons for fluid leaks and rectify if necessary. Also inspect the brake disc as described later in this Section.

15 Thoroughly inspect the pads. All four pads must be renewed even if

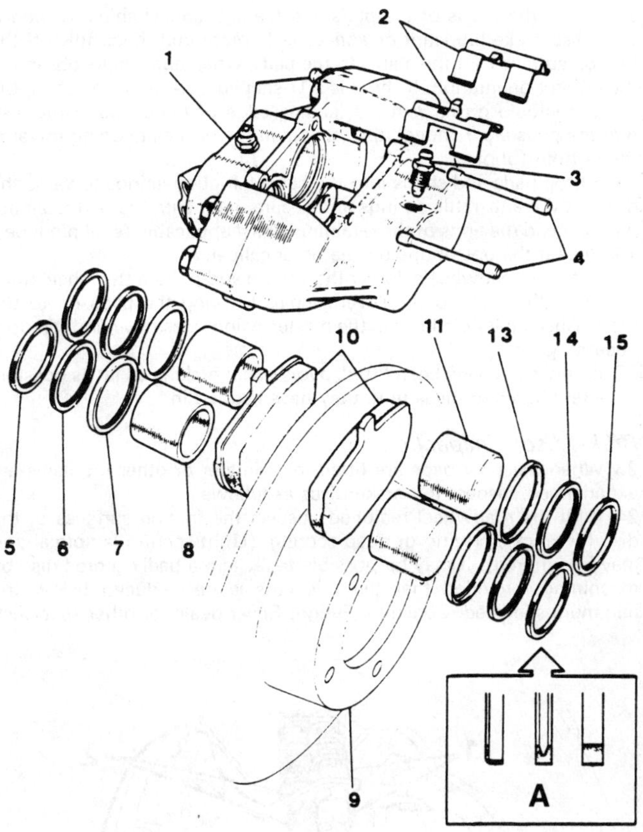

Fig. 13.38 Disc brake components. Four-piston caliper shown – others only have two (Sec 16)

1 Caliper
2 Anti-rattle springs
3 Bleed nipple
4 Pad retaining pins
5 Piston seal
6 Wiper seal
7 Wiper seal retainer
8 Piston
9 Disc
10 Pads
11 Piston
13 Wiper seal retainer
14 Wiper seal
15 Fluid seal
A Retainer and seal profiles

only one is worn to the limit. Do not interchange pads in an attempt to even out wear.

16 If new pads are to be fitted, the caliper pistons must be pushed back into their bores to make room for the thickness of the unworn linings. Do this with a tyre lever or a similar blunt instrument. As the pistons are pushed back, the fluid level in the master cylinder reservoir will rise. Syphon out some fluid if necessary, using an old poultry baster or battery hydrometer. Do not syphon the fluid by mouth, it is poisonous.

16.11 Brake pad upper retaining pin (split pin type) and anti-rattle spring in position

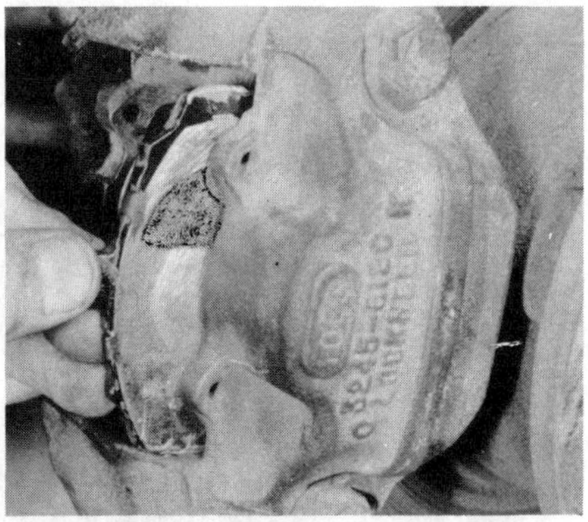

16.12 Removing a brake pad. Cut-out in shim (shaded) is uppermost

16.17 Pad and shim ready for refitting. Cut-out (arrowed) in pad goes at bottom, cut-out in shim at top

17 Smear the backs of the pads and the anti-squeal shims (if fitted) with disc brake lubricant or anti-squeal compound. Keep this off the friction surfaces. Fit the shims to the pads, when applicable, observing the correct orientation. If there is a D-shaped cut-out in the shim, the cut-out should go nearest the leading edge of the caliper. Note also that the pads are 'handed', the cut-out in the backplate going towards the bottom (photo).
18 Fit the pads and shims into the caliper, friction linings towards the disc. Fit the anti-rattle springs and secure them with the pad retaining pins. Spread the ends of the retaining pins if applicable (split pin type).
19 Repeat the operations on the other caliper.
20 Refit the roadwheels, lower the vehicle and tighten the wheel nuts.
21 Press the brake pedal several times to bring the pads up to the discs. Check the fluid level in the master cylinder reservoir and top up if necessary.
22 If new pads have been fitted, avoid harsh braking as far as possible for a few hundred miles until they have bedded in.

Brake disc – inspection

23 Whenever new pads are fitted, or if judder or other problems are experienced, inspect the brake discs as follows.
24 With the roadwheel removed, inspect the friction surfaces of the disc for cracks, crazing or deep scoring. (Light scoring is normal and may be ignored). It may be possible to reclaim a badly scored disc by machining, provided that the thickness is not reduced below the minimum specified. Consult a Freight Rover dealer or other specialist.

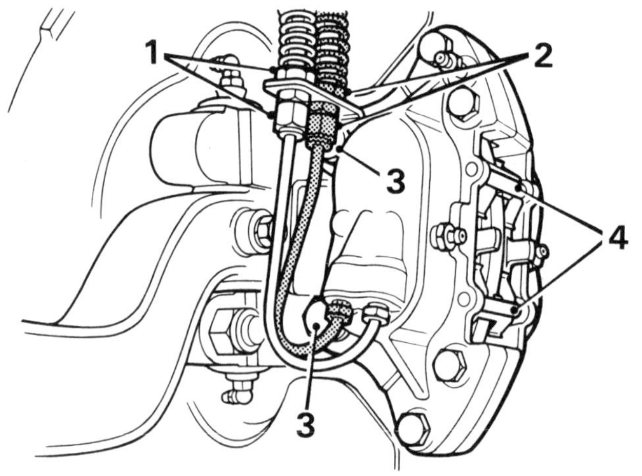

Fig. 13.39 Fitting details of four-piston caliper (Sec 16)

1 *Primary hose and pipe*
2 *Secondary hose and pipe*
3 *Caliper mounting bolts*
4 *Pad retaining pins*

25 Use a micrometer to measure the thickness of the disc in six different places across the friction area. Thickness variation in excess of 0.001 in (0.025 mm) can cause judder and should be corrected by machining, or by renewal of the disc.
26 Position a dial gauge or fixed pointer against the outer circumference of the friction area. Rotate the disc and measure the maximum run-out, either by reading the gauge or by inserting feeler blades between the pointer and the disc. Run-out in excess of 0.009 in (0.23 mm) is normally reckoned to cause judder. Excessive run-out may be due to distortion of the disc, or to excessive play in the hub bearings. Sometimes an improvement can be made by re-positioning the disc on the hub.
27 Ideally both discs should be renewed or machined together, in order to maintain balanced braking characteristics.

Brake disc – removal and refitting

28 Remove the brake caliper without disconnecting the hydraulic hose(s) and tie it up out of the way.
29 Remove the hub bearing grease cap. Extract the split pin, remove the nut retainer and unscrew and remove the nut. Recover the bearing retaining washer.
30 Pull the hub/disc assembly off the stub axle. Recover any loose bearing races; do not mix them up.
31 Clamp the disc in a vice with protected jaws. Remove the nuts which secure the disc to the hub and pull the hub off the disc.
32 Before refitting, clean the hub and disc mating surfaces.
33 Refit by reversing the removal operations. Tighten the disc-to-hub nuts to the specified torque. Adjust the hub bearings before fitting the brake caliper, and check the disc run-out as previously described.

Brake caliper – removal and refitting

34 Remove the brake pads as previously described.
35 Clean around the hydraulic union(s) on the caliper and at the other end of the flexible hose(s). Deal with the union(s) according to number and type as follows:

- (a) *Banjo union – clamp the hose, remove the union bolt and recover the washers*
- (b) *Single hose – clamp the hose, slacken the caliper union and unscrew the caliper after removing the mounting bolts. Recover the washer (when fitted) (photo)*
- (c) *Twin hoses – disconnect the hoses at the support bracket*

36 Plug or cap open hoses and unions to minimise fluid loss and to keep dirt out.
37 Remove the two caliper mounting bolts (photo). These are on the inboard side of the caliper. **Do not** remove the two bolts (on the outboard side) which hold the caliper halves together.
38 Lift the caliper off the stub axle and (when applicable) unscrew it from the flexible hose (photo).
39 Refit by reversing the removal operations. Tighten the caliper mounting bolts to the specified torque and use new sealing washers on the hydraulic unions (when applicable).
40 Make sure that the routing and 'set' of the flexible hose(s) is satisfactory at all angles of lock. See the end of this Section for further details.
41 Bleed the appropriate hydraulic circuit(s) on completion.

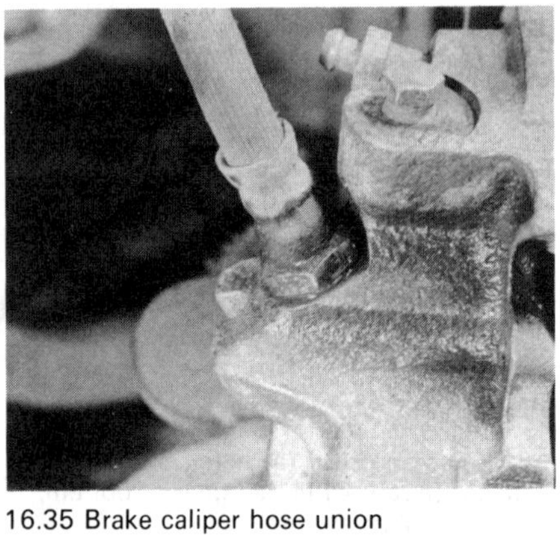

16.35 Brake caliper hose union

16.37 Removing a caliper mounting bolt

16.38 Removing a brake caliper

16.45A Removing the wiper seal and retainer ...

16.45B ... and the piston seal

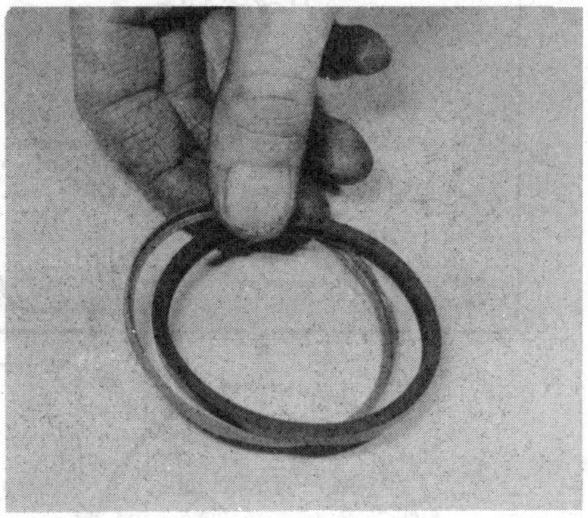

16.48 Wiper seal and retainer

Brake caliper – overhaul

42 Remove the caliper as previously described. (If it is proposed to use hydraulic pressure to remove the caliper pistons, the caliper may be left connected to the hydraulic system, but take care not to strain the hoses or pipes).

43 Clean the caliper with a soft cloth or brush and methylated spirit.

44 Restrain all but one of the caliper pistons by inserting or clamping a suitably shaped piece of wood. (See Fig. 13.40 for the dimensions suitable for four-piston calipers). Apply low air pressure (from a foot pump), or depress the brake pedal gently, to eject the free piston. **Caution:** The piston may be ejected with some force. If the piston is seized in its bore, renew the caliper.

45 Remove the wiper seal and retainer from the mouth of the caliper bore. Hook out the piston seal from the groove in the bore, using a blunt or non-metallic instrument (photos).

46 Clean the piston and bore using clean rags and methylated spirit or special brake cleaning solvent. Do not use abrasives. If corrosion or scoring is evident, the caliper must be renewed.

47 Dip a new piston seal in clean hydraulic fluid and fit it to the groove in the bore, using the fingers only. The shape of the groove will cause the edge of the seal furthest from the mouth of the bore to stand proud.

48 Insert a new wiper seal into its retainer (photo). Press the retainer and seal into the mouth of the bore until the retainer is flush.

49 Lubricate the piston with clean hydraulic fluid. Push the piston into the bore, leaving about 0.16 in (4 mm) protruding (photo). This may be easier if the bleed screw is slackened.

50 Reposition the wooden block and repeat the above operations on the remaining piston(s). Refit the caliper and bleed the appropriate hydraulic circuit(s).

16.49 Fitting a caliper piston

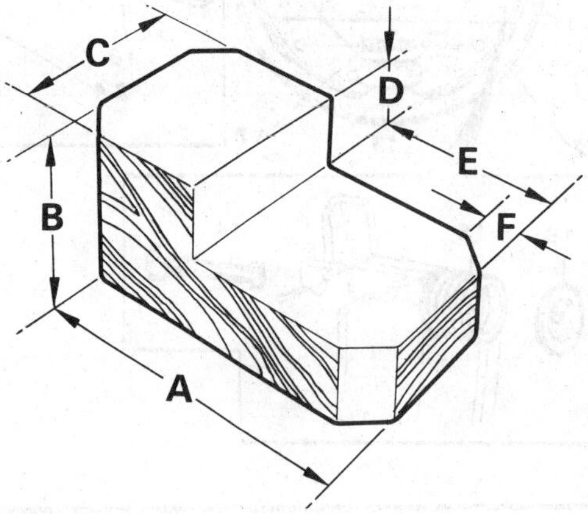

Fig. 13.40 Dimensions of the wooden block needed to restrain three caliper pistons (Sec 16)

A	*3.5 in (89 mm)*	*D*	*0.8 in (20 mm)*
B	*1.75 in (45 mm)*	*E*	*1.75 in (45 mm)*
C	*2.0 in (51 mm)*	*F*	*0.4 in (10 mm)*

Rear brake shoes (200 Series, 1985 on) – removal and refitting

52 Chock the front wheels and slacken the rear wheel nuts. Release the handbrake. Raise and support the rear of the vehicle and remove the rear wheels.

53 Remove the two screws which secure the brake drum to the hub.

54 Pull off the brake drum. If it is stuck, first check that the handbrake is released, then remove the rubber bung from the backplate for access to the adjuster mechanism. Depress the adjuster with a screwdriver to release it (photos). The drum should now pull off without difficulty.

55 Remove the shoe steady pins, springs and washers by depressing the washers and turning them 90° (photo).

56 Note how the springs are fitted to the shoes (photos). Unhook and remove the bottom spring.

57 Disengage the shoes from the bottom pivot and the wheel cylinder. Release the strut from between the shoes.

58 Unhook the top springs from the shoes and strut. Remove the leading (front) shoe complete with adjuster mechanism. Recover the strut.

59 Release the handbrake cable from the lever on the trailing shoe. Remove the trailing shoe and lever.

60 Clean and inspect the brake components as described in Chapter 9, Section 4, paragraphs 7 to 11. Note the following points:

- *(a) The use of special brake lubricant ot anti-seize compound is to be preferred to grease (paragraph 11)*
- *(b) Do not lubricate the mating faces or the edges of the adjuster plates*
- *(c) When assessing lining wear, remember that the lining on the leading shoe is originally twice as thick as that on the trailing shoe*

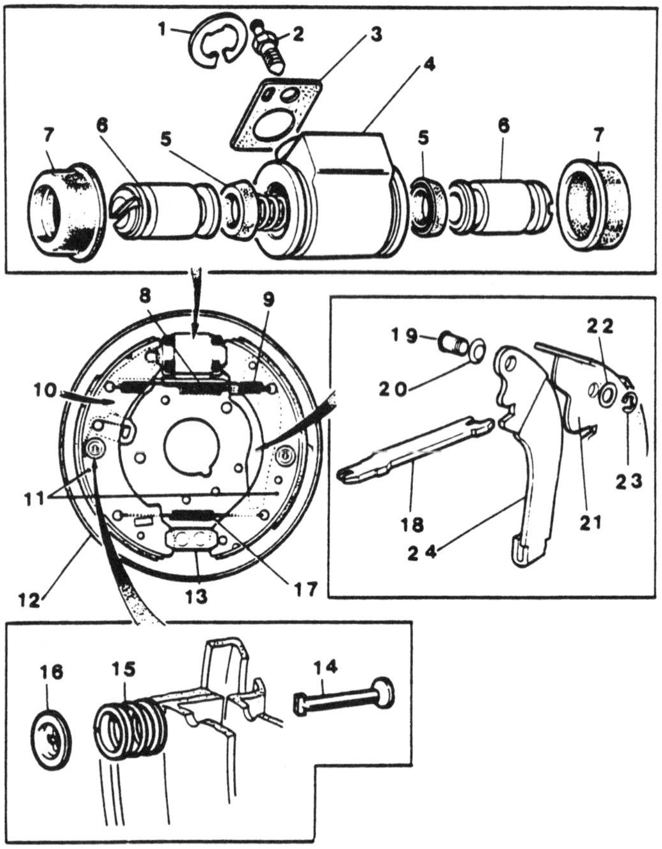

Fig. 13.41 Rear components (200 Series, 1985 on) (Sec 16)

1 *Circlip*
2 *Bleed nipple*
3 *Gasket*
4 *Wheel cylinder*
5 *Piston seals*
6 *Pistons*
7 *Boots*
8 *Strut spring*
9 *Pull-off spring*
10 *Adjuster mechanism (behind shoe)*
11 *Shoes*
12 *Backplate*
13 *Bottom pivot*
14 *Steady pin*
15 *Spring*
16 *Steady pin washer*
17 *Pull-off spring*
18 *Strut*
19 *Pivot pin*
20 *Spring washer*
21 *Trailing shoe*
22 *Plain washer*
23 *Circlip*
24 *Handbrake lever*

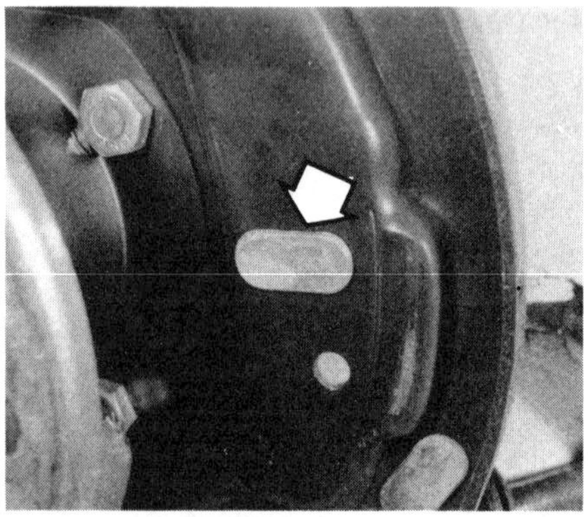
16.54A Remove the access bung (arrowed) from the other side of the backplate ...

16.54B ... and release the adjuster with a screwdriver. (Brake shoes removed for clarity)

16.55 Shoe steady pin and washer

16.56A Correct fitting of bottom spring ...

16.56B ... and top springs

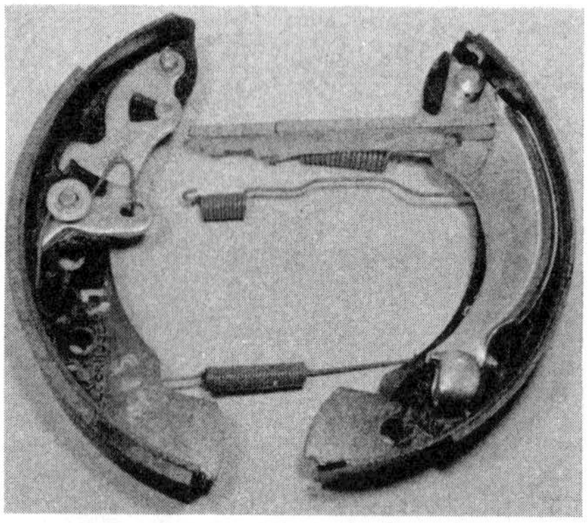
16.61 Brake shoes with handbrake lever and adjuster mechanism attached

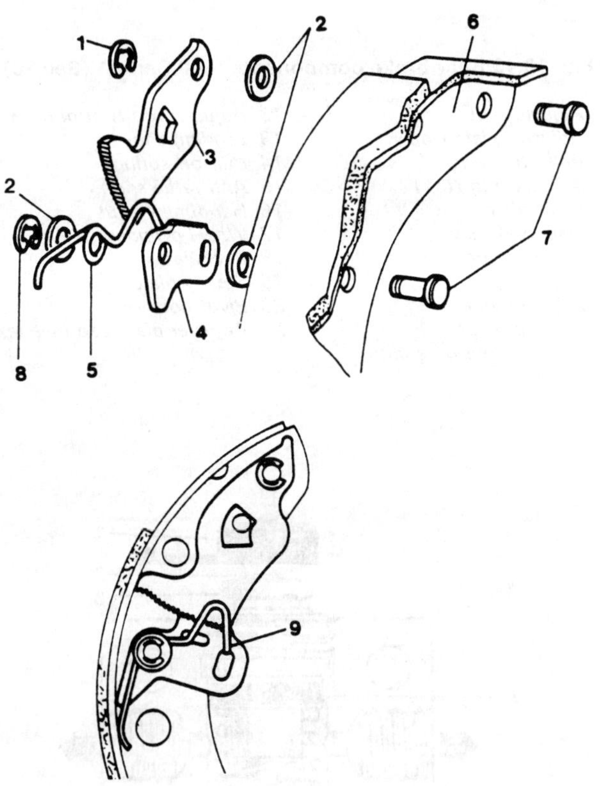

Fig. 13.42 Brake self-adjusting mechanism (200 Series) (Sec 16)

1	*Circlip*	*6*	*Leading shoe*
2	*Plain washers*	*7*	*Pivot pins*
3	*Adjuster plate (large)*	*8*	*Circlip*
4	*Adjuster plate (small)*	*9*	*End of spring*
5	*Spring*		

61 The adjuster mechanism and handbrake lever can be transferred to new shoes if desired. Observe the correct positioning of the washers, and the fitted position of the adjuster mechanism spring (photo).
62 Offer the trailing shoe and handbrake lever to the backplate. Engage the handbrake cable in the lever.
63 Fit the leading shoe, engaging the strut between the leading and trailing shoes. Fit the top springs to the shoes and strut.
64 Fit the bottom spring and engage the shoes in the bottom pivot and wheel cylinder pistons.
65 Refit the steady pins, springs and washers. Secure the washers by turning them 90°.
66 Refit the brake drum. (If difficulty is experienced, collapse the adjuster via the access bung, and make sure that the shoes are centralised). Secure the brake drum with the two screws.
67 Press the brake pedal several times. The adjuster should be heard to operate and the pedal travel should decrease as the shoes are brought towards the drum.
68 Repeat the operations on the other rear brake. Check the operation of the handbrake and adjust if necessary.
69 Refit the wheels, lower the vehicle and tighten the wheel nuts.
70 If new shoes have been fitted, avoid harsh braking as far as possible for the first few hundred miles until the linings have bedded in.

Rear brake shoes (300 Series) – removal and refitting

71 Chock the front wheels, release the handbrake and slacken the rear wheels nuts. Raise and support the rear of the vehicle and remove the rear wheels.
72 Remove the countersunk screw which secures the brake drum. Pull off the drum. If the brake shoes prevent removal, back off the adjuster by turning the adjuster wheel through the access hole at the bottom of the backplate.
73 Remove the steady pin washers by depressing them with pliers and turning them 90°. Remove the springs and steady pins.
74 Pull the shoes apart at the bottom and remove the adjuster strut, at the same time unhooking the adjuster tension spring from the bottom of the leading shoe. Remove the bottom adjuster plate and rod.
75 Pull the shoes apart at the top and remove the handbrake strut and anti-rattle spring.
76 Remove the leading shoe by swivelling it outwards and upwards, at the same time unhooking the pull-off spring.
77 Similarly remove the trailing shoe, pull-off spring and handbrake lever. Unhook the handbrake cable from the lever. Also unhook the top adjuster rod.
78 Retain the wheel cylinder pistons with a spring clamp or a stout rubber band. Do not press the brake pedal whilst the drum is removed.
79 Clean and inspect the brake components as described in Chapter 9, Section 4, paragraphs 7 to 11. The use of special brake lubricant or anti-seize compound is to be preferred to grease, however (paragraph 11).
80 Commence refitting by removing the spring clip or rubber band from the wheel cylinder. Connect the handbrake cable to the handbrake lever. Offer up the trailing shoe and engage its pull-off spring and the adjuster rod.
81 Fit the leading shoe and pull-off spring.
82 Fit the handbrake strut and anti-rattle spring between the shoe webs at the top. The flat edge of the strut faces the backplate, and the spring goes next to the leading shoe. Make sure that the shoes are correctly engaged with the wheel cylinder pistons.
83 Back off the adjuster rod to its shortest setting. Fit the rod between the shoes, with the adjuster wheel at the trailing shoe end.
84 Fit the adjuster plate, rod and spring, engaging them in their respective holes.
85 Fit the shoe steady pins, springs and washers. Centralise the shoes as far as possible by eye.
86 Turn the adjuster wheel until there is a clearance of 0.010 to 0.030 in (0.25 to 0.7 mm) between the bottom tip of the adjuster plate and the trailing shoe (gap 'X' in Fig. 13.43).
87 Fit the brake drum, altering the position of the shoes or backing off the adjuster a little more if necessary. Press the brake pedal firmly several times to centralise the shoes.
88 Remove the drum and recheck the adjuster plate clearance (paragraph 86). Readjust if necessary, then refit the drum, applying a little anti-seize compound to the drum and hub contact areas. Refit the countersunk screw.
89 Repeat the operations on the other rear brake, then refit the roadwheels and lower the vehicle. Tighten the wheel nuts.
90 Drive the vehicle in reverse and brake hard several times to operate the adjusters. Footbrake and handbrake travel should be normal when adjustment is complete.
91 If new shoes have been fitted, avoid harsh braking as far as possible for the first few hundred miles until the new linings have bedded in.

Servo assembly (later models) – overhaul

92 Later models with two-piston caliper disc brakes have a Lockheed type 50 servo. This can be overhauled as described for the type 38 (Chapter 9, Section 18).
93 Models with four-piston caliper disc brakes have a Lockheed type 65 servo. This cannot be overhauled, apart from renewing the non-return valve and the air filter.

Hydraulic system (later models) – description

94 All models have a dual circuit brake hydraulic system. The way in which the two circuits are divided varies as follows.
95 On models with a single bleed screw on each caliper, the primary hydraulic circuit operates the front brakes and the secondary circuit operates the rear brakes.
96 On models with three bleed screws on each caliper, the primary hydraulic circuit operates the rear brakes and one pair of pistons in each front caliper. The secondary circuit operates the other pair of pistons in each caliper.
97 A delay valve (when fitted) in the front brake circuit(s) provides the correct synchronisation between front and rear brakes. On some models a load sensing valve reduces the hydraulic pressure to the rear brakes when the vehicle is lightly laden (photo).

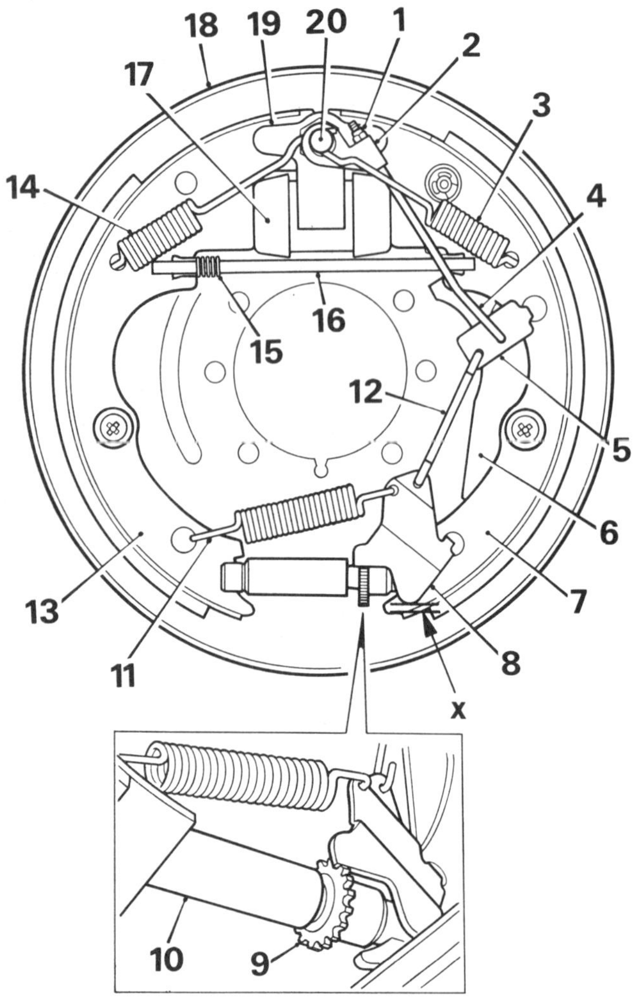

Fig. 13.43 Rear brake components (300 Series) (Sec 16)

1 *Adjuster nut*
2 *Adjuster plate (top)*
3 *Pull-off spring*
4 *Adjuster rod (top)*
5 *Pivot plate*
6 *Handbrake lever*
7 *Trailing shoe*
8 *Adjuster plate (bottom)*
9 *Adjuster wheel*
10 *Adjuster strut*
11 *Adjuster tension spring*
12 *Adjuster rod (bottom)*
13 *Leading shoe*
14 *Pull-off spring*
15 *Anti-rattle spring*
16 *Handbrake strut*
17 *Wheel cylinder*
18 *Backplate*
19 *Steady plate*
20 *Pivot post*
X *Adjuster plate gap (see text)*

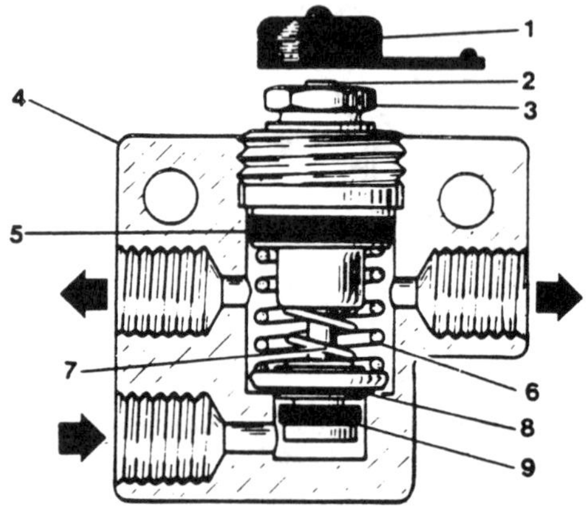

Fig. 13.44 Sectional view of a pressure delay valve (Sec 16)

1 *Cover*
2 *Plunger*
3 *Cap*
4 *Body*
5 *Stem seal*
6 *Spring*
7 *Spring*
8 *Main seal*
9 *Bypass seal*

16.97 Brake load sensing valve

16.98 Delay valve plunger (arrowed) may need to be depressed when bleeding

Bleeding the hydraulic system – later models

98 Models with a single bleed screw on each caliper are bled as described in Chapter 9, Section 3. If fluid will not pass the delay valve, remove the plunger cover and depress the plunger(s) when bleeding (photo).

99 Models with three bleed screws per caliper must be bled in the sequence shown in Fig. 13.45 when it is necessary to bleed the whole system. Note that the two primary circuit bleed screws on each caliper are bled together, so two bleed tubes will be needed. Again, depress the delay valve plunger(s) if necessary.

100 The rules concerning partial bleeding of the three bleed screw system are as follows:

(a) If the front part of the primary circuit has been broken, the complete primary circuit must be bled

(b) If the rear part of the primary circuit has been broken and no air has entered the front part, only the rear part need be bled

(c) If any part of the secondary circuit has been broken, the complete secondary circuit must be bled.

Bleeding equipment – all models

101 Various brake bleeding kits are available from motor accessory shops. The aim of the kits is to simplify the bleeding process, usually allowing 'one-man' bleeding instead of having to call on an assistant (photo). Follow the maker's instructions when using such a kit.

102 The use of 'pressure bleeding' equipment is particularly recommended for routine fluid renewal. DIY versions of this equipment use air pressure from the spare tyre to pressurize a large reservoir of hydraulic fluid. When the bleed screws are opened, hydraulic fluid under pressure flows out.

Delay valve – removal and refitting

103 The delay valve is located under the bonnet, near the brake master cylinder. It is not fitted to all models.

104 Clean around the hydraulic pipe unions on the valve, then disconnect and cap them. Be prepared for fluid spillage.

105 Remove the two nuts or bolts which secure the valve. Remove the valve.

106 A defective delay valve must be renewed.

107 Refit by reversing the removal operations. Bleed the appropriate hydraulic circuit(s) on completion.

Load sensing valve – adjustment

108 The tyre pressures must be correct and the vehicle must be unladen for this adjustment. The rear axle weight must also be known, so have the rear of the vehicle weighed if necessary.

109 Consult the load sensing valve setting table (in the right-hand door stop) and note the setting dimension 'Z' corresponding to the rear axle weight (photo).

110 Raise the rear of the vehicle on ramps so that the weight is still on the wheels. Disconnect the load sensing valve spring by removing the clevis pin.

111 Hold the spring vertically below the load sensing valve arm. Measure the distance between the hole in the arm and the spring eye. This should correspond to the setting dimension 'Z'.

112 If adjustment is necessary, slacken the clamp bolt which secures

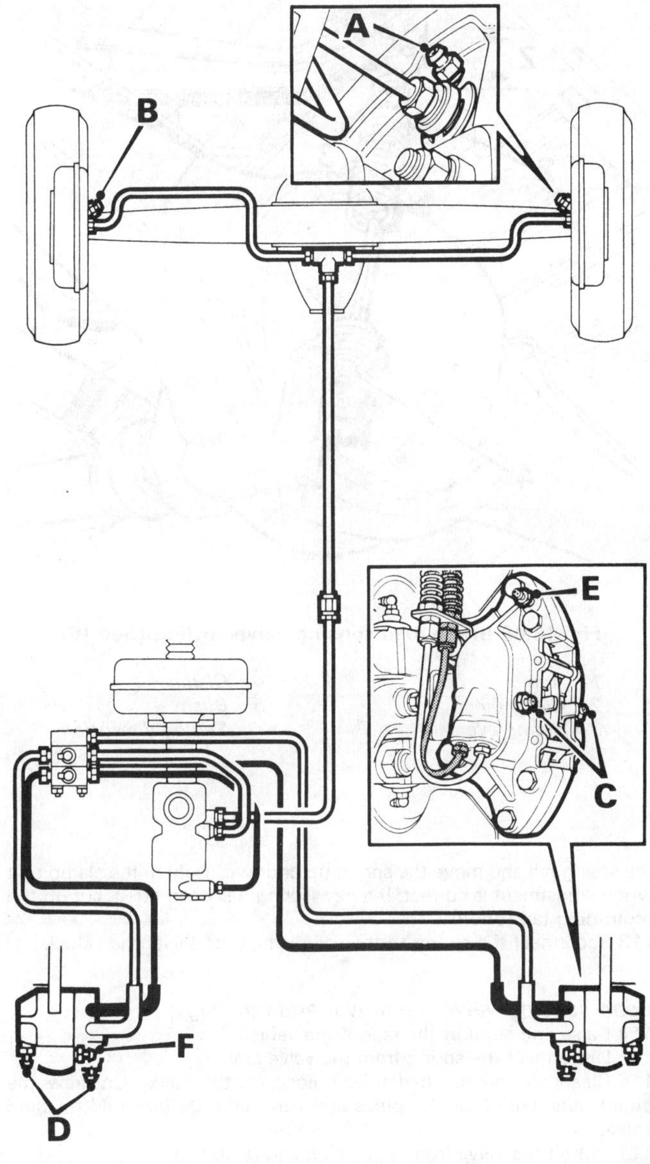

Fig. 13.45 Brake bleeding sequence for models with three bleed screws per caliper (Sec 16)

Primary circuit – A-B-C-D
Secondary circuit – E-F

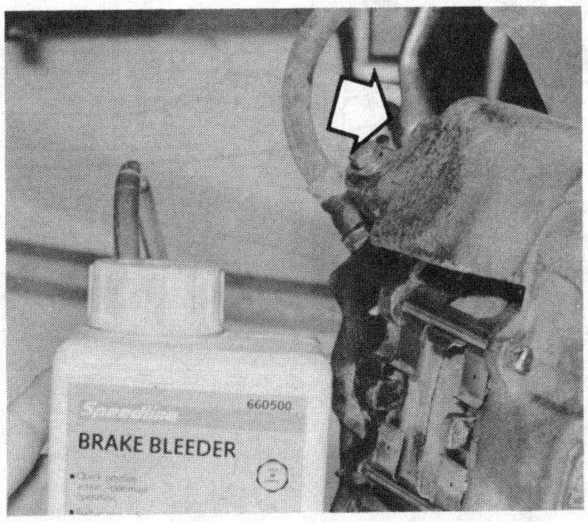

16.101 One-man brake bleeding kit. Tube is connected to bleed screw (with spanner on it – arrowed)

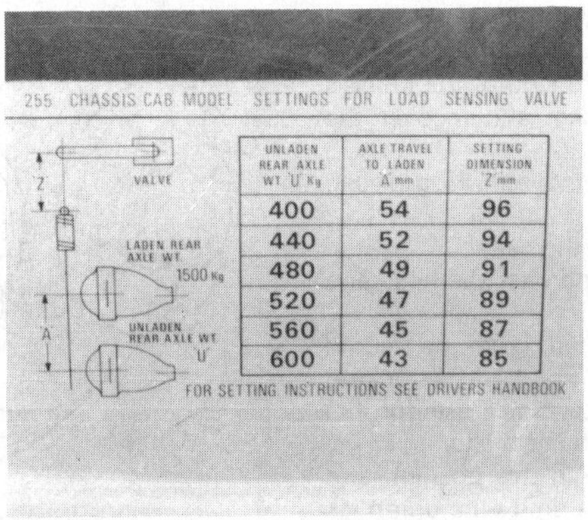

UNLADEN REAR AXLE WT 'U' Kg	AXLE TRAVEL TO LADEN 'A' mm	SETTING DIMENSION 'Z' mm
400	54	96
440	52	94
480	49	91
520	47	89
560	45	87
600	43	85

16.109 Typical load sensing valve setting table

16.112 Load setting valve spring tail clamp

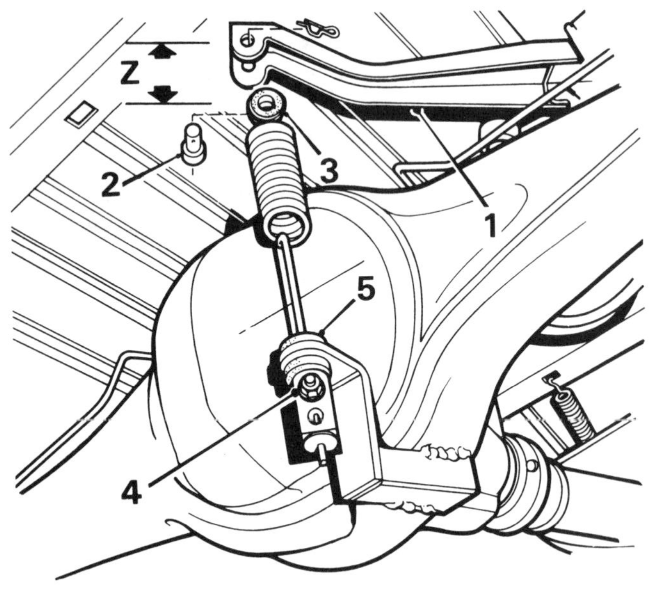

Fig. 13.46 Brake load sensing valve setting (Sec 16)

1 Arm
2 Clevis pin
3 Spring eye
4 Clamp
5 Bush
Z Setting dimension

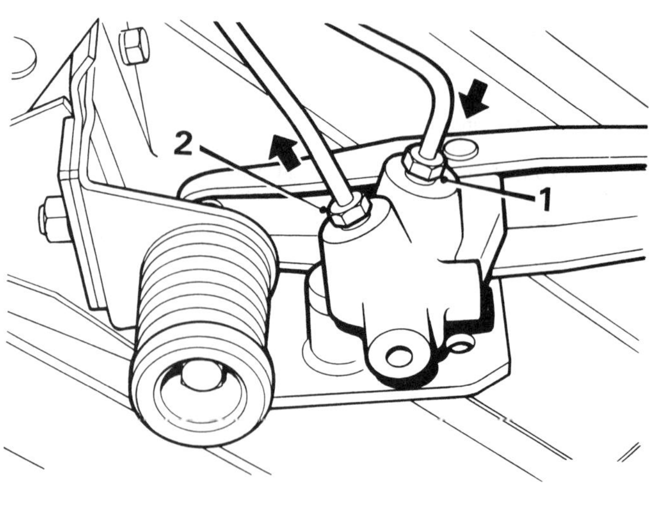

Fig. 13.47 Load sensing valve hydraulic connections (Sec 16)

1 From master cylinder *2 To rear wheel*

the spring tail and move the spring up or down. Tighten the clamp bolt when adjustment is correct. If a new spring has been fitted, cut off the protruding tail (photo).
113 Reconnect the spring to the valve arm, then lower the vehicle.

Load sensing valve – removal and refitting

114 Raise and support the rear of the vehicle.
115 Disconnect the spring from the valve arm.
116 Clean around the hydraulic unions on the valve. Unscrew the union nuts, withdraw the pipes and cap them. Be prepared for fluid spillage.
117 Unbolt the valve from its bracket and remove it.
118 Refit by reversing the removal operations. Check the adjustment of the valve spring before connecting it, as just described.
119 Bleed the appropriate hydraulic circuit on completion.

Front flexible hose (disc brake models) – removal and refitting

120 Clean around the hose unions. Disconnect the hose union at the rigid pipe; be prepared for fluid spillage.
121 Remove the nut and washer from the inboard side of the chassis bracket. Withdraw the hose end from the bracket and recover the locking plate.
122 Disconnect the hose from the caliper and remove it. Plug or cap open unions.
123 Commence refitting by securing the hose to the caliper.
124 Fit the locking plate to the chassis bracket and pass the hose end through the plate and bracket.
125 The locking plate and the hose end are serrated. From the neutral position (hose not twisted) turn the hose by one notch of the locking plate to give it the correct 'set'. Turn the hose clockwise for the left-hand brake and anti-clockwise for the right-hand brake.
126 Engage the hose end with the locking plate in this position and secure it with the washer and nut. Reconnect the rigid pipe union.
127 Check that the position of the hose if satisfactory at all angles of lock.
128 Bleed the appropriate hydraulic circuit on completion.

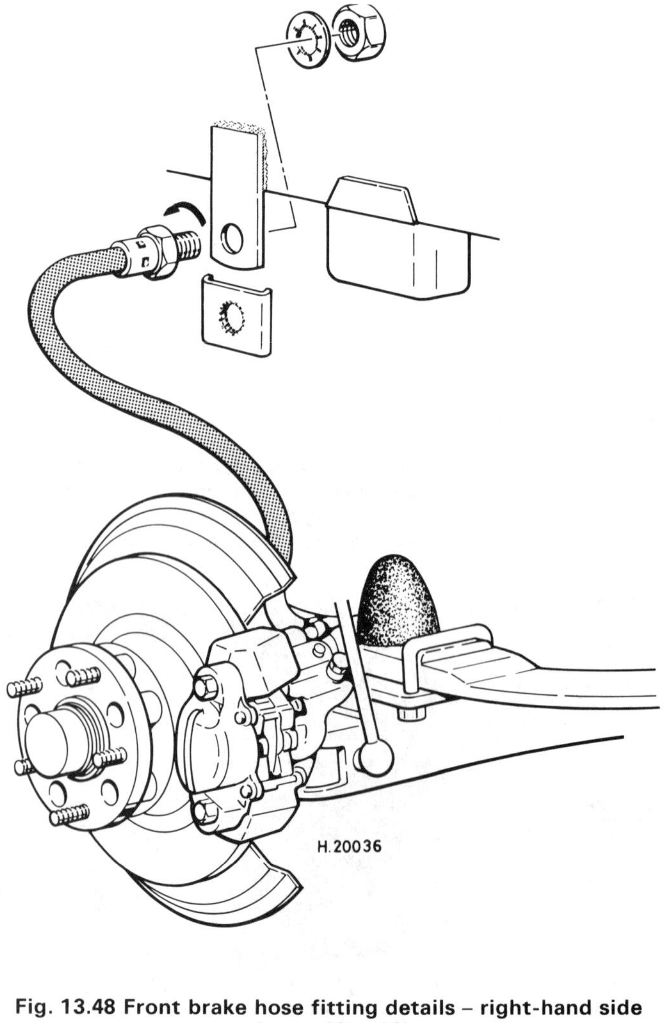

Fig. 13.48 Front brake hose fitting details – right-hand side shown (Sec 16)

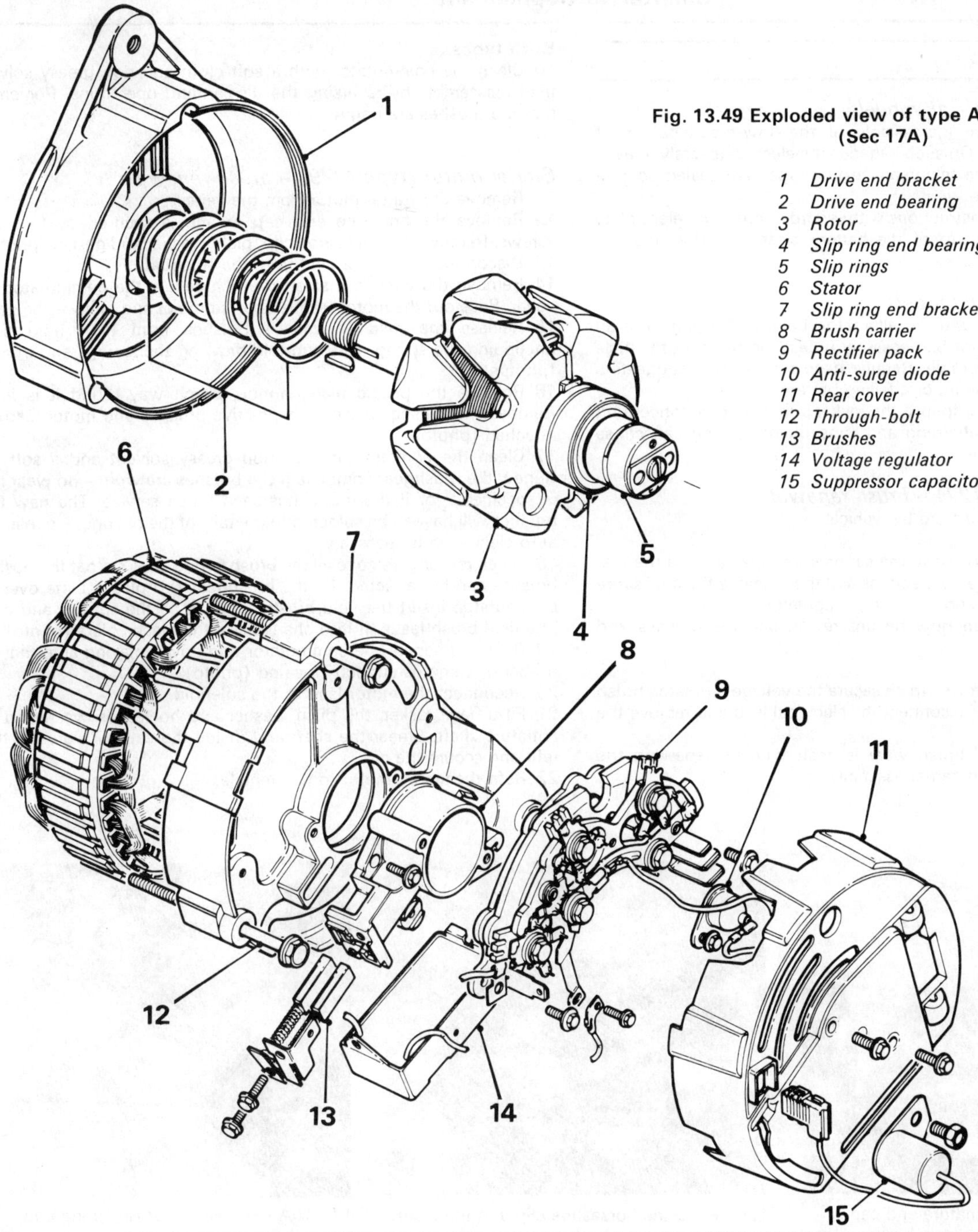

Fig. 13.49 Exploded view of type A115 alternator (Sec 17A)

1 Drive end bracket
2 Drive end bearing
3 Rotor
4 Slip ring end bearing
5 Slip rings
6 Stator
7 Slip ring end bracket
8 Brush carrier
9 Rectifier pack
10 Anti-surge diode
11 Rear cover
12 Through-bolt
13 Brushes
14 Voltage regulator
15 Suppressor capacitor

17A.6A Remove the three screws ...

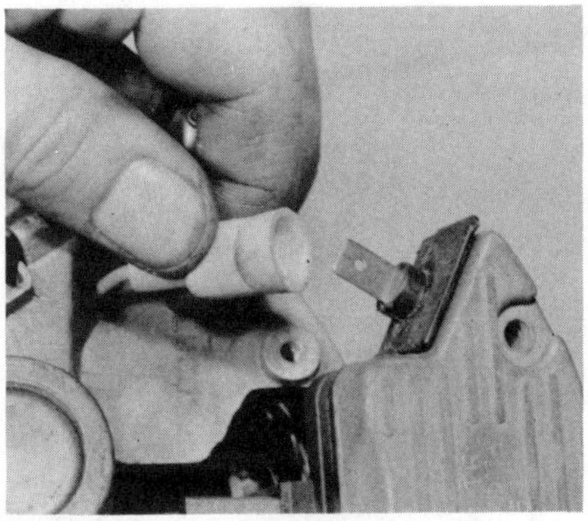

17A.6B ... disconnect the lead ...

17A.6C ... and remove the voltage regulator/bush carrier

17A Electrical system

Battery maintenance – all models

1 Modern batteries are increasingly of the 'low-maintenance' or 'maintenance-free' type. On such batteries the electrolyte rarely, if ever, needs topping-up. Some maintenance-free types are sealed so that topping-up is impossible.

2 Follow the maker's instructions with regard to checking electrolyte level. Always keep the top of the battery clean and dry, and the terminals clean and tight.

Battery charging – all models

3 On modern vehicles with an alternator, battery charging from an external source should not be necessary unless the pattern of vehicle use is extremely unfavourable. Regular need for charging suggests a fault in the electrical system, or a failing battery.

4 With sealed batteries, follow the maker's instructions concerning the charging rate. Overcharging and consequent gassing can cause such batteries to explode.

Alternator (A115/A127) – brush renewal

5 Remove the alternator from the vehicle.

Type A115

6 Refer to Fig. 13.49. Remove the suppressor capacitor and the rear cover. Make a note of the connections and then remove the anti-surge diode, the rectifier pack and the voltage regulator.

7 The brush carrier can now be unscrewed and the brushes and springs renewed.

Type A127

8 Remove the three screws which secure the voltage regulator/brush carrier to the alternator. Disconnect the electrical lead and remove the assembly (photos).

9 As far as is known, brush wear is dealt with by renewing the complete regulator/brush carrier assembly.

Both types

10 Clean the commutator with a soft cloth and non-greasy solvent, then reassemble by reversing the dismantling operations. Remember that the brushes are fragile.

Starter motor (type M79) – brush renewal

11 Remove the starter motor from the vehicle.

12 Remove the armature end cap, which is secured by two Torx screws. Remove the horseshoe clip, plain washer and gasket (photos).

13 Disconnect the motor lead from the solenoid.

14 Remove the two Torx screws which secure the commutator end cover. Prise out the motor lead rubber grommet and remove the cover.

15 Release the field (negative) brushes from their holders by unclipping the spring caps and withdrawing the springs and brushes (photos).

16 Remove the plastic plate, noting which way round it is fitted. Remove the brushgear with the positive brushes and motor lead still attached (photo).

17 Clean the commutator with non-greasy solvent and a soft rag. Renew the brushgear complete if the brushes are worn – no wear limit is specified – or if the motor has seen much service. The new field brushes will have to be soldered to the tails of the old ones – consult an auto electrician if necessary.

18 When refitting, remove all the brush spring caps so that the positive brushes can be retracted. Fit the brushgear and plastic plate over the commutator, insert the field brushes and fit all the springs and caps. The field brush leads fit into channels in the plastic plate (photo).

19 Refit and secure the commutator end cover, making sure that the rubber grommet is correctly located (photo).

20 Reconnect the motor lead to the solenoid.

21 Fit a new gasket, the plain washer and horseshoe washer to the armature shaft. Smear the shaft and inside the cap with grease, then refit and secure the cap.

22 Refit the starter motor to the vehicle.

17A.12A Remove the armature end cap ...

17A.12B ... the horseshoe clip and the plain washer

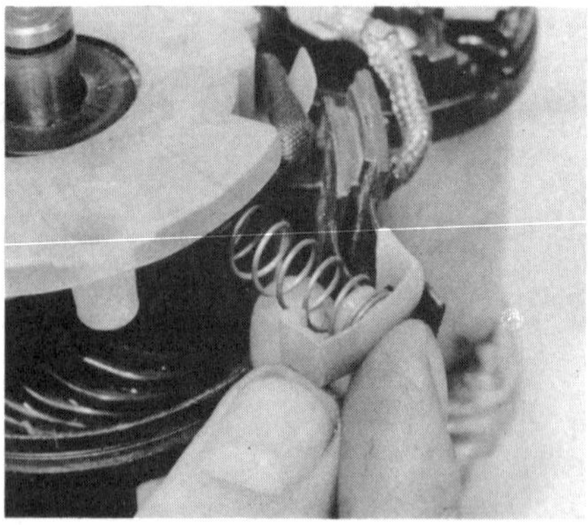

17A.15A Removing a brush spring and cap ...

17A.15B ... followed by a field brush

17A.16 Brushgear removed from motor

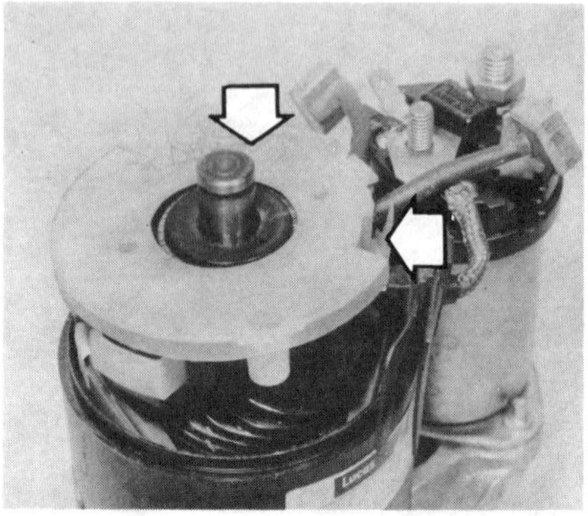

17A.18 Field brush leads go into channels (arrowed) in plastic plate

17A.19 Refitting the commutator end cover

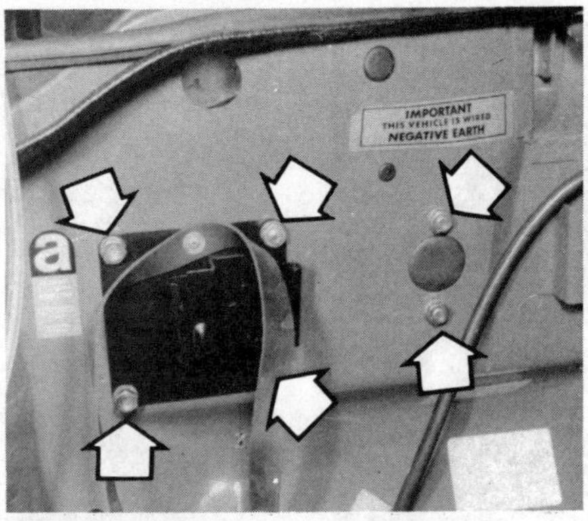

17A.26 Six nuts (arrowed) which secure the wiper motor and the washer bottle bracket

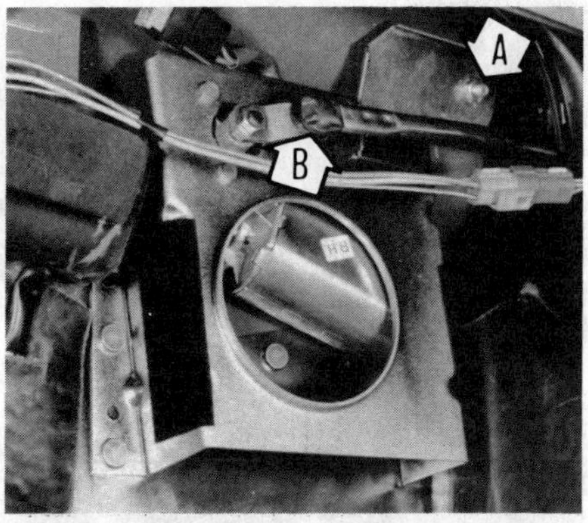

17A.28 One further nut (A) secures the wiper motor bracket. Link rod (B) must be freed from crankarm first

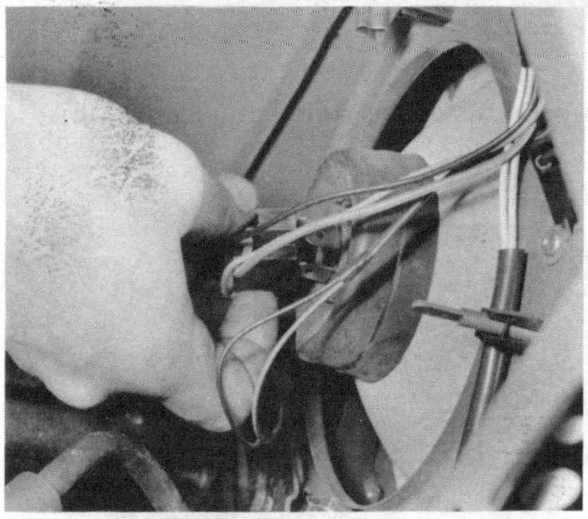

17A.32 Disconnecting a headlight bulb

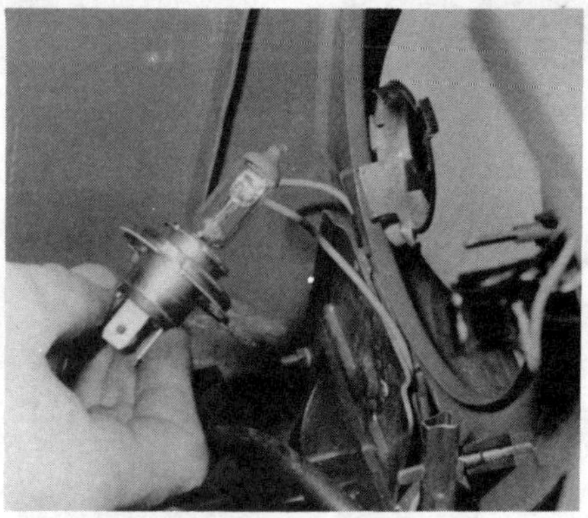

17A.33 Removing a headlight bulb

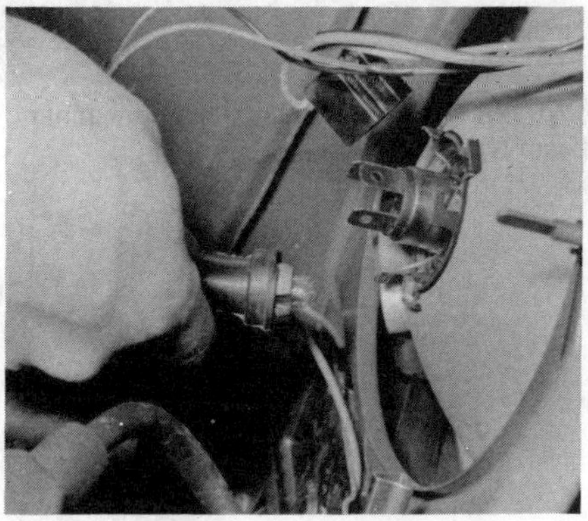

17A.39A Removing a sidelight bulb and holder – bulb type headlight

Windscreen wiper motor (later models) – removal and refitting

23 On later models the wiper motor drive is by means of rods instead of a flexible rack. The motor is located above the left-hand parcel shelf and is removed as follows.

24 Disconnect the battery negative lead.

25 Remove the left-hand parcel shelf.

26 Working under the bonnet, unclip the windscreen washer bottle and move it aside. Remove the six nuts which secure the wiper motor and the washer bottle bracket (photo).

27 Inside the cab, free the link rod from the wiper motor crankarm by prising it with a large screwdriver.

28 Remove the remaining nut which secures the wiper motor bracket to the bulkhead (photo). Pull the motor and bracket into the left-hand footwell and unplug the electrical connector.

29 The motor can now be unbolted from the bracket if required.

30 Refit by reversing the removal operations.

Headlight (later models) – bulb renewal

31 Access to the headlight bulbs is from under the bonnet. There is no need to remove the light unit.

32 Unplug the wiring connector from the bulb, then remove the rubber boot (photo).

33 Release the spring clip and withdraw the bulb (photo). **Caution:** If the bulb has just been in use, it may be very hot.

34 Do not touch the glass of the new bulb with the fingers. If it is accidentally touched, clean it with a soft rag and methylated spirit.

35 Fit the new bulb, making sure that the lugs and pips on the bulb and reflector match up. Secure it with the spring clip.

36 Refit the rubber boot, plug in the wiring connector and check for correct operation.

Headlight beam (200/300 Series) – adjustment

37 Adjustment is made from under the bonnet, by turning one or more of the 'T-bar' adjusters which can be seen at the rear of the headlight unit.

38 As with the earlier models, adjustment is best left to a garage having optical alignment equipment.

Front sidelight (in headlight) – bulb renewal

39 Open and prop the bonnet. Remove the sidelight bulbholder from the rear of the headlight bowl. In the case of sealed beam headlight units, the bulbholder must be twisted to free it from its mounting (photos).

40 Renew the bulb, refit the bulholder and check for correct operation. Close the bonnet.

Rear light cluster (200/300 Series chassis cab) – bulb renewal

41 The rear light units fitted to chassis cab models in production differ from those on other models. They will not necessarily be retained when some bodywork is attached to the chassis.

42 Unclip the lens surround and hinge it away from the lens (photo). Remove the lens.

43 The bulbs are now exposed and may be renewed as required (photo).

44 Refit the lens, making sure that it fits into the groove in the rubber. Clip the lens surround in place to secure it.

Direction indicator side repeater – bulb renewal

45 Open and prop the bonnet. Reach for the rear of the side repeater and pull the bulb and holder out of it.

46 Pull out the wedge-base bulb and push in the new one (photo).

47 Refit the bulb and holder, check for correct operation and close the bonnet.

Direction indicator/hazard warning flasher units (later models) – location

48 Some time after 1980 a combined direction indicator/hazard warning flasher was adopted. At first it was located in the steering column cowl.

49 Around 1985 the flasher unit was relocated behind a cover at the left-hand end of the left-hand parcel shelf.

17A.39B Sidelight bulb, holder and mounting removed from sealed beam type headlight

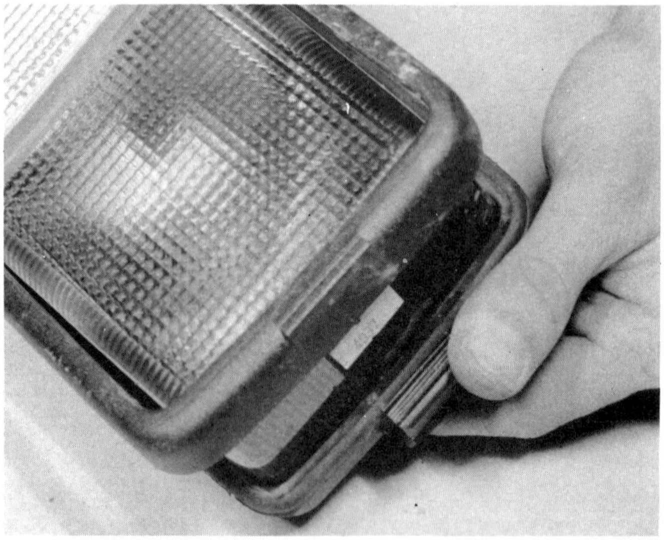
17A.42 Unclipping a rear light cluster lens surround (chassis cab)

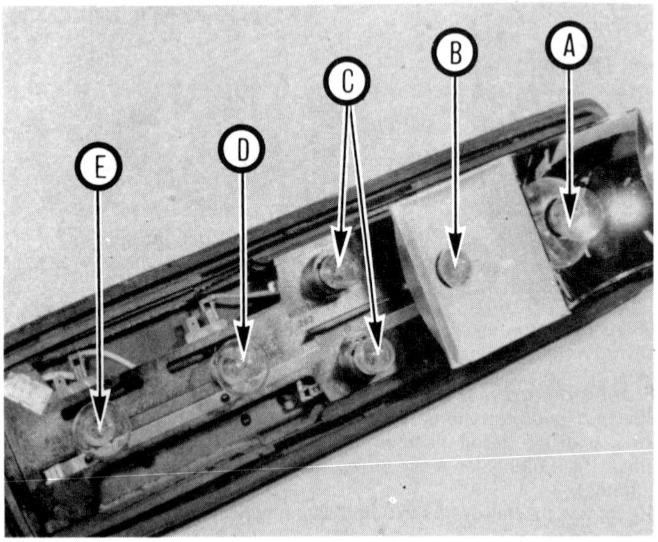

17A.43 Rear light cluster bulbs (chassis cab)

A Rear fog
B Reversing
C Tail/number plate
D Stop
E Direction indicator

17A.46 Direction indicator side repeater bulb

Switch illumination bulbs – renewal

50 Carefully prise or pull off the switch cover.

51 Unscrew or pull out the bulb, according to type, using the fingers or a length of rubber tube.

52 Fit the new bulb, check for correct operation and refit the switch cover.

Ignition/starter switch (later models) – removal and refitting

53 Proceed as described in Chapter 10, Section 38, but note that on some models the switch is secured by a single grub screw at the side of the steering lock/switch housing, instead of by two screws on the face of the housing.

Steering column switches (later type) – removal and refitting

54 Disconnect the battery negative lead.

55 Remove the steering column shroud. This is secured by four screws, of which three are not accessible until the left-hand half of the shroud has been removed (photos).

Fig. 13.50 Switch illumination bulb renewal – push-push type (Sec 17A)

1 Cover end lens
2 Bulb

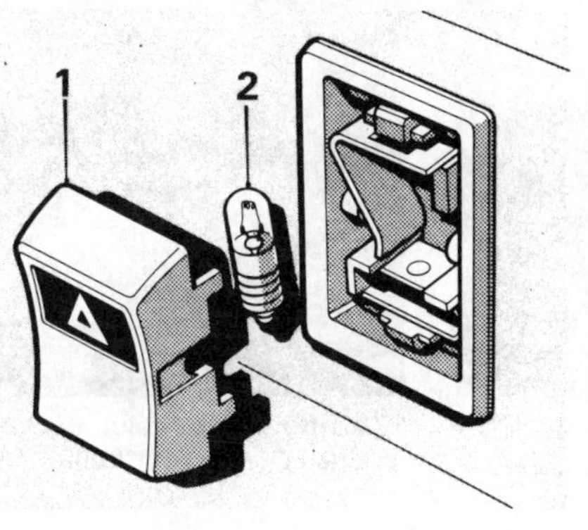

Fig. 13.51 Hazard warning switch (early type) bulb renewal (Sec 17A)

1 Cover 2 Bulb

Fig. 13.52 Hazard warning switch (later type) bulb renewal (Sec 17A)

1 Cover 2 Bulb

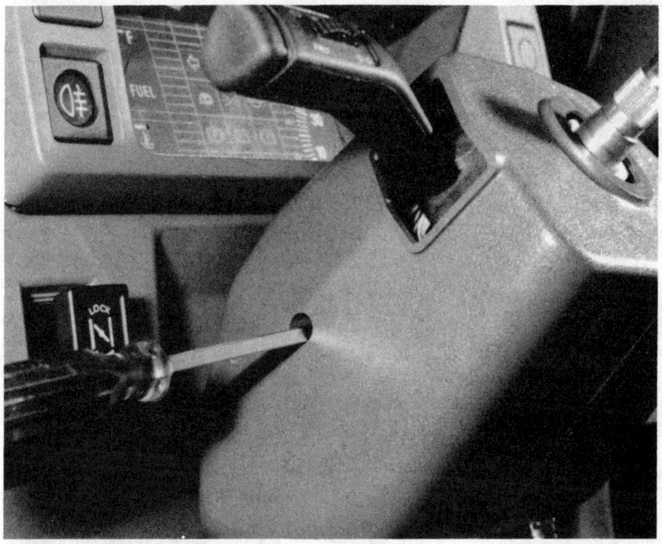

17A.55A Removing the left-hand shroud screw

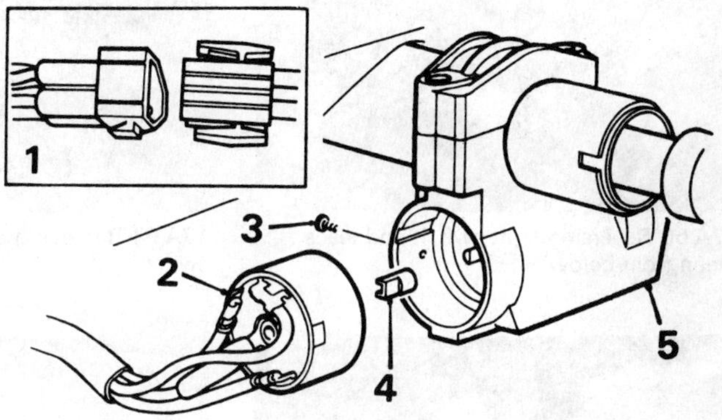

Fig. 13.53 Ignition/starter switch – later models (Sec 17A)

1 Multi plug
2 Switch
3 Grub screw
4 Switch driver
5 Steering lock/switch housing

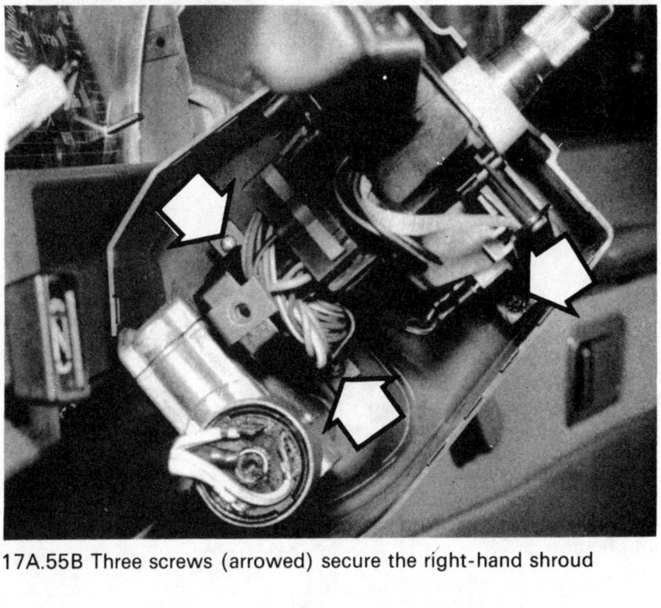

17A.55B Three screws (arrowed) secure the right-hand shroud

56 Depress the switch securing lugs and slide the switch from its location. Disconnect the multi-plug and (when applicable) the fibre optic 'bootlace' which carries the switch illumination (photos).
57 Refit by reversing the removal operations.

Reversing light switch (5-speed gearbox) – adjustment

58 Remove the six screws which secure the gear lever gaiter. Pull back the gaiter.
59 Switch on the ignition and engage reverse gear.
60 Screw the reversing light switch inwards or outwards until the lights just come on, then screw it a further half-turn inwards from this point. It may be necessary to disconnect and reconnect the switch leads to prevent excessive twisting (photo).
61 Select neutral and reverse several times and check that the operation of the switch is satisfactory.
62 Switch off the ignition and refit the gear lever gaiter.

Interior light door switch – removal and refitting

63 Later models are fitted with door switches which operate the interior light.
64 Remove the switch securing screw and pull the switch out of its location in the door pillar (photo).
65 Unplug the wiring connector. Secure the wire with a clothes peg or a piece of string so that it does not fall back into the pillar.
66 Refit by reversing the removal procedures.

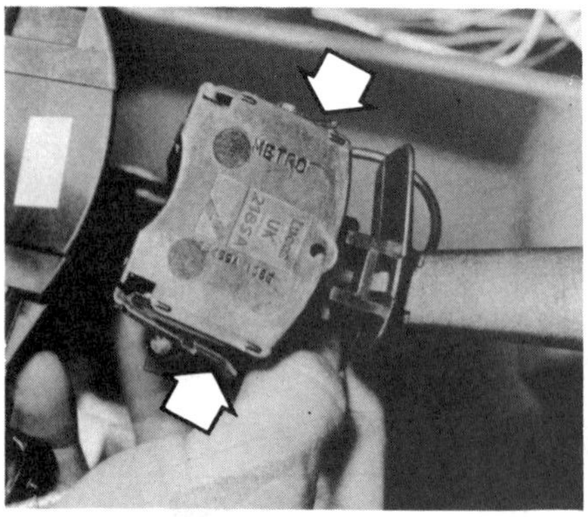

17A.56A Depress the lugs (arrowed) and slide out the switch ...

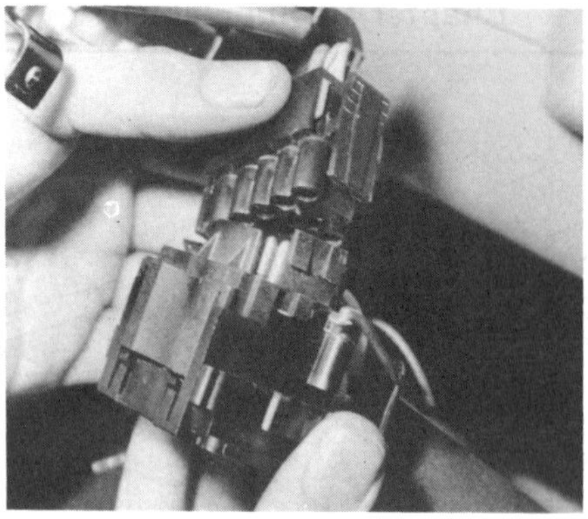
17A.56B ... disconnect the multi-plug ...

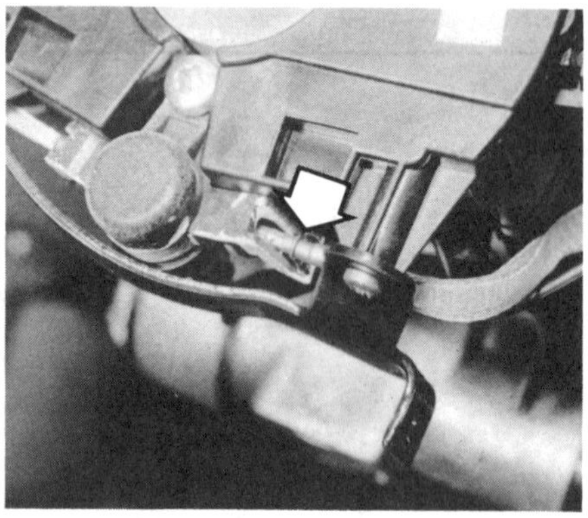
17A.56C ... and unplug the fibre-optic 'bootlace' (arrowed) from the bulbholder

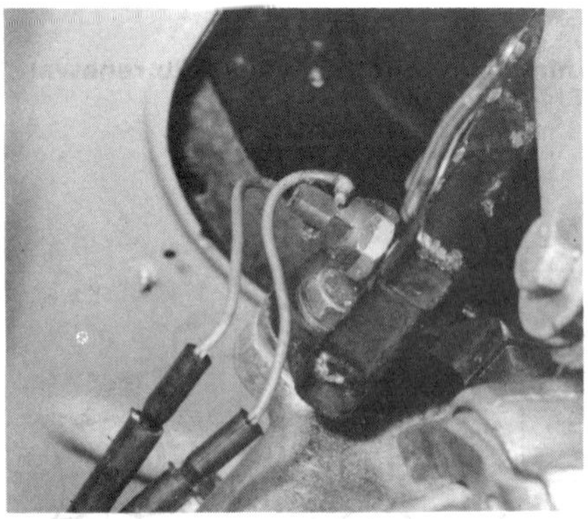
17A.60 Reversing light switch and leads (seen from below)

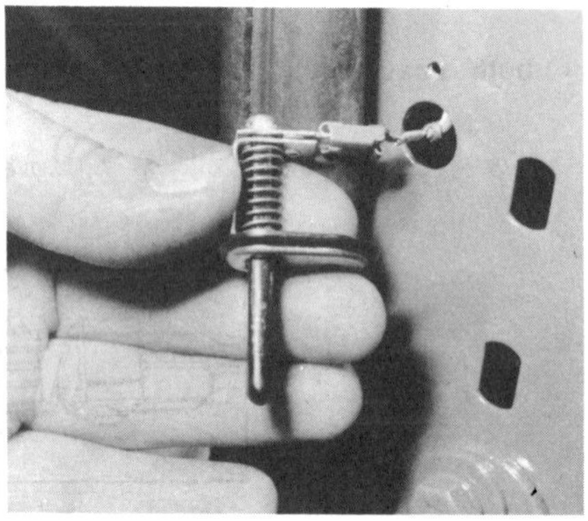
17A.64 Removing an interior light door switch

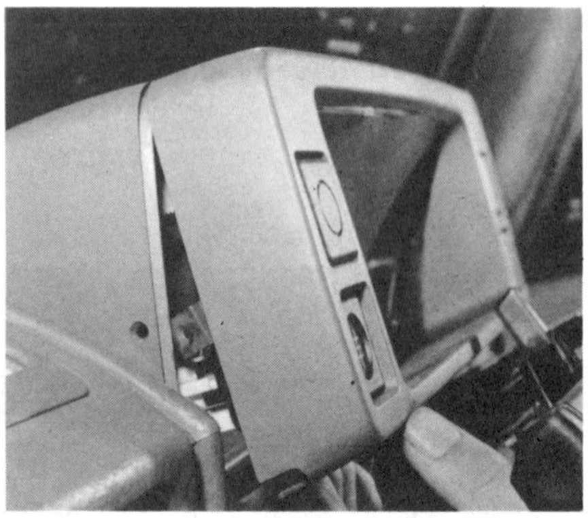
17A.68 Removing the instrument panel shroud

17A.69A Two of the instrument panel securing screws (arrowed)

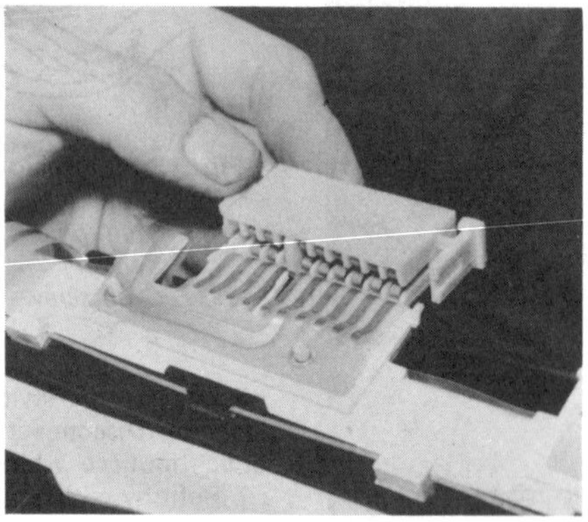
17A.69B Disconnecting an instrument panel multi-plug

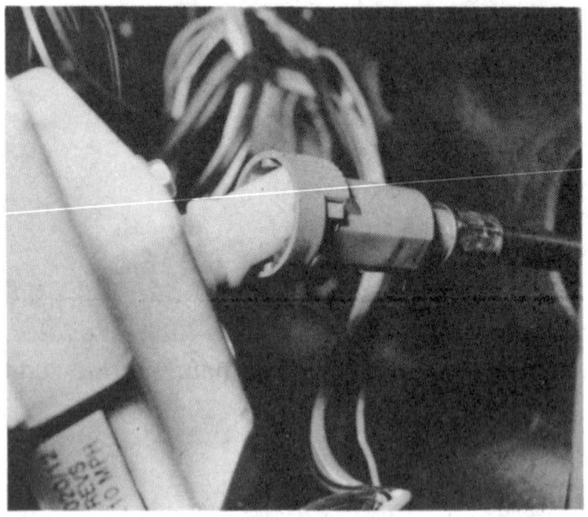
17A.69C Disconnecting the speedometer cable

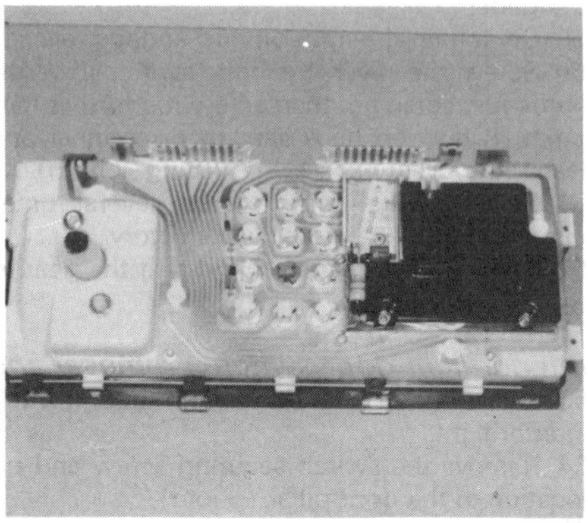
17A.70 Rear view of the instrument panel

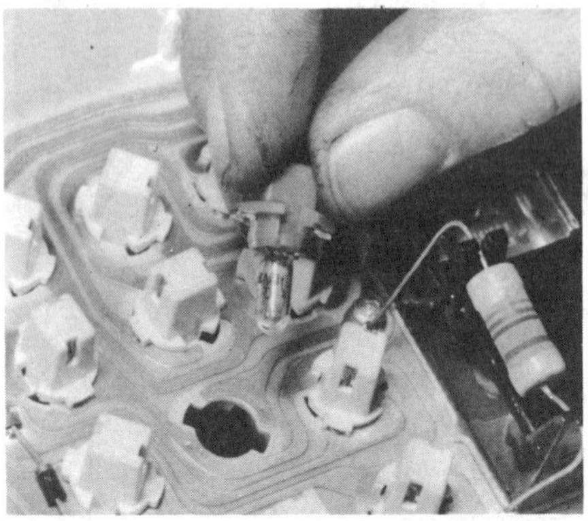
17A.71 Removing the alternator warning light bulb

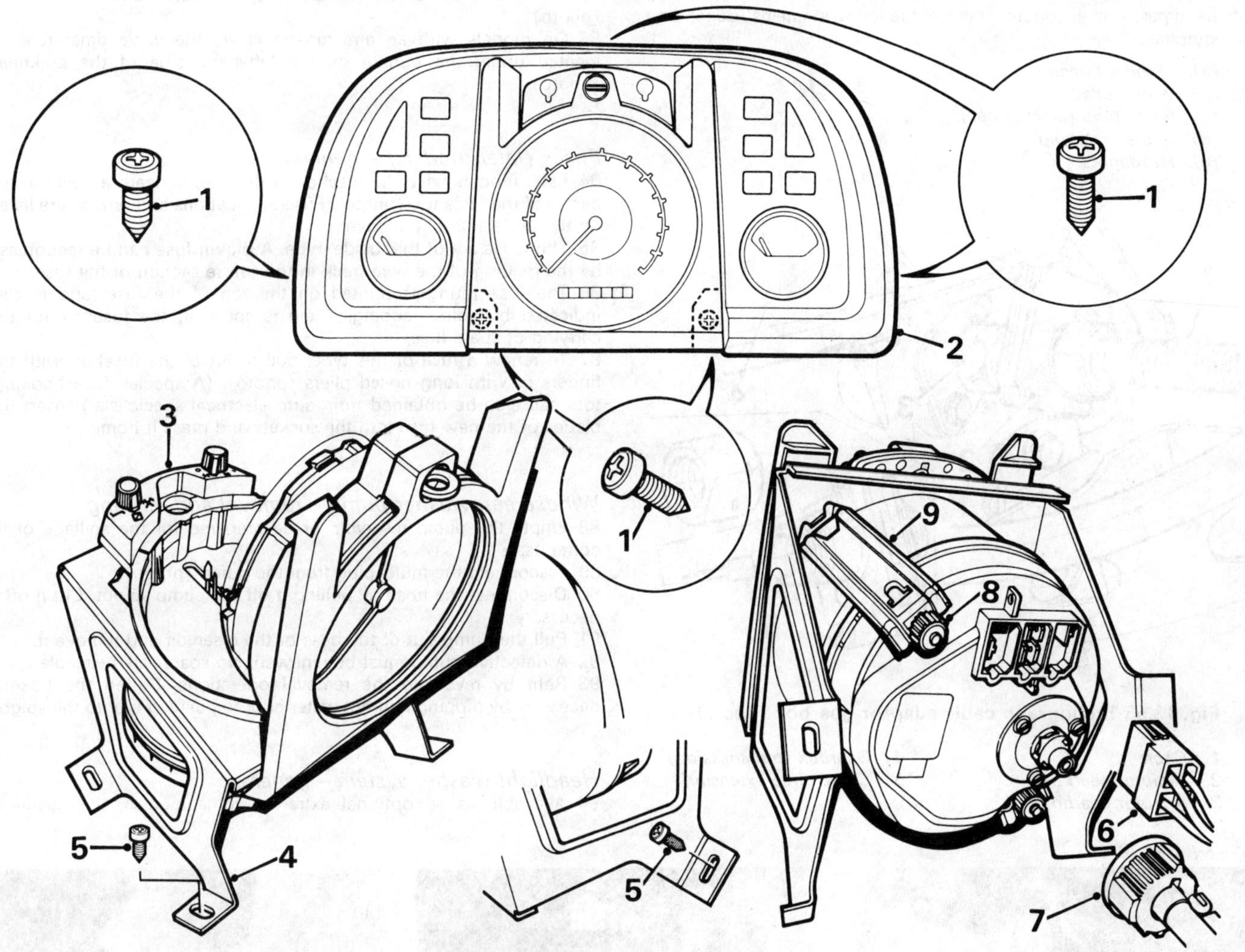

Fig. 13.54 Tachograph fitting details (Sec 17A)

1 *Instrument panel screws*
2 *Instrument panel*
3 *Tachograph*
4 *Mounting plate*
5 *Mounting screws*
6 *Multi-plug*
7 *Drive cable*
8 *Knurled nut*
9 *Bracket*

Instrument panel (later type) – removal and refitting

67 Disconnect the battery negative lead.
68 Remove the two screws which secure the instrument panel shroud. Remove the shroud, disconnecting any switches contained in it (photo).
69 Remove the four screws which secure the instrument panel. Withdraw the panel, disconnecting the wiring multi-plugs and the speedometer cable (photos).
70 The various instruments, bulbs etc can now be renewed as required. Be careful when handling the printed circuit, it is fragile (photo).
71 If renewing the alternator (no-charge) warning light bulb, which is identified by its red holder, note that the bulb rating is 2W (photo).
72 Refit by reversing the removal operations.

Tachograph – general

73 Some models are fitted with a tachograph to comply with regulations controlling drivers' working hours and the operation of the vehicle.
74 Follow the maker's instructions for day-to-day use of the tachograph, insertion and removal of charts, etc.
75 If the tachograph ever needs to be removed, this is done in a similar way to any other instrument. However, anti-tamper seals are fitted to the tachograph drive cable; if these seals are broken, the tachograph must be calibrated and new seals fitted by an authorised agent after refitting.
76 At the gearbox end of the tachograph cable is an adaptor gearbox, also fitted with anti-tamper seals. The nut which secures the adaptor gearbox has a left-hand thread at the adaptor gearbox end, and a normal thread at the drive pinion end.

Dim-dip system – description

77 On 1987 models the headlights and sidelights are wired in such a way that it is not possible to drive on sidelights alone. If the ignition and sidelights are switched on together, a relay supplies the headlight dipped beam filaments via a dropper resistor. This is to comply with lighting regulations for 1987 and later vehicles.
78 The components of the system are a relay, located under the ashtray, and a dropper resistor located under the left-hand headlight unit (photo).
79 No circuit diagram was available at the time of writing.

Relays (later models) – description

80 Various relays are found in the electrical system of later models. A relay is an electrically-operated switch, sometimes with a timer or delay mechanism incorporated. If a circuit or system which includes a relay malfunctions, remember that it could be the relay at fault.

81 Relays located behind the panel at the left-hand end of the left-hand parcel shelf (photos) control the following items (depending on equipment):

(a) Manifold heater
(b) Wiper delay
(c) Fuel pump (electric pump)
(d) Starter solenoid
(e) Headlights

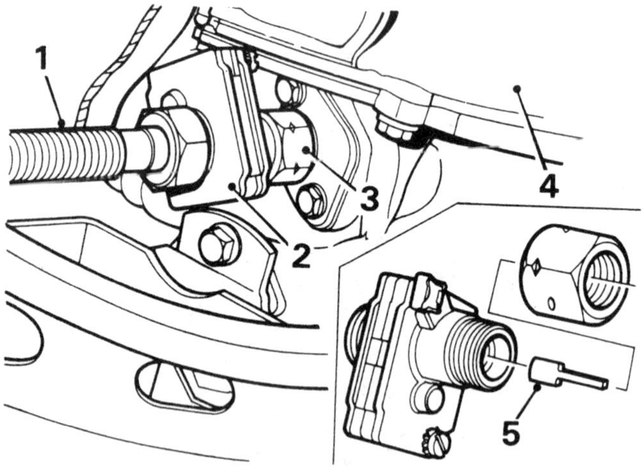

Fig. 13.55 Tachograph cable adaptor gearbox (Sec 17A)

1 Cable
2 Adaptor gearbox
3 Adaptor gearbox nut
4 Gearbox (transmission)
5 Cable drive extension

82 On 1987 models, the dim-dip relay is located under the ashtray (photo).
83 On models with an anti-run-on valve, the valve timer relay is located under the bonnet on the left-hand side of the bulkhead (photo).

Fuses (later models) – general

84 Later models have the fusebox in the cab, beneath a cover in the centre of the facia top (photo). Fuse applications and ratings are listed on the underside of the cover.
85 The fuses are of the 'blade' type. A blown fuse can be recognised by the melting of the wire track in the centre section of the fuse.
86 The fuse rating is printed on the top of the fuse, and is also indicated by colour coding. There is space in the fusebox for the carrying of spare fuses.
87 To renew a fuse of this type, pull it out of the fusebox with the fingers or with long-nosed pliers (photo). (A special 'fuse hooking' tool can also be obtained from auto electrical specialists.) Insert the blades of the new fuse into the sockets and press it home.

Windscreen washer pump – removal and refitting

88 Empty the pump reservoir, or be prepared for the spillage of its contents.
89 Disconnect the multi-plug from the pump (photo).
90 Disconnect the hose by pulling it off the pump spigot. Cut it off if necessary.
91 Pull the pump out of the base of the reservoir and remove it.
92 A defective pump must be renewed. No spares are available.
93 Refit by reversing the removal operations. Soften the hose if necessary by dipping it in hot water prior to pushing it onto the spigot.

Headlight washer system – general

94 Available as an optional extra on some models, the headlight

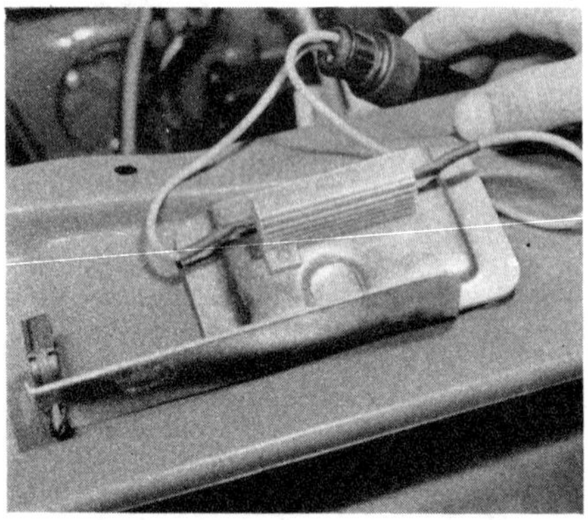

17A.78 Dim-dip resistor removed from below the left-hand headlight

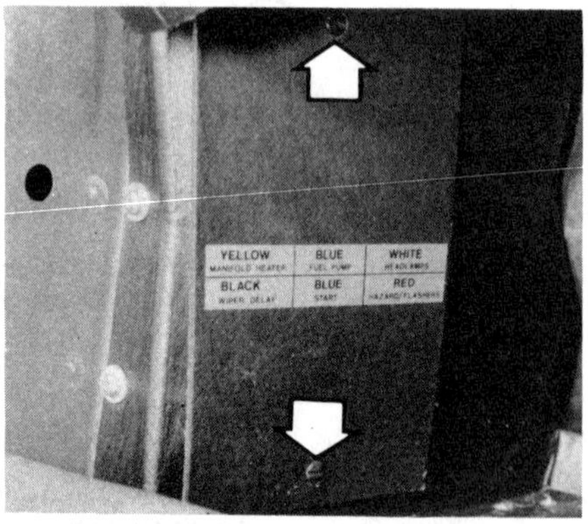

17A.81A Remove the panel (securing screws arrowed) ...

17A.81B ... to reveal the relays and the flasher unit

17A.82 Dim-dip relay under the ashtray

17A.83 Anti-run-on valve timer relay

17A.84 Fusebox and cover

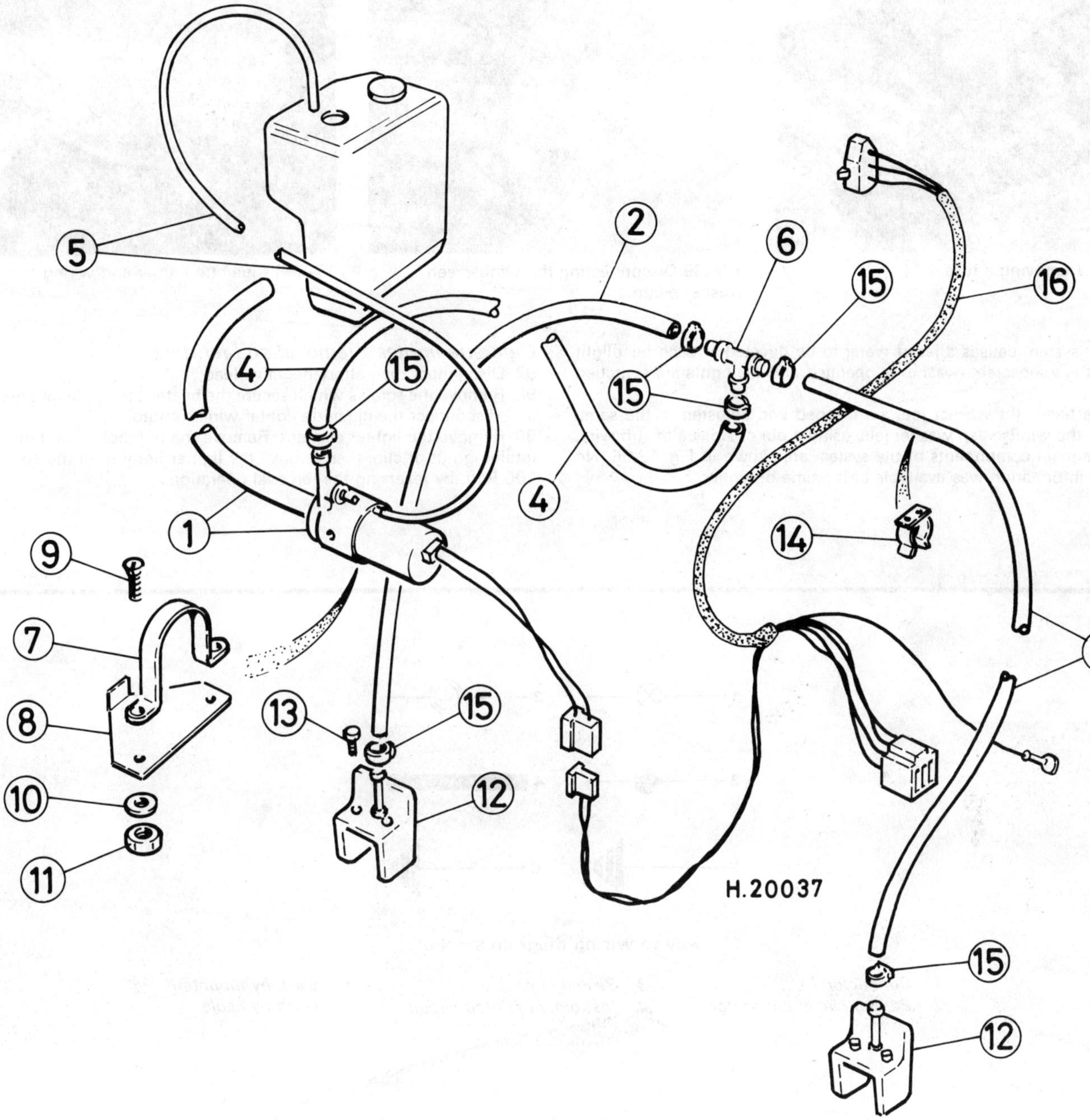

Fig. 13.56 Headlight washer system components (Sec 17A)

1 Pump and inlet tube
2 Pressure tube to nozzle
3 Pressure tube to nozzle
4 Pressure tube from pump
5 Air bleed tube
6 T-piece
7 Pump bracket
8 Pump mounting plate
9 Screw
10 Washer
11 Nut
12 Nozzles
13 Screw
14 Clip
15 Hose clips

17A.87 Removing a fuse

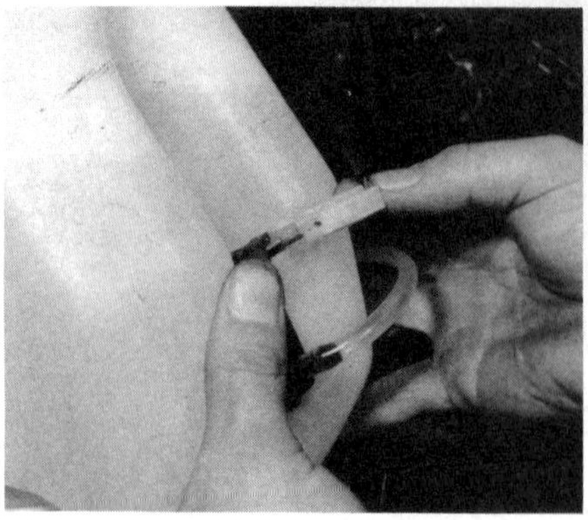

17A.89 Disconnecting the windscreen washer pump

17A.98 Cigarette lighter and wiring

washer system causes a jet of water to be directed at each headlight when the windscreen washer is operated and the lights are switched on.

95 The headlight washer jets are cleaned and adjusted in the same way as the windscreen washer jets, using a pin or a piece of stiff wire.

96 The main components of the system are shown in Fig. 13.56. No further information was available at the time of writing.

Cigarette lighter – removal and refitting

97 Disconnect the battery negative lead.

98 Remove the screws which secure the heater cover. Lower the cover and disconnect the cigarette lighter wires (photo).

99 Remove the lighter element. Remove the bulbholder and the body retaining nut or clip, then remove the lighter body from the cover.

100 Refit by reversing the removal operations.

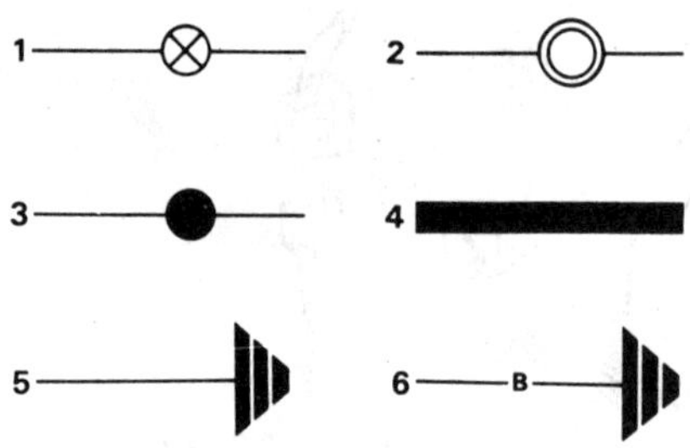

Key to wiring diagram symbols

1 Connector
2 Printed circuit connector
3 Sealed joint
4 Instrument printed circuit
5 Earth by mounting
6 Earth by cable

Colour codes

B Black
G Green
K Pink
L Light
N Brown
O Orange
P Purple
R Red
S Slate
U Blue
W White
Y Yellow

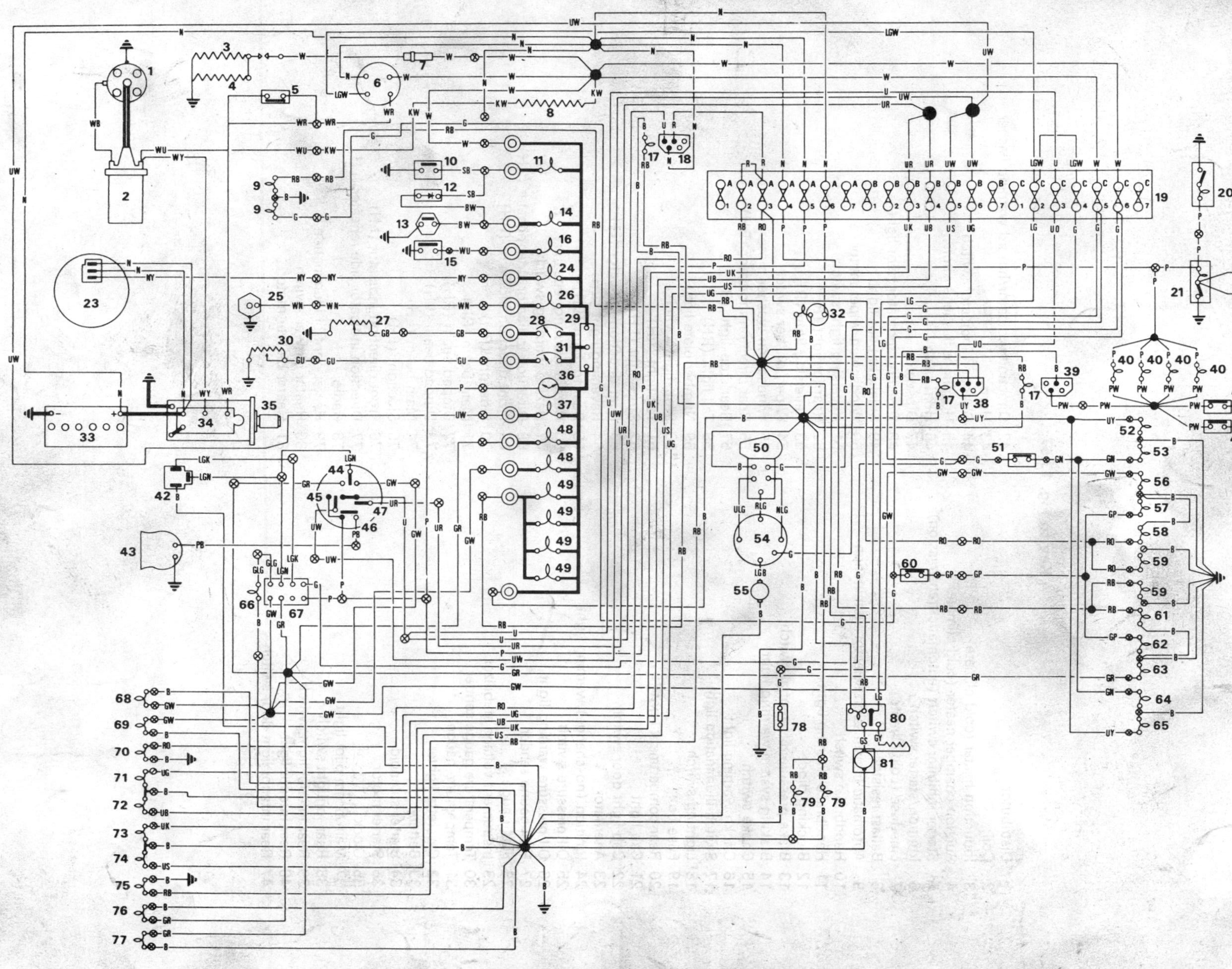

Fig. 13.57 Wiring diagram – 200 Series without tachograph. Not all items are fitted to all models

Key to Fig. 13.57

1 Distributor
2 Coil
3 Induction heater (cold climate)
4 Suction chamber heater (cold climate)
5 Starter inhibitor switch (automatic transmission)
6 Ignition/starter switch
7 Line fuse (cold climate)
8 Ballast resistor
9 Automatic transmission selector illumination
10 Handbrake switch
11 Handbrake warning light
12 Blocking diode
13 Brake pressure differential switch
14 Braking system warning light
15 Choke switch
16 Choke warning light
17 Switch illumination light
18 Lighting switch
19 Fuse box
20 Rear compartment light
21 Cab light
22 Cab light door switch
23 Alternator
24 Ignition (no-change) warning light
25 Oil pressure switch
26 Oil pressure warning light
27 Fuel gauge sender
28 Fuel gauge
29 Instrument voltage stabilizer
30 Temperature gauge sender
31 Temperature gauge
32 Cigarette lighter
33 Battery
34 Starter solenoid
35 Starter motor
36 Clock
37 Main beam pilot light
38 Rear foglight switch
39 Rear interior light switch
40 Rear interior light
47 Rear interior light door switch
42 Direction indicator/hazard warning flasher
43 Horn
44 Direction indicator switch
45 Headlight flasher switch
46 Horn push
47 Dipswitch
48 Direction indicator pilot light
49 Sidelight pilot light
50 Windscreen wiper motor
51 Reversing light switch
52 Rear foglight (RH)
53 Reversing light (RH)
54 Wiper/washer switch
55 Windscreen washer motor
56 Rear direction indicator (RH)
57 Rear stop-light (RH)
58 Tail light (RH)
59 Number plate light
60 Stop-light switch
61 Tail light (LH)
62 Stop-light (LH)
63 Rear direction indicator (LH)
64 Reversing light (LH)
65 Rear foglight (LH)
66 Hazard warning repeater
67 Hazard warning switch
68 Direction indicator side repeater (RH)
69 Front direction indicator (RH)
70 Sidelight (RH)
71 Main beam (RH)
72 Dipped beam (RH)
73 Dipped beam (LH)
74 Main beam (LH)
75 Sidelight (LH)
76 Front direction indicator (LH)
77 Direction indicator side repeater (LH)
78 Radio
79 Heater control illumination
80 Heater blower switch
81 Heater blower motor

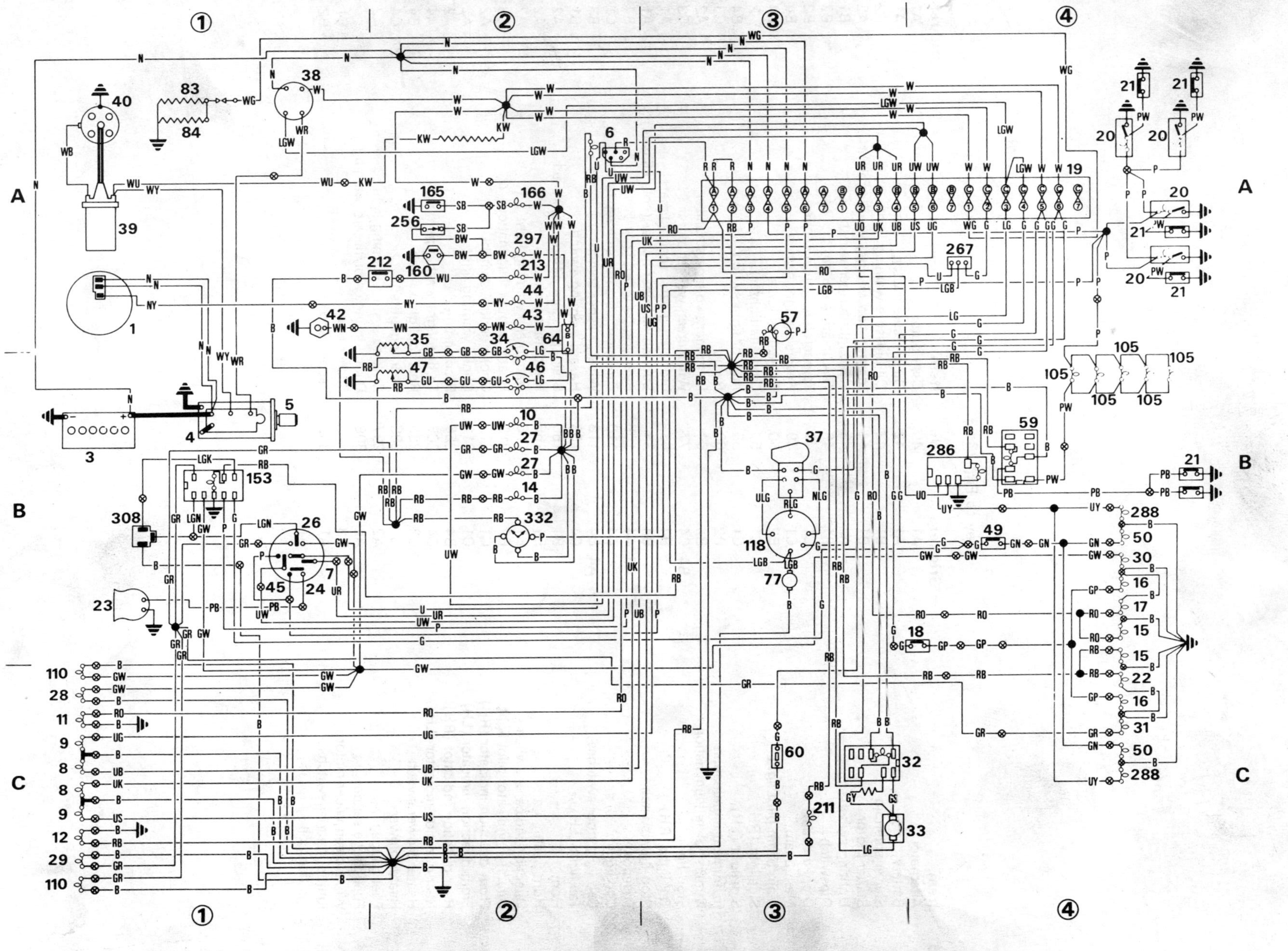

Fig. 13.58 Wiring diagram – 300 Series with tachograph. Not all items are fitted to all models

Key to Fig. 13.58

No	Item	Grid ref
1	Alternator	A1
3	Battery	A1
4	Starter solenoid	B1
5	Starter motor	B1
6	Lighting switch	A2
7	Dipswitch	B1
8	Dipped beam	C1
9	Main beam	C1
10	Main beam pilot light	B2
11	Sidelight (RH)	C1
12	Sidelight (LH)	C1
14	Instrument illumination	B2
15	Number plate illumination	B4
16	Stop-lights	B4, C4
17	Tail light (RH)	B4
18	Stop-light switch	B4
19	Fusebox	A3, A4
20	Cab light	A4
21	Cab light door switches	A4, B4
22	Tail light (LH)	C4
23	Horn	B1
24	Horn push	B1
26	Direction indicator switch	B1
27	Direction indicator pilot light	B2
28	Front direction indicator (RH)	C1
29	Front direction indicator (LH)	C1
30	Rear direction indicator (RH)	B4
31	Rear direction indicator (LH)	C4
32	Heater blower switch	C3
33	Heater blower	C3
34	Fuel gauge	A2
35	Fuel gauge sender	A2
37	Windscreen wiper motor	B3
38	Ignition/starter switch	A1
39	Ignition coil	A1
40	Distributor	A1
42	Oil pressure switch	A1
43	Oil pressure warning light	A2
44	Ignition (no-charge) warning light	A2
45	Headlight flasher switch	B1
46	Temperature gauge	B2
47	Temperature gauge sender	B2
49	Reversing light switch	B4
50	Reversing lights	C4
57	Cigarette lighter	A2
59	Interior light switch	B4
60	Radio	C3
64	Instrument voltage stabiliser	A2
77	Windscreen washer motor	B3
83	Induction heater (cold climate)	A1
84	Suction chamber heater (cold climate)	A1
105	Rear compartment light	B4
110	Direction indicator side repeaters	C1
118	Wiper/washer switch	B3
153	Hazard warning switch	B1
160	Brake pressure differential switch	A2
165	Handbrake switch	A2
166	Handbrake warning light	A2
211	Heater control illumination	C3
212	Choke control switch	A2
213	Choke warning light	A2
256	Blocking diode	A2
267	To headlight washer pump	A4
286	Rear foglight switch	B4
288	Rear foglights	B4, C4
297	Braking system warning light	A2
308	Direction indicator/hazard warning flasher unit	B1
332	Tachograph	B2

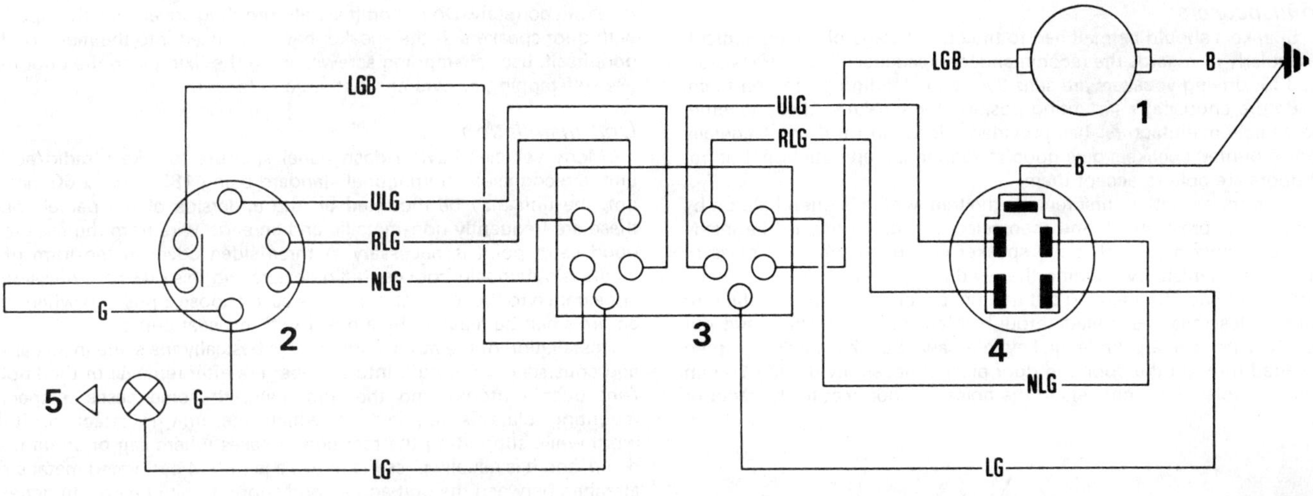

Fig. 13.59 Supplementary wiring diagram – wiper delay circuit

1 Windscreen washer motor
2 Wash/wipe switch
3 Wiper delay unit
4 Wiper motor
5 To fuse No 5C

17B Mobile radio equipment – interference-free installation

Aerials – selection and fitting

The choice of aerials is now very wide. It should be realised that the quality has a profound effect on radio performance, and a poor, inefficient aerial can make suppression difficult.

A wing-mounted aerial is regarded as probably the most efficient for signal collection, but a roof aerial is usually better for suppression purposes because it is away from most interference fields. Stick-on wire aerials are available for attachment to the inside of the windscreen, but are not always free from the interference field of the engine and some accessories.

Motorised automatic aerials rise when the equipment is switched on and retract at switch-off. They require more fitting space and supply leads, and can be a source of trouble.

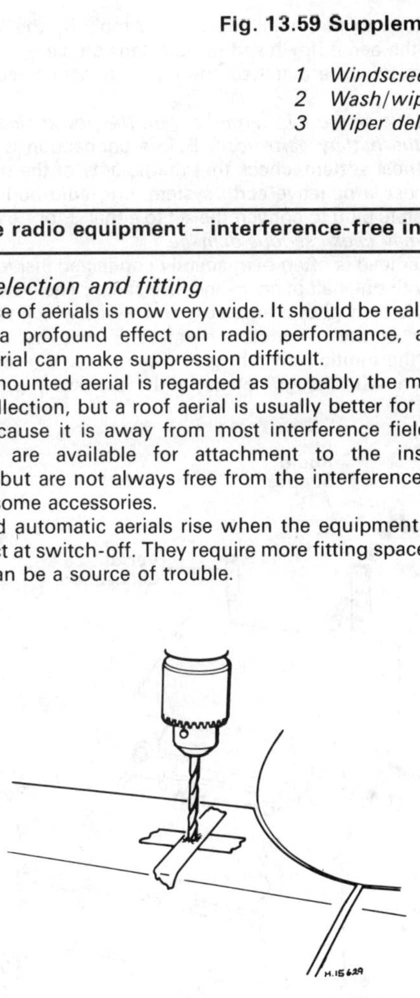

Fig. 13.60 Drilling the bodywork for aerial mounting (Sec 17B)

There is no merit in choosing a very long aerial as, for example, the type about three metres in length which hooks or clips on to the rear of the car, since part of this aerial will inevitably be located in an interference field. For VHF/FM radios the best length of aerial is about one metre. Active aerials have a transistor amplifier mounted at the base and this serves to boost the received signal. The aerial rod is sometimes rather shorter than normal passive types.

A large loss of signal can occur in the aerial feeder cable, especially over the Very High Frequency (VHF) bands. The design of feeder cable is invariably in the co-axial form, ie a centre conductor surrounded by a flexible copper braid forming the outer (earth) conductor. Between the inner and outer conductors is an insulator material which can be in solid or stranded form. Apart from insulation, its purpose is to maintain the correct spacing and concentricity. Loss of signal occurs in this insulator, the loss usually being greater in a poor quality cable. The quality of cable used is reflected in the price of the aerial with the attached feeder cable.

The capacitance of the feeder should be within the range 65 to 75 picofarads (pF) approximately (95 to 100 pF for Japanese and American equipment), otherwise the adjustment of the car radio aerial trimmer may not be possible. An extension cable is necessary for a long run between aerial and receiver. If this adds capacitance in excess of the above limits, a connector containing a series capacitor will be required, or an extension which is labelled as 'capacity-compensated'.

Fitting the aerial will normally involve making a $^{7}/_{8}$ in (22 mm) diameter hole in the bodywork, but read the instructions that come with the aerial kit. Once the hole position has been selected, use a centre punch to guide the drill. Use sticky masking tape around the area for this helps with marking out and drill location, and gives protection to the paintwork should the drill slip. Three methods of making the hole are in use:

(a) Use a hole saw in the electric drill. This is, in effect, a circular hacksaw blade wrapped round a former with a centre pilot drill.
(b) Use a tank cutter which also has cutting teeth, but is made to shear the metal by tightening with an Allen key.
(c) The hard way of drilling out the circle is using a small drill, say $^{1}/_{8}$ in (3 mm), so that the holes overlap. The centre metal drops out and the hole is finished with round and half-round files.

Whichever method is used, the burr is removed from the body metal and paint removed from the underside. The aerial is fitted tightly ensuring that the earth fixing, usually a serrated washer, ring or clamp, is making a solid connection. *This earth connection is important in reducing interference.* Cover any bare metal with primer paint and topcoat, and follow by underseal if desired.

Aerial feeder cable routing should avoid the engine compartment and areas where stress might occur, eg under the carpet where feet will be located. Roof aerials require that the headlining be pulled back and that a path is available down the door pillar. It is wise to check with the vehicle dealer whether roof aerial fitting is recommended.

Loudspeakers

Speakers should be matched to the output stage of the equipment, particularly as regards the recommended impedance. Power transistors used for driving speakers are sensitive to the loading placed on them.

Before choosing a mounting position for speakers, check whether the vehicle manufacturer has provided a location for them. Generally door-mounted speakers give good stereophonic reproduction, but not all doors are able to accept them.

For door mounting, first remove the trim, which is often held on by 'poppers' or press studs, and then select a suitable gap in the inside door assembly. Check that the speaker would not obstruct glass or winder mechanism by winding the window up and down. A template is often provided for marking out the trim panel hole, and then the four fixing holes must be drilled through. Mark out with chalk and cut cleanly with a sharp knife or keyhole saw. Speaker leads are then threaded through the door and door pillar, if necessary drilling 10 mm diameter holes. Fit grommets in the holes and connect to the radio or tape unit correctly. Do not omit a waterproofing cover, usually supplied with door speakers. If the speaker has to be fixed into the metal of the door itself, use self-tapping screws, and if the fixing is to the door trim use self-tapping screws and flat spire nuts.

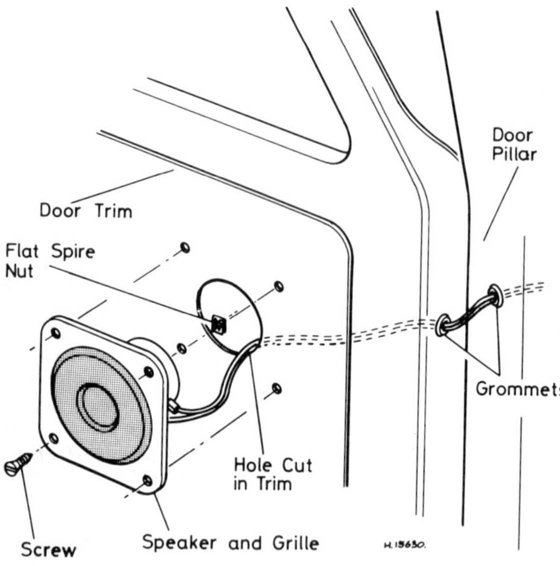

Fig. 13.61 Door-mounted speaker installation (Sec 17B)

Unit installation

Many vehicles have a dash panel aperture to take a radio/audio unit, a recognised international standard being 189.5 mm x 60 mm. If not, the unit may be mounted on the underside of the parcel shelf; these are frequently non-metallic and an earth wire from the case to a good earth point is necessary. A three-sided cover in the form of a cradle is obtainable from car radio dealers and this gives a professional appearance to the installation; in this case choose a position where the controls can be reached by a driver with his seat belt on.

Installation of the radio/audio unit is basically the same in all cases, and consists of offering it into the aperture after removal of the knobs (*not* push buttons) and the trim plate. In some cases a special mounting plate is required to which the unit is attached. It is worthwhile supporting the rear end in cases where sag or strain may occur, and it is usually possible to use a length of perforated metal strip attached between the unit and a good support point nearby. In general it is recommended that tape equipment should be installed at or nearly horizontal.

Connections to the aerial socket are simply by the standard plug terminating the aerial downlead or its extension cable. Speakers for a stereo system must be matched and correctly connected, as outlined previously.

Note: *While all work is carried out on the power side, it is wise to disconnect the battery earth lead.* Before connection is made to the vehicle electrical system, check that the polarity of the unit is correct. All Sherpas use a negative earth system, but radio/audio units often have a reversible plug to convert the set to either + or – earth. *Incorrect connection may cause serious damage.*

The power lead is often permanently connected inside the unit and terminates with one half of an in-line fuse carrier. The other half is fitted with a suitable fuse (3 or 5 amperes) and a wire which should go to a power point in the electrical system. This may be the accessory terminal on the ignition switch, giving the advantage of power feed with ignition or with the ignition key at the 'accessory' position. Power

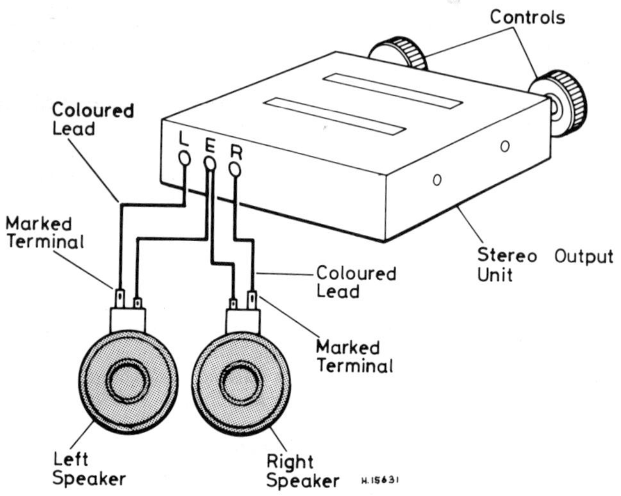

Fig. 13.62 Speaker connections must be correctly made as shown (Sec 17B)

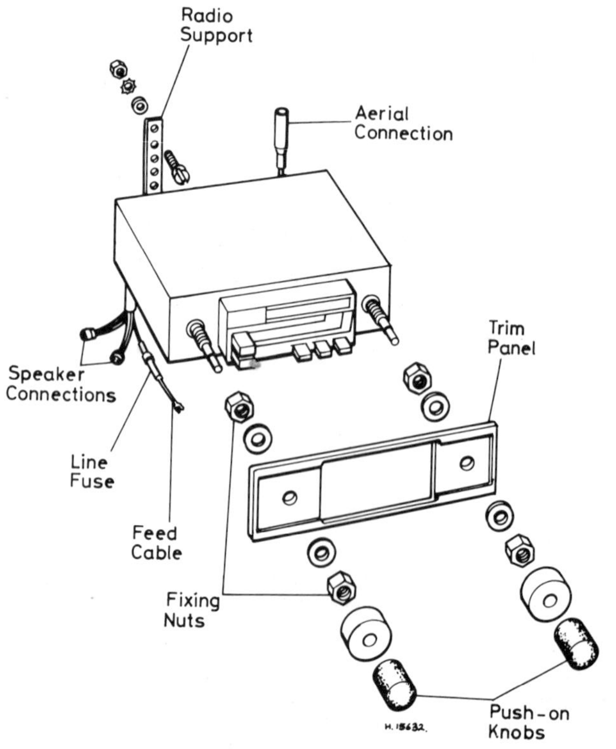

Fig. 13.63 Mounting component details for radio/cassette unit (Sec 17B)

to the unit stops when the ignition key is removed. Alternatively, the lead may be taken to a live point at the fusebox with the consequence of having to remember to switch off at the unit before leaving the vehicle.

Before switching on for initial test, be sure that the speaker connections have been made, for running without load can damage the output transistors. Switch on next and tune through the bands to ensure that all sections are working, and check the tape unit if applicable. The aerial trimmer should be adjusted to give the strongest reception on a weak signal in the medium wave band, at say 200 metres.

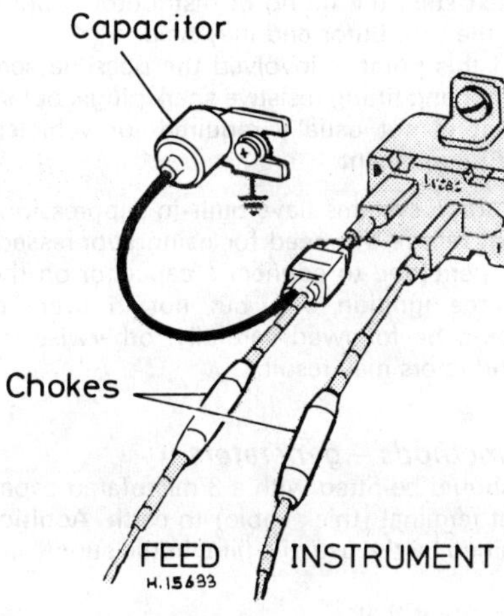

Fig. 13.64 Voltage stabiliser interference suppression (Sec 17B)

Interference

In general, when electric current changes abruptly, unwanted electrical noise is produced. The motor vehicle is filled with electrical devices which change electric current rapidly, the most obvious being the contact breaker.

When the spark plugs operate, the sudden pulse of spark current causes the associated wiring to radiate. Since early radio transmitters used sparks as a basis of operation, it is not surprising that the car radio will pick up ignition spark noise unless steps are taken to reduce it to acceptable levels.

Interference reaches the car radio in two ways:

(a) by conduction through the wiring.
(b) by radiation to the receiving aerial.

Initial checks presuppose that the bonnet is down and fastened, the radio unit has a good earth connection (*not* through the aerial downlead outer), no fluorescent tubes are working near the car, the aerial trimmer has been adjusted, and the vehicle is in a position to receive radio signals, ie not in a metal-clad building.

Switch on the radio and tune it to the middle of the medium wave (MW) band off-station with the volume (gain) control set fairly high. Switch on the ignition (but do not start the engine) and wait to see if irregular clicks or hash noise occurs. Tapping the facia panel may also produce the effects. If so, this will be due to the voltage stabiliser, which is an on-off thermal switch to control instrument voltage. It is located usually on the back of the instrument panel, often attached to the speedometer. Correction is by attachment of a capacitor and, if still troublesome, chokes in the supply wires.

Switch on the engine and listen for interference on the MW band. Depending on the type of interference, the indications are as follows.

A harsh crackle that drops out abruptly at low engine speed or when the headlights are switched on is probably due to a voltage regulator.

A whine varying with engine speed is due to the dynamo or alternator. Try temporarily taking off the fan belt – if the noise goes this is confirmation.

Regular ticking or crackle that varies in rate with the engine speed is due to the ignition system. With this trouble in particular and others in general, check to see if the noise is entering the receiver from the wiring or by radiation. To do this, pull out the aerial plug, (preferably shorting out the input socket or connecting a 62 pF capacitor across it). If the noise disappears it is coming in through the aerial and is *radiation noise.* If the noise persists it is reaching the receiver through the wiring and is said to be *line-borne.*

Interference from wipers, washers, heater blowers, turn-indicators, stop lamps, etc is usually taken to the receiver by wiring, and simple treatment using capacitors and possibly chokes will solve the problem. Switch on each one in turn (wet the screen first for running wipers!) and listen for possible interference with the aerial plug in place and again when removed.

Electric petrol pumps are now finding application again and give rise to an irregular clicking, often giving a burst of clicks when the ignition is on but the engine has not yet been started. It is also possible to receive whining or crackling from the pump.

Note that if most of the vehicle accessories are found to be creating interference all together, the probability is that poor aerial earthing is to blame.

Component terminal markings

Throughout the following sub-sections reference will be found to various terminal markings. These will vary depending on the manufacturer of the relevant component. If terminal markings differ from those mentioned, reference should be made to the following table, where the most commonly encountered variations are listed.

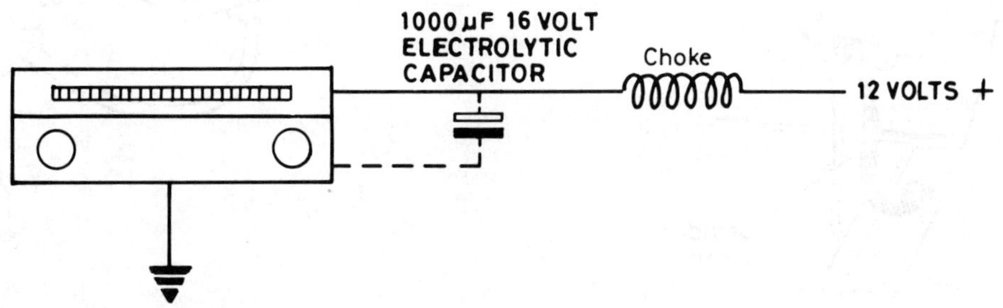

Fig. 13.65 Line-borne interference suppression (Sec 17B)

Alternator	Alternator terminal (thick lead)	Exciting winding terminal
DIN/Bosch	B+	DF
Delco Remy	+	EXC
Ducellier	+	EXC
Ford (US)	+	DF
Lucas	+	F
Marelli	+B	F

Ignition coil	Ignition switch terminal	Contact breaker terminal
DIN/Bosch	15	1
Delco Remy	+	–
Ducellier	BAT	RUP
Ford (US)	B/+	CB/–
Lucas	SW/+	–
Marelli	BAT/+B	D

Voltage regulator	Voltage input terminal	Exciting winding terminal
DIN/Bosch	B+/D+	DF
Delco Remy	BAT/+	EXC
Ducellier	BOB/BAT	EXC
Ford (US)	BAT	DF
Lucas	+/A	F
Marelli		F

Suppression methods – ignition

Suppressed HT cables are supplied as original equipment by manufacturers and will meet regulations as far as interference to neighbouring equipment is concerned. It is illegal to remove such suppression unless an alternative is provided, and this may take the form of resistive spark plug caps in conjunction with plain copper HT cable. For VHF purposes, these and 'in-line' resistors may not be effective, and resistive HT cable is preferred. Check that suppressed cables are actually fitted by observing cable identity lettering, or measuring with an ohmmeter – the value of each plug lead should be 5000 to 10 000 ohms.

A 1 microfarad capacitor connected from the LT supply side of the ignition coil to a good nearby earth point will complete basic ignition interference treatment. *NEVER fit a capacitor to the coil terminal to the contact breaker – the result would be burnt out points in a short time.*

If ignition noise persists despite the treatment above, the following sequence should be followed:

(a) Check the earthing of the ignition coil; remove paint from fixing clamp.

(b) If this does not work, lift the bonnet. Should there be no change in interference level, this may indicate that the bonnet is not electrically connected to the car body. Use a proprietary braided strap across a bonnet hinge ensuring a first class electrical connection. If, however, lifting the bonnet increases the interference, then fit resistive HT cables of a higher ohms-per-metre value.

(c) If all these measures fail, it is probable that re-radiation from metallic components is taking place. Using a braided strap between metallic points, go round the vehicle systematically – try the following: engine to body, exhaust system to body, front suspension to engine and to body, steering column to body (especially French and Italian cars), gear lever to engine and to body (again especially French and Italian cars), Bowden cable to body, metal parcel shelf to body. When an offending component is located it should be bonded with the strap permanently.

(d) As a next step, the fitting of distributor suppressors to each lead at the distributor end may help.

(e) Beyond this point is involved the possible screening of the distributor and fitting resistive spark plugs, but such advanced treatment is not usually required for vehicles with entertainment equipment.

Electronic ignition systems have built-in suppression components, but this does not relieve the need for using suppressed HT leads. In some cases it is permitted to connect a capacitor on the low tension supply side of the ignition coil, but not in every case. Makers' instructions should be followed carefully, otherwise damage to the ignition semiconductors may result.

Suppression methods – generators

Alternators should be fitted with a 3 microfarad capacitor from the B+ main output terminal (thick cable) to earth. Additional suppression may be obtained by the use of a filter in the supply line to the radio receiver.

It is most important that:

(a) Capacitors are never connected to the field terminal.

(b) Alternators must not be run without connection to the battery.

Suppression methods – voltage regulators

Alternator regulators come in three types:

(a) Vibrating contact regulators separate from the alternator. Used extensively on continental vehicles.

(b) Electronic regulators separate from the alternator.

(c) Electronic regulators built-in to the alternator.

In case (a) interference may be generated on the AM and FM (VHF) bands. For some cars a replacement suppressed regulator is

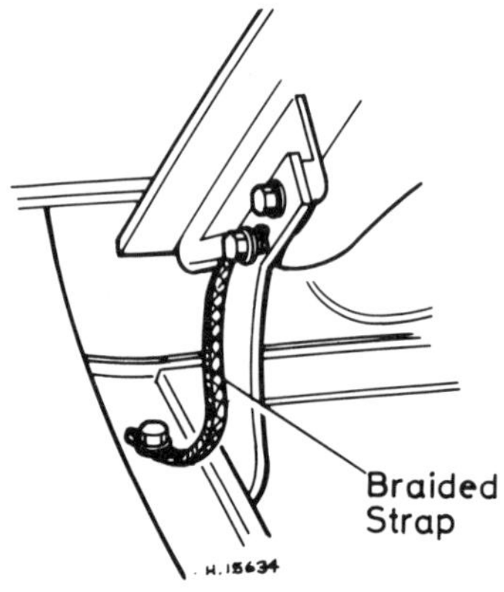

Fig. 13.66 Braided earth strap between bonnet and body (Sec 17B)

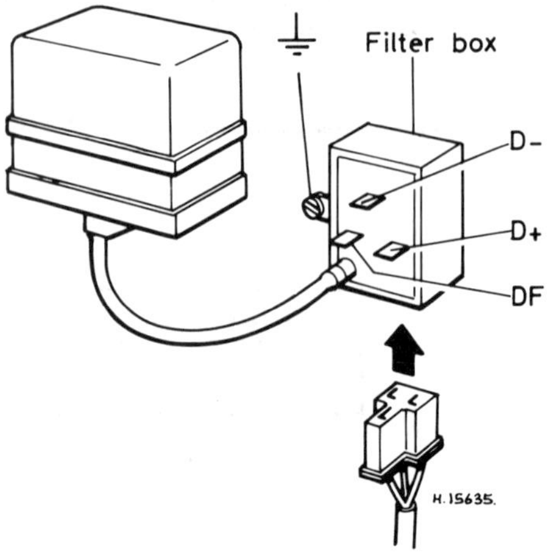

Fig. 13.67 Typical filter box for vibrating contact voltage regulator (alternator equipment) (Sec 17B)

available. Filter boxes may be used with non-suppressed regulators. But if not available, then for AM equipment a 2 microfarad or 3 microfarad capacitor may be mounted at the voltage terminal marked D+ or B+ of the regulator. FM bands may be treated by a feed-through capacitor of 2 or 3 microfarad.

Electronic voltage regulators are not always troublesome, but where necessary, a 1 microfarad capacitor from the regulator + terminal will help.

Integral electronic voltage regulators do not normally generate much interference, but when encountered this is in combination with alternator noise. A 1 microfarad or 2 microfarad capacitor from the warning lamp (IND) terminal to earth for Lucas ACR alternators and Femsa, Delco and Bosch equivalents should cure the problem.

Suppression methods – other equipment

Wiper motors – Connect the wiper body to earth with a bonding strap. For all motors use a 7 ampere choke assembly inserted in the leads to the motor.

Heater motors – Fit 7 ampere line chokes in both leads, assisted if necessary by a 1 microfarad capacitor to earth from both leads.

Electronic tachometer – The tachometer is a possible source of ignition noise – check by disconnecting at the ignition coil CB terminal. It usually feeds from ignition coil LT pulses at the contact breaker terminal. A 3 ampere line choke should be fitted in the tachometer lead at the coil CB terminal.

Horn – A capacitor and choke combination is effective if the horn is directly connected to the 12 volt supply. The use of a relay is an alternative remedy, as this will reduce the length of the interference-carrying leads.

Electrostatic noise – Characteristics are erratic crackling at the receiver, with disappearance of symptoms in wet weather. Often shocks may be given when touching bodywork. Part of the problem is the build-up of static electricity in non-driven wheels and the acquisition of charge on the body shell. It is possible to fit spring-loaded contacts at the wheels to give good conduction between the rotary wheel parts and the vehicle frame. Changing a tyre sometimes helps – because of tyres' varying resistances. In difficult cases a trailing flex which touches the ground will cure the problem. If this is not acceptable it is worth trying conductive paint on the tyre walls.

Fuel pump – Suppression requires a 1 microfarad capacitor between the supply wire to the pump and a nearby earth point. If this is insufficient a 7 ampere line choke connected in the supply wire near the pump is required.

Fluorescent tubes – Vehicles used for camping/caravanning frequently have fluorescent tube lighting. These tubes require a relatively high voltage for operation and this is provided by an inverter (a form of oscillator) which steps up the vehicle supply voltage. This can give rise to serious interference to radio reception, and the tubes themselves can contribute to this interference by the pulsating nature of the lamp discharge. In such situations it is important to mount the aerial as far away from a fluorescent tube as possible. The interference problem may be alleviated by screening the tube with fine wire turns spaced an inch (25 mm) apart and earthed to the chassis. Suitable chokes should be fitted in both supply wires close to the inverter.

Radio/cassette case breakthrough

Magnetic radiation from dashboard wiring may be sufficiently intense to break through the metal case of the radio/cassette player. Often this is due to a particular cable routed too close and shows up as ignition interference on AM and cassette play and/or alternator whine on cassette play.

The first point to check is that the clips and/or screws are fixing all parts of the radio/cassette case together properly. Assuming good earthing of the case, see if it is possible to re-route the offending cable – the chances of this are not good, however, in most cars.

Next release the radio/cassette player and locate it in different positions with temporary leads. If a point of low interference is found, then if possible fix the equipment in that area. This also confirms that local radiation is causing the trouble. If re-location is not feasible, fit the radio/cassette player back in the original position.

Alternator interference on cassette play is now caused by radiation from the main charging cable which goes from the battery to the output terminal of the alternator, usually via the + terminal of the starter motor relay. In some vehicles this cable is routed under the dashboard, so the solution is to provide a direct cable route. Detach the

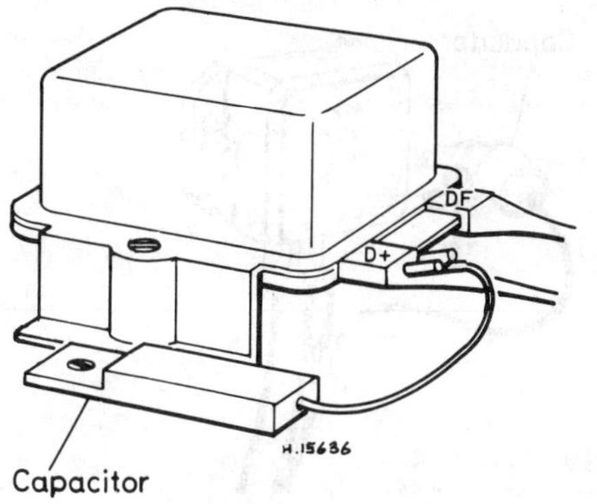

Fig. 13.68 Suppression of AM interference by vibrating contact voltage regulator (alternator equipment) (Sec 17B)

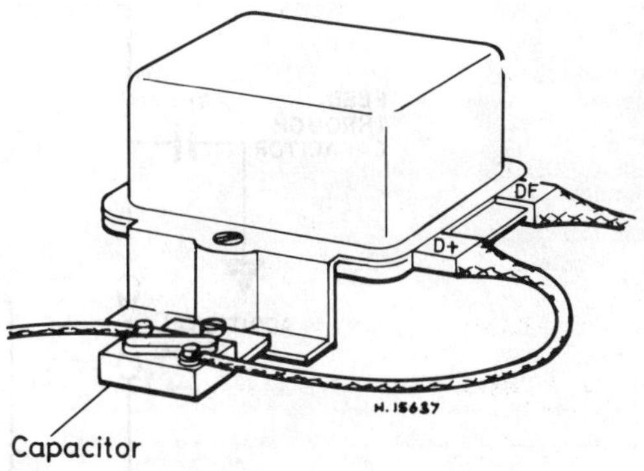

Fig. 13.69 Suppression of FM interference by vibrating contact voltage regulator (alternator equipment) (Sec 17B)

original cable from the alternator output terminal and make up a new cable of at least 6 mm² cross-sectional area to go from alternator to battery with the shortest possible route. *Remember – do not run the engine with the alternator disconnected from the battery.*

Ignition breakthrough on AM and/or cassette play can be a difficult problem. It is worth wrapping earthed foil round the offending cable run near the equipment, or making up a deflector plate well screwed down to a good earth. Another possibility is the use of a suitable relay to switch on the ignition coil. The relay should be mounted close to the ignition coil; with this arrangement the ignition coil primary current is not taken into the dashboard area and does not flow through the ignition switch. A suitable diode should be used since it is possible that at ignition switch-off the output from the warning lamp alternator terminal could hold the relay on.

Connectors for suppression components

Capacitors are usually supplied with tags on the end of the lead, while the capacitor body has a flange with a slot or hole to fit under a nut or screw with washer.

Connections to feed wires are best achieved by self-stripping connectors. These connectors employ a blade which, when squeezed down by pliers, cuts through cable insulation and makes connection to the copper conductors beneath.

Chokes sometimes come with bullet snap-in connectors fitted to the wires, and also with just bare copper wire. With connectors, suitable female cable connectors may be purchased from an auto-accessory shop together with any extra connectors required for the cable ends after being cut for the choke insertion. For chokes with bare wires, similar connectors may be employed together with insulation sleeving as required.

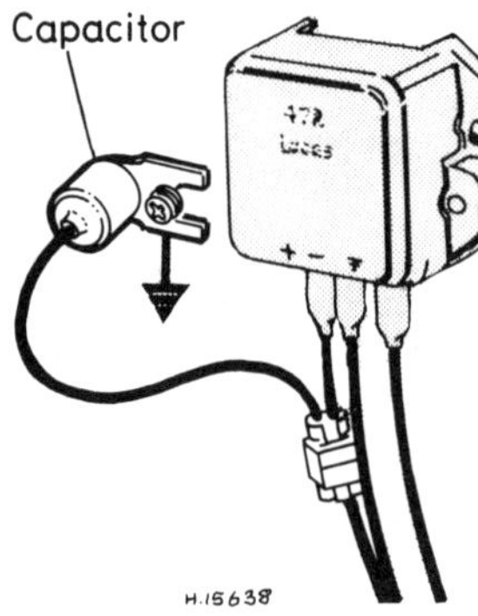

Fig. 13.70 Electronic voltage regulator suppression (Sec 17B)

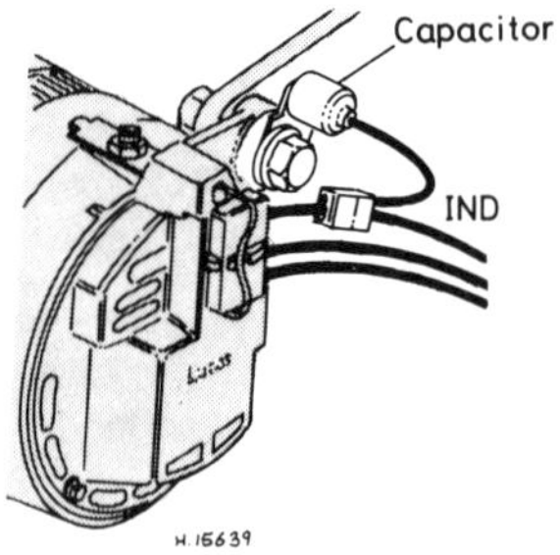

Fig. 13.71 Suppression of interference from electronic voltage regulator when integral with alternator (Sec 17B)

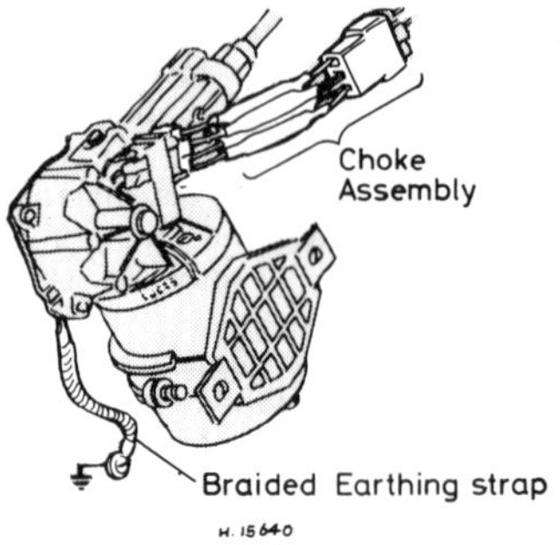

Fig. 13.72 Wiper motor interference suppression (Sec 17B)

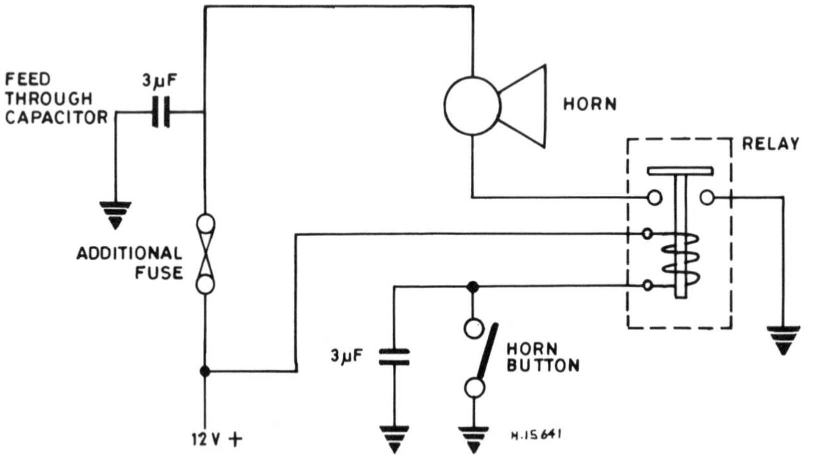

Fig. 13.73 Use of relay to reduce horn interference (Sec 17B)

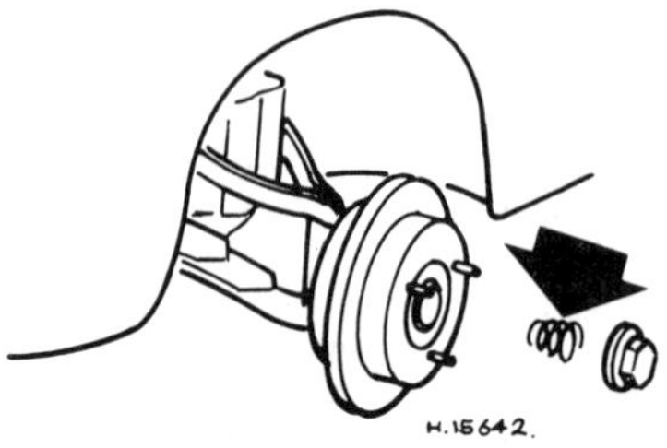

Fig. 13.74 Use of spring contacts at wheels (Sec 17B)

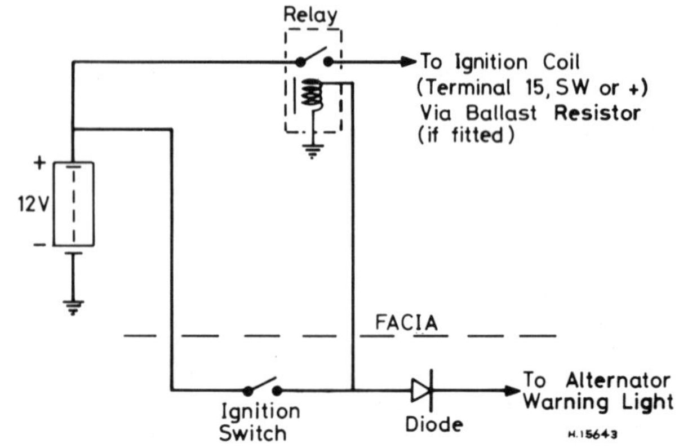

Fig. 13.75 Use of ignition coil relay to suppress case breakthrough (Sec 17B)

VHF/FM broadcasts

Reception of VHF/FM in an automobile is more prone to problems than the medium and long wavebands. Medium/long wave transmitters are capable of covering considerable distances, but VHF transmitters are restricted to line of sight, meaning ranges of 10 to 50 miles, depending upon the terrain, the effects of buildings and the transmitter power.

Because of the limited range it is necessary to retune on a long journey, and it may be better for those habitually travelling long distances or living in areas of poor provision of transmitters to use an AM radio working on medium/long wavebands.

When conditions are poor, interference can arise, and some of the suppression devices described previously fall off in performance at very high frequencies unless specifically designed for the VHF band. Available suppression devices include reactive HT cable, resistive distributor caps, screened plug caps, screened leads and resistive spark plugs.

For VHF/FM receiver installation the following points should be particularly noted:

(a) Earthing of the receiver chassis and the aerial mounting is important. Use a separate earthing wire at the radio, and scrape paint away at the aerial mounting.
(b) If possible, use a good quality roof aerial to obtain maximum height and distance from interference generating devices on the vehicle.
(c) Use of a high quality aerial downlead is important, since losses in cheap cable can be significant.
(d) The polarisation of FM transmissions may be horizontal, vertical, circular or slanted. Because of this the optimum mounting angle is at 45° to the vehicle roof.

Citizens' Band radio (CB)

In the UK, CB transmitter/receivers work within the 27 MHz and 934 MHz bands, using the FM mode. At present interest is concentrated on 27 MHz where the design and manufacture of equipment is less difficult. Maximum transmitted power is 4 watts, and 40 channels spaced 10 kHz apart within the range 27.60125 to 27.99125 MHz are available.

Aerials are the key to effective transmission and reception. Regulations limit the aerial length to 1.65 metres including the loading coil and any associated circuitry, so tuning the aerial is necessary to obtain optimum results. The choice of a CB aerial is dependent on whether it is to be permanently installed or removable, and the performance will hinge on correct tuning and the location point on the vehicle. Common practice is to clip the aerial to the roof gutter or to employ wing mounting where the aerial can be rapidly unscrewed. A popular solution is to use the 'magmount' – a type of mounting having a strong magnetic base clamping to the vehicle at any point, usually the roof.

Aerial location determines the signal distribution for both transmission and reception, but it is wise to choose a point away from the engine compartment to minimise interference from vehicle electrical equipment.

The aerial is subject to considerable wind and acceleration forces. Cheaper units will whip backwards and forwards and in so doing will alter the relationship with the metal surface of the vehicle with which it forms a ground plane aerial system. The radiation pattern will change correspondingly, giving rise to break-up of both incoming and outgoing signals.

Interference problems on the vehicle carrying CB equipment fall into two categories:

(a) Interference to nearby TV and radio receivers when transmitting.
(b) Interference to CB set reception due to electrical equipment on the vehicle.

Problems of break-through to TV and radio are not frequent, but can be difficult to solve. Mostly trouble is not detected or reported because the vehicle is moving and the symptoms rapidly disappear at the TV/radio receiver, but when the CB set is used as a base station any trouble with nearby receivers will soon result in a complaint.

It must not be assumed by the CB operator that his equipment is faultless, for much depends upon the design. Harmonics (that is, multiples) of 27 MHz may be transmitted unknowingly and these can fall into other user's bands. Where trouble of this nature occurs, low pass filters in the aerial or supply leads can help, and should be fitted in base station aerials as a matter of course. In stubborn cases it may be necessary to call for assistance from the licensing authority, or, if possible, to have the equipment checked by the manufacturers.

Interference received on the CB set from the vehicle equipment is, fortunately, not usually a severe problem. The precautions outlined previously for radio/cassette units apply, but there are some extra points worth noting.

It is common practice to use a slide-mount on CB equipment enabling the set to be easily removed for use as a base station, for example. Care must be taken that the slide mount fittings are properly earthed and that first class connection occurs between the set and slide-mount.

Vehicle manufacturers in the UK are required to provide suppression of electrical equipment to cover 40 to 250 MHz to protect TV and VHF radio bands. Such suppression appears to be adequately effective at 27 MHz, but suppression of individual items such as alternators/dynamos, clocks, stabilisers, flashers, wiper motors, etc, may still be necessary. The suppression capacitors and chokes available from auto-electrical suppliers for entertainment receivers will usually give the required results with CB equipment.

Other vehicle radio transmitters

Besides CB radio already mentioned, a considerable increase in the use of transceivers (ie combined transmitter and receiver units) has taken place in the last decade. Previously this type of equipment was fitted mainly to military, fire, ambulance and police vehicles, but a large business radio and radio telephone usage has developed.

Generally the suppression techniques described previously will suffice, with only a few difficult cases arising. Suppression is carried out to satisfy the 'receive mode', but care must be taken to use heavy duty chokes in the equipment supply cables since the loading on 'transmit' is relatively high.

18 Suspension and steering

Introduction

1 Few changes have been made to the suspension and steering systems. The most significant are the introduction of telescopic front shock absorbers, needle roller bearing swivel pins and (on some models) composite springs.

Front hub bearings (disc braked models) – removal and refitting

2 Proceed as for brake disc removal (Section 16) but do not separate the brake disc from the hub.

3 Renew the bearings as described in Chapter 11, Section 2, paragraphs 6 to 16.

4 Adjust the bearings, refit the brake caliper and the roadwheel and lower the vehicle. Tighten the wheel nuts.

Front hub bearings (disc braked models) – adjustment

5 Proceed as described in Chapter 11, Section 3.

Swivel pins (needle roller type) – general

6 Some later models have swivel pins which run in needle roller races instead of plain bushes.

7 Overhaul procedures are similar to those described in Chapter 11, Section 4, making due allowance for detail differences. Unlike the bushes, the needle roller races will not require reaming.

8 The clearance between the top of the axle beam and the stub axle is the same for both types of assembly.

Composite springs – description

9 From 1986 model year, 200 Series vehicle springs are made of a composite plastic/glass-fibre material instead of metal. It is anticipated that this type of spring will eventually be fitted to all models.

10 The advantages of the composite spring include lower weight, greater strength and improved damping characteristics. Road noise transmission is also said to be reduced.

11 In an emergency a metal spring can be fitted in place of a composite spring, but the converse is not possible because of the different eye fittings (photo).

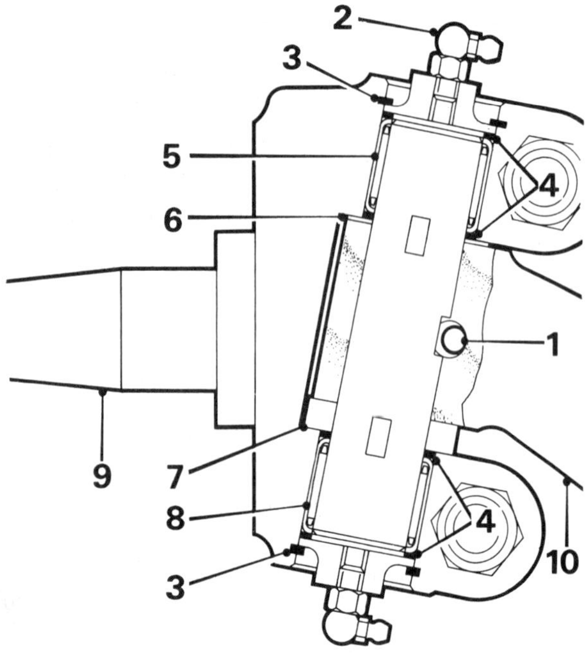

Fig. 13.76 Sectional view of needle roller type swivel pin (Sec 18)

1 Cotter pin
2 Grease nipple
3 Circlip
4 O-rings
5 Upper bearing
6 Shim
7 Thrust bearing
8 Lower bearing
9 Stub axle
10 Axle beam

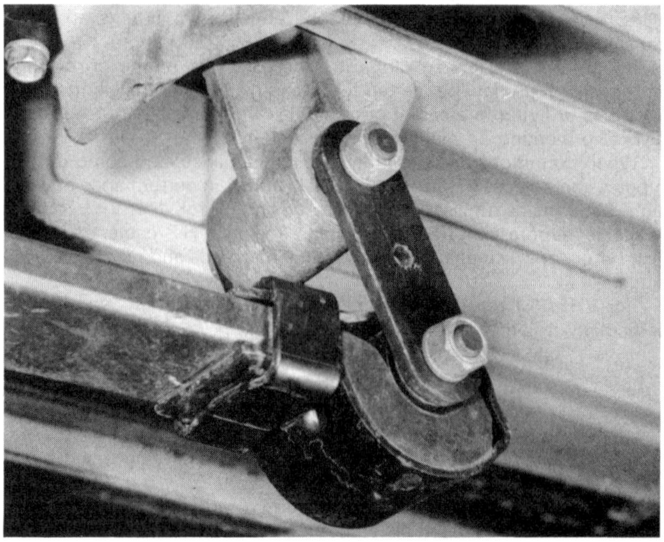

18.11 Composite spring eye and shackle

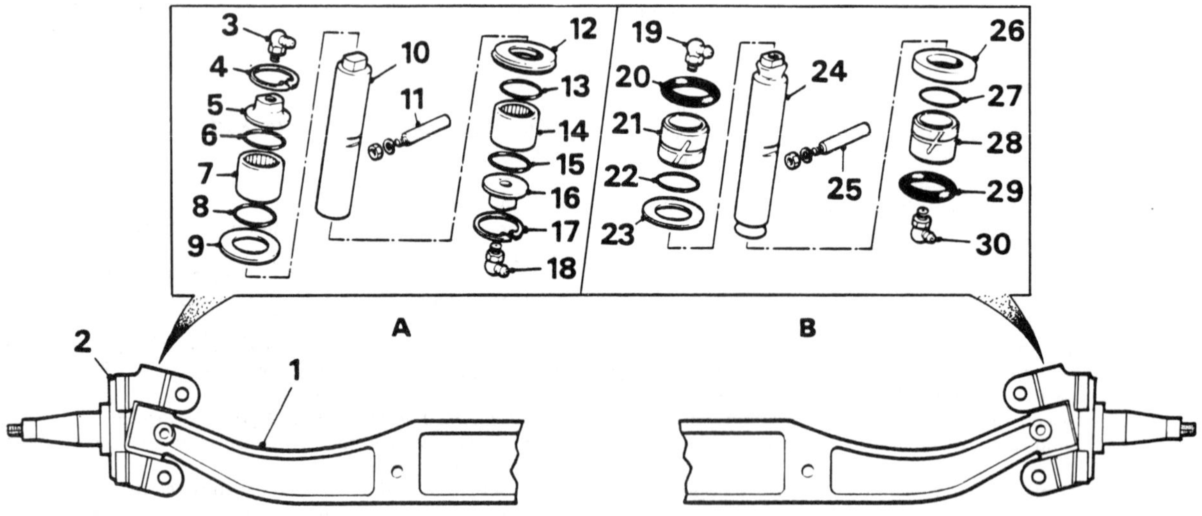

Fig. 13.77 Comparison of needle roller (A) and plain bush (B) swivel pin components (Sec 18)

1 Axle beam
2 Stub axle
3 Grease nipple
4 Circlip
5 Retainer
6 O-ring
7 Upper bearing
8 O-ring
9 Shim
10 Swivel pin
11 Cotter pin
12 Thrust bearing
13 O-ring
14 Lower bearing
15 O-ring
16 Retainer
17 Circlip
18 Grease nipple
19 Grease nipple
20 O-ring
21 Upper bush
22 O-ring
23 Shim
24 Swivel pin
25 Cotter pin
26 Thrust bearing
27 O-ring
28 Lower bush
29 O-ring
30 Grease nipple

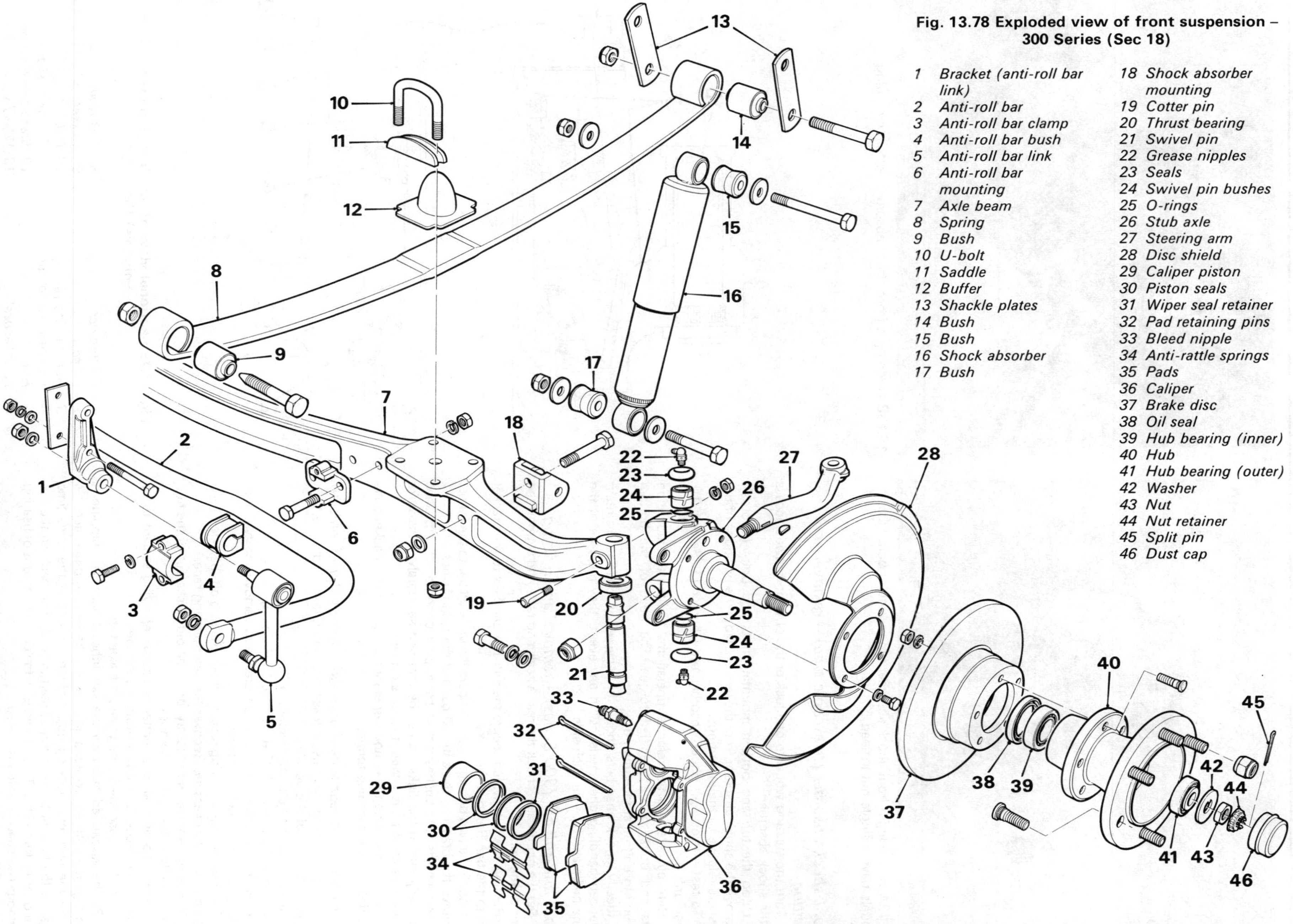

Fig. 13.78 Exploded view of front suspension – 300 Series (Sec 18)

1 *Bracket (anti-roll bar link)*
2 *Anti-roll bar*
3 *Anti-roll bar clamp*
4 *Anti-roll bar bush*
5 *Anti-roll bar link*
6 *Anti-roll bar mounting*
7 *Axle beam*
8 *Spring*
9 *Bush*
10 *U-bolt*
11 *Saddle*
12 *Buffer*
13 *Shackle plates*
14 *Bush*
15 *Bush*
16 *Shock absorber*
17 *Bush*
18 *Shock absorber mounting*
19 *Cotter pin*
20 *Thrust bearing*
21 *Swivel pin*
22 *Grease nipples*
23 *Seals*
24 *Swivel pin bushes*
25 *O-rings*
26 *Stub axle*
27 *Steering arm*
28 *Disc shield*
29 *Caliper piston*
30 *Piston seals*
31 *Wiper seal retainer*
32 *Pad retaining pins*
33 *Bleed nipple*
34 *Anti-rattle springs*
35 *Pads*
36 *Caliper*
37 *Brake disc*
38 *Oil seal*
39 *Hub bearing (inner)*
40 *Hub*
41 *Hub bearing (outer)*
42 *Washer*
43 *Nut*
44 *Nut retainer*
45 *Split pin*
46 *Dust cap*

18.13A Removing a front shock absorber top mounting bolt. Later models have a single nut instead

18.13B Undoing a front shock absorber bottom mounting

Front shock absorber (telescopic type) – removal and refitting

12 Turn the steering wheel to one lock or the other to provide access to the shock absorber.
13 Undo the top and bottom mounting nuts and bolts (photos). Remove the shock absorber from the vehicle. Note the positions of the washers relative to the mountings.
14 To test a telescopic shock absorber, see Chapter 11, Section 10, paragraph 3.
15 Refit by reversing the removal operations. Tighten the mounting nuts and bolts to the specified torque. The head of the bottom mounting bolt must face inwards.
16 Ideally, shock absorbers should be renewed in pairs to maintain balanced handling characteristics. In any event, they must be of the same make and type on each side.

Front anti-roll bar (300 Series) – removal and refitting

17 Set the handbrake, chock the rear wheels and raise and support the front of the vehicle.
18 Remove the nuts which secure the anti-roll bar links to the bracket and to the bar. Remove the links.
19 Unbolt the two brackets which secure the anti-roll bar to the frame. Remove the brackets, the anti-roll bar and the split bushes.
20 Renew the split bushes if they are in poor condition. The bushes and balljoints in the links cannot be renewed separately – the links themselves must be renewed if wear is evident.
21 Refit by reversing the removal operations. Tighten the bracket bolts and links to the specified torque.

Rear hub (twin rear wheel models) – removal and refitting

22 Chock the front wheels, release the handbrake and raise and support the rear of the vehicle.
23 Remove the brake drum (Section 16).
24 Remove the half shaft (Section 15, 16 HA axle).
25 Relieve the lockwasher tabs. Unscrew and remove the two hub locking nuts and the lockwasher.
26 Remove the outer bearing race, then pull the hub off the axle. Use a slide hammer if necessary. Recover the inner bearing race.
27 If bearing renewal is necessary, drive the old tracks out of the hub and carefully drive in the new ones.
28 Renew the oil seal as a matter of course, applying sealant to its non-moving faces. Similarly seal the oil thrower.
29 Liberally grease the bearings and the axle tube.
30 Fit the assembled hub over the axle tube, if necessary tapping it home gently with a soft-faced hammer. Fit the bearing outer race.
31 Fit the hub inner locking nut. Tighten the nut, at the same time rotating the hub, until the bearings start to bind, then back off the nut by a quarter-turn (90°). This should give the correct amount of bearing endfloat (no specific value is stated).
32 Fit a new lockwasher and the hub outer locking nut. Tighten the

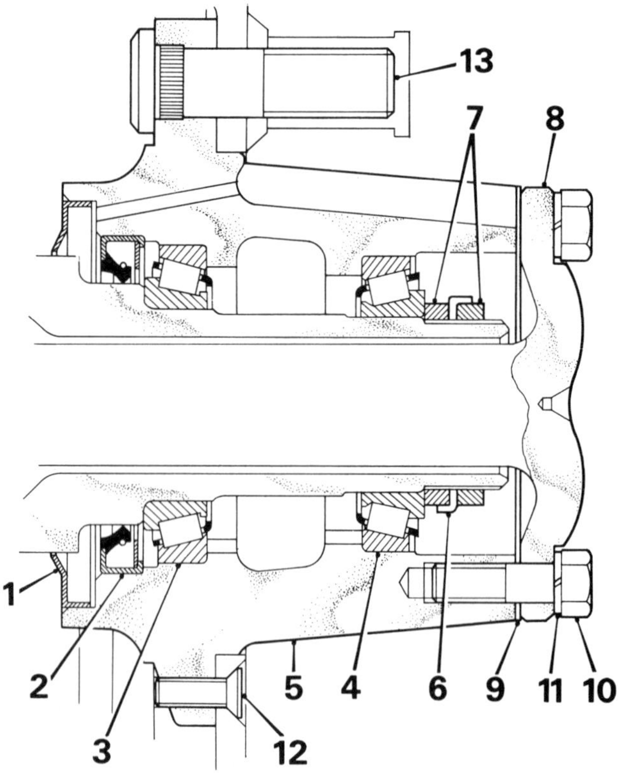

Fig. 13.79 Sectional view of rear hub (twin rear wheel models) (Sec 18)

1 Oil thrower
2 Oil seal
3 Hub bearing (inner)
4 Hub bearing (outer)
5 Hub
6 Lockwasher
7 Locking nuts
8 Half shaft
9 Gasket
10 Bolt
11 Spring washer
12 Brake down screw
13 Wheel stud

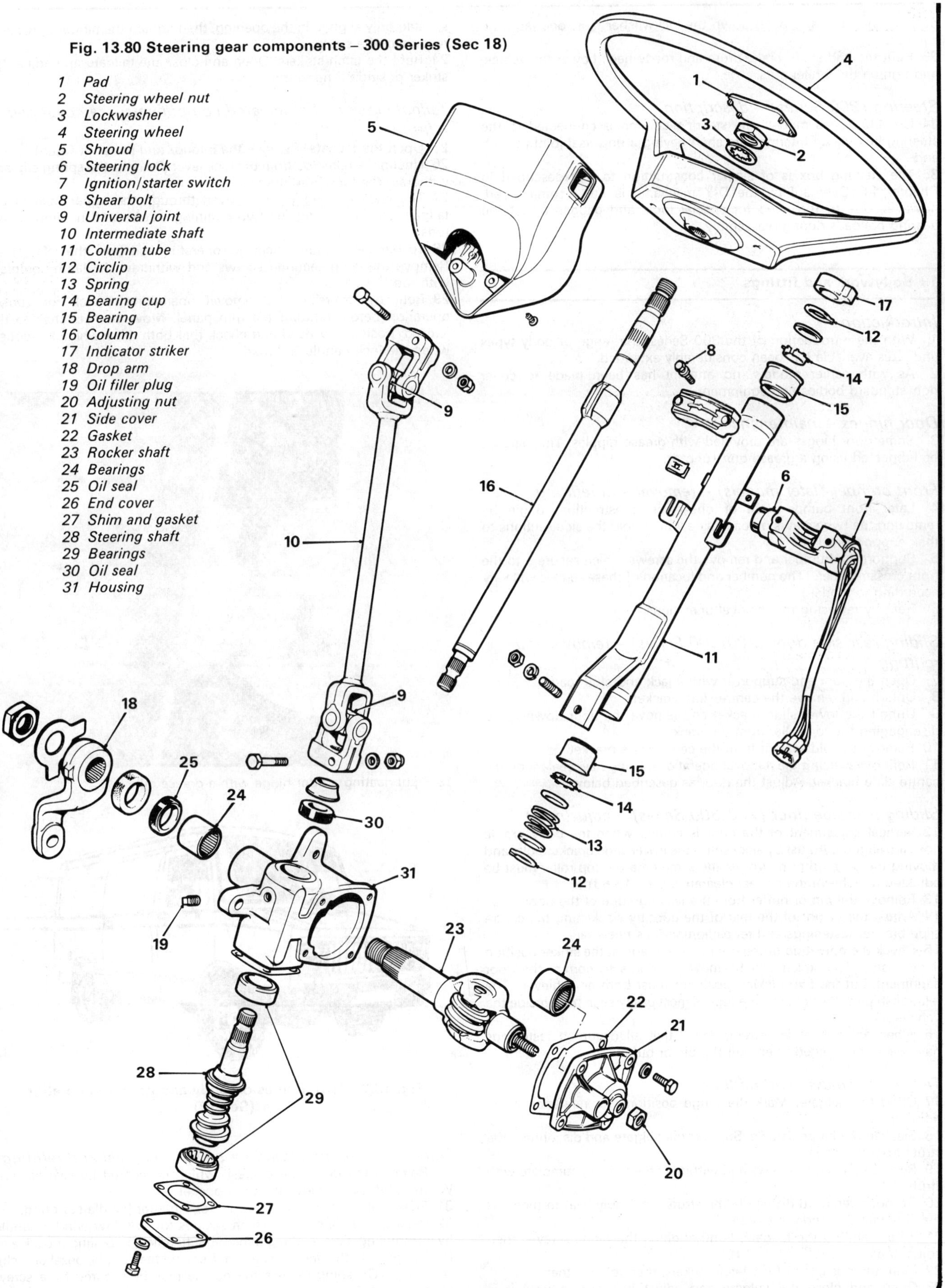

Fig. 13.80 Steering gear components – 300 Series (Sec 18)

1 Pad
2 Steering wheel nut
3 Lockwasher
4 Steering wheel
5 Shroud
6 Steering lock
7 Ignition/starter switch
8 Shear bolt
9 Universal joint
10 Intermediate shaft
11 Column tube
12 Circlip
13 Spring
14 Bearing cup
15 Bearing
16 Column
17 Indicator striker
18 Drop arm
19 Oil filler plug
20 Adjusting nut
21 Side cover
22 Gasket
23 Rocker shaft
24 Bearings
25 Oil seal
26 End cover
27 Shim and gasket
28 Steering shaft
29 Bearings
30 Oil seal
31 Housing

outer locking nut and bend down the lockwasher tabs, one tab onto each nut.
33 Refit the half shaft, brake drum and roadwheel. Lower the vehicle and tighten the wheel nuts.

Steering (300 Series) – description
34 On 300 Series models the steering column is connected to the steering box via an intermediate shaft having a universal joint at each end.
35 The steering box is of similar construction to that described in Chapter 11. Overhaul by the DIY mechanic is not recommended, however, due to the need for special tools and gauges to set the bearing preloads during reassembly.

19 Bodywork and fittings

Introduction
1 With the introduction of the 300 Series, the range of body types and sizes available has been considerably extended.
2 As with earlier models, no attempt has been made to cover non-standard bodies and equipment.

Door hinges – maintenance
3 Some door hinges are provided with grease nipples. They should be lubricated using a grease gun (photo).

Front bumper (later models) – removal and refitting
4 Later front bumpers are of one-piece construction. Begin by removing the two screws on each side which hold the side sections to their support brackets.
5 Support the bumper and remove the screws which secure it to the front crossmember. (The number and location of these screws will vary according to model.)
6 Refit by reversing the removal operations.

Sliding rear side door (200/300 Series) – removal and refitting
7 Open the door and support it with a jack and cloth pad.
8 Unbolt and remove the centre slide bracket.
9 Unbolt the lower slide bracket and remove the door downwards, disengaging the top roller from the track.
10 Remove the old sealant from the centre slide bracket.
11 Refit by reversing the removal operations, using new sealant on the centre slide bracket. Adjust the door as described below.

Sliding rear side door (200/300 Series) – adjustment
12 Vertical adjustment of the door is correct when the body profile lines are aligned. Adjust by slackening the lower slide bracket bolts and moving the door up or down. At the same time the top roller must be adjusted to achieve the correct clearance (Fig. 13.83).
13 Remove the pin or buffer from the leading edge of the door.
14 Adjust the height of the rear of the door by slackening the centre slide bracket fastenings and repositioning it as necessary.
15 Check the operation of the door lock and adjust the striker position if necessary. The striker can be moved inwards to correct the door alignment, but the two locking positions must both be achieved. The relationship of the striker to the lock is most easily seen from inside the vehicle.
16 When adjustment is correct, check that all disturbed fastenings have been retightened, then refit the pin or buffer.

Tailgate – removal and refitting
17 Open the tailgate. Mark the hinge positions for reference when refitting.
18 Slacken the hinge screws. Support the tailgate and disconnect the struts from the body.
19 Remove the hinge screws and withdraw the tailgate complete with struts.
20 Do not attempt to dismantle the struts, nor apply heat to them, as they contain gas under pressure.
21 When refitting the tailgate, lightly tighten the hinge screws, then adjust it as follows.
22 Mark the positions of the latch strikers, then remove them.
23 Open and close the tailgate and adjust its position until it is satisfactorily aligned in the opening, then tighten the hinge screws and reconnect the struts.
24 Refit the latch strikers. Open and close the tailgate and adjust the striker position if necessary.

Tailgate lock and associated components – removal and refitting
25 Open the tailgate. Remove the interior trim panel (if fitted).
26 Unclip the relay rod from the lock lever. Remove the spring clip and withdraw the lock from the tailgate.
27 The relay lever may be removed through the upper hole in the tailgate after removing the two retaining screws and unclipping the rods.
28 To remove a latch, unclip its operating rod from the relay lever. Remove the latch securing screws and withdraw the latch complete with rod.
29 Refit by reversing the removal operations. Check for correct operation before refitting the trim panel. Move both latches to the 'closed' position by hand and check that both release simultaneously when the lock handle is turned.

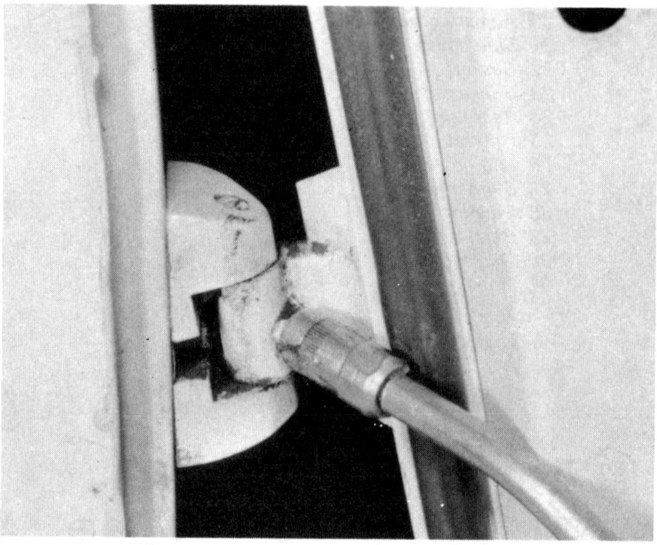

19.3 Lubricating a door hinge with a grease gun

Fig. 13.81 Profile lines on body and door should align (Sec 19)

Door trim (winding window type) – removal and refitting
30 Remove the door pull/armrest, which is secured by two screws. When fitted, also remove the door pocket.
31 Prise free the two halves of the door interior handle surround.
32 Release the spring clip which secures the window winder handle by introducing a piece of rag behind the handle and pulling it back and forth to pick up the ends of the clip. Remove the handle, bezel and clip (photos). (On some models the handle may be secured by a screw instead of a clip.)

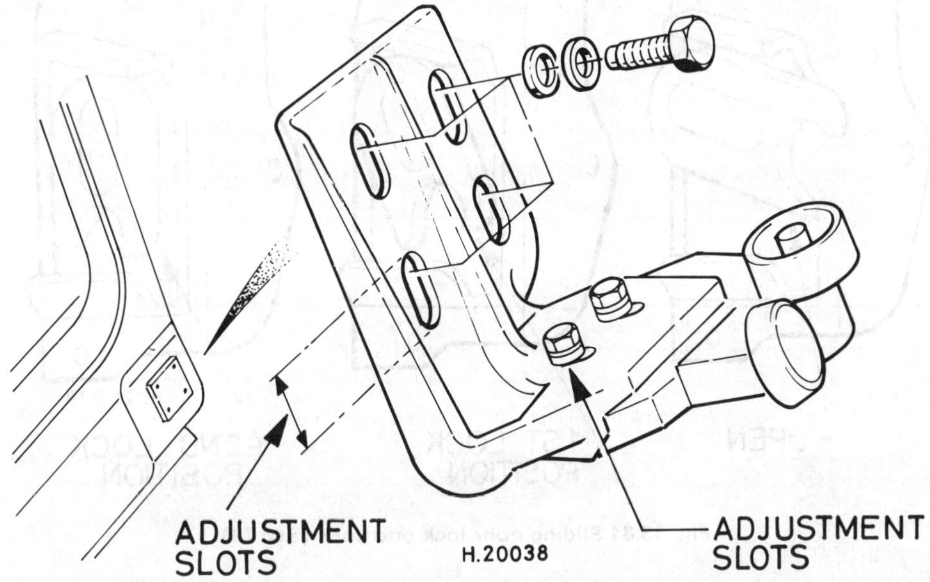

Fig. 13.82 Sliding door lower bracket adjustment (Sec 19)

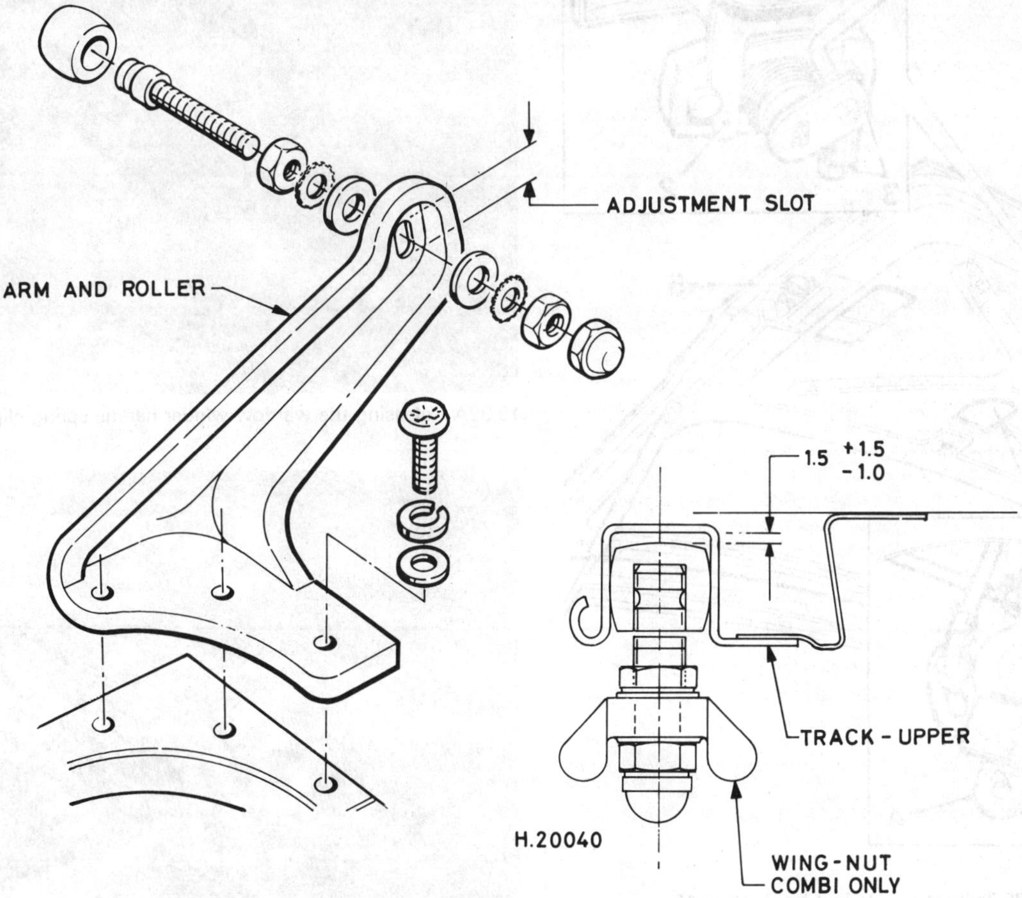

Fig. 13.83 Sliding door upper arm and roller. Sectional view shows clearance (in mm) of roller in track (Sec 19)

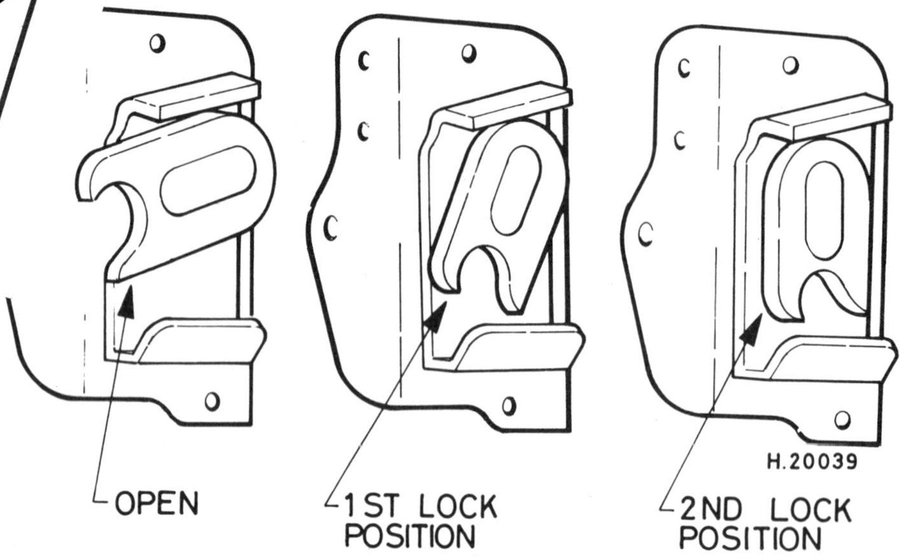

Fig. 13.84 Sliding door lock positions (Sec 19)

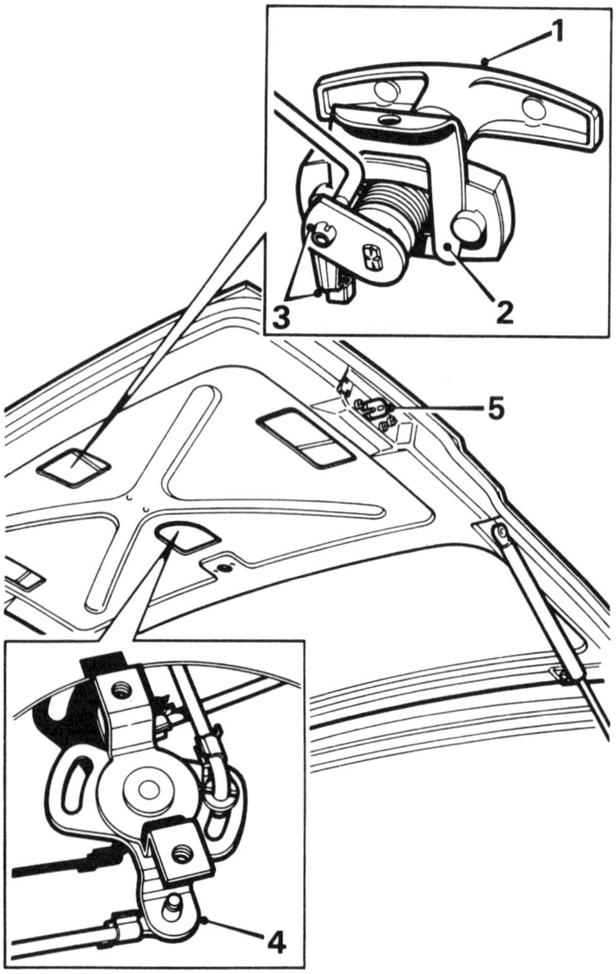

Fig. 13.85 Tailgate lock and latch (Sec 19)

1 Handle
2 Spring clip
3 Relay rod bush and clip
4 Relay lever
5 Latch

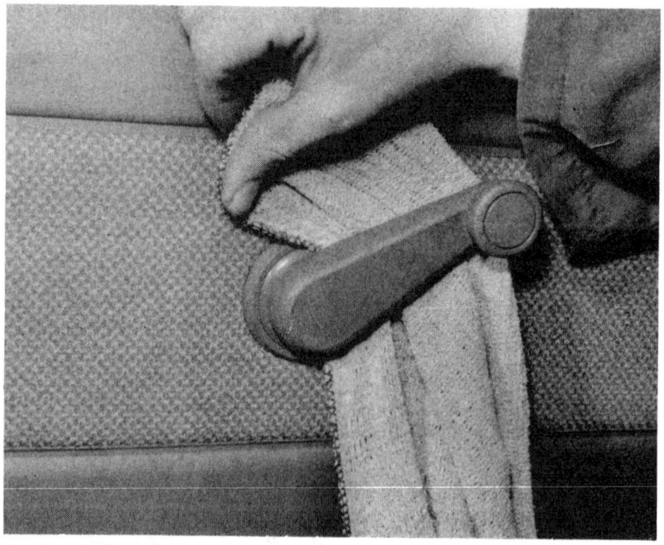

19.32A Releasing the window winder handle spring clip

19.32B Removing the window winder handle

33 Unclip the trim panel, using a palette knife or a wide-bladed screwdriver to release the panel clips from the door.
34 Peel the waterproof sheet off the inside of the door (photo).
35 Refit by reversing the removal operations, renewing broken or damaged components as necessary. Fit the spring clip to the window winder handle before pressing the handle home.

Winding window glass and regulator – removal and refitting

36 Remove the door trim panel as previously described.
37 Raise the window glass and wedge or tape it in position.
38 Remove the single screw which secures the bottom of the front channel, and the four screws which secure the regulator mechanism (photo).
39 Free the regulator arm from the window bottom channel. Remove the regulator from the door (photo).
40 Carefully lower the window glass into the door. Free the weatherstrip from the window aperture, removing the rigid waist sections completely.
41 Drill out the blind rivet which secures the top of the front channel. Free the channel from the fixed quarter-light and remove the quarter-light (photo).
42 Remove the front channel downwards through the door (photo).
43 Lift the window glass out through the slot in the door (photo).
44 Refit by reversing the removal operations. Use a new blind rivet to secure the top of the front channel.

Door/wing mirror – removal and refitting

45 The type of mirror and associated fitting will vary according to model year and door configuration. Typically the mirror arm bracket is secured to the door or body by three screws concealed beneath a plastic cover (photo).
46 Remove the three screws to remove the mirror, arm and bracket together; remove the arm-to-bracket nut if it is just wished to remove the mirror and arm. The mirror and arm cannot always be separated.
47 Refit by reversing the removal operations.

Heater cover (300 Series) – removal and refitting

48 Disconnect the battery negative lead.
49 Remove the screws which secure the cover to the parcel shelves and facia brackets.
50 Disconnect the cigarette lighter and illumination wiring (as applicable).
51 Remove the heater cover and lower console together. The two may then be separated if desired.
52 Refit by reversing the removal operations. Do not fully tighten the parcel shelf and facia bracket screws until the alignment of the cover is satisfactory.

Heater controls (300 Series) – adjustment

53 Remove the heater cover, then adjust the controls as follows. (The order of adjustment is not important.)

Face level air control

54 Move the control to the 'off' position. Release the screw which clamps the face level air flap lever to the control rod. Move the lever fully clockwise so that the flap is closed, hold it in this position and tighten the clamp screw.

Air temperature control

55 Move the control to the 'cold' position. Release the screw which clamps the air temperature flap lever to the control rod. Move the lever fully anti-clockwise so that the flap is closed, hold it in this position and tighten the clamp screw.

Air distribution control

56 Move the control to direct air to the windscreen. Release the screw which clamps the distribution flap lever to the control rod. Move the lever fully clockwise and tighten the clamp screw.

All controls

57 Check for correct operation, then refit the heater cover.

Heater unit (300 Series) – removal and refitting

58 Drain the cooling system.
59 Remove the heater cover.
60 Position a drain pan to catch any spillage, then disconnect the coolant hoses from the matrix pipes. Have a couple of bungs ready to plug the pipes.

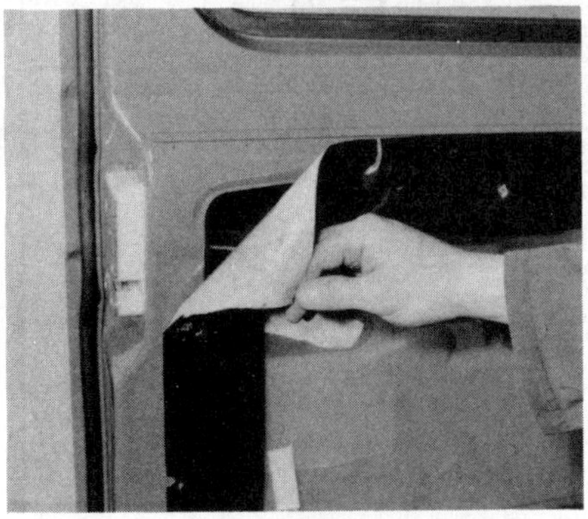

19.34 Peeling off the waterproof sheet

19.38 Window regulator mechanism, showing securing screws

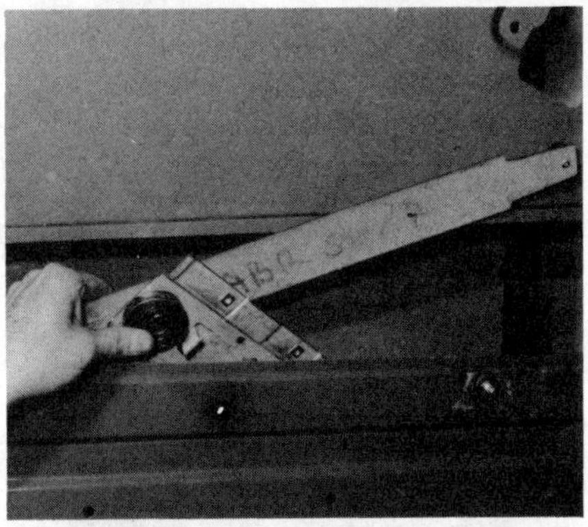

19.39 Removing the regulator mechanism

19.41 Removing the quarter-light

19.42 Removing the window front channel

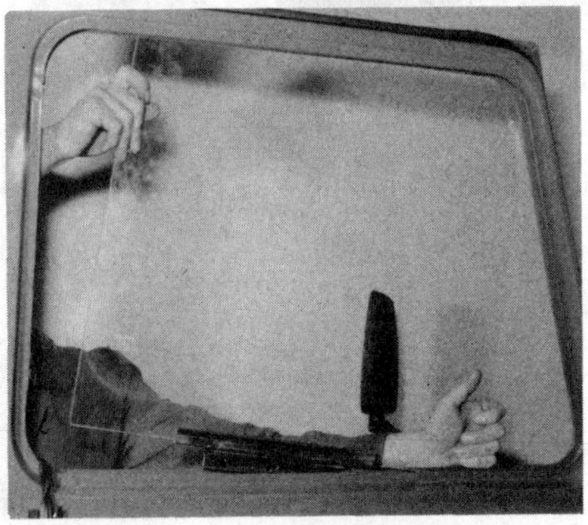

19.43 Removing the window glass

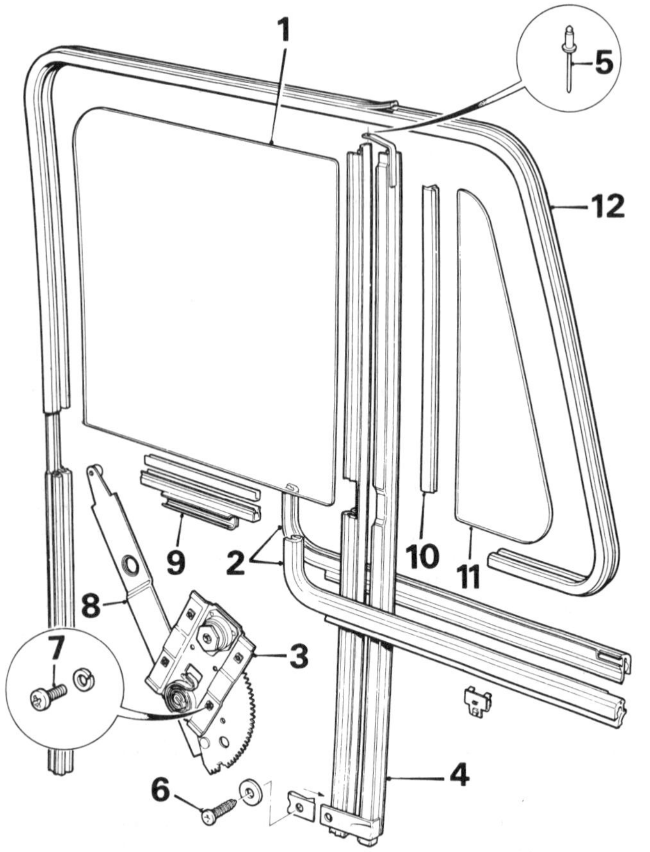

Fig. 13.86 Front door winding window glass components (Sec 19)

1 *Window*
2 *Waist weatherstrips*
3 *Regulator*
4 *Front channel*
5 *Blind rivet*
6 *Front channel screw*
7 *Regulator screw*
8 *Regulator arm*
9 *Bottom channel*
10 *Quarter-light rubber*
11 *Quarter-light (fixed)*
12 *Weatherstrip*

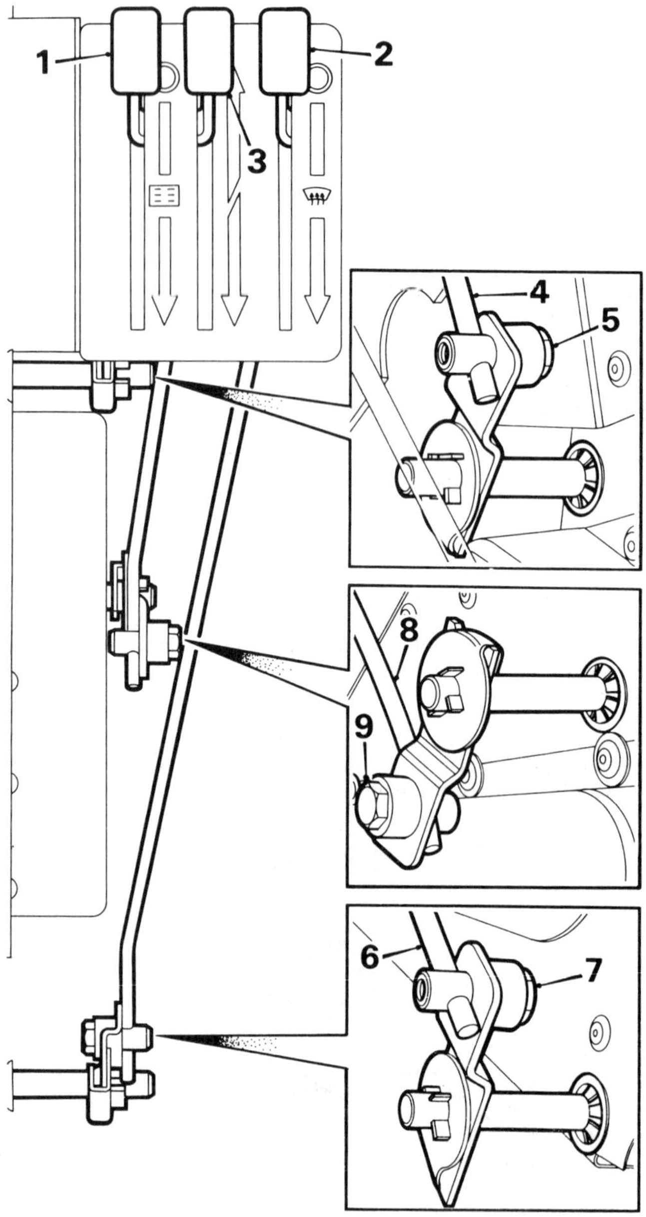

Fig. 13.87 Heater controls – 300 Series (Sec 19)

1 *Face level air control*
2 *Temperature control*
3 *Air distribution control*
4 *Face level control rod*
5 *Clamp screw*
6 *Temperature control rod*
7 *Clamp screw*
8 *Distribution control rod*
9 *Clamp screw*

19.45 Wing mirror mounting bracket

61 Remove the nuts and bolts which secure the air inlet to the bulkhead. An assistant will be needed since it is necessary to work on both sides of the bulkhead at once.
62 Remove the bolts which secure the heater unit brackets to the facia rail.
63 Free the demister tubes from the heater and withdraw it, complete with controls.
64 Refit by reversing the removal operations.

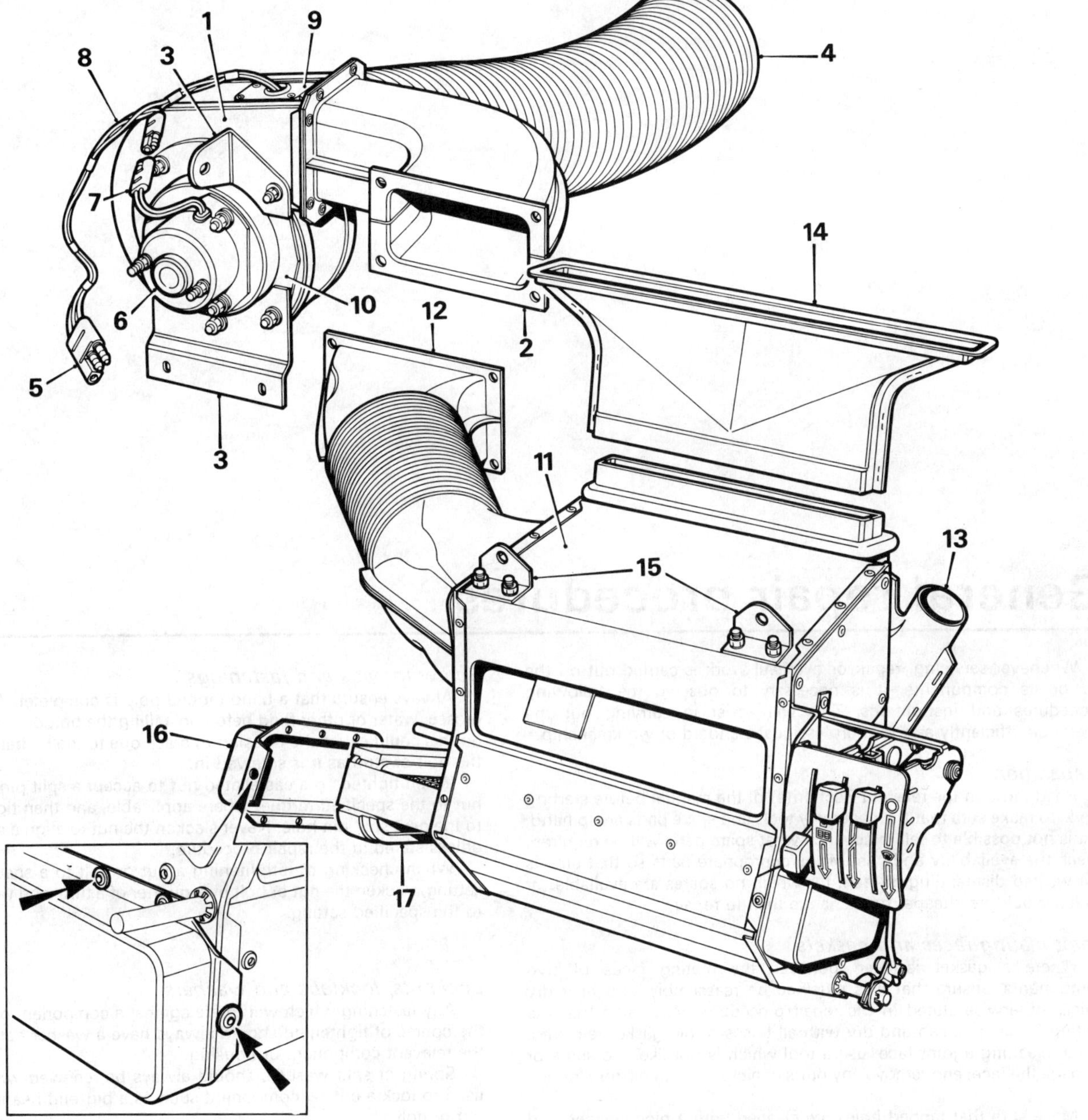

Fig. 13.88 Heater unit – 300 Series. Blind rivets (arrowed) secure matrix cover (Sec 19)

1 Blower unit
2 Blower outlet
3 Brackets
4 Inlet trunking
5 Multi-plug
6 Blower motor
7 Multi-plug
8 Wiring harness
9 Resistor
10 Cover
11 Heater assembly
12 Heater air inlet
13 Side demist tube
14 Demister duct
15 Brackets
16 Matrix cover
17 Matrix

Heater unit (300 Series) – overhaul

65 Check the availability of parts before becoming committed to overhauling the heater.
66 Release the control rod clamp screws and disconnect the rods from the levers. Grease the articulating joints of the controls and adjust the controls as previously described.
67 Remove the matrix cover by drilling out the rivets which secure it to the heater unit case. Remove the seal and matrix.
68 Flush out or renew the matrix, then refit it using new pipe seals.
69 Refit the matrix cover and seal. Secure the cover using new blind rivets.

Heater blower (300 Series) – removal and refitting

70 Disconnect the battery negative lead.
71 Remove the nuts and bolts which secure the blower outlet to the bulkhead. Two people will be needed, one working in the cab and the other under the bonnet.
72 Unbolt the blower brackets from the bulkhead and from the bonnet hinge bracket.
73 Free the blower from the inlet trunking, disconnect the multi-plug and remove it. Recover the sealer from the blower outlet.
74 The blower may be dismantled for renewal of the motor, the resistor or the fan, but check first that parts are available.
75 Refit by reversing the removal operations.

General repair procedures

Whenever servicing, repair or overhaul work is carried out on the car or its components, it is necessary to observe the following procedures and instructions. This will assist in carrying out the operation efficiently and to a professional standard of workmanship.

Preparation

Read through the relevant Section(s) of the manual before starting work, to make sure that the necessary tools and spare parts are to hand. If it is not possible to tell in advance what spare parts will be required, check the availability and cost of the appropriate parts so that time is not wasted dismantling an item for which no spares are available, or which would be cheaper to exchange than to repair.

Joint mating faces and gaskets

Where a gasket is used between the mating faces of two components, ensure that it is renewed on reassembly, and fit it dry unless otherwise stated in the repair procedure. Make sure that the mating faces are clean and dry with all traces of old gasket removed. When cleaning a joint face, use a tool which is not likely to score or damage the face, and remove any burrs or nicks with an oilstone or fine file.

Make sure that tapped holes are cleaned with a pipe cleaner, and keep them free of jointing compound if this is being used unless specifically instructed otherwise.

Ensure that all orifices, channels or pipes are clear and blow through them, preferably using compressed air.

Oil seals

Whenever an oil seal is removed from its working location, either individually or as part of an assembly, it should be renewed.

The very fine sealing lip of the seal is easily damaged and will not seal if the surface it contacts is not completely clean and free from scratches, nicks or grooves. If the original sealing surface of the component cannot be restored, the component should be renewed.

Protect the lips of the seal from any surface which may damage them in the course of fitting. Use tape or a conical sleeve where possible. Lubricate the seal lips with oil before fitting and, on dual lipped seals, fill the space between the lips with grease.

Unless otherwise stated, oil seals must be fitted with their sealing lips toward the lubricant to be sealed.

Use a tubular drift or block of wood of the appropriate size to install the seal and, if the seal housing is shouldered, drive the seal down to the shoulder. If the seal housing is unshouldered, the seal should be fitted with its face flush with the housing top face.

Screw threads and fastenings

Always ensure that a blind tapped hole is completely free from oil, grease, water or other fluid before installing the bolt or stud. Failure to do this could cause the housing to crack due to the hydraulic action of the bolt or stud as it is screwed in.

When tightening a castellated nut to accept a split pin, tighten the nut to the specified torque, where applicable, and then tighten further to the next split pin hole. Never slacken the nut to align a split pin hole unless stated in the repair procedure.

When checking or retightening a nut or bolt to a specified torque setting, slacken the nut or bolt by a quarter of a turn, and then retighten to the specified setting.

Locknuts, locktabs and washers

Any fastening which will rotate against a component or housing in the course of tightening should always have a washer between it and the relevant component or housing.

Spring or split washers should always be renewed when they are used to lock a critical component such as a big-end bearing retaining nut or bolt.

Locktabs which are folded over to retain a nut or bolt should always be renewed.

Self-locking nuts can be reused in non-critical areas, providing resistance can be felt when the locking portion passes over the bolt or stud thread.

Split pins must always be replaced with new ones of the correct size for the hole.

Special tools

Some repair procedures in this manual entail the use of special tools such as a press, two or three-legged pullers, spring compressors etc. Wherever possible, suitable readily available alternatives to the manufacturer's special tools are described, and are shown in use. In some instances, where no alternative is possible, it has been necessary to resort to the use of a manufacturer's tool and this has been done for reasons of safety as well as the efficient completion of the repair operation. Unless you are highly skilled and have a thorough understanding of the procedure described, never attempt to bypass the use of any special tool when the procedure described specifies its use. Not only is there a very great risk of personal injury, but expensive damage could be caused to the components involved.

Professional motor mechanics are trained in safe working procedures. However enthusiastic you may be about getting on with the job in hand, do take the time to ensure that your safety is not put at risk. A moment's lack of attention can result in an accident, as can failure to observe certain elementary precautions.

There will always be new ways of having accidents, and the following points do not pretend to be a comprehensive list of all dangers; they are intended rather to make you aware of the risks and to encourage a safety-conscious approach to all work you carry out on your vehicle.

Essential DOs and DON'Ts

DON'T rely on a single jack when working underneath the vehicle. Always use reliable additional means of support, such as axle stands, securely placed under a part of the vehicle that you know will not give way.

DON'T attempt to loosen or tighten high-torque nuts (e.g. wheel hub nuts) while the vehicle is on a jack; it may be pulled off.

DON'T start the engine without first ascertaining that the transmission is in neutral (or 'Park' where applicable) and the parking brake applied.

DON'T suddenly remove the filler cap from a hot cooling system – cover it with a cloth and release the pressure gradually first, or you may get scalded by escaping coolant.

DON'T attempt to drain oil until you are sure it has cooled sufficiently to avoid scalding you.

DON'T grasp any part of the engine, exhaust or catalytic converter without first ascertaining that it is sufficiently cool to avoid burning you.

DON'T allow brake fluid or antifreeze to contact vehicle paintwork.

DON'T syphon toxic liquids such as fuel, brake fluid or antifreeze by mouth, or allow them to remain on your skin.

DON'T inhale dust – it may be injurious to health (see *Asbestos* below).

DON'T allow any spilt oil or grease to remain on the floor – wipe it up straight away, before someone slips on it.

DON'T use ill-fitting spanners or other tools which may slip and cause injury.

DON'T attempt to lift a heavy component which may be beyond your capability – get assistance.

DON'T rush to finish a job, or take unverified short cuts.

DON'T allow children or animals in or around an unattended vehicle.

DO wear eye protection when using power tools such as drill, sander, bench grinder etc, and when working under the vehicle.

DO use a barrier cream on your hands prior to undertaking dirty jobs – it will protect your skin from infection as well as making the dirt easier to remove afterwards; but make sure your hands aren't left slippery. Note that long-term contact with used engine oil can be a health hazard.

DO keep loose clothing (cuffs, tie etc) and long hair well out of the way of moving mechanical parts.

DO remove rings, wristwatch etc, before working on the vehicle – especially the electrical system.

DO ensure that any lifting tackle used has a safe working load rating adequate for the job.

DO keep your work area tidy – it is only too easy to fall over articles left lying around.

DO get someone to check periodically that all is well, when working alone on the vehicle.

DO carry out work in a logical sequence and check that everything is correctly assembled and tightened afterwards.

DO remember that your vehicle's safety affects that of yourself and others. If in doubt on any point, get specialist advice.

IF, in spite of following these precautions, you are unfortunate enough to injure yourself, seek medical attention as soon as possible.

Asbestos

Certain friction, insulating, sealing, and other products – such as brake linings, brake bands, clutch linings, torque converters, gaskets, etc – contain asbestos. *Extreme care must be taken to avoid inhalation of dust from such products since it is hazardous to health.* If in doubt, assume that they *do* contain asbestos.

Fire

Remember at all times that petrol (gasoline) is highly flammable. Never smoke, or have any kind of naked flame around, when working on the vehicle. But the risk does not end there – a spark caused by an electrical short-circuit, by two metal surfaces contacting each other, by careless use of tools, or even by static electricity built up in your body under certain conditions, can ignite petrol vapour, which in a confined space is highly explosive.

Always disconnect the battery earth (ground) terminal before working on any part of the fuel or electrical system, and never risk spilling fuel on to a hot engine or exhaust.

It is recommended that a fire extinguisher of a type suitable for fuel and electrical fires is kept handy in the garage or workplace at all times. Never try to extinguish a fuel or electrical fire with water.

Note: *Any reference to a 'torch' appearing in this manual should always be taken to mean a hand-held battery-operated electric lamp or flashlight. It does NOT mean a welding/gas torch or blowlamp.*

Fumes

Certain fumes are highly toxic and can quickly cause unconsciousness and even death if inhaled to any extent. Petrol (gasoline) vapour comes into this category, as do the vapours from certain solvents such as trichloroethylene. Any draining or pouring of such volatile fluids should be done in a well ventilated area.

When using cleaning fluids and solvents, read the instructions carefully. Never use materials from unmarked containers – they may give off poisonous vapours.

Never run the engine of a motor vehicle in an enclosed space such as a garage. Exhaust fumes contain carbon monoxide which is extremely poisonous; if you need to run the engine, always do so in the open air or at least have the rear of the vehicle outside the workplace.

If you are fortunate enough to have the use of an inspection pit, never drain or pour petrol, and never run the engine, while the vehicle is standing over it; the fumes, being heavier than air, will concentrate in the pit with possibly lethal results.

The battery

Never cause a spark, or allow a naked light, near the vehicle's battery. It will normally be giving off a certain amount of hydrogen gas, which is highly explosive.

Always disconnect the battery earth (ground) terminal before working on the fuel or electrical systems.

If possible, loosen the filler plugs or cover when charging the battery from an external source. Do not charge at an excessive rate or the battery may burst.

Take care when topping up and when carrying the battery. The acid electrolyte, even when diluted, is very corrosive and should not be allowed to contact the eyes or skin.

If you ever need to prepare electrolyte yourself, always add the acid slowly to the water, and never the other way round. Protect against splashes by wearing rubber gloves and goggles.

When jump starting a car using a booster battery, for negative earth (ground) vehicles, connect the jump leads in the following sequence: First connect one jump lead between the positive (+) terminals of the two batteries. Then connect the other jump lead first to the negative (–) terminal of the booster battery, and then to a good earthing (ground) point on the vehicle to be started, at least 18 in (45 cm) from the battery if possible. Ensure that hands and jump leads are clear of any moving parts, and that the two vehicles do not touch. Disconnect the leads in the reverse order.

Mains electricity and electrical equipment

When using an electric power tool, inspection light etc, always ensure that the appliance is correctly connected to its plug and that, where necessary, it is properly earthed (grounded). Do not use such appliances in damp conditions and, again, beware of creating a spark or applying excessive heat in the vicinity of fuel or fuel vapour. Also ensure that the appliances meet the relevant national safety standards.

Ignition HT voltage

A severe electric shock can result from touching certain parts of the ignition system, such as the HT leads, when the engine is running or being cranked, particularly if components are damp or the insulation is defective. Where an electronic ignition system is fitted, the HT voltage is much higher and could prove fatal.

Conversion factors

Length (distance)

Inches (in)	X 25.4	= Millimetres (mm)	X 0.0394	= Inches (in)
Feet (ft)	X 0.305	= Metres (m)	X 3.281	= Feet (ft)
Miles	X 1.609	= Kilometres (km)	X 0.621	= Miles

Volume (capacity)

Cubic inches (cu in; in^3)	X 16.387	= Cubic centimetres (cc; cm^3)	X 0.061	= Cubic inches (cu in; in^3)
Imperial pints (Imp pt)	X 0.568	= Litres (l)	X 1.76	= Imperial pints (Imp pt)
Imperial quarts (Imp qt)	X 1.137	= Litres (l)	X 0.88	= Imperial quarts (Imp qt)
Imperial quarts (Imp qt)	X 1.201	= US quarts (US qt)	X 0.833	= Imperial quarts (Imp qt)
US quarts (US qt)	X 0.946	= Litres (l)	X 1.057	= US quarts (US qt)
Imperial gallons (Imp gal)	X 4.546	= Litres (l)	X 0.22	= Imperial gallons (Imp gal)
Imperial gallons (Imp gal)	X 1.201	= US gallons (US gal)	X 0.833	= Imperial gallons (Imp gal)
US gallons (US gal)	X 3.785	= Litres (l)	X 0.264	= US gallons (US gal)

Mass (weight)

Ounces (oz)	X 28.35	= Grams (g)	X 0.035	= Ounces (oz)
Pounds (lb)	X 0.454	= Kilograms (kg)	X 2.205	= Pounds (lb)

Force

Ounces-force (ozf; oz)	X 0.278	= Newtons (N)	X 3.6	= Ounces-force (ozf; oz)
Pounds-force (lbf; lb)	X 4.448	= Newtons (N)	X 0.225	= Pounds-force (lbf; lb)
Newtons (N)	X 0.1	= Kilograms-force (kgf; kg)	X 9.81	= Newtons (N)

Pressure

Pounds-force per square inch (psi; lbf/in^2; lb/in^2)	X 0.070	= Kilograms-force per square centimetre (kgf/cm^2; kg/cm^2)	X 14.223	= Pounds-force per square inch (psi; lbf/in^2; lb/in^2)
Pounds-force per square inch (psi; lbf/in^2; lb/in^2)	X 0.068	= Atmospheres (atm)	X 14.696	= Pounds-force per square inch (psi; lbf/in^2; lb/in^2)
Pounds-force per square inch (psi; lbf/in^2; lb/in^2)	X 0.069	= Bars	X 14.5	= Pounds-force per square inch (psi; lbf/in^2; lb/in^2)
Pounds-force per square inch (psi; lbf/in^2; lb/in^2)	X 6.895	= Kilopascals (kPa)	X 0.145	= Pounds-force per square inch (psi; lbf/in^2; lb/in^2)
Kilopascals (kPa)	X 0.01	= Kilograms-force per square centimetre (kgf/cm^2; kg/cm^2)	X 98.1	= Kilopascals (kPa)
Millibar (mbar)	X 100	= Pascals (Pa)	X 0.01	= Millibar (mbar)
Millibar (mbar)	X 0.0145	= Pounds-force per square inch (psi; lbf/in^2; lb/in^2)	X 68.947	= Millibar (mbar)
Millibar (mbar)	X 0.75	= Millimetres of mercury (mmHg)	X 1.333	= Millibar (mbar)
Millibar (mbar)	X 0.401	= Inches of water (inH$_2$O)	X 2.491	= Millibar (mbar)
Millimetres of mercury (mmHg)	X 0.535	= Inches of water (inH$_2$O)	X 1.868	= Millimetres of mercury (mmHg)
Inches of water (inH$_2$O)	X 0.036	= Pounds-force per square inch (psi; lbf/in^2; lb/in^2)	X 27.68	= Inches of water (inH$_2$O)

Torque (moment of force)

Pounds-force inches (lbf in; lb in)	X 1.152	= Kilograms-force centimetre (kgf cm; kg cm)	X 0.868	= Pounds-force inches (lbf in; lb in)
Pounds-force inches (lbf in; lb in)	X 0.113	= Newton metres (Nm)	X 8.85	= Pounds-force inches (lbf in; lb in)
Pounds-force inches (lbf in; lb in)	X 0.083	= Pounds-force feet (lbf ft; lb ft)	X 12	= Pounds-force inches (lbf in; lb in)
Pounds-force feet (lbf ft; lb ft)	X 0.138	= Kilograms-force metres (kgf m; kg m)	X 7.233	= Pounds-force feet (lbf ft; lb ft)
Pounds-force feet (lbf ft; lb ft)	X 1.356	= Newton metres (Nm)	X 0.738	= Pounds-force feet (lbf ft; lb ft)
Newton metres (Nm)	X 0.102	= Kilograms-force metres (kgf m; kg m)	X 9.804	= Newton metres (Nm)

Power

Horsepower (hp)	X 745.7	= Watts (W)	X 0.0013	= Horsepower (hp)

Velocity (speed)

Miles per hour (miles/hr; mph)	X 1.609	= Kilometres per hour (km/hr; kph)	X 0.621	= Miles per hour (miles/hr; mph)

*Fuel consumption**

Miles per gallon, Imperial (mpg)	X 0.354	= Kilometres per litre (km/l)	X 2.825	= Miles per gallon, Imperial (mpg)
Miles per gallon, US (mpg)	X 0.425	= Kilometres per litre (km/l)	X 2.352	= Miles per gallon, US (mpg)

Temperature

Degrees Fahrenheit = (°C x 1.8) + 32

Degrees Celsius (Degrees Centigrade; °C) = (°F - 32) x 0.56

**It is common practice to convert from miles per gallon (mpg) to litres/100 kilometres (l/100km), where mpg (Imperial) x l/100 km = 282 and mpg (US) x l/100 km = 235*

Index

D

E

O

P